ORGANIC AND PHYSICAL CHEMISTRY OF POLYMERS

ORGANIC AND PHYSICAL CHEMISTRY OF POLYMERS

Yves Gnanou
Michel Fontanille

WILEY-
INTERSCIENCE

A JOHN WILEY & SONS, INC., PUBLICATION

Library of Congress Cataloging-in-Publication Data:

Gnanou, Yves.
    Organic and physical chemistry of polymers / by Yves Gnanou and Michel Fontanille.
        p. cm.
    Includes index.
    ISBN 978-0-471-72543-5 (cloth)
1.   Polymers. 2.   Polymers—Synthesis. 3.   Chemistry, Physical and theoretical.   I. Fontanille, M. (Michel), 1936–   II. Title.
    QD381.G55 2008
    547′.7—dc22

                                                                2007029090

Printed in the United States of America

10 9 8 7 6 5 4 3 2 1

CONTENTS

FOREWORD

Polymers, commonly known as plastics, are perhaps the most important materials for society today. They are employed in nearly every device. The interior of every automobile is essentially entirely made of polymers; polymers are also used for body parts and for under-the-hood applications. Progress in the aerospace industry has been aided by new light, strong nanocomposite polymeric materials. Many construction materials (e.g., insulation, pipes) and essentially all adhesives, sealants, and coatings (paints) are made from polymers. The computer chips used in our desktops, laptops, cell phones, Ipods, or Iphones are enabled by polymers used as photoresists in microlithographic processes. Many biomedical applications require polymers for tissue or bone engineering, drug delivery, and also for needles, tubing, and containers for intravenous delivery of medications. Some new applications call for smart or "intelligent" polymers that can respond to external stimuli and change shape and color to be used as artificial muscles or sensors.

Thus, it is not surprising that the annual production of polymers approaches 200 million tons and 50% of the chemists in USA, Japan or Western Europe work in one way or other with polymeric materials. However, polymer awareness has not yet reached the appropriate level, for many of those chemists do not fully comprehend nor do they take advantage of concepts of free volume, glass transition, and microphase separation; consequently they do not know how to precisely control polymer synthesis. One may also argue that some polymer scientists do not sufficiently appreciate most recent developments in organic and physical chemistry, although polymer science has a very interdisciplinary character and bridges synthetic chemistry with precise characterization techniques offered by the methodologies of physical chemistry.

Organic and Physical Chemistry of Polymers by Yves Gnanou and Michel Fontanille provides a unique approach to combine fundamentals of organic and physical chemistry and apply them to explain complex phenomena in polymer science. The authors employ a very methodical way, straightforward for polymer science novices and at the same time, attractive for more experienced polymer scientists. On reading this book, one can easily comprehend not only how to make conventional and new polymeric materials, but also how to characterize them and use them for classic and new advanced applications.

I read the book with a great interest, and I am convinced that this book will become an excellent polymer science textbook for senior undergraduate and graduate students.

KRZYSZTOF MATYJASZEWSKI

J.C. Warner University Professor of Natural Sciences
Carnegie Mellon University
Fall 2007, Pittsburgh, USA

PREFACE

Although the uses of polymers in miscellaneous applications are as old as humanity, polymer science began only in the 1920s, after Staudinger conclusively proved to sceptics the concept of long chain molecules consisting of atoms covalently linked one to another. Then came the contributions of physicists: Kuhn first accounted for the flexibility of certain polymers and understood the role of entropy in the elasticity of rubber. Flory subsequently explained most of the physical properties of polymers using very simple ideas, and Edwards found a striking analogy between the conformation of a polymer chain and the trajectory of a quantum mechanical particle.

The aim of this textbook is to do justice to the interdisciplinary nature of polymer science and to break the traditional barriers between polymer chemistry and the physical chemistry and physics of polymers. Through the description of the structures found in polymers and the reactions used to synthesize them, through the account of their dynamics and their energetics, are conveyed the basic concepts and the fundamental principles that lay the foundations of polymer science. We tried to keep in view this primary emphasis throughout most of the book, and chose not to elaborate on applicative and functional aspects of polymers.

At the core of this book lie three main ideas:

1. —the synthesis of polymer chains requires reactions exhibiting high selectivity, including regio-, chemo- and sometimes stereoselectivity. Mother Nature also produces macromolecules that are useful for life (proteins, DNA, RNA) but with a much higher selectivity;

2. —polymers represent a class of materials that are by essence ambivalent, exhibiting at the same time viscous and elastic behaviors. Indeed, a polymer chain never behaves as a purely elastic material or as an ideal viscous liquid. Depending upon the temperature and the polymer considered, the time scale of the stress applied, either the viscous or the elastic component dominates in its response;

3. —an assembly of polymer chains can adopt a variety of structures and morphologies and self-organize in highly crystalline lamellae or exist as a totally disordered amorphous phase and intermediately as mesomorphic structures.

Polymers are thus materials with peculiar physical properties which are controlled by their methods of synthesis and their internal structure. The first chapters (I to III) introduce the notions of configuration and conformation of polymers, their dimensionality, and how their multiple interactions contribute to their overall cohesion. The three next chapters are concerned with physical chemistry, namely the thermodynamics of polymer solutions (IV), the structures typical of polymer assemblies (V), and the experimental methods used to characterize the size, the shape and the structures of polymers (VI). Four chapters (VII to IX) then follow that elaborate on the methods of synthesis and modification of polymers, and the engineering of complex architectures (X). Chapters XI to XIII subsequently describe the thermal transitions and relaxations of polymers, their mechanical properties and their rheology. These thirteen chapters are rounded off by monographs (chapters XIV to XVI) of natural polymers and of some common monodimensional and tridimensional polymers.

Since the 1920s, polymer science has moved on at a dramatic rate. Significant advances have been made in the synthesis and the applications of polymeric materials, paving the way for the award of the Nobel Prize in five instances to polymer scientists. Staudinger in 1953, Ziegler and Natta in 1963, Flory in 1974, de Gennes in 1991, and more recently McDiarmid, Shirakawa and Heeger in 2000 indeed received this distinction. Their contributions and the many developments witnessed in the area of specialty polymers have made necessary to write a book that provides the basics of polymer science and a bridge to an understanding of the huge primary literature now available. This book is intended for students with no prior knowledge or special background in mathematics and physics; it can serve as a text for a senior-level undergraduate or a graduate-level course.

In spite of our efforts, some mistakes certainly remain; we would appreciate reports about these from readers.

Last but not least, we wish to mention our debt and express our gratitude to Professors Robert Pecora (Standford University), Marcel van Beylen (Leuven University) and colleagues from our University who read and checked most of the chapters. We are also indebted to Professor K. Matyjaszewski for accepting to write the foreword of this book.

YVES GNANOU
MICHEL FONTANILLE

Summer 2007, Bordeaux, France.

1

INTRODUCTION

1.1. HISTORY

The term *polymer* is quite old and has been used since 1866 after Berthelot mentioned that "When styrolene (now called *styrene*) is heated up to 200°C for several hours, it is converted into a resinous polymer...." Is it the first synthetic polymer recognized as such? Probably, yes. However, the concept of polymeric chain as we understand it today had to wait for the work of Staudinger (Nobel Prize laureate in 1953) before being fully accepted. It is only from that time onward—approximately the 1920s—that the "macromolecular" theory ultimately prevailed over the opposite "micellar" theory.

Meanwhile, artificial and synthetic polymers had acquired due acceptance and began to be utilized as substitutes for rare substances (celluloid in lieu of ivory, artificial silk, etc.) or in novel applications (bakelite, etc.) due to their peculiar properties.

The variety of synthetic polymers discovered by Staudinger is impressive, and a number of today's polymeric substances were prepared for the first time by this outstanding scientist. His work soon attracted the keen interest and attention of the chemical industry, and as soon as 1933 the ICI company obtained a grade of polyethylene whose world production is still several tens of million tons per annum. A little later (1938), and after some failures in the field of polyesters, scientists

Organic and Physical Chemistry of Polymers, by Yves Gnanou and Michel Fontanille
Copyright © 2008 John Wiley & Sons, Inc.

headed by Carothers at DuPont de Nemours discovered the polyamides (known as "nylons"). This breakthrough illustrated the ability of polymer chemists to design and invent materials with mechanical characteristics surpassing those of materials originating from the vegetable or animal worlds.

By the end of the Second World War, polymers had shown their ability to replace many traditional materials, but were somehow plagued by a reputation of affording only poor-quality products. From the research work carried out in both academic laboratories and industrial research centers since then, many unexpected improvements have been accomplished in terms of processes and properties, so that today's polymers are present in most advanced sectors of technology.

It is no surprise that the name of several Nobel laureates appear on the list of scientists who have contributed the most to polymer science. In addition to Staudinger, these include Ziegler, Natta, Flory, de Gennes, McDiarmid, Shirakawa, Heeger, and, recently, Chauvin, Grubbs, and Schrock. There are also many scientists whose names are known only to experts and whose contributions were instrumental in the development of the polymer field. Owing to the economic significance of polymer materials, industry has also been keen on supporting research work in the field of polymers. They are indeed present everywhere and appear in almost all aspects of our daily life. With the continuous improvement of their properties, the old tendency to look down on polymers has given way to attention and consideration; more than ever, the current perception is: "There are no bad polymers but only bad applications."

Table 1.1 contains important dates that have marked the progress witnessed in the field of polymers throughout the last 150 years or so. Most of them correspond to the discovery of new methodologies and materials, followed by their industrial development. These successes have been possible because of a sustained investment in basic research and the surge of knowledge that has resulted from it.

1.2. SEVERAL DEFINITIONS

What is a polymer? Several answers can be given, but, for the moment, the most common and generally accepted definition is: a system formed by an assembly of macromolecules—that is, a system of molecular entities with large dimension, which are obtained by the covalent linking of a large number of constitutional repeat units, more commonly called monomeric units. The macromolecular structures corresponding to this definition have molecular dimensions (characterized by their molar mass) much larger than those of the simple molecules. This, in turn, provides the polymer considered with properties of practical application—in particular, in the field of materials.

It is difficult to precisely define the change induced by the transition from the simple molecular level to the macromolecular one. Depending upon the property considered, the macromolecular effect will be indeed perceptible at a lower or higher threshold of molar mass; for example, the majority of industrially produced linear polymers used in daily life are in the range of $\sim 10^5 \, \text{g·mol}^{-1}$.

Table 1.1. Main dates in the history of polymers

1838: A. Payen succeeded in extracting from wood a compound with the formula $(C_6H_{10}O_5)_n$, which he called *cellulose*.

1844: C. Goodyear developed the vulcanization of natural rubber.

1846: C. Schonbein obtained nitrocellulose (which was the first "artificial" polymer) by action of a sulfo-nitric mixture on cellulose.

1866: M. Berthelot discovered that upon heating "styrolene" up to $200°C$ for several hours, the latter is converted into a "resinous polymer."

1883: H. de Chardonnet obtained "artificial silk" by spinning a collodion (concentrated solution) of nitrocellulose.

1907: A. Hofmann prepared the first synthetic rubber by polymerization of conjugated dienes.

1910: L. Baekeland developed the first industrial process for the production of a synthetic polymer; formo-phenolic resins were produced under the name of "bakelite."

1919: H. Staudinger introduced the concept of macromolecule and then carried out the polymerization of many vinyl and related monomers. He can be viewed as the father of macromolecular science.

1925: Th. Svedberg presented experimental evidence of the existence of macromolecules by measuring their molar mass using ultracentrifugation.

1928: K. Meyer and H. Mark established the relationship between the chemical and crystallographic structures of polymers.

1933: E. Fawcett and R. Gibson, working for I.C.I., carried out the free radical polymerization of ethylene under high pressure.

1938: W. Carothers (of DuPont de Nemours) and his team prepared the first synthetic polyamides (known under the "nylon" tradename).

1942: P. Flory and M. Huggins proposed a theory accounting for the behavior of macromolecular solutions.

1943: O. Bayer synthesized the first polyurethane.

1947: T. Alfrey and C. Price proposed a theory of chain copolymerization.

1953: F. Crick and J. Watson identified the double helix structure of DNA using X-ray crystallography. They shared the Nobel Prize in 1962.

1953: K. Ziegler discovered the polymerization of ethylene under low pressure, using a catalyst generated from $TiCl_4$ and AlR_3.

1954: G. Natta obtained and identified isotactic polypropene.

1955: M. Williams, R. Landel, and J. Ferry proposed a relation (WLF equation) between the relaxation time of polymer chains at a certain temperature and that measured at the glass transition temperature.

1956: M. Szwarc established the principles of "living" polymerizations based on his work on the anionic polymerization of styrene.

1957: A. Keller obtained and characterized the first macromolecular monocrystal.

1959: J. Moore developed size exclusion chromatography as a technique to fractionate polymers.

1960: Discovery of thermoplastic elastomers and description of the corresponding morphologies.

1970–1980: P.-G. de Gennes formulated the scaling concepts which accounted for the variation of the characteristic sizes of a polymer with its concentration. He introduced with Doi and Edwards the concept of reptation of polymer chains in the molten state.

(*continued overleaf*)

Table 1.1. (*continued*)

1974: Development of aromatic polyamides by DuPont de Nemours.

1980: W. Kaminsky and H. Sinn discovered the effect of aluminoxanes on the polymerization of olefins catalyzed by metallocenes.

1982: A DuPont de Nemours team working under O. Webster and D. Sogah discovered the group transfer polymerization of acrylic monomers and initiate various research works related to the controlled polymerization of these monomers.

1982: T. Otsu introduced the concept of controlled radical polymerization. This concept was applied by E. Rizzardo and D. Solomon (1985) then by M. George (1992) to the controlled radical polymerization of styrene.

1986: D. Tomalia described the synthesis of the first dendrimers.

1992: D. Tirrell synthesized the first perfectly uniform polymer using methods of genetic engineering.

1994: M. Sawamoto and K. Matyjaszewski developed a new methodology of controlled radical polymerization by atom transfer.

2000: After more than 20 years of work on intrinsically conducting polymers, H. Shirakawa, A. Heeger, and A. McDiarmid were awarded the Nobel Prize in Chemistry.

2005: Y. Chauvin, R. Grubbs, and R. Schrock have been awarded the 2005 Nobel Prize in Chemistry for improving the metathesis reaction, a process used in making new polymers.

Remark. The terms *polymer* and *macromolecule* are often utilized without discrimination. Some specialists prefer using the term *macromolecule* for compounds of biological origin, which often have more complex molecular structure than synthetic polymers. For our part, we will utilize the two terms interchangeably.

The number of monomer units constituting a polymer chain is called the *degree of polymerization** (DP); it is directly proportional to the molar mass of the polymer. An assembly of a small number of monomer units within a macromolecular chain is called *sequence* and the first terms of the series of sequences are referred to as *dyad, triad, tetrad, pentad*, and so on. Chains made up of a small number of monomer units are called *oligomers*; typically, the degrees of polymerization of oligomers vary from 2 to a few tens. Synthetic polymers are obtained by reactions known as polymerization reactions, which transform simple molecules called monomer molecules (or monomers) into a covalent assembly of monomer units or polymer. When a polymer is obtained from the polymerization of different monomer molecules (indicated in this case by *comonomers*) exhibiting different molecular structure, it is called a *copolymer*.

---

*The symbol recommended by IUPAC for the average number of monomeric units in a polymeric chain is $\overline{X}$, DP being the abbreviation for the degree of polymerization.

Monomeric units that are part of a polymer chain can be linked one to another by a varying number of bonds; we suggest to call this number *valence*.[†] This term should be preferred to *functionality*, which can be misleading (see page 216). Thus, monomeric units can be mono-, di-, tri-, tetra-, or plurivalent and so are the corresponding monomer molecules.

The average valence of monomeric units in a macromolecular chain determines its *dimensionality* (see Section 1.4.3).

1.3. REPRESENTATION OF POLYMERS

Depending upon the level of precision and the type of information required, one has at one's disposal different adequate representations of the polymer structure. To represent the macromolecular nature of a linear polymer, a mere continuous line as shown in Figure 1.1 is perfectly relevant. Representations appearing in Figures 1.3 and 3.1 (see the corresponding paragraphs) illustrate more complex architectures and for the first one of higher dimensionality.

The most suitable representation of the chemical structure of a macromolecular compound is a monomeric unit flanked by two brackets and followed by a number, n, appearing as an index to indicate the degree of polymerization. Such a representation disregards the chain ends, which are necessarily different from the main chain, as well as possible defects along the polymer backbone (Section 3.2). This is illustrated in the following three examples, which are based on conventions borrowed from organic chemistry.

Poly(vinylidene chloride)

Poly(methyl methacrylate)

cis-1,4-Polyisoprene

To address configurational aspects, one generally relies on the Fischer projections used in organic chemistry, with a rotation $\pi/2$ of the line representing the main chain. However, in the case of polymers, it is the relative configuration of

---

[†]The term valence of monomers or of monomeric units is proposed by anology with the valence of atoms which corresponds to the number of orbitals available for bonding. The valence of a monomer thus corresponds to the number of covalent bonds that it forms with the nearest monomeric units.

a sequence of monomer units that is considered, which implies that several such units are represented. The two following examples take into consideration these conventions:

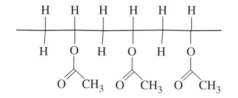

Sequence of 3 successive units *(triad)* of poly(vinyl acetate) presenting the same configuration

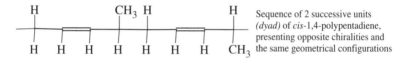

Sequence of 2 successive units *(dyad)* of *cis*-1,4-polypentadiene, presenting opposite chiralities and the same geometrical configurations

This method of representation is certainly easier to use than the one based on the principles established by Cram, which is illustrated below:

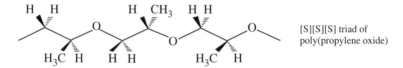

[S][S][S] triad of poly(propylene oxide)

1.4. CLASSIFICATION OF ORGANIC POLYMERS

1.4.1. Depending upon their **origin**, one can classify polymers into three categories:

- **Natural polymers** are obtained from vegetable or animal sources. Their merits and utility are considerable, but they will be only briefly described in the first part of this work. To this category belong all families of polysaccharides (cellulose, starch, etc.), proteins (wool, silk, etc.), natural rubber, and so on;
- **Artificial polymers** are obtained by chemical modification of natural polymers in order to transform some of their properties; some of them, such as cellulose esters (nitrocellulose, cellulose acetate, etc.), have been economically important for a long time;
- **Synthetic polymers** are exclusively the result of human creation; they are obtained by polymerization of monomer molecules. There exists a large variety of such polymers, and henceforth they will be described in detail.

1.4.2. A classification by **applications** would not be exhaustive because of the extreme variability of the polymer properties and the endless utilization of polymers, particularly in the field of materials. However, one can identify three main categories of polymers as a function of the application contemplated:

- **Large-scale** polymers (also called **commodity** polymers), whose annual production is in the range of millions of tons, are used daily by each of us. Polyethylene, polystyrene, poly(vinyl chloride), and some other polymers are included in this category of polymers of great economic significance;
- **Technical** polymers (also called **engineering** plastics) exhibit mechanical characteristics that enable them to replace traditional materials (metals, ceramics, etc.) in many applications; polyamides, polyacetals, and so on, are part of this family;
- **Functional** polymers are characterized by a specific property that has given rise to a particular application. Conducting polymers, photoactive polymers, thermostable polymers, adhesives, biocompatible polymers, and so on, belong to this category.

Depending on whether they are producers, formulators, or users of polymers, experts do not assign the same definition to each of these categories even if they broadly agree on the terms.

1.4.3. Polymers can also be classified into three categories as a function of their **structure** (dimensionality):

- **Linear** (or **monodimensional**) polymers, which consist of a (possibly) high (but finite) number of monomeric units; such systems are obtained by the polymerization of bivalent monomers, and a linear macromolecule can be schematically represented by a continuous line divided into intervals to indicate the monomer units (Figure 1.1); an assembly of polymer chains consists of entities with variable length, a characteristic designated by the term *dispersity*;[‡]
- **Two-dimensional** polymers are mainly found in Nature (graphite, keratin, etc.); two-dimensional synthetic polymers are objects that have not yet crossed the boundaries of laboratories. They appear in the form of two-dimensional layers with a thickness comparable to that of simple molecules (Figure 1.2);

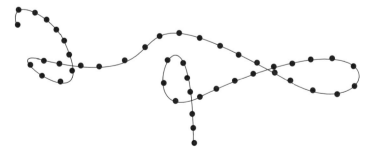

Figure 1.1. Representation of the chain of a linear polymer.

[‡]term recommended in 2007 by the IUPAC Subcommittee on Macromolecular Nomenclature to replace *polydispersity*.

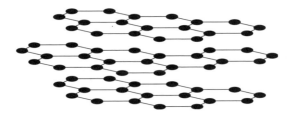

Figure 1.2. Schematic representation of a two-dimensional polymer, here carbon graphite.

- **Three-dimensional** polymers result either from the polymerization of mono-mers whose average valence is higher than two or from the cross-linking of linear polymers (formation of a three-dimensional network) through physical or chemical means. Their molecular dimension can be regarded as infinite for all covalently linked monomeric units of the sample are part of only one simple macromolecule. Chains grow at the same time in the three dimensions of space, and a volume element of such a system can be represented as shown in Figure 1.3.

This last mode of classification is extremely useful since all the properties of the macromolecular systems—mechanical properties in particular—are very strongly affected by the dimensionality of the polymer systems. Monographs on the various families of synthetic polymers will be presented in two different chapters to highlight this point.

Remark. Irrespective of their dimensionality and/or their topology, synthetic polymers can be classified as homopolymers and copolymers, depending on their molecular structure (see Section 3.2).

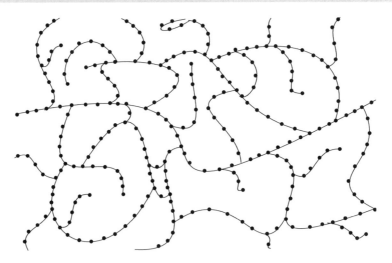

Figure 1.3. Schematic representation of a three-dimensional polymer.

1.5. NOMENCLATURE OF POLYMERS

There are three ways to name polymers.

The **first one**, which is official, follows the recommendations of the International Union of Pure and Applied Chemistry (IUPAC). It consists in naming the monomer unit according to the rules used for small organic molecules and, after insertion between brackets, in appending the prefix *poly* before it.

Poly(1-phenylethylene)

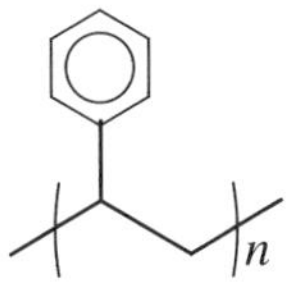

Poly(methylene) $— (CH_2)— n$

Poly(oxyethylene) $— (O\text{-}CH_2\text{-}CH_2)— n$

Poly[1-(methoxycarbonyl)-1-methylethylene]

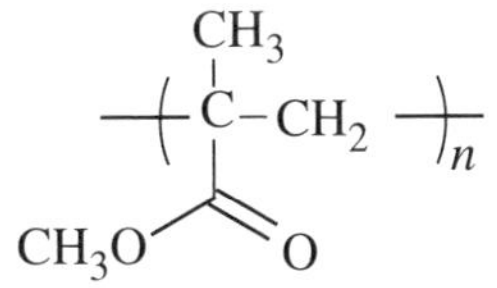

Poly[imino(1-oxohexamethylene)]

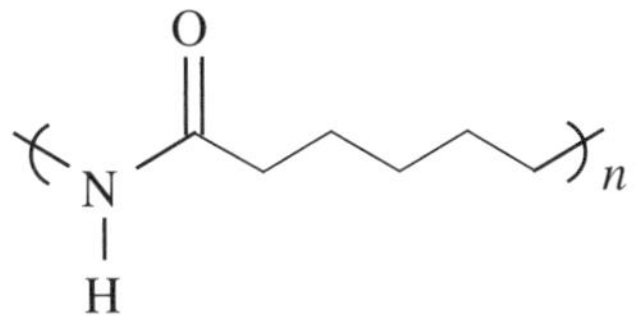

This method is based on the structure of polymer irrespective of the method of preparation.

The **second one**, which is the most frequently used, refers to the polymerization of a particular monomer and may reflect the process used. For example, poly(ethylene oxide) results from the polymerization of ethylene oxide:

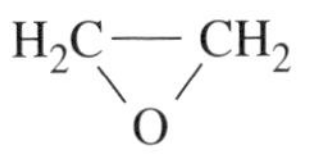

Polyethylene $-(CH_2-CH_2)_n-$ is obtained by polymerization of ethylene $H_2C = CH_2$ (which should be called ethene). Polypropylene and poly(vinyl chloride) are obtained from the polymerization of propylene (which should be called propene) and vinyl chloride, respectively:

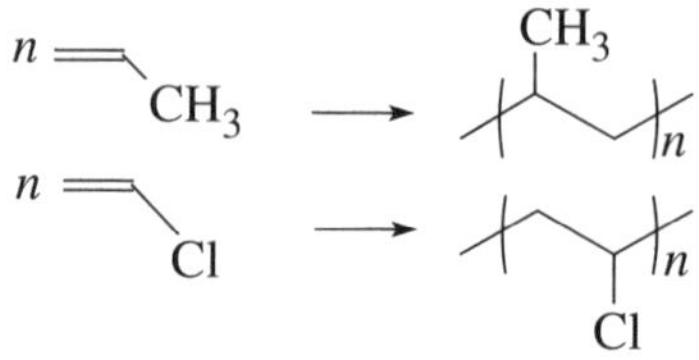

Remarks

(a) When the monomer name consists of several words, it is inserted between brackets and the prefix "poly" is added before it.

(b) The same polymer can have several names if it can be prepared by different methods. For instance, the polyamide shown below, whose acronym is PA-6, can be called polycaprolactam or poly(ε-capramide) whether it is obtained by chain polymerization of ε-caprolactam or by self-polycondensation of ε-aminocaproic acid:

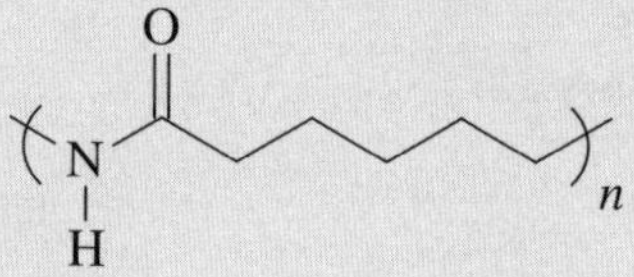

(c) Each natural polymer has its own name: cellulose, starch, keratin, lignin, and so on.

For the most commonly used polymers, a **third method**, based on acronyms, is widespread; these acronyms can designate either

- a particular polymer: PVC for poly(vinyl chloride), PS for polystyrene, and so on, or
- a family of polymers: PUR for polyurethanes, UP for unsaturated polyesters, and so on.

Acronyms can be also utilized to emphasize a structural characteristic; for instance, UHMWPE indicates a polyethylene with ultra-high molar mass, whereas "generic" polyethylene is simply designated by PE. Other examples of designations will be given later on—in particular in Chapter 3, which addresses the molecular structure of polymers. Table 1.2 gives the three types of naming for the most important and/or significant polymers.

Table 1.2. Designation of several common polymers

Structure of Monomeric Unit	IUPAC Designation	Common Designation	Acronym
	Poly(methylene)	Polyethylene	PE
	Poly(1-methylethylene)	Polypropylene	PP
CN	Poly(1-cyanoethylene)	Polyacrylonitrile	PAN
	Poly(oxyethylene-oxyterephthaloyl)	Poly(ethylene terephthalate)	PET
$\left(\!O-CH_2\!\right)_n$	Poly(oxymethylene)	Polyformaldehyde	POM
	Poly(1-acetoxyethylene)	Poly(vinyl acetate)	PVAC
OH	Poly(1-hydroxyethylene)	Poly(vinyl alcohol)	PVAL
F F / F F	Poly(difluoromethylene)	Polytetrafluoro-ethylene	PTFE
	Poly[imino (1,6-dioxohexamethylene) iminohexamethylene]	Poly(hexamethylene adipamide)	PA-6,6
H_3C	Poly(1-methylbut-1-enylene)	1,4- *cis*-Polyisoprene	NR
CH_3 / CH_3	Poly(1,1-dimethylethylene)	Polyisobutene	PIB

(continued overleaf)

Table 1.2. (*continued*)

Structure of Monomeric Unit	IUPAC Designation	Common Designation	Acronym
	Poly(1-vinylethylene)	1,2-Polybutadiene	1,2-PBD
	Poly(butenylene)	1,4-*cis*-Polybutadiene	PBD or BR

Remarks

(a) The acronyms BR and NR, which refer to polybutadiene and natural polyisoprene, correspond to the abbreviation of butadiene rubber and natural rubber, respectively.

(b) In general, chains of synthetic polydienes contain variable proportions of 1,2-, 1,4-, and 3,4-type monomer units.

(c) Designations of polymers other than linear homopolymers are the subject of specific rules. Some of them will be indicated while presenting the corresponding structure.

LITERATURE

G. Allen, *Perspectives*, in *Comprehensive Polymer Science*, Vol. 1, Pergamon Press, Oxford, 1989.

J. Bandrup, E. H. Immergut, E. A. Grulke, *Polymer Handbook*, 4th edition, Wiley, New York, 1999.

W. V. Metanomski, *Compendium of Macromolecular Nomenclature*, Blackwell Scientific Publishers, Oxford, 1991.

2

COHESIVE ENERGIES OF POLYMERIC SYSTEMS

Most of the properties of polymers, which are used in a very large variety of applications, are closely related to their cohesion. The cohesion energy, above all, depends on the strength of molecular interactions that develop between molecular groups.

Considered individually, these interactions are not stronger than those observed in a system composed of simple molecules. However, in polymeric systems, the multiplicity of interactive groups and the forces resulting from their repetition along the same macromolecular chain lead to considerable cohesion energies that are in turn responsible for the peculiar mechanical properties of polymeric materials.

2.1. MOLECULAR INTERACTIONS

Three types of interactions are responsible for the cohesion observed in polymers.

2.1.1. Van der Waals Interactions

These are attraction forces between dipoles, which can have various origins.

Keesom forces correspond to the mutual attraction between two permanent dipoles. The energy of interaction (ϵ_K) is given by the relation

$$\epsilon_K = -(2\mu^4/3RT)r^{-6}$$

where μ represents the dipole moment of the polarized molecular group and r represents the distance between dipoles, R and T being the gas constant and absolute

Organic and Physical Chemistry of Polymers, by Yves Gnanou and Michel Fontanille
Copyright © 2008 John Wiley & Sons, Inc.

Figure 2.1. Keesom interaction in a linear polyester.

temperature, respectively. Such interactions are formed in polymers having polar groups such as poly(alkyl acrylate)s, cellulose esters, and so on. The corresponding cohesion energy varies from $\sim$0.5 to 3 kJ·mol^{-1}. Figure 2.1 shows how such an interaction is established.

Debye forces (or *induction* forces) correspond to the mutual attraction of a permanent dipole and the dipole that it induces on a nearby polarizable molecular group:

$$\epsilon_D = -2\alpha\mu^4 r^{-6}$$

where α represents the polarizability of the polarizable molecular group. The cohesion energy corresponding to this type of molecular interaction varies from 0.02 to 0.5 kJ·mol^{-1}. Figure 2.2 gives an example of such an interaction.

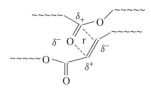

Figure 2.2. Debye interaction in an unsaturated polyester.

London forces (or *dispersion* forces) result from the asymmetric nature of the instantaneous electronic configuration of atoms. The energy developed between two instantaneous dipoles is given by the following relation:

$$\epsilon_L = -3/2[\alpha_1\alpha_2 I_1 I_2/(I_1 + I_2)]r^{-6}$$

where α_1 and α_2 denote the polarizabilities of the interactive groups, and I_1 and I_2 denote the corresponding ionization energies. These are low-energy forces for organic molecules with small atomic number (0.5 to 2 kJ·mol^{-1}) and have important effects mainly in the case of the compounds that do not have polar groups (polyethylene, polybutadiene, etc.).

Whatever the type of interaction, one has to bear in mind that the energy produced by van der Waals interactions scales with r^{-6}, which explains that both intra- and intermacromolecular interactions contribute to the cohesion of polymeric systems.

2.1.2. Hydrogen Bonds

Hydrogen bonds differ from van der Waals interactions by their strength. They arise from electrostatic or ionic interactions and, in certain cases, even from covalent bonds. Hydrogen bonds are formed between a hydrogen atom carried by a strongly electronegative atom (F, O or N) and another molecular group containing a strongly electronegative atom (O, N, F, etc., and sometimes Cl).

$$R_1-A-H--B-R_2 \qquad \text{(A and B are strongly electronegative elements.)}$$

Whatever their origin, these H bonds produce an energy that can attain $40\,kJ \cdot mol^{-1}$, a high value that results from the strong polarity of the bonds involved and the small size of the hydrogen atom, which can come very close to interacting groups. H bonds induce particularly high cohesion in the polymeric materials that contain them. Such interactions are found in proteins, and chemists copied Nature with the synthesis of polyamides (Figure 2.3). The presence of these H bonds explains the high tenacity of cellulose-based fibers and their high hydrophilicity even though they are insoluble in water.

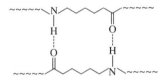

Figure 2.3. Hydrogen bonds in polycaprolactam (PA-6).

2.1.3. Ionic Bonds

Bonds of this type are sometimes generated to increase cohesion in polymers. Such polymers are called *ionomers*. When anions (carboxylates, sulfonates, etc.) carried by the polymeric chain are associated with monovalent cations, they form ion pairs that are assembled in aggregates, thus leading to a physical cross-linking of the macromolecular systems. When the same anions are associated with bivalent cations (Ca^{2+}, Zn^{2+}), the latter establish, in addition to the aggregates, bridges between chains. For example, acrylic acid can be copolymerized with a (meth)acrylic ester to give, after treatment with a zinc salt (Figure 2.4), an ionic bridging between chains.

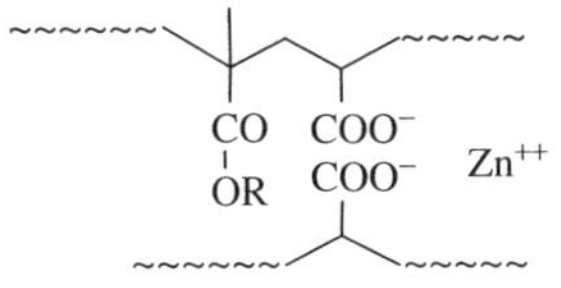

Figure 2.4. Ionic bonds in a (meth)acrylic copolymer with zinc carboxylate groups.

2.2. COHESION ENERGY IN POLYMERS

Many physical and mechanical properties of condensed state matter are determined by the strength of its internal molecular interactions. To quantitatively treat their effects, it is useful to define the notion of *cohesive energy*.

For a liquid, the *molar cohesive energy* (E_{co}) can be defined as the molar energy required to disrupt all molecular interactions; it is then possible to relate it with the heat of evaporation ΔH_{vap} by means of the following expression:

$$E_{co} = \Delta H_{vap} - RT$$

where RT corresponds to the work of the pressure forces.

The quality of the molecular interactions can be evaluated by means of the *specific cohesion* or *cohesion energy density,*

$$e = E_{co}/V \qquad (V = \text{molar volume in } cm^3 \cdot mol^{-1})$$

or even by the *solubility parameter* δ (Hildebrand theory),

$$\delta = (E_{co}/V)^{1/2} = e^{1/2}$$

For simple compounds, E_{co} can be calculated either from the heat of evaporation or from the variation of the vapor pressure with temperature. For macromolecular compounds, vapor pressure can be neglected and the transition to a gaseous state upon increasing the temperature could occur only by rupture and degradation of covalent bonds and formation of small molecules. The measurement of E_{co} requires the use of indirect methods such as the comparison of swelling or dissolution in liquids of known solubility parameter.

If one assumes that the cohesion energy is an additive parameter, then E_{co} is equal to the sum of the contributions of the various groups constituting the monomer unit. Hence, knowing the cohesion energy of each group, one should be able to calculate the actual value of E_{co}. As a matter of fact, the molar cohesion energies are not strictly additive, unlike what is called the *molar attraction constant* (F):

$$F = (E_{co} \cdot V_{298 \ K})^{1/2}$$

For example:

$$F_{CH_2} = 263 \ J^{1/2} \cdot mol^{-1} \cdot cm^{3/2}$$
$$F_{C=O} = 526 \ J^{1/2} \cdot mol^{-1} \cdot cm^{3/2}$$
$$F_{C\equiv N} = 708 \ J^{1/2} \cdot mol^{-1} \cdot cm^{3/2}, \text{ etc.}$$

Table 2.1. Molar cohesive energy of several important polymers

Polymer	Formula	V (cm^3·mol^{-1})	δ_{exp} (J$^{1/2}$·cm$^{3/2}$)	E_{co} (J·mol^{-1}) (calculated from δ)	E_{co} (J·mol^{-1}) (calculated from F)
Polyethylene		33.0	16.5	9,000	8,500
Polyisobutene		66.8	16.3	17,800	18,200
Polystyrene		98.0	18.2	32,000	36,000
Poly(vinyl chloride)		45.2	21.3	19,300	17,800
1,4-*cis*-Polybutadiene		60.7	17.1	17,500	17,600
Poly(ethylene terephthalate)		143.2	20.5	62,000	60,500
Poly(hexamethylene adipamide)		208.3	28.0	161,000	150,000

One can then deduce E_{co} for a polymer of known molecular structure and compare the obtained value with that experimentally determined (Table 2.1).

Compounds that exhibit a low cohesion energy density—polyalkadienes, EPDM rubber, polyisobutene, polysiloxanes, and so on—can be used as elastomers, provided that they are vulcanizable and noncrystalline in the absence of stress.

Materials having a higher cohesion can be used as plastics [polystyrene, poly (methyl methacrylate), etc.].

Finally, compounds whose cohesion is very high can be used for the manufacture of textile fibers for which the mechanical properties must be excellent in order to ensure a high tenacity (polyamides, polyacrylonitrile, etc.).

The cohesion of a polymer determines also its capacity to dissolve in a solvent of a given cohesion. Polymer dissolution in solvent entails the replacement of polymer–polymer interactions by polymer–solvent interactions (see Chapter 4).

LITERATURE

D. W. van Krevelen, *Properties of Polymers*, 3rd edition, Elsevier, Amsterdam, 1990.

J. Brandrup and E. J. Immergut, *Polymer Handbook*, 3rd edition, Wiley, New York, 1989.

A. F. M. Barton, *Handbook of Solubility Parameters and Other Cohesion Parameters*, CRC Press, Boca Raton, (FL), 1983.

J. Bicerano, *Prediction of Polymer Properties*, Marcel Dekker, New York, 1996.

3

MOLECULAR STRUCTURE
OF POLYMERS

Even though the main thrust of this textbook is to focus on organic polymers, it should be noted that there is no fundamental difference between organic and inorganic polymers. Indeed the nature of the atoms that constitute polymeric chains has relatively little effect on the basic properties of the polymer, the latter being mainly governed by the macromolecular character of these substances.

The term "structure" has quite different meanings in the field of polymers; it can refer to a sequence of atoms, a sequence of monomeric units, a chain as a whole, or an assembly of a more or less large number of chains. This is why it is necessary to propose a specific denomination for each one of them. One major difference with other fields of chemistry is the fact that synthetic polymers are far from being perfect structurewise, owing to the methods commonly used to produce them; whatever the level considered (molecular or higher), flaws always exist which affect most of the properties of the resulting materials in proportion to their occurrence.

3.1. TOPOLOGY AND DIMENSIONS

For the sake of simplicity, a polymer chain can be visualized as a very long noodle or a worm whose dots would correspond to successive monomeric units (Figure 3.1). The chain drawn in Figure 3.1 consists of two ends and is called linear or *monodimensional*; its molar mass, which is related to its size, has a finite value. All polymers having a finite size irrespective of their topology (or

Organic and Physical Chemistry of Polymers, by Yves Gnanou and Michel Fontanille
Copyright © 2008 John Wiley & Sons, Inc.

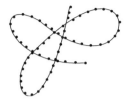

Figure 3.1. Worm-like structure of a linear polymer.

structures)—branched polymers, stars, combs, ladders, and macrocycles (Figure 3.2)—belong to this category.

Another way to account for the dimensionality of a macromolecular system consists in assigning a valence to each monomeric unit, which corresponds to the number of covalent bonds that it forms with the nearest monomeric units. Thus, the average valence of a polymer sample ($\bar{v}$) can be defined using the relation

$$\bar{v} = \frac{\sum_i n_i v_i}{\sum_i n_i}$$

where n_i is the number of monomeric units of valence v_i (the bar over $\bar{v}$ symbolizes the averaging process).

In the case of a monodimensional polymer, $\bar{v}$ is equal to $(2 - \varepsilon)$; ε corresponds to the monovalence of the two chain-ends and is equal to $2/(\bar{X}_n)$, where $\bar{X}_n$ is the number-average degree of polymerization (see Section 3.4.2).

Remarks

(a) In the case of a macrocyclic polymer, $\varepsilon = 0$.

(b) For polymers of high average degree of polymerization, ε is generally neglected.

When the average valence is higher than 2, this implies that all monomeric units of a sample are connected to each other by covalent bonds and that a unique polymer chain of macroscopic dimension and same size as that of the sample is eventually obtained. One can then regard its molecular size as "infinitely large," and the system is known as three-dimensional (see Figure 1.3).

The higher the average valence of a system, the greater its density of crosslinking. The description and the characterization of polymer networks will be described in Section 3.5.

Remark.   Because reactive groups have less opportunity to react with each other in a dense network, it is difficult to obtain polymeric systems with $\bar{v} > 3$ and hence, generally, $2 < \bar{v} < 3$.

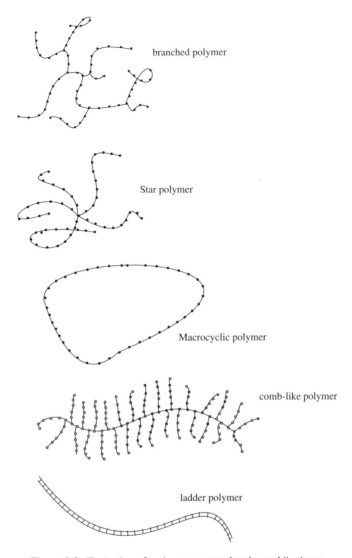

branched polymer

Star polymer

Macrocyclic polymer

comb-like polymer

ladder polymer

Figure 3.2. Illustration of various macromolecular architectures.

The concepts that will be developed hereafter are mainly relative to linear polymers; they can also be applied to chains linking two junctions of a polymer network, provided that they are sufficiently loose.

3.2. ARRANGEMENT OF MONOMERIC UNITS IN POLYMERS

Two main categories of polymers can be distinguished by whether they result from the polymerization of one or several monomers.

3.2.1. Homopolymers

When monomeric units are linked one after another in accordance with a regiose-lective process, the polymer is called *regular*. In the case of a vinyl polymer, this kind of structure is called *head-to-tail*:

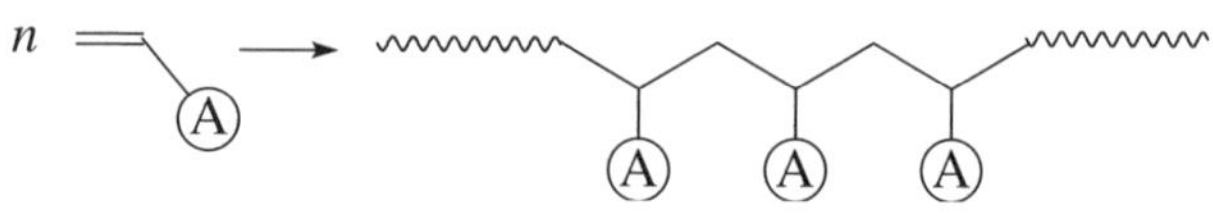

In the opposite case, it is called *irregular*. The following scheme shows successively head-to-head and tail-to-tail structures; this arrangement of monomeric units forms an irregular *triad* (i.e., constituted by three successive monomeric units with an irregular linking):

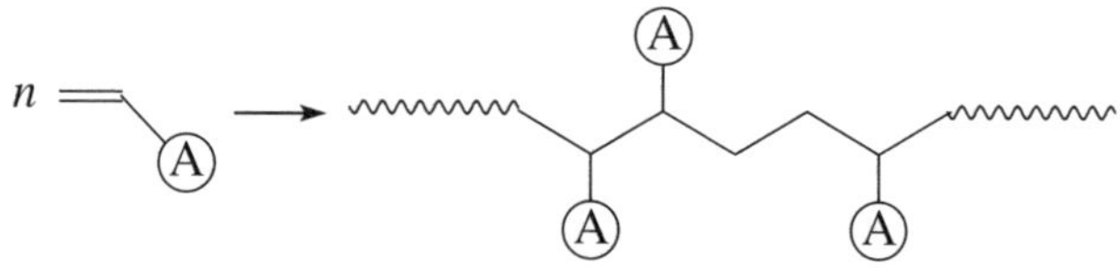

In synthetic polymers obtained by chain polymerization, most of the dyads are regular because active centers occur at the substituted carbon atom due to the possibility of stabilization by resonance and also due to steric effects (see Section 8.5.4).

Besides the positional isomerism described above, one can encounter another form of isomerism in homopolymers as illustrated by the polymerization of butadiene. Either 1,2 or 1,4 monomeric units may be incorporated in the polybu-tadiene chain, depending upon the experimental conditions used:

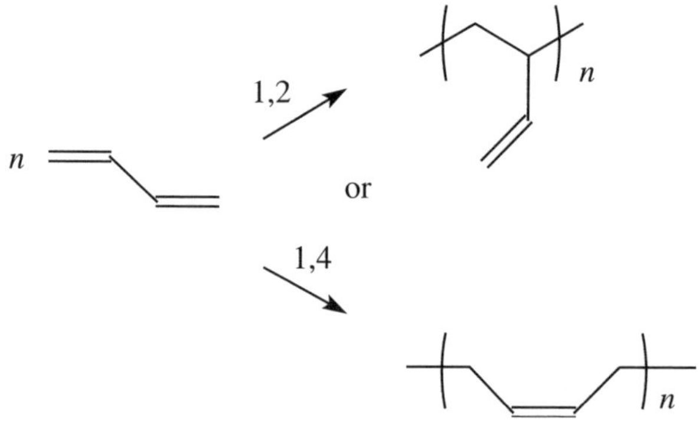

The relative proportion of 1,2 *versus* 1,4 units determines the properties of the resulting polymer.

In general, polymers obtained by step-growth polymerizations exhibit perfect regularity due either to the symmetry of the monomers used or to the fact that the condensation reaction considered cannot generate irregularity.

3.2.2. Copolymers

Copolymers are obtained from the polymerization of two different *comonomers*.

When more than two monomers are polymerized together: a *terpolymer* is generated from three comonomers, a *quaterpolymer* from four comonomers, and so on. For the sake of simplicity, only the various sequences found in *bipolymers* and resulting from the copolymerization of two different comonomers A and B will be described hereafter. They are designated as poly(A-*co*-B).

Statistical copolymers correspond to a certain statistics of the distribution of the comonomeric units along the chain backbone. This statistic is determined by the composition of the reaction mixture (content in each comonomer) during the polymerization and by the value of the reactivity ratios of a pair of comonomers (see Section 8.5.3). The two sequences shown hereafter, each one comprising eight consecutive monomeric units *(octad)*, have the same composition but a different number of alternations (-AB- or -BA-):

$$\sim\!\sim\!\sim\!\sim\!\sim\!\sim\text{-ABBABAAB-}\sim\!\sim\!\sim\!\sim\!\sim\!\sim\qquad \text{(5 alternations)}$$

$$\sim\!\sim\!\sim\!\sim\!\sim\!\sim\text{-AAABABBB-}\sim\!\sim\!\sim\!\sim\!\sim\!\sim\qquad \text{(3 alternations)}$$

The properties of statistical copolymers are determined, first of all, by their composition and, second, by the frequency of alternations; their properties are generally intermediate between those of the corresponding homopolymers (Figure 3.3).

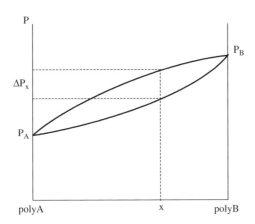

Figure 3.3. Variation of a given property (*P*) for a poly(A-*stat*-B) copolymer as a function of its composition *x*; ΔP_x mirrors the incidence of the frequency of alternations (AB or BA) on the property (*P*).

Remarks

(a) A particular case among statistical copolymers is that of *random* copolymers (Bernouillian statistics) in which the frequency of alternations is determined only by the composition of the reaction medium.

(b) When the detailed structure of a copolymer is not specified and is named poly(A-*co*-B), it is generally a statistical copolymer that should be denoted poly(A-*stat*-B).

Alternating copolymers, for which the number of alternations is maximum, are referred to as (-AB-)$_n$ and correspond to the equimolecularity of their composition. They could even be regarded as homopolymers, the "monomeric" unit of which would be (-AB-). For example, polyhexamethyleneadipamide, usually called *polyamide-6,6*,

and obtained from the perfect alternating copolymerization of hexamethylenediamine with adipic acid (polyamidification), can be considered as a homopolymer or a copolymer indifferently. Alternating copolymers are designated as, poly(A-*alt*-B) with, for example, the case of poly[styrene-*alt*-(maleic anhydride)]:

Block copolymers are a category of copolymers that exhibit very long sequences of each comonomer; these sequences have a macromolecular size and are referred to as blocks:

$$\underbrace{\text{A}\sim\text{AAAA}\sim\sim\sim\text{AAAA}}_{\text{PolyA}}\underbrace{\text{BBBB}\sim\sim\sim\text{BBBB}\sim\sim\text{B}}_{\text{PolyB}}$$

The copolymers with two blocks have only one alternation, -AB-; those with three blocks two alternations; and so on. One designates them as poly A-block-poly B for a copolymer with two blocks of poly(A) and poly(B).

Instead of a single alternation from the poly(A) block to the poly(B) block, certain copolymers exhibit a progressive transformation from A to B as indicated below:

~~~~~~~~AAAAAAABAAAABAAAABBAABBAABBBABBBBBBBB~~~~~~

They are called *tapered block copolymers*.

Copolymers referred to as *segmented* are copolymers with short and numerous blocks. They are generally obtained by a *chain extension* process from difunctional oligomers.

Due to the nonmiscibility between blocks of different chemical nature (see Section 4.4), block copolymers are generally segregated in multiphase systems in which each phase preserves its properties. In that respect, they are definitely different from statistical copolymers having the same overall composition.

Graft copolymers consist of a backbone of poly(A) carrying side chains of poly(B), also called *branches* or *grafts*. Grafts can be regularly or randomly distributed along the skeleton:

```
~~~~ AAAAAAAA~~~~~~~AAAAAAAAAAAAA~~~~
         B              B             B
         B              B             B
         B              B             B
         B              B             B
         B                            B
         B                            B
                                      B
                                      B
```

Graft copolymers are also subject to phase segregation and give rise to morphologies that are different from those found in block copolymers. When each monomeric unit of the backbone carries a graft of the same length, the corresponding copolymers are called *comb-like copolymers*:

PolyA

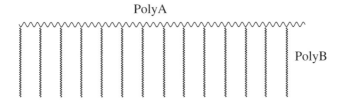

PolyB

3.2.3. Characterization of the Arrangement of Monomeric Units

Most of the methods commonly used for the characterization of the molecular structure of organic molecules can be applied to the identification of polymers—in particular, conventional absorption spectroscopies (NMR, IR, UV–visible). However, it is important to point out that the determination of the copolymer composition by UV spectroscopy can lead to erroneous values; as a matter of fact, interactions between local chromophoric groups can modify the values of the overall molar extinction coefficients as compared to those of simple molecules taken as models.

There are also a number of methods that are specific to the characterization of the unit arrangement; only two characteristic examples will be given here to illustrate them.

(a) *Determination of Structural Defects in Poly(vinyl acetate).* A very sensitive method is required in the latter case because the percentage of these defects is generally low and that of two successive defects is even lower.

Regular dyad   Irregular dyad

After measurement of its number-average molar mass (Section 4.3), this polymer is subjected to hydrolysis to give poly(vinyl alcohol) (PVAL), before being selectively oxidized by periodic acid (HIO_4). The regular dyads of PVAl corresponding to a 1,3-diol-type structure are not oxidized, whereas the irregular dyads of 1,2-diol type are cleaved to give aldehyde functional groups:

Regular dyad   Irregular dyad

Each defect is the cause of the chain cleavage, and, for an average of two defects per chain, the initial $\overline{M}_n$ will be divided by a factor of 3. From the number-average molar masses before and after oxidation, it is possible to calculate the average number of defects per chain.

(b) *Measurement of the Composition of a Poly[(vinylidene chloride)-stat-(vinyl chloride)] Copolymer and Determination of the Number of Alternations.* Poly(vinyl chloride) (PVC) decomposes upon heating to give benzene in two steps:

Poly(vinyl chloride) triad

As for poly(vinylidene chloride) (PVDC), it decomposes upon heating by the same mechanism and leads to the formation of trichlorobenzene:

Poly(vinylidene chloride) triad   1,3,5-Trichlorobenzene

Mixed triads of such a copolymer give monochlorobenzene and dichlorobenzene respectively upon decomposition.

A technique such as *pyrolytic chromatography*, which brings about the thermal decomposition of a polymer and the subsequent analysis of decomposed products by gas-phase chromatography, is useful in the latter case. It allows: (a) the identification of the four benzenic derivatives formed by decomposition of the various types of triads; and (b) the quantitative determination of the concentration of these compounds and their respective proportions. From such characterization it is possible to establish the molecular structure of the original copolymer.

Due to their extremely low concentration, the identification and quantitative determination of the molecular groups located at the chain ends is very difficult. Mass spectrometry by ablation of a matrix in which the polymer is dispersed by a laser radiation (MALDI) has opened very promising prospects for the resolution of this problem (see Section 6.3).

3.3. CONFIGURATIONAL STRUCTURES

Rules that are used to represent and designate simple organic molecules apply to macromolecular systems; however, because of the repetition of sites giving potentially rise to configurational isomerism, new structural features have to be considered.

3.3.1. Tacticity

Tacticity mirrors the existence of a relative configurational regularity of successive monomeric units along the macromolecular chain.

3.3.1.1. Simple Tacticity. This notion is particularly interesting in the case of vinyl and related polymers. If the polymerization reaction is stereospecific and generates successive units that are inserted into the growing chain with the same absolute configuration as that of the preceding unit, the polymer is described as *isotactic*. To represent it, one uses the method recommended by Fischer for simple organic molecules with a rotation of $\pi/2$ in the plane of the sheet as per convention. Hence, an isotactic vinyl polymer can be represented as

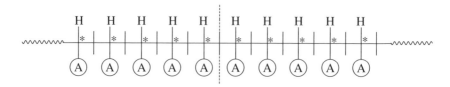

Due to its symmetry and intramolecular compensations, the resulting perfect macro-molecule does not exhibit any optical activity even if each tertiary carbon atom is intrinsically asymmetrical. The situation is analogous to that of simple organic molecules such as D, L-tartaric acid. However, with synthetic polymers, one cannot expect a perfect arrangement of the successive configurational units, and therefore the compensation observed reflects the average configuration. The isotactic content is defined as the proportion of triads comprising monomer units having the same configuration ([R] or [S]):

$$\sim\sim\sim\sim\text{-[R]-[R]-[R]-}\sim\sim\sim\sim\text{-[S]-[S]-[S]-}\sim\sim\sim\sim$$

For example, commercial isotactic polypropylene exhibits an isotactic content higher than 96%.

Another type of stereoregularity results from the alternation of configurations of successive monomeric units. It can be schematized as

$$\sim\sim\sim\sim\text{[R]-[S]-[R]-[S]-[R]-[S]-[R]-[S]}\sim\sim\sim\sim$$

and, for vinyl polymers, represented by

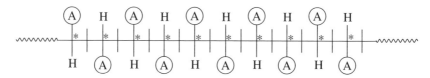

In this case, the polymer is known as *syndiotactic* and the syndiotactic content is defined as the proportion of -[S]-[R]-[S]- or -[R]-[S]-[R]- triads in the polymer under consideration.

When the polymerization reaction is not stereospecific, the resulting polymer is known as *atactic*. In the case of a total configurational disorder,

25% of triads are isotactic -[R]-[R]-[R]- or -[S]-[S]-[S]-,

25% of triads are syndiotactic -[S]-[R]-[S]- or -[R]-[S]-[R]-, and

50% of triads are heterotactic -[R]-[R]-[S]- or -[R]-[S]-[S]- or
-[S]-[S]-[R]- or -[S]-[R]-[R]-.

Hence, a proportion of isotactic or syndiotactic triads higher than 25% indicates a tendency to *stereoregulation* in the polymerization reaction.

It is interesting to consider the configurational and tactic arrangement of sequences larger than triads. This is possible upon using the two configurational possibilities offered by a dyad:

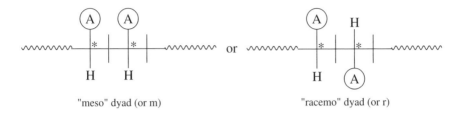

       "meso" dyad (or m)   "racemo" dyad (or r)

Thus, the three types of triads with different relative configurational structure can be distinguished on the basis of m and r dyads. To be meaningful, such a representation requires that the last monomer unit of the preceding dyad be taken into consideration in the definition of the next dyad; it is just a consequence of the covalent linking of the constituting units:

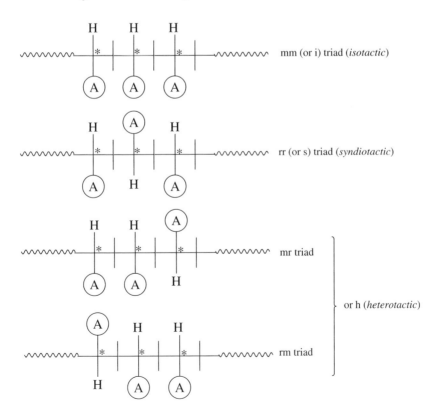

> **Remark.** Since only the relative configurations are considered, it is not necessary to differentiate mr and rm triads in the case of vinyl and related polymers. It is different for polymers exhibiting a "true" asymmetry for each monomer unit (see Section 3.3.2).

By the same principle, longer sequences can be described; for example, those corresponding to the arrangement of seven consecutive monomer units are shown below:

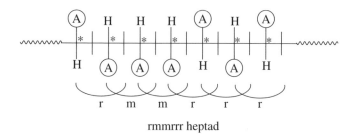

rmmrrr heptad

3.3.1.2. Ditacticity.
Ditactic polymers possess two centers of asymmetry per monomeric unit; it is the case of 1,2-disubstituted vinyl polymers whose two chiral centers are generated upon polymerization, and it is also the case of heterocycles carrying two centers of asymmetry. The three possible tacticities are schematically represented as follows:

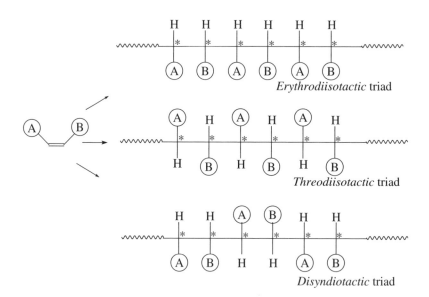

Erythrodiisotactic triad

Threodiisotactic triad

Disyndiotactic triad

3.3.2. Optical Activity

Two cases must be distinguished.

The first one is that of polymers obtained from optically active monomers for which the optical purity of the polymer is that of the original monomer. In this case, the polymerization reaction does not modify the chirality of the monomer unit as compared to that of the initial monomer. Macromolecules obtained from the polymerization of ethylenic monomers having an optically active lateral group correspond to this case; for example, optically active 4-methylhex-1-ene generates an optically pure polymer whose chiral centers are located on the side-chain:

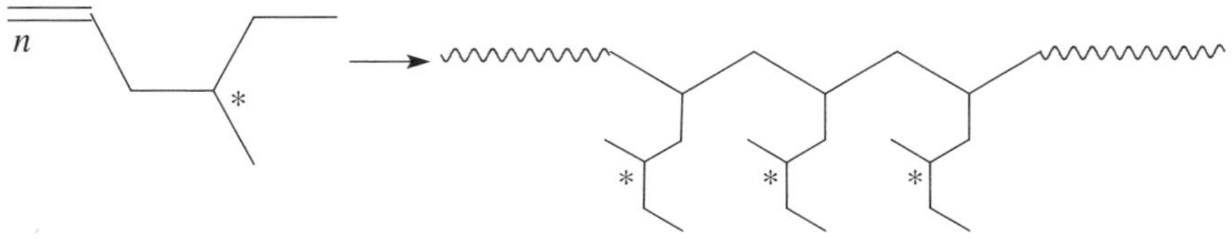

Poly([R] or [S]-4-methylhex-1-ene) triad

Polymers obtained from certain substituted heterocycles are also optically pure, provided that each monomeric unit possesses a true asymmetry. It is the case of the polymethyloxirane (PPOx) whose mm triad is represented in the following example:

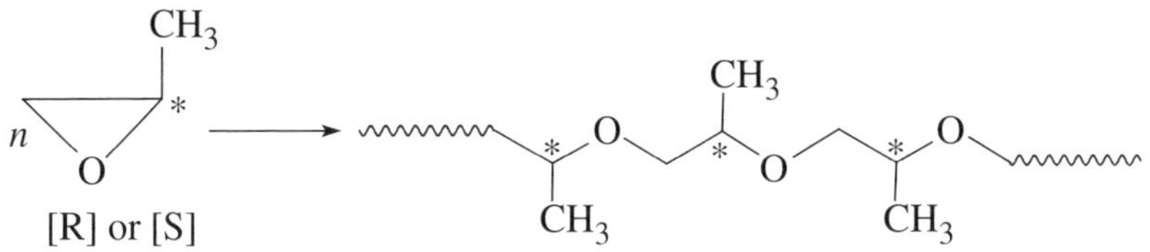

[R] or [S]

Polycondensates obtained from an optically active monomer enter also in this category:

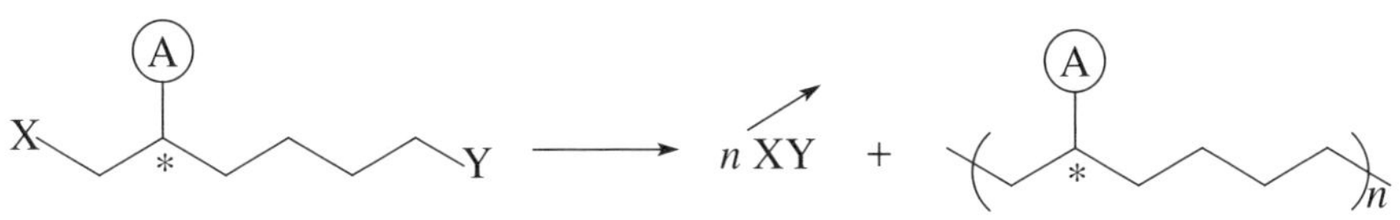

The second case is that of polymers obtained from pro-chiral monomers and whose optical activity is the result of a stereoregulating polymerization process. The two examples shown below illustrate this situation:

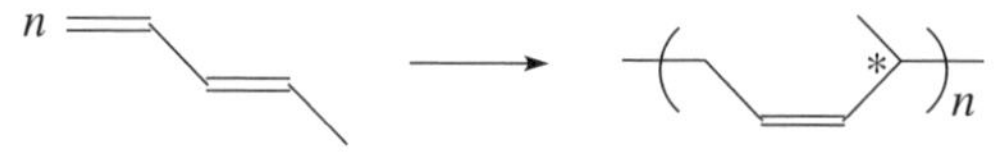

Poly(penta-1,3-diene)

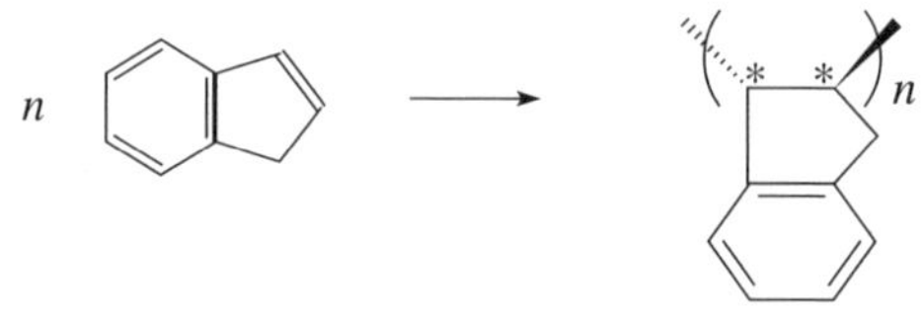

Polyindene

In the latter case, the polymerization generates two asymmetric centers per monomeric unit, which also leads to *ditacticity*.

3.3.3. Geometrical Isomerism

Geometrical isomerism is analogous to that found in simple organic molecules, and the relative configurations of the successive units are not to be taken into consideration. Two kinds of geometrical isomerism can then be distinguished:

- *Z/E Isomerism (generally indicated by cis/trans).* This is found in polymers obtained from conjugated dienes or by ring-opening metathesis polymerization of cycloalkenes (polyalkenamers). The example illustrated hereafter is that of 1,4-polyisoprene (or 2-methylbuta-1,3-diene). It is important from an economical point of view—in particular, in the industry of natural and synthetic elastomers.

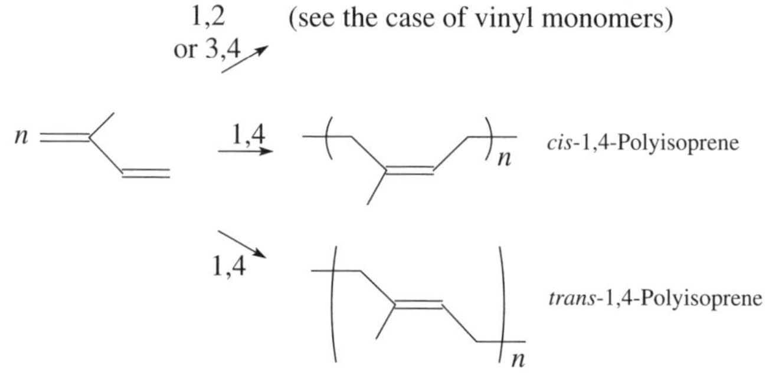

cis-1,4-Polyisoprene

trans-1,4-Polyisoprene

Remark.   Natural rubber is almost perfectly regular 1,4-cis-polyisoprene; the mechanism of its formation is complex. Gutta-percha is the natural trans isomer. One can also obtain synthetic 1,4-polyisoprenes with a great configurational regularity by polymerization of isoprene.

- *cis/trans Isomerism (for polycyclic compounds)*. It is seldom considered because of the characterization difficulties. As a matter of fact, very few polymers are subject to cis/trans isomerism. For example, in saturated polynorbornene,

Norbornene   Polynorbornene

the a and b carbon atoms can give rise to cis/trans isomerism and the saturated polynorbornene shown is the trans isomer.

Polynorbornenamer obtained by ring-opening metathesis polymerization (ROMP) of norbornene can also produce two types of geometrical isomerism. The monomeric unit represented hereafter has a "Z" configuration relative to the double bond and a "trans" configuration relative to the cyclopentylene ring.

Tacticity based on relative geometrical configurations of successive monomeric units can be regarded and treated as that observed in the sequences of monomeric units carrying an asymmetrical carbon atom; however, it has little usefulness.

3.3.4. Determination of the Configurational Structures

To characterize the optical activity and the geometrical isomerism of polymers, one can rely on the conventional methods used in organic chemistry. The characterization of tacticity is almost exclusively the domain of NMR spectroscopy; the particular case of vinyl and related polymers will be discussed in detail further in this textbook.

It was previously shown why vinyl and related polymers do not exhibit optical activity even if they are strongly isotactic: they are pseudo-chiral with respect to the whole chain, but tacticity takes only into account the relative configurations and consequently the immediate neighbors of the site under consideration.

3.3.4.1. ¹H NMR. In vinyl and related polymers, two types of sites can be considered stereosensitive.

First, **methylene protons** ($-H_aCH_b-$) whose two immediate neighbors are pseudo-asymmetrical carbon atoms. With these first neighbors, two cases can arise:

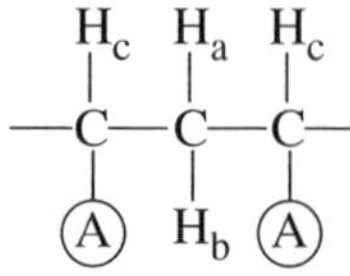

corresponding to a meso dyad

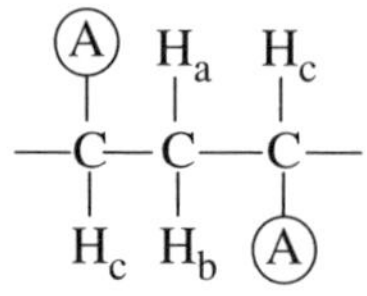

corresponding to a racemic dyad

H_a and H_b protons of the m dyad are not in chemically equivalent environments and will thus display two different frequencies of resonance, producing two different NMR signals. On the other hand, in the r dyad the two methylenic protons have an identical environment and will give only one NMR signal. Thus, in a polymer containing the two types of dyads the methylene protons are to produce three signals whose intensity will be proportional to the content in m and r dyads, respectively:

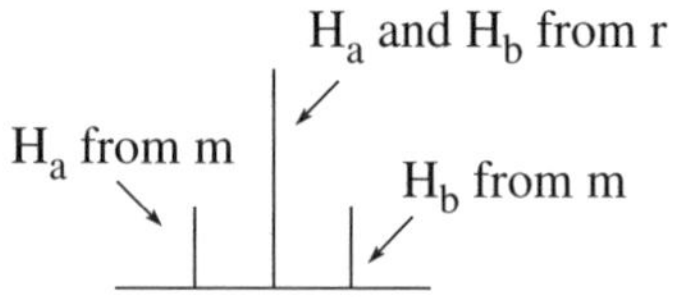

With NMR spectrometers functioning at higher magnetic fields, spectral features arising from still longer sequences consisting of an even number of monomer units (tetrads, hexads, octads) can be detected. For example, the mmm tetrad shown hereafter will give two signals for $-H_aCH_b-$, whereas the mrr tetrad will produce only one:

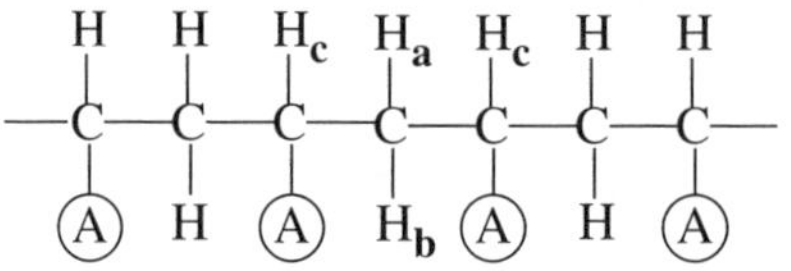

mmm tetrad

or

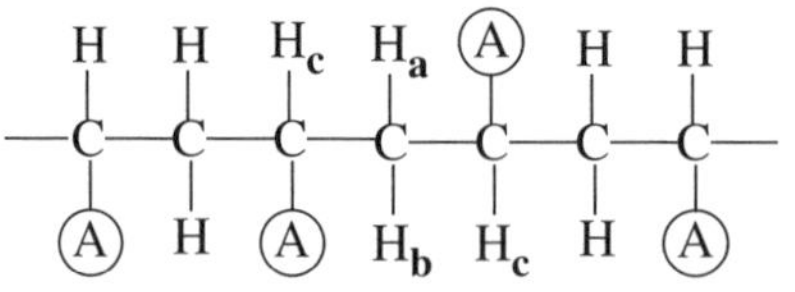

mrr tetrad

Thus, the signal (or the two signals) corresponding to each type of dyad is split into as many secondary signals as there exist different tetrads; their intensity is proportional to their relative concentration. These multiple signals still correspond to the resonance of H_a and H_b protons, but their magnetic environment can be disturbed by the relative configuration of more distant asymmetrical carbon atoms. In the case of certain polymers (PMMA, etc.) the stereosensitivity of H_a and H_b protons is very high and by using high-resolution NMR spectrometers and high temperature, it is possible to identify very long even sequences (hexads, octads, etc.).

Methyne protons $-CH_cA-$ also have a strong stereosensitivity. These groups are flanked by two methylene groups, which are symmetrical and therefore have no stereochemical effect on the resonance of H_c protons. In contrast, protons on neighboring tertiary carbon atoms (which are asymmetrical) do affect the resonant frequency of H_c. Three different situations corresponding to three types of triads can be singled out:

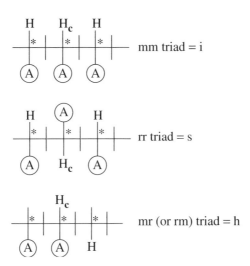

The frequency of the magnetic resonance of H_c is thus affected by the relative configurational structure of the monomeric unit under consideration and that of its two neighbors, these three units corresponding to a triad whose central position is occupied by H_c. For a polymer containing three types of triads, one obtains three main signals for each group of nonequivalent protons, whose relative intensity is proportional to the relative concentration of the triads they represent:

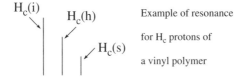

For vinyl polymers it should be emphasized that mr and rm triads correspond to the same magnetic environment for H_c and thus resonate as a single signal.

Figure 3.4 shows in the same NMR spectrum, the signals arising from various groups of protons contained in m dyads ($-CH_2-$) and in mm triads ($-HC-$ and $-CH_3$) of isotactic polypropene. Their position and multiplicity resulting from spin–spin

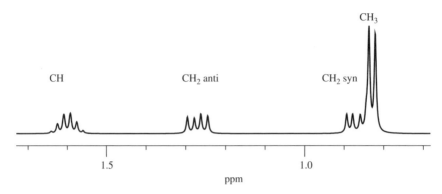

Figure 3.4. ¹H NMR spectrum (simulated at 400 MHz) of isotactic polypropylene. Parameters which are used are the following ones: $\delta_{CH_3} = 0.80$ ppm, $\delta_{CH_2} = 1.20-1.35$ ppm, $\delta_{CH} = 1.6$ ppm. $^3J_{[CH-CH_3]} = 6.5$ Hz, $\delta_{CH_3} = 0.80$ ppm, $\delta_{CH_2} = 1.20-1.35$ ppm, $\delta_{CH} = 1.6$ ppm. $^3J_{[CH-CH_3]} = 6.5$ Hz, $^3J_{[CH-CH_2]} = 8$ Hz, $^2J_{[CH_2]} = -10$ Hz $^3J_{[CH-CH_2]} = 8$ Hz, $^2J_{[CH_2]} = -10$ Hz.

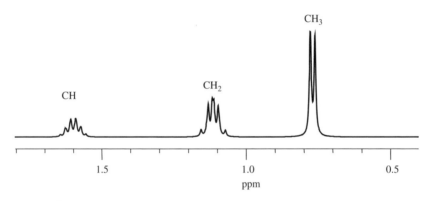

Figure 3.5. ¹H NMR spectrum (simulated at 400 MHz) of syndiotactic polypropylene. Parameters that are used are the same as for Figure 3.4.

couplings between protons were simulated. In the case of syndiotactic polypropene, the signal corresponding to H_a and H_b is single (Figure 3.5).

As in the case of methylene protons, an increase in the magnetic field of an NMR spectrometer can reveal effects and consequently more subtle long-range magnetic influences; the relative configuration of the immediate neighbors can be identified under these conditions. These systems form sequences with an odd number of monomeric units whose relative configuration of the constituting units is mirrored by the resonant frequency of H_c. Starting from central mm triads (i) which produces a main signal, one can detect the effect of the mmmm, rmmm (or mmmr) rmmr pentads: the main signal is split into three secondary signals. Similarly, the signal of mr (or rm) triads (h) is split into four secondary signals corresponding to mmrm (or mrmm), mmrr (or rrmm), rmrm (or mrmr), and rmrr (or rrmr) pentads. Finally, the signal corresponding to rr triads (s) is split into three signals corresponding to mrrm, rrrm (or mrrr), and rrrr. Hence, the observation of pentads can lead to 10 different signals whose intensity gives information about the configurational statistics of the polymerization reaction (see Section 8.3.4).

The number N_x of the different x-ads is given by the following relation:

$$N_x = 2^{(x-2)} + 2^{(y-1)}$$

where $y = x/2$ if x is even and $y = (x - 1)/2$ if x is odd. The assignment of the signals requires statistical considerations, which will be presented in Section 8.3.

3.3.4.2. ^{13}C NMR. Unlike ^{1}H, ^{13}C is a rare nucleus (only 1% in abundance) and therefore ^{13}C–^{13}C couplings do not usually show up in the spectrum. Through spin decoupling by irradiation of the protons nuclei with a strong radio-frequency field, ^{13}C–^{1}H couplings can also be eliminated, so that each resonance exists as a single narrow line. This simplifies the spectrum, makes the assignment easier, and provides better resolution (over 200 ppm) so that minor changes in structure cause a shift of carbon atoms that are a few bonds away.

However, the structural characterization of polymers by ^{13}C NMR does not lend itself to an easy quantitative determination of the polymers being studied. Indeed, it is well known that the carbon atoms exist in different hybridization states with different relaxation times. Moreover, due to the Overhauser effect, decoupling by irradiation of the protons modifies the population of the excited ^{13}C atoms. Consequently, it is difficult to consider that the intensity of the signals is directly proportional to the population of carbon atoms which they represent. Nevertheless, for the determination of tacticity, one generally considers that all stereosensitive carbon atoms are in the same hybridization state and that the intensities of their signals are an accurate measure of the content in different x-ads. Figure 3.6 shows

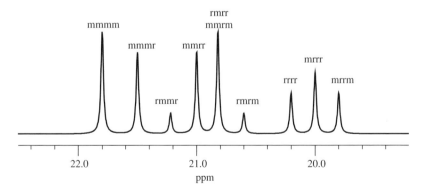

Figure 3.6. Simulated ^{13}C NMR spectrum (at 100.6 MHz and with protons decoupling) in the range of CH_3 resonance. This spectrum points out the different pentads of an atactic polypropylene.

an example of such a spectrum: the proportion of m dyads seems to be definitely higher than that of the r dyads; one can deduce from this spectrum that the corresponding polypropene exhibits a tendency to isotacticity.

3.4. DISPERSITY OF MOLAR MASSES – AVERAGE MOLAR MASSES

3.4.1. Dispersity*

Except for certain natural polymers (enzymes) and those recently obtained by genetic engineering (perfectly defined polymers), the immense majority of macromolecular systems are disperse. This means that they consist of chains having different sizes (and consequently different molar masses), such a fluctuation arising from the random nature of the polymerization reactions.

The most practical representation of the dispersity of molar masses is shown in Figure 3.7, which consists in plotting either the number (N_i) of moles of species corresponding to a degree of polymerization (X_i) or their mass (N_iM_i), *versus* either the degree of polymerization (X_i) or the corresponding molar mass (M_i). This representation of $N_i = f(X_i$ or $M_i)$ affords a curve that gives information about the numeral distribution of chains.

A slightly different representation consists in plotting the mass of the chains corresponding to a degree of polymerization (X_i)—that is, N_iM_i—*versus* X_i (or M_i); one then obtains a curve that gives the distribution of chains according to their mass. Such curves can be established by means of any method that allows one to fractionate a broadly distributed sample into a large number of low dispersity fractions whose mass and molar mass can be measured.

Complementary to differential distribution functions, it might be useful to establish integral (cumulative) distribution functions (Figure 3.8)—in particular, when one is interested in the proportion of the population to which a property value applies or not. The most probable molar mass (molar mass at the peak in Figure 3.7) corresponds to the inflection point of the curve (Figure 3.8).

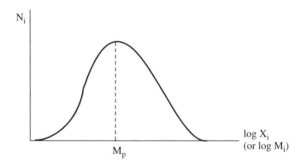

Figure 3.7. Diagram representing the chain-length distribution of a polymer sample.

*The term "Dispersity" was recently recommended by the IUPAC Subcommittee on Macromolecumar Terminology.

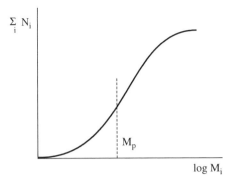

Figure 3.8. Diagram representing the integral (cumulative) chain-length distribution of a polymer sample.

After defining the average molar masses in the next section (Section 3.4.2), the various methods of quantifying the molar-mass dispersity will be presented.

> **Remark.** Although it is not recommended by the competent committees, the terms *polydispersity* and *monodispersity* (as well as the corresponding adjectives: *polydisperse* and *monodisperse*, respectively) are commonly used instead of *high dispersity* and *low dispersity*. For reasons of etymology, the terms mono- and polydispersity are to be rejected.

3.4.2. Average Molar Masses

They are defined using N_i and M_i, which are a number of moles of species (i) and their molar mass, respectively.

The *number-average molar mass* $(\overline{M}_n)$ is defined as the sum $(\sum_i)$ of the molar masses (M_i) of all the i families of species present in the system times their number fraction, that is, $N_i/\sum_i N_i$:

$$\overline{M}_n = \sum_i M_i \frac{N_i}{\sum_i N_i} = \frac{\sum_i N_i M_i}{\sum_i M_i}$$

The bar over $\overline{M}$ symbolizes the averaging process, but the IUPAC Nomenclature Subcommittee recently suggested $<M>_n$ as symbol of an average number, and it is easier to type.

Because $N_i M_i$ denotes the total mass of each family of species (i), their sum $\sum_i N_i M_i$ represents the total mass of the sample. Thus $\overline{M}_n$ is equal to the total mass of the sample divided by the total number of moles of polymeric species present. For the determination of $\overline{M}_n$, methods that count the number of molecules forming the sample are appropriate; these methods will be described in detail in Section 4.3.

The *mass-average molar mass* $(\overline{M}_w)$* is defined as the sum $\sum_i$ of the molar masses (M_i) of the i families of species present in the sample times their mass fraction, that is, $N_i M_i / \sum_i N_i M_i$:

$$\overline{M}_w = \sum_i M_i \frac{N_i}{\sum_i N_i M_i} = \frac{\sum_i N_i M_i^2}{\sum_i N_i M_i}$$

$\overline{M}_w$ of a polymer is then the ratio of the second moment of the number distribution of molar masses to the corresponding first moment.

According to the latter definition, longer chains have a higher statistical weight than that of shorter chains. The measurement of $\overline{M}_w$ is based on phenomena whose intensity is proportional to the size of the particles to be measured (see Section 4.3).

The *Z-average molar mass* $(\overline{M}_z)$ is seldom used to characterize synthetic polymers; it can be defined by a similar logic as above:

$$\overline{M}_z = \frac{\sum_i N_i M_i^3}{\sum_i N_i M_i^2}$$

The three preceding average molar masses result from logical definitions and are measured in $g \cdot mol^{-1}$.

Viscometric-average molar masses $(\overline{M}_v)$ cannot be defined with the same rigor since they result from an empirical expression, the Mark–Houwink relation correlating the *intrinsic viscosity* (or *limit index of viscosity*) $[\eta]$ to the molar mass M of a nondispersed (*isomolecular*) fraction of a polymer sample. The way to establish the relation

$$[\eta] = K M^\alpha$$

is discussed in Section 4.3.

In the case of a dispersed system, every family (i) is characterized by its own intrinsic viscosity,

$$[\eta_i] = K M_i^\alpha$$

and for an assembly of macromolecules of a sample, $[\eta]$ is equal to the sum of $[\eta_i]$ multiplied by their mass statistical weight:

$$[\eta] = \sum_i [\eta_i] \cdot \frac{N_i M_i}{\sum_i N_i M_i}$$

---

*This quantity, often miscalled *weight-average molecular weight*, reflects the chemist's legendary carelessness to distinguish between *mass* and *weight*.

If one sets $[\eta]$ is equal to

$$[\eta] = K\overline{M}_v^\alpha = K\frac{\sum_i N_i M_i^{(1+\alpha)}}{\sum_i N_i M_i}$$

then the definition of $\overline{M}_v$ can be deduced:

$$\overline{M}_v = \left[\frac{\sum_i N_i M_i^{(1+\alpha)}}{\sum_i N_i M_i}\right]^{1/\alpha}$$

For $\alpha \neq 1$, which is generally the case, $\overline{M}_v$ cannot be expressed in $g\cdot mol^{-1}$ because it is a relative average molar mass.

Remarks

(a) Generally, $0.50 < \alpha < 0.90$ and thus $\overline{M}_n < \overline{M}_v < \overline{M}_w < \overline{M}_z$.
(b) The number-average, the mass-average, and so on, degrees of polymerization ($\overline{X}$) can be deduced from the corresponding average molar masses. For example, $\overline{X}_n = \overline{M}_n/m_0$, in which m_0 is the molar mass of the monomeric unit.

There are two ways of **characterizing the dispersity**.

The first one consists in calculating the deviation ε from the number-average molar mass for every family of polymer chains exhibiting a molar mass M_i, that is $\varepsilon = |M_i - \overline{M}_n|$.

The number-standard deviation σ is the average of the square of deviations

$$\sigma^2 = \sum_i \frac{N_i}{\sum_i N_i}(M_i - \overline{M}_n)^2$$

$$\sigma^2 = \sum_i \frac{N_i}{\sum_i N_i}(M_i^2 - \overline{M}_n^2 - 2M_i\overline{M}_n)$$

$$\sigma^2 = \frac{\sum_i N_i M_i^2}{\sum_i N_i} - 2\overline{M}_n\frac{\sum_i N_i M_i}{\sum_i N_i} + \overline{M}_n^2$$

$$\sigma^2 = \frac{\sum_i N_i M_i^2}{\sum_i N_i} - 2\overline{M}_n^2 = \overline{M}_w\overline{M}_n - \overline{M}_n^2$$

which is generally used in the form

$$\sigma = [\overline{M}_n(\overline{M}_w - \overline{M}_n)]^{1/2}$$

Although very logical by its definition, the standard deviation σ is rarely used to measure the dispersity of polymer samples. The *dispersity index* is generally used:

$$D_M = \overline{M}_w / \overline{M}_n \qquad \text{(often called } I_p)$$

The dispersity index can vary from unity for perfectly nondispersed samples ("isometric" systems) up to several tens for samples characterized by a strong dispersion in the size or molar mass of the constituting macromolecules (highly dispersed systems).

3.5. POLYMER NETWORKS

3.5.1. Description of the Networks

Polymer chains can form a three-dimensional network when each one of them is connected to more than two of its homologs *through* monomeric units whose valence is equal to or higher than 3. These units are also called *junction points*. In contrast to crystal lattices, the term *polymer network* does not imply the idea of long-range order; it is used to indicate that the macromolecular structure created in this manner ranges over a three-dimensional space until it occupies the volume of its container—that is, the dimension of a macroscopic object. One can then speak of an infinite network whose ends are connected one to another through a single macromolecule of macroscopic dimensions. Such networks are insoluble and nonfusible; they can, at most, swell in a good solvent.

The mass of a mole of such a macroscopic molecule is obviously "infinitely large," but by no means infinite, which is often stated as a figure of speech. If *n* grams of monomer form one single macromolecule after cross-linking, the molar mass of the latter would be $M = n \times 6.02 \times 10^{23}$ g·mol^{-1} (according to Avogadro), which is an extremely large value of molar mass but certainly not an infinite one.

Just before cross-linking, an abrupt transition occurs during which the reaction medium passes from the state of a solution to that of an elastic gel, undergoing a *sol–gel transition* (which is also called *gel point*): at this stage, which corresponds to a specific conversion of the reactive functional groups (p_{gel}), the mass-average molar mass of the polymer as well as its viscosity diverge abruptly toward infinity.

Before the gel point the chains present in the reaction medium have finite size; that is, they are still soluble and fusible. Hence, the sol–gel transition is a critical phenomenon, which corresponds to a transition of connectivity that can be analyzed within the framework of the theory of *percolation*.

The first condition to obtain an infinite three-dimensional macromolecule and consequently a network is to introduce in the polymerizing system a branching agent whose valence is at least equal to three. Its role is to allow a nonlinear growth of the chains. However, this condition is not sufficient to generate an infinite macromolecule.

We note that the probability for two branching units to be connected by the sequence of units

$$-A + A-\diagdown\substack{A \\ \diagup \\ A} + B-B \longrightarrow \substack{\diagdown \\ \diagup}-A\!-\![B\text{-}BA\text{-}A]_x\!B\text{-}BA\!-\diagdown$$

is

$$p_a[p_b(1 - \rho)p_a]^x \cdot p_b\rho$$

where ρ is the proportion of a function carried by the branching unit, and p_a and p_b are the probabilities that A and B, respectively, have reacted. Then the general case of a branching unit being linked to another one by an elastic chain of any size can be expressed by the following probability $(\bar{\alpha})$:

$$\bar{\alpha} = \sum_{x=0}^{\infty}[p_b(1 - \rho)p_a]^x \cdot p_a p_b$$

which can also be written

$$\bar{\alpha} = \frac{p_a p_b \rho}{1 - [p_a p_b(1 - \rho)]}$$

If all A functions belong to the branching units and for stoichiometric conditions ($\rho = 1$, $p_a = p_b = p$), the previous equation reduces to

$$\bar{\alpha} = p^2$$

The critical value $(\bar{\alpha}_c)$ at which the system grows toward an infinite network can be deduced as follows: for a system with trivalent branching units ($v = 3$), each elastic chain reacting with one of these units has the possibility to be succeeded by two more chains. For the branching of successive chains to continue indefinitely, $\bar{\alpha}$ should be higher than 1/2; under these conditions, only the expected number of elastic chains in succeeding generations can be greater than the number of chains in the preceding ones.

In other words, n chains can give rise to $2n\bar{\alpha}$ chains with $2n\bar{\alpha} > n$ only for $\bar{\alpha} > 1/2$. For a system whose branching units is of valence v, gelation will occur for

$$\bar{\alpha}_c = \frac{1}{v - 1}$$

and at the critical conversion $p_c = \left(\frac{1}{v-1}\right)^{1/2}$.

Polymer networks can be obtained in various ways:

- By step-growth polymerizations between monomers (or precursors) carrying antagonistic functional groups and having an average valence higher than two.
- By chain copolymerization of a divalent vinyl or related monomer with a multi-unsaturated comonomer.
- By random cross-linking of chains carrying reactive functional groups (vulcanization, etc.).

In each of the cases, cross-linking occurs only under certain conditions that depend on the proportion and valence of the agent responsible for branching. For a three-dimensional step-growth polymerization involving antagonistic molecules A and B with the valence distribution

$$A_{v_1} + A_{v_2} + \cdots + A_{v_k} + B_{g_1} + B_{g_2} + \cdots + B_{g_j}$$

the gel point (p_{gel}) occurs at the following critical conversion:

$$p_{gel}^2 = \frac{1}{(v_e - 1)(g_e - 1)}$$

where v_e and g_e denote the average valence of monomers A and B, respectively.

$$f_e = \sum_i f_i a_{fi}$$

and

$$g_e = \sum_j g_j b_{gj}$$

a_{fi} and b_{gj} being the molar fractions of the various species present in the reaction medium and v_i and g_j their valence. If the ratio $1/(v_e - 1)(g_e - 1)$ is higher than 1, the system cannot generate an infinite macromolecule.

For chain copolymerization reactions involving a tetravalent monomer (e.g., a bis-unsaturated monomer $[CH_2 = CH(A) + CH_2 = CH(A) - (A)HC = CH_2]$), the conversion at the gel point can be written as

$$p_{gel} = \frac{1 - q}{(v - 2)aq}$$

where a is the molar fraction of the tetravalent monomer and q is the probability of chain growth [$q = R_p/(R_p + R_t)$, R_p is the rate of propagation, and R_t is the rate of termination].

In the case of chain vulcanization, where the latter react through side groups carried by the monomeric units or through unsaturations present in the backbone, the gel point (p_{gel}) occurs for

$$p_{gel} = \frac{1}{X_{n-1}} \approx \frac{1}{X_n}$$

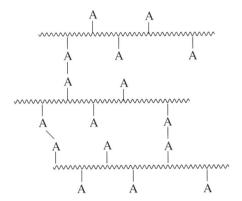

Beyond the gel point the cross-linked fraction increases at the expense of the *sol* fraction as the reaction proceeds (see Figure 3.9).

In addition to the chemical nature of its constituting monomeric units, a polymer network can be defined through two essential parameters, namely, the number (ν) of its elastic chains and the number (μ) of its *junction points* or *cross-links*. According to their definition, elastic chains are connected by their two ends to junction points from which emanate at least two other elastic chains. However, a network not only is a collection of ν chains connecting μ junction points, but also contains also a fraction of loose ends, or dangling chains, loops, and entanglements. A model network is defined as a defect-free one where all cross-links have the same functionality and all elastic chains are of the same size (identical number of monomeric units).

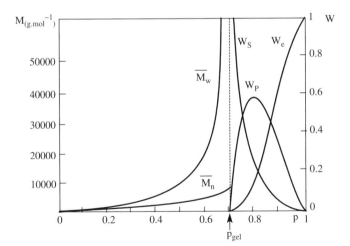

Figure 3.9. Variation of average molar masses and of mass fractions related to extractable chains (W_s), elastic chains (W_e), and dangling chains (W_p), *versus* conversion (p) for a cross-linking step-growth reaction.

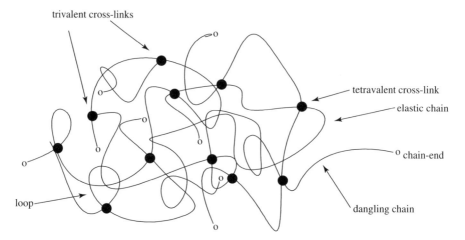

Figure 3.10. Elastic chains, cross-links, and defects in a polymeric network.

From a practical and experimental point of view, one resorts to the notions of cross-links, degree of cross-linking, or molar mass (M_c) of the elastic chains (Figure 3.10). The density of cross-linking (γ) corresponds to the number of moles of cross-links per unit of volume (or per unit of mass), whereas μ represents their molar fraction with respect to the monomeric units of the chains. μ, γ, and M_c are interrelated to one another by

$$\rho = \gamma / N_0$$

$$M_c = M_0 N_0 / \gamma$$

where N_0 is the number of moles of the monomeric units per unit of volume and M_0 is their molar mass.

Practically, the chemical properties of polymer networks depend on the chemical nature of elastic chains and on the type of cross-links. The mechanical properties are, in contrast, essentially governed by the cross-linking density and the mobility of the elastic chains. Polymer networks may thus be soft or hard and exhibit rubbery or brittle behavior.

3.5.2. Characterization of the Networks

Due to their infinitely large size, which confers insolubility and nonfusibility to them, polymer networks cannot be characterized by the traditional methods used for linear and soluble polymers. It is mainly through the study of their mechanical properties (experiments of traction and compression) that one can attain a better knowledge of the structure of networks. As for other solid materials, polymer networks behave like Hookean bodies (within the limit of moderate deformations); that is the deformation is directly proportional to the applied stress. At this stage, it is necessary to make a distinction between rigid networks, made up of crystallized

Table 3.1. Comparison between respective characteristics of flexible and rigid polymeric networks

Characteristic	Rigid Networks (Elasticity from Enthalpic Origin)	Elastomeric Networks (Elasticity from Entropic Origin)
Elastic modulus	High (2–3 GPa)	low (1×10^{-3} GPa)
Reversible strain	Low (0.1%)	High (100% and more)
Variation of the temperature while stretching	Decreasing	Increasing
Variation of the length while heating	Stretching	Shrinking

chains or whose glass transition temperature is higher than the service temperature (T_s), and elastomeric networks whose chains are characterized by a glass transition temperature lower than T_s. The marked differences that characterize these two types of networks are due to the nature of their elasticity; the latter has (a) an enthalpic origin in networks made up of rigid chains and (b) an entropic one for those that consist of flexible chains (Table 3.1).

The mechanical properties of elastomeric networks are described within the theory of rubber elasticity, which accounts for the behavior of a network—in fact, its elastic modulus—as a function of its molecular parameters (number of elastic chains and cross-links).

Valuable information regarding the structure of networks can also be obtained from swelling measurements. The swelling ratio is directly related to the number of elastic chains by the Flory–Rehner equation (see paragraph on rubber elasticity). Networks that swell in a solvent are called *gels*; the elasticity of gels—even if they result from rigid networks in a dry state—can be analyzed using the theory of rubber elasticity. From a simple experiment such as the extraction of the soluble fraction (ω_s), one can obtain useful information regarding the conversion ratio (p) or the fraction of dangling chains thanks to relations that connect ω_s to these two structural parameters.

In addition, more sophisticated techniques have been developed for a more precise analysis of the structure and the behavior of networks: dynamic light scattering proved to be useful to determine the fluctuation of density in networks, and neutron scattering allows one to understand the mechanism of deformation of chains when submitting the network to a macroscopic deformation.

3.5.3. Physical Networks

In addition to networks comprising covalent cross-links (irreversible by nature), there are also transient networks known as "physical networks." The mechanism of formation of these physical networks is somehow similar to the chain vulcanization previously described, but in this case the cross-links formed have a reversible character. Chain "bridging" can occur through the establishment of either weak molecular interactions (van der Waals bonds) or moderate interactions (hydrogen bonds) or by self-assembly of rigid polymer blocks in ABA-type block copolymers. Such physical networks are schematized in Figure 3.11.

Nature also provides a number of examples of physical networks. The most common example is that of aqueous gelatin solutions whose cross-linking corresponds to a transition: from coil to helix of polypeptide chains and to their partial *renaturation* into native collagen. In such networks, some polypeptide chains self-assemble and adopt triple-helix organization while others remain as statistical coils.

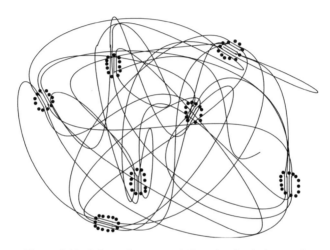

Figure 3.11. Schematic representation of a physical network.

LITERATURE

H. G. Barth and J. W. Mays, *Modern Methods of Polymer Characterization*, Wiley, New York, 1991.

F. A. Bovey and L. W. Jelinski, *Chain Structure and Conformation of Macromolecules*, Academic Press, New York, 1982.

T. R. Crompton, *Analysis of Polymers: An Introduction*, Pergamon Press, Oxford, 1989.

R. N. Ibbnett (Ed.), *NMR Spectroscopy of Polymers*, Chapman & Hall, London, 1993.

K. Matsuzaki, U. Toshiyuki, and T. Asakura, *NMR Spectroscopy and Stereoregularity of Polymers*, Karger, Basel, 1996.

4

THERMODYNAMICS OF MACROMOLECULAR SYSTEMS

4.1. GENERAL CHARACTERISTICS OF POLYMER SOLUTIONS

By definition, a solution contains more than one component. A solution can be gaseous, liquid, or solid. The term *macromolecular* (or *polymer*) *solution* will be used to indicate a mixture of a polymer with a small-molecule solvent and *polymer blend* when solvent and solute are both polymers. In this chapter the thermodynamics of polymer solutions and of solid polymer blends will thus be discussed separately.

The study of the behavior of polymer solutions is important for several reasons. One of the most important is that most methods used to characterize polymers are usually applicable to liquid solutions. An exception to this is neutron scattering, which is often applied to polymers that are condensed as liquids or solids. Polymers are best characterized in solutions with small-molecule solvents. Their molar mass and "statistical" molecular dimensions can be determined using osmometry, size exclusion chromatography, viscometry, and/or light scattering—that is, all of which methods require dilute polymer solutions. From these methods, information about the polymer radius of gyration, the distribution of the molar masses, their average value, and so on, can be obtained. In addition to their importance for polymer characterization, polymer solutions have wide applications (paints, varnishes, oils for engine lubrication, etc.) that take advantage of their particular properties (high viscosity, etc.).

The true nature of polymers was initially revealed from investigations of their solution properties. These studies established that macromolecular chains are made of repetitive units connected by covalent bonds.

Organic and Physical Chemistry of Polymers, by Yves Gnanou and Michel Fontanille
Copyright © 2008 John Wiley & Sons, Inc.

49

Like simple organic molecules, polymers can be dispersed or dissolved in a solvent but their properties differ completely from those exhibited by common solutions of simple molecules. For the same range of concentration, the viscosity of a solution of macromolecules is much higher (due to the chain entanglement) than that of a solution of simple homologous molecules. In addition, marked deviations in their colligative properties (i.e., properties related to the number of species present in the system) are observed compared to the ideal behavior or even to the behavior of real systems consisting of small molecules. This experimental fact—the unusual deviations from Raoult's law—and also the stability of macromolecular solutions led to the final demise of the theory of molecular aggregates glued together by weak interactions (the micellar theory as opposed to the macromolecular theory).

Mixing two chemical compounds, whether simple molecules or macromolecular chains, leads to a change in the entropy (S), the enthalpy (H) and even the volume of the solution. For an isothermic mixing, these changes cause a variation of the free energy, which can be expressed as

$$\Delta G_{mix} = \Delta H_{mix} - T \Delta S_{mix}$$

where T is the absolute temperature. The miscibility of two compounds is thermodynamically favored when $\Delta G_{mix} < 0$.

Solutions can be classified into five main categories—ideal, athermic, regular, irregular, and "theta":

- An **ideal solution** is characterized by an enthalpy of mixing equal to zero and an entropy of mixing equal to the conformational entropy (or *combinatorial entropy*);
- An **athermal solution** is also characterized by an enthalpy of mixing equal to zero but its entropy of mixing is higher than the conformational entropy. It thus exhibits an excess entropy;
- A **regular solution** is characterized by a nonzero enthalpy of mixing and an entropy of mixing equal to the conformational entropy;
- An **irregular** (or **real**) **solution** corresponds to a solution whose enthalpy of mixing does not equal zero and whose entropy of mixing comprises an excess entropy in addition to the conformational term;
- A polymer solution is in **"theta" conditions** when the enthalpy of mixing compensates the excess entropy of mixing at a given temperature. Thus, at this temperature, the solution can be regarded as *ideal*.

Remark. The combinatorial entropy related to the multiplicity of conformational arrangements, which can be adopted by a macromolecule, is sometimes erroneously called *configurational*, because the configuration of a polymer refers to different levels of structure.

4.2. FLORY–HUGGINS THEORY

Statistical thermodynamics can be used to calculate both the enthalpic and the entropic contributions to the free energy of mixing by means of statistical calculations. Attempts were first made to apply a theory appropriate to simple regular solutions, the Hildebrand model, to the case of macromolecular solutions. This model does not adequately account for the specific behavior of polymer solutions primarily because the entropy of mixing in a polymer solution is strongly affected by the connectivity of the polymer—that is, by the existence of covalent bonds between the repetitive units.

This prompted Flory and Huggins to propose a model specific to polymers, which is an extension of the lattice fluid theory but better considers the specificity of polymers; it compares the free energy of polymer–solvent systems before and after mixing.

4.2.1. Entropy of Mixing

In the Flory–Huggins theory, also referred to as the Flory–Huggins mean-field theory, solutions are depicted as three-dimensional crystalline lattices comprising n_t cells (Figure 4.1). Each solvent molecule occupies one cell of this lattice while the macromolecules occupy neighboring cells in a number equal to that of the degree of polymerization (X). It is assumed that each repetitive unit occupies one cell of the lattice, and thus a volume identical to that of a solvent molecule. Each molecule of solvent or solute possesses a number of nearest neighbors (z)—called the *coordination number*—which is also the number of possible interactions for the resident of the cell. Moreover, the volume of such cells is considered to remain unchanged after mixing of the two components.

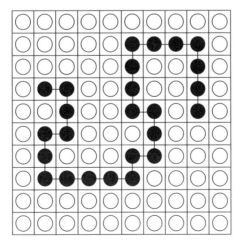

Figure 4.1. Representation of a polymer solution by a lattice. Flory–Huggins model.

Using this model, it is possible to calculate the conformational entropy (ΔS_{co}) resulting from the mixing of n_1 molecules of solvent with n_2 macromolecules having the same degree of polymerization X, considering that the total number of elements (n_t) is equal to ($n_1 + n_2 X$). As the solution is relatively dilute and as the solvent molecules and the macromolecules are randomly distributed, it is assumed that both the conformational entropy (which reflects all possible positional combinations of solvent molecules and of the macromolecules) and the mixing entropy are identical.

The calculation of the entropy of mixing of a macromolecular solution begins with the determination of the number of ways of placing n_1 solvent molecules and n_2 macromolecules in a total of n_t spaces. In a system where j macromolecules have already been placed, let us introduce the $(j + 1)$th macromolecule. At this stage, the number of vacant cells is equal to $n_t - jX$, a number that also corresponds to the number of ways (number of combinations) of placing the first monomer unit of the $(j + 1)$th macromolecule. The number of ways of placing the next monomer unit is equal to the product of the coordination number (z) of the lattice times the fraction $(1 - f_j)$ of the vacant cells—that is, $z(1 - f_j)$.

Remark. Equations are numbered in the following part only when necessary. Each chapter has its own independent numbering.

The third monomer unit and all subsequent units will find one of the adjacent cells occupied by the preceding unit so that only $(z - 1)(1 - f_j)$ vacant cells will remain for them. Hence, the number of ways of placing this $(j + 1)$th macromolecule can be expressed as

$$v_{j+1} = (n_t - jX)z(z - 1)^{X-2}(1 - f_j)^{X-1} \tag{4.1}$$

The number of ways of arranging all the macromolecules of the solution which correspond to the number of possible combinations can therefore be written as

$$P_2 = \frac{v_1 v_2 \ldots v_n}{n_2!} = \frac{1}{n_2!} \prod_{j=0}^{n_2-1} v_{j+1} = \frac{1}{n_2!} \prod_{j=1}^{n_2} v_j \tag{4.2}$$

The presence of $n_2!$ in this expression as denominator is essential for the probability P_2 to be independent of the order of introduction of the n_2 macromolecules.

Now it remains to place the n_1 molecules of solvent: the solvent molecules being indistinguishable, their number of combinations is equal to 1. Thus, expression (4.2) represents the number of combinations of the solution.

The fraction of vacant cells $(1 - f_i)$ is expressed as

$$(1 - f_j) = \frac{n_t - jX}{n_t} \tag{4.3}$$

and the number of combinations is given by

$$P_2 = \frac{1}{n_2!} \prod_{j=1}^{n_2} z(z-1)^{X-2} \frac{[n_t - X(j-1)]^X}{n_t^{X-1}} \qquad (4.4)$$

considering (4.1), (4.2) and (4.3).

After regrouping the terms independent of j, this leads to

$$P_2 = \frac{z^{n_2}(z-1)^{(X-2)n_2}}{n_2! n_t^{(X-1)n_2}} \prod_{j=1}^{n_2} [n_t - X(j-1)]^X \qquad (4.5)$$

Such a product can be written in terms of factorials:

$$\prod_{j=1}^{n_2} [n_t - X(j-1)]^X = X^{n_2 X} \prod_{j=1}^{n_2} \left[\frac{n_t}{X} - j + 1 \right]^X \sim X^{n_2 X} \left[\frac{(n_t/X)!}{(n_1/X)!} \right]^X$$

Now equation (4.5) becomes

$$P_2 = \frac{z^{n_2}(z-1)^{(X-2)n_2} X^{n_2 X}}{n_2! n_t^{(X-1)n_2}} \left[\frac{(n_t/X)!}{(n_1/X)!} \right]^X \qquad (4.6)$$

Using Stirling's approximation ($\ln a! \sim a \ln a - a$) to remove factorials, (4.6) can be formulated as follows:

$$\ln P_2 = -n_2 \ln \left[\frac{X n_2}{n_t} \right] - n_1 \ln \left[\frac{n_1}{n_t} \right]$$

$$+ n_2 [\ln X - X + 1 + \ln z + (X-2) \ln(z-1)] \qquad (4.7)$$

The conformational entropy of the solution can then be derived from the Boltzmann equation ($S = k \ln P_2$):

$$S_{\text{co}} = -k \left[n_2 \ln \frac{X n_2}{n_t} + n_1 \ln \frac{n_1}{n_t} \right]$$

$$+ k n_2 [\ln X - X + 1 + \ln z + (X-2) \ln(z-1)] \qquad (4.8)$$

where k is the Boltzmann constant.

The conformational entropies of pure solvent and solute can be easily calculated from (4.8), by setting

$$n_2 = 0 \quad \text{leading to} \quad (S_{\text{co}})_1 = 0 \qquad (4.9)$$

and

$$n_1 = 0 \quad \text{leading to} \quad (S_{co})_2 = kn_2[\ln X - X + 1 + \ln z + (X - 2)\ln(z - 1)]$$
$$(4.10)$$

Hence the entropy of mixing (ΔS_{co}) for mixtures with low ΔH_{mix} is simply expressed as

$$\Delta S_{mix} = S_{co} - [(S_{co})_1 + (S_{co})_2]$$

$$\Delta S_{mix} = -k\left[n_1 \ln \frac{n_1}{n_t} + n_2 \ln \frac{n_2 X}{n_t}\right] \qquad (4.11)$$

The same expression can be rewritten as a function of the volume fractions of the two components, with Φ_1 being the volume fraction of the solvent and Φ_2 being that of the polymer:

$$\Phi_1 = \frac{n_1}{n_t} \quad \text{and} \quad \Phi_2 = \frac{n_2 X}{n_t}$$

which results in a very simple expression for ΔS_{mix}:

$$\Delta S_{mix} = -k(n_1 \ln \Phi_1 + n_2 \ln \Phi_2)$$

Substituting in this expression of ΔS_{mix} the number of moles for the number of molecules gives (with $R = kN_a$, R being the gas constant)

$$\Delta S_{mix} = -R(N_1 \ln \Phi_1 + N_2 \ln \Phi_2) \qquad (4.12)$$

where N_1 and N_2 are the number of moles of solvent and macromolecules, respectively.

It should be emphasized that this expression of the entropy of mixing is applicable only to athermic systems or to mixtures exhibiting only weak interactions between molecules—that is, solutions with low enthalpy of mixing. Deviations from ideality could arise in particular in the following situations, which will be discussed later on:

- When molecules interact strongly, the assumption of a random placement of the components in the solution is not realistic. Strong interactions induce a short-range order, which leads to a lower entropy of mixing;
- In dilute solutions, polymers are subjected to *excluded volume*. Excluded volume prohibits the access of any other homolog into the vicinity of a chain segment and thus causes a lower entropy of mixing;
- At high temperature, the density of a mixture can considerably decrease and the contribution of the "free volume" to the entropy of mixing cannot be neglected.

4.2.2. Enthalpy and Free Energy of Mixing

Macromolecular solutions deviate generally from ideality and are characterized by a nonzero enthalpy of mixing. In the Flory–Huggins theory the calculation of ΔH_{mix} is inspired by that of the enthalpy of mixing of regular solutions. Three types of interactions between nearest neighbor pairs of molecules are considered: solvent–solvent, solvent–monomeric segment, and segment–segment interactions characterized by ε_{11}, ε_{12}, ε_{22}; ε_{ij} corresponds to the potential energy of an ij pair contact or the energy to dissociate it. The proportion of the various interactions depends on the relative proportions of solvent and solute. The enthalpy of mixing of such a system can be calculated starting from the relation:

$$\Delta H_{\text{mix}} = H - (H_1 + H_2)$$

where H, H_1, and H_2 are the energies of the interactions which develop within the mixture and in the pure components (solvent and polymer), respectively.

The energy required to break the $n_1/2$ solvent–solvent interactions that occur in a lattice exclusively constituted of n_1 molecules of solvent is equal to

$$H_1 = z(n_1/2)\varepsilon_{11}$$

where z is the number of immediate neighbors. In the same way, the enthalpy required to break segment–segment interactions in a medium containing n_2 macro-molecules having degree of polymerization X can be written as

$$H_2 = z(n_2 X/2)\varepsilon_{22}$$

In the case of a binary mixture, each solvent molecule is surrounded by $zn_1/(n_1 + n_2 X)$ molecules of solvent and $zn_2 X/(n_1 + n_2 X)$ repetitive units. The energy corresponding to the interactions of the n_1 solvent molecules involved in solvent–solvent and solvent–segment contacts is given by

$$\frac{1}{2} z \left[\frac{n_1^2}{n_1 + n_2 X} \right] \varepsilon_{11} + \frac{1}{2} z \left[\frac{n_1 n_2 X}{n_1 + n_2 X} \right] \varepsilon_{12}$$

the factor of 1/2 is necessary in the above expression since each solvent–solvent contact is counted twice. In the same manner, the energy of segment–segment and segment–solvent contacts involving the polymer repeating units can be expressed as

$$\frac{1}{2} z \left[\frac{n_2^2 X^2}{n_1 + n_2 X} \right] \varepsilon_{22} + \frac{1}{2} z \left[\frac{n_1 n_2 X}{n_1 + n_2 X} \right] \varepsilon_{12}$$

The sum of all these energies H is equal to

$$H = \frac{1}{2} z \left[\frac{n_1^2}{n_1 + n_2 X} \right] \varepsilon_{11} + \frac{1}{2} z \left[\frac{n_2^2 X^2}{n_1 + n_2 X} \right] \varepsilon_{22} + z \left[\frac{n_1 n_2 X}{n_1 + n_2 X} \right] \varepsilon_{12}$$

and the enthalpy of mixing is established as follows:

$$\Delta H_{\text{mix}} = z \left[\left(\frac{n_1^2 \varepsilon_{11}}{2(n_1 + n_2 X)} + \frac{n_2^2 X^2 \varepsilon_{22}}{2(n_1 + n_2 X)} + \frac{n_1 n_2 X \varepsilon_{12}}{(n_1 + n_2 X)} \right) - \left(\frac{n_1 \varepsilon_{11}}{2} + \frac{n_2 X \varepsilon_{22}}{2} \right) \right]$$

or

$$\Delta H_{\text{mix}} = z X \frac{n_1 n_2}{(n_1 + n_2 X)} \Delta \varepsilon_{12}$$

with

$$\Delta \varepsilon_{12} = \varepsilon_{12} - \left(\frac{\varepsilon_{11}}{2} + \frac{\varepsilon_{22}}{2} \right)$$

Strong interactions result in negative values of ε. If interactions of 1–1 and 2–2 types are stronger than 1–2 type, $\Delta \varepsilon_{12}$ and ΔH_{mix} are positive and the mixture is then endothermic. The mixing will be exothermic in the opposite case.

ΔH_{mix} can also be written as a function of the volume fraction of polymer (Φ_2):

$$z n_1 \Phi_2 \Delta \varepsilon_{12}$$

This corresponds to ΔH_{mix} per unit of volume; to obtain the molar enthalpy of mixing, it must be multiplied by V_1, the molar volume. Defining $\chi_{12} = z \frac{\Delta \varepsilon_{12} V_1}{RT}$, the expression for the enthalpy of mixing becomes

$$\Delta H_{\text{mix}} = RT \chi_{12} N_1 \Phi_2 \tag{4.13}$$

where χ_{12} is called *polymer–solvent interaction parameter*. The free energy of mixing ΔG_{mix} is then easily established as

$$\Delta G_{\text{mix}} = RT(N_1 \ln \Phi_1 + N_2 \ln \Phi_2 + \chi_{12} N_1 \Phi_2) \tag{4.14}$$

χ_{12} is also frequently associated with the Hildebrand solubility parameters through the relation

$$\chi_{12} \equiv V_m (\delta_1 - \delta_2)^2 / RT$$

where δ_1 and δ_2 are the solubility parameters of the two components and V_m is their molar volume taken identical for solvent molecules and monomer units.

4.2.3. Miscibility Conditions and Phase Separation

In the case of an athermic solution, the replacement of a contact between similar species by a "hetero-contact" ($\Delta \varepsilon_{12}$) between solvent and repetitive unit (segment) does not cause any modification of ΔG_{mix}, since, in this case, χ_{12} is 0. Except for some rare cases of athermic solutions, solvent–polymer mixtures are generally endothermic, characterized by a positive enthalpy of mixing. The interactions developed within such solutions are intermolecular repulsive forces of the van der Waals type.

In the case of nonpolar polymer–solvent mixtures, which are the only ones considered here, these interactions are indeed controlled by the polarizability of the components and are described by the relation

$$\varepsilon_{ij} = -3/2[I_i I_j \alpha_i \alpha_j/(I_i + I_j)]r^{-6}$$

where I_i and I_j are the ionization potentials and α_i, α_j are the polarizabilities of components i and j. Hence, the interaction parameter that reflects the whole of these interactions can be written as

$$\chi_{12} = A(\alpha_1 \alpha_2)^2$$

where A is a constant and indices 1 and 2 correspond to solvent and polymer, respectively. This expression, which is established considering only London-type van der Waals interactions, shows that interactions between dissimilar units are necessarily repulsive (or zero); hence, χ_{12} should be positive.

Even in the case of toluene–polystyrene solutions, χ_{12} is in the range 0.3–0.4. Thus a positive enthalpy of mixing tends to oppose the polymer dissolution in a solvent ($\Delta G > 0$). In the Flory–Huggins model, two components can mix with each other only if the positive enthalpy term is compensated by the entropy term ($-T\Delta S$), which is always negative.

In the particular case of specific interactions of higher energy—such as hydrogen bonding—the interaction parameter can take negative values. Solvent and polymer are then miscible in all proportions, but the Flory–Huggins theory does not account for this case, which implies a completely different calculation of the entropy of mixing.

Hence, the interaction parameter χ_{12} is a measure of the quality of a solvent, and its knowledge is essential to the prediction of the domains of concentration corresponding either to the miscibility or to the phase separation of the components. As a matter of fact, it is possible, using the Flory–Huggins theory, to delimit these domains as a function of the concentration of the species and of the interaction parameter.

Figure 4.2 shows the variation of the free energy of mixing (ΔG_{mix}) as a function of the volume fraction of component 2 for various values of the interaction parameter χ_{12}. A symmetrical variation of ΔG_{mix} is observed when the components 1 and 2 have the same size.

The situation for a solution containing a polymer with a degree of polymerization X is different. The variation of ΔG_{mix} becomes strongly asymmetrical in this case.

The effect of χ_{12} can be seen in the form of the curves depicting the variation of ΔG_{mix} with Φ_2 (Figure 4.2). When concave, these curves indicate a total miscibility of the two components. This occurs for χ_{12} values lower than 0.5. When χ_{12} is

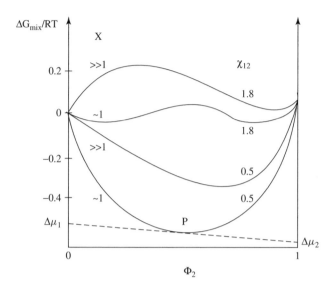

Figure 4.2. Variation of the average free energy of mixing (in *RT* units) as the function of volume fraction of the aqueous solution for various values of χ_{12} and the degree of polymerization *X*.

higher than 0.5, these curves show a maximum and two minima and therefore two inflection points, characteristic of the presence of two phases in equilibrium.

These inflection points, also called *spinodal points*, correspond to

$$\partial^2 (\Delta G_{mix}) / \partial \Phi_2{}^2 = 0$$

They define the thermodynamic limits of metastability. For concentrations corresponding to the spinodal points, the system is unstable and demixes spontaneously into two distinct continuous phases which form an "interpenetrating system." This type of phase separation characteristic of spinodal regions, is also called *spinodal decomposition*.

As for the minima, they are called *binodal points* and a common tangent line passes through them. The chemical potential of a component at these binodal points is the same in each of the two phases in equilibrium (called *prime* and *double prime*):

$$\mu_1' = \mu_1'' \quad \text{et} \quad \mu_2' = \mu_2''$$

which gives

$$\Delta \mu_1' = \mu_1'' - \mu_1^\circ = \mu_1'' - \mu_1^\circ = \Delta \mu_1''$$

and

$$\Delta \mu_2' = \mu_2' - \mu_2^\circ = \mu_2'' - \mu_2^\circ = \Delta \mu_2''$$

The chemical potential (μ_i) of a component i is by definition the variation of the free energy of mixing ΔG_{mix} resulting from the introduction of N moles of i:

$$(\partial \Delta G_{mix} / \partial N)_{T,P} = \mu_i \tag{4.15}$$

According to the Gibbs–Duhem relation

$$\sum_i N_i d\mu_i = 0 \tag{4.16}$$

and for a mixture of components 1 and 2, one obtains

$$\Delta G_{mix} = N_1(\mu_1 - \mu_1^\circ) + N_2(\mu_2 - \mu_2^\circ)$$

a relation that can also be written as

$$\Delta G_{mix} = [\Delta\mu_1 + (\Delta\mu_2 - \Delta\mu_1)\Phi_2]N_t$$

where

$$N_t = N_1 + N_2$$

ΔG_{mix} thus varies linearly with Φ_2 with a slope equal to $(\Delta\mu_2 - \Delta\mu_1)$, and the chemical potentials are given by the intercepts of the function $\Delta G_{mix} = f(\Phi_2)$ for $\Phi_2 \to 0$ and $\Phi_2 \to 1$; this function also corresponds to the tangent (P) to the curve shown in Figure 4.2 for a given composition Φ_2.

When a polymer of degree of polymerization X is one of the two components, one obtains

$$\Delta G_{mix} = \left[\Delta\mu_1 - \left(\Delta\mu_1 - \frac{\Delta\mu_2}{X}\right)\Phi_2\right]N_t \tag{4.17}$$

The slope of the common tangent that passes through the two binodal points is equal to $(\Delta\mu_1 - \Delta\mu_2/X)$, and its intercept for $\Phi_2 = 0$ is equal to $\Delta\mu_1$. Insofar as the two binodal points possess a common tangent, the chemical potentials of the two components are identical in both phases for p' and p'' compositions.

As for the compositions located between the spinodal and binodal points, the free energy of mixing of the corresponding systems, albeit negative, is higher than those of bimodal compositions. These systems will thus demix into two phases with compositions equal to those of the binodal points in order to minimize their free energy. Indeed, even a negative energy of mixing is not necessarily synonymous with miscibility: should a lower free energy be accessible, a system will tend to it even if it requires that it demixes into two phases.

To summarize, three areas can be distinguished at a given temperature:

- Between $\Phi_2 = 0$ or 1 and the binodal points, a system forms homogeneous solutions and only one stable phase.

- Between the binodal points, two phases coexist whose composition is given by the contact points of the tangent to the curve; the regions between binodal and spinodal points form *metastable* solutions; in this case, the phase separation is kinetically controlled by the nucleation and the growth of nuclei leading to the dispersion of one phase into the other.
- The regions between the spinodal points lead to unstable solutions that demix spontaneously into two phases.

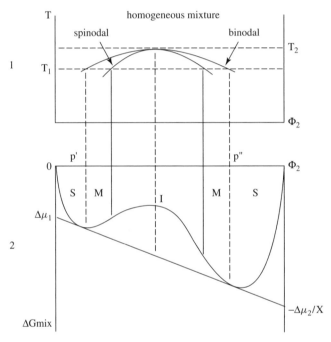

Figure 4.3. 1: Phase diagram of a macromolecular solution whose phase separation occurs through a decrease of temperature (UCST). 2: Variation of the average free energy of mixing as a function of the volume fraction of the solute: (a) formation of a homogeneous solution at T_2; (b) demixing in two phases for compositions between p' and p'' at T_1. S, M, I indicate the regions of stability, metastability, and instability, respectively.

Curves of binodal and spinodal points can be drawn as a function of the temperature up to the critical temperature (T_c) $(T_2$ in Figure 4.3) where these two curves meet. Beyond T_c, the system forms only one phase. At this critical temperature, the partial first- and second-order derivatives of the chemical potential $(\Delta\mu_1)$ are equal to zero; the chemical potential is the derivative of the free energy of mixing (ΔG_{mix}) relative to the number of moles (N_1):

$$\Delta\mu_1 = \frac{\partial\Delta G_{\text{mix}}}{\partial N_1} = RT\left[\ln(1-\Phi_2) + \left(1-\frac{1}{X}\right)\Phi_2 + \chi_{1,2}\Phi_2^2\right] \qquad (4.18)$$

$$\partial \Delta \mu_1 / \partial \Phi_2 = RT \left[2\chi_{1,2} \Phi_2 - (1 - \Phi_2)^{-1} + \left(1 - \frac{1}{X} \right) \right] = 0 \qquad (4.19)$$

$$\partial \Delta^2 \mu_1 / \partial \Phi_2 = RT 2\chi_{1,2} - (1 - \Phi_2)^{-2} = 0 \qquad (4.20)$$

which gives the following relation for the *critical volume fraction* of phase 2:

$$\Phi_{2,\mathrm{crit}} = \frac{1}{1 + \sqrt{X}} \qquad (4.21)$$

The *critical parameter of interaction of demixing* is given by

$$\chi_{1,2,\mathrm{crit}} = \frac{1}{2} \left(1 + \frac{1}{\sqrt{X}} \right)^2 \sim \frac{1}{2} + \frac{1}{\sqrt{X}} \qquad (4.22)$$

For values of the degree of polymerization tending to infinity, $\Phi_{2,\mathrm{crit}}$ thus tends to 0 and $\chi_{1,2,\mathrm{crit}}$ tends to 0.5.

4.2.4. Determination of the Interaction Parameter (χ_{12})

χ_{12} can be determined through osmometry measurements (see Section 6.1.2.1). The osmotic pressure, which is the pressure to apply to stop the flow of the solvent molecules through a semipermeable membrane (permeable to the solvent and impermeable to the macromolecules), is related to the solution activity and to the chemical potential by the relation

$$-\Pi V_1 = RT \ln a_1 = \Delta \mu_1 \qquad (4.23)$$

V_1 being the molar volume of the pure solvent; taking into account (4.18), one obtains the following relation for the solvent activity:

$$\ln \left(\frac{a_1}{\Phi_1} \right) - \left(1 - \frac{1}{X} \right) \Phi_2 = \chi_{12} \Phi_2^2 \qquad (4.24)$$

χ_{12} is the slope of the variation of the term of the left member of the aforementioned equation versus Φ_2^2.

The osmotic pressure can be easily deduced from equations (4.20) and (4.23) after expressing them as functions of the concentration C_2. Considering that $\Phi_1 = (1 - \Phi_2)$ and that $\ln (1 - \Phi_2)$ can be developed into a series,

$$\left[-\Phi_2 - \frac{1}{2} \Phi_2^2 - \frac{1}{3} \Phi_2^3 - \cdots \right]$$

equation (4.18) becomes

$$\Delta \mu_1 = -RT \left[\frac{\Phi_2}{X} + \left(\frac{1}{2} - \chi_{12} \right) \Phi_2^2 + \cdots \right] \qquad (4.25)$$

After observing that Φ_2 can also be written as $C_2\overline{V}_2$ or $C_2V_2{}^0/M_2$, where C_2 is the concentration of polymer, $\overline{V}_2$ the partial specific volume of the polymer, and $V_2{}^0$ its molar volume, and considering that X is also the ratio of the molar volume of polymer to that of the solvent ($V_2{}^0/V_1{}^0$), one obtains

$$\Delta\mu_1 = -RTV_1^0\left[\frac{C_2}{M_2} + \left(\frac{1}{2} - \chi_{12}\right)\frac{\overline{V}_2^{\,2}}{V_1^0}C_2^2 + \frac{\overline{V}_2^{\,3}}{3V_1^0}C_2^3 + \cdots\right] \qquad (4.25a)$$

The expression for the osmotic pressure can be deduced as follows:

$$\Pi = RT\left[\frac{C_2}{M_2} + \left(\frac{(1/2 - \chi_{12})\overline{V}_2^2}{V_1^0}\right)C_2^2 + \frac{\overline{V}_2^3}{3V_1^0}C_2^3 + \cdots\right]$$

$$= RT(A_1C_2 + A_2C_2^2 + A_3C_2^3 + \cdots) \qquad (4.26)$$

where A_1, A_2, and A_3, the virial coefficients, correspond to

$$A_1 = \frac{1}{M_2}, \qquad A_2 = \frac{[1/2 - \chi_{12}]\overline{V}_2^2}{V_1^0}, \qquad A_3 = \frac{\overline{V}_2^3}{3V_1^0} \qquad (4.27)$$

Knowing the osmotic pressure (Π) and hence the chemical potential, one can determine the interaction parameter (χ_{12}).

4.2.5. Real Macromolecular Solutions

As already shown above, the Flory–Huggins theory appears particularly well-suited to the case of regular macromolecular solutions whose components are nonpolar; indeed, their enthalpy of mixing is slightly positive and their entropy of mixing has mainly a conformational origin. Solutions of polyisobutene in benzene or natural rubber in benzene belong to such category. The phase separation in such systems occurs upon decreasing the temperature (upper critical solution temperature (UCST)) (Figure 4.4), which means that their enthalpy of mixing is independent of the temperature and χ_{12} is inversely proportional to the temperature. These are two basic assumptions of the Flory–Huggins theory.

On the other hand, the same model does not predict the lower critical solution temperature case (LCST), which is observed in macromolecular solutions with polar components. Indeed, demixing upon an increase of the temperature (Figure 4.4) is a well-known phenomenon for solutions of polar polymers that are characterized by a high enthalpy of mixing. Conscious of this shortcoming, Flory modified the initial version of his model and reconsidered the assumptions of an enthalpy of mixing and of an energy of contact independent of the temperature.

Observing that in solutions containing polar components, segment–segment interactions can be favored and perturb a random distribution of macromolecular chains, Flory proposed to take into account the existence of such interactions

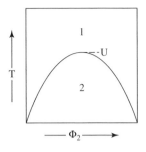 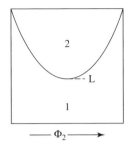 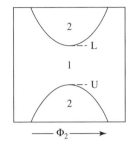

Figure 4.4. Phase diagrams showing the temperatures of demixing (T) as a function of the volume fraction of polymer, with one-phase region (1) and two-phase regions (2). U is used for "UCST"; it means that phase separation occurs upon decreasing the temperature. L is used for "LCST"; in the latter case, phase separation occurs upon increasing the temperature.

in the calculation of the entropy of mixing. In addition to the traditional conformational component, he introduced a term reflecting such interactions or associations in the expression of the entropy of mixing. In such an event, the expression of the exchange (or contact) energy ($\Delta \varepsilon_{12}$) also necessitates a reformulation with an entropy term introduced in complement to the enthalpy term, the latter reflecting the variation of enthalpy due to solvent–solute contact:

$$\Delta\varepsilon_{12} = \Delta\varepsilon_{12,H} - T\Delta\varepsilon_{12,S} \qquad (4.28)$$

The parameter of interaction in such a case becomes

$$\chi_{12} = (z-2)\frac{\Delta\varepsilon_{12,H}}{RT} - (z-2)\frac{\Delta\varepsilon_{12,S}}{R} = \chi_{12,H} + \chi_{12,S} \qquad (4.29)$$

where $\chi_{12,H}$ and $\chi_{12,S}$ represent the enthalpic and entropic contributions, respectively. The entropy and the enthalpy of mixing can be expressed as

$$\Delta S_{\text{mix}} = -R[N_1 \ln \Phi_1 + N_2 \ln \Phi_2 + \frac{\partial(\chi_{12}T)}{\partial T}N_1\Phi_2] \qquad (4.30)$$

$$\Delta H_{\text{mix}} = -RT^2 N_1 \Phi_2 \left[\frac{\partial \chi_{12}}{\partial T}\right] \qquad (4.31)$$

Only for contact energies really independent of the temperature and χ_{12} varying in a proportional manner to temperature—which are two assumptions of the theory of Flory–Huggins—is the traditional expression (4.13) for ΔH_{mix} valid.

One can also observe that the term 1/2 in the expression (4.25) for the chemical potential ($\Delta \mu_1$) originates from the entropic term $\ln(1 - \Phi_2)$ and results from the development into a power series of the latter. Since χ_{12} can also be written as ($\chi_{12,H} + \chi_{12,S}$), this last term can be regrouped with 1/2 to define an entropic parameter: $\Psi = \frac{1}{2} - \chi_{12,S}$. Because $\chi_{12,H}$ represents the enthalpic contribution—and thus has the dimension of an energy—the Ψ term should also have the dimension of an energy and must be divided by T. A temperature θ thus exists at which the variation of the chemical potential resulting from solvent–solute interactions and from the variation of the free energy of mixing are equal to 0 ($\Delta G = \Delta H - \theta\,\Delta S = 0$):

$$\chi_{12,H} - \theta(\Psi/T) = 0 \quad \text{and thus} \quad \chi_{12,H} = \theta(\Psi/T)$$

which gives the relation

$$\frac{1}{2} - \chi_{12} = \Psi\left[1 - \frac{\theta}{T}\right] \tag{4.32}$$

This expression shows that at the said θ temperature, the parameter of interaction (χ_{12}) is equal to 1/2, and the second virial coefficient is equal to 0. Under θ *conditions*, the excluded volume (u) is also equal to 0, which will be shown later on. The system is exactly at the boundary between the "good" and "bad" solvent regimes; polymer segments do not exhibit a particular preference for the molecules of solvent or for other segments of the chain under θ conditions, and they behave like "phantoms." A solvent is said to behave as a good solvent for a particular polymer for values of χ_{12} in the range 0–0.3; it is a poor solvent if χ_{12} values lie between 0.4 and 0.5 and a non-solvent for higher values. From the expressions (4.22) and (4.32), the critical temperature of demixing (T_c) and θ temperature can be correlated through

$$\frac{1}{T_c} = \frac{1}{\theta}\left[1 + \frac{1}{\Psi}\left(\frac{1}{\sqrt{X}} + \frac{1}{2X}\right)\right] \cong \frac{1}{\theta}\left[1 + \frac{1}{\Psi}\frac{1}{\sqrt{X}}\right] \tag{4.33}$$

While examining the variation of Tc as a function of X, Flory observed that the entropic parameter (Ψ) can take negative values for highly polar systems. As Ψ and χ_{12} are always of the same sign, negative values of Ψ and χ_{12} reflect an exothermic mixing. In its second version the Flory–Huggins theory predicts the demixing of a macromolecular solution when the temperature is raised and hence the existence of a lower critical solution temperature (LCST). Such behavior is observed whenever polymer (usually polar) and solvent exchange strong interactions.

4.2.6. Kinetics of Phase Separation

Phase separation occurs in a macromolecular solution by one of the two following mechanisms:

- *Binodal transition* (implying nucleation and growth) or *spinodal* decomposition. Nucleation and growth are associated with the metastability zone between the binodal and spinodal curves. This process implies the existence of an energy barrier and the occurrence of large composition fluctuations. Upon raising or reducing the temperature, when the bimodal curve is crossed, spherical domains of a minimum size, also called *critical nuclei*, are formed and grow with time. The growth of these domains is accomplished by diffusion of the one of the two components (Figure 4.5a).

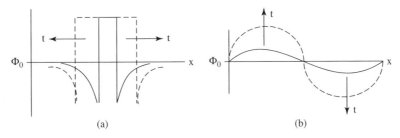

Figure 4.5. Schematic representation of the phase separation processes: (a) by nucleation and growth and (b) by spinodal decomposition.

- The region enclosed by the spinodal curve is that of the *spinodal decomposition* which refers to a phase separation of negligible energy barrier. Overlapping worms are formed and grow through small fluctuations in composition (Figure 4.5b), affording purer phases with time. These fluctuations in composition occur with a certain periodicity, and their amplitude increases with time until reaching the gradient corresponding to phase separation. The domains eventually formed are of about the same size as the original periods of the fluctuations observed in the early stages of phase separation. Cahn and Hillard have described in detail the kinetics of such phase separation.

4.3. DILUTE MACROMOLECULAR SOLUTIONS

The expression of the entropy of mixing proposed by the regular Flory–Huggins model is not appropriate to the case of dilute macromolecular solutions. Its assumption that the polymer chains are randomly distributed in the lattice is indeed untenable in dilute solutions in which polymers instead exist as isolated rafts surrounded by a sea of solvent. In such a case the local density in segments can be high in the vicinity of long chains, whereas the overall concentration in segments in the solution can be very small. As a matter of fact, the assumption of the Flory–Huggins theory of a local concentration in segments identical to the average concentration applies only to concentrated or fairly concentrated media and is erroneous at large dilutions.

A tangible manner to perceive the shortcoming of the Flory–Huggins model is to measure the osmotic pressure whose expression at low concentrations can be written as

$$\frac{\Pi}{RTC_2} = \frac{1}{M_2} + A_2C_2 + A_3C_2^2 + \cdots \tag{4.26a}$$

The second virial coefficient (A_2), which reflects the quality of solvent and is measured by taking the slope of the right-hand side of the equation $\Pi/RTC_2 = f(C_2)$, is assumed to be independent of the molar mass of the sample. Experimentally, one observes in contrast a decrease of A_2 with the sample molar mass, which is not accounted for by the Flory–Huggins model due to its oversimplification—in particular, in the case of low concentration regions. The concept of excluded volume that applies to dilute macromolecular solutions was introduced to overcome this shortcoming.

4.3.1. Concept of Excluded Volume: Case of Compact Molecules

In an ideal dilute medium, the chains are isolated and they are in contact only with solvent molecules. In other words, chains and even segments of a same chain exclude each other from the volume they occupy, and the expulsion of alien chains or segments of a certain volume is called *"excluded volume"*; this leads to long-range interactions of a steric nature. Steric exclusion affects only the entropy and not the enthalpy of the system, and therefore interactions of enthalpic nature can be neglected in a first approximation. The thermodynamics of such a system can thus be treated within the framework of ideal dilute solutions obeying the Henry law.

As macromolecules are isolated, the lattice model used for concentrated solutions is inappropriate to describe their behavior.

The calculation of the entropy of mixing in such a case is based on the assumption that the contribution of each macromolecule to the entropy of the system depends on the number of ways of placing it in the solution. This number is proportional to the difference between the total volume of the solution and the volume inaccessible to this macromolecule due to the presence of other macromolecules.

Let V and R be the volume and the radius of a sphere (radius of the *equivalent sphere*) containing a compact macromolecule of spherical form (Figure 4.6).

Because centers of gravity of two such spheres cannot approach each other beyond a distance $d = 2R$, a certain volume of the sphere is excluded to the other one. Hence, the volume excluded by a compact spherical macromolecule

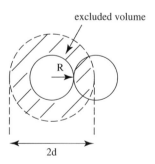

excluded volume

2d

Figure 4.6. Representation of the excluded volume in the case of two spherical macromolecules of identical volume.

corresponds to

$$u = \frac{4\pi d^3}{3} = 8\left(\frac{4}{3}\pi R^3\right) = 8V$$

which is equal to eight times the volume V containing the macromolecule under consideration.

In the case of a rod, it is more difficult to evaluate the excluded volume because the longitudinal axes may well be oriented randomly ($\gamma = 0$ if they are parallel). The excluded volume by a rigid rod of length L, circumference U and volume $V = LU$ for others is equal to

$$U_r = 8V(1 + L/U \sin \gamma)$$

The calculation of the entropy of mixing of a dilute macromolecular solution made up of compact spheres may be developed as follows: the number of ways of placing the first macromolecule is proportional to the volume V available ($v_1 = AV$, where A is a constant of proportionality), the number of ways of placing the jth chain can be written as

$$v_j = A[V - (j-1)u] \tag{4.34}$$

The total number of possible ways (P_2) to arrange n_2 macromolecules can be deduced easily:

$$P_2 = \prod_{j=1}^{n_2} v_i/n_2! = (AV)^{n_2} \prod_{j=1}^{n_2}[1 - (j-1)u/V]/n_2! \tag{4.35}$$

The $n_2!$ factor takes into consideration the number of ways of permuting n_2 indistinguishable chains.

From the relation between P_2 and the conformational entropy (S_{co}),

$$S_{co} = k \ln P_2$$

the following relation can be established:

$$S_{co} = k[n_2 \ln AV + \sum_{j=1}^{n_2} \ln[1 - (j-1)\frac{u}{V}] - \ln n_2!] \tag{4.36}$$

Since $\sum_{j=1}^{n_2} \ln[1 - (j-1)\frac{u}{V}]$ can also be written in the form

$$\sum_{j=1}^{n_2} \ln\left(1 - j\frac{u}{V}\right) \sim -\frac{u}{V}\sum_{j=1}^{n_2} j \tag{4.37}$$

and since the sum of the first integers is equal to $\frac{1}{2}n_2(n_2 - 1)$, S_{co} is equal to

$$S_{co} = kn_2 \ln A + kn_2 \ln V - k \ln n_2! - \frac{k}{2}\frac{u}{V}n_2^2 \tag{4.36a}$$

Expressed as a function of the numbers of moles of solvent (N_1) and polymer (N_2) ($V \cong V_1^0(N_1 + XN_2)$), S_{co} becomes

$$S_{co} = k\mathcal{N}_a N_2 \ln A + k\mathcal{N}_a N_2 \ln[V_1{}^0(N_1 + XN_2)] - k\ln(\mathcal{N}_a N_2!)$$

$$-\frac{k}{2}\frac{u}{V_1^0}\frac{N_2^2}{(N_1 + XN_2)}\mathcal{N}_a^2 \tag{4.36b}$$

where $V_1{}^0$ is the molar volume of pure solvent and N_a is Avogadro's number.

For the calculation of the entropy of mixing, it is necessary to consider the conformational entropies of pure solvent, of pure polymer, and of the solute previously calculated [equation (4.12)]:

$$\Delta S_m = S_{co,\,solute} - (S_{co})_1 - (S_{co})_2 \tag{4.38}$$

$(S_{co})_2$ can be deduced from the expression (4.36b) with $N_1 = 0$; as for $(S_{co})_1$, it is equal to 0, and ΔS_m can be written as

$$\Delta S_m = -k\mathcal{N}_a N_2 \ln\frac{XN_2}{N_1 + XN_2} + \frac{k}{2}\frac{u}{V_1^0}\frac{N_2}{X}\frac{N_1}{N_1 + XN_2}\mathcal{N}_a^2 \tag{4.38a}$$

Expressed as a function of the volume fractions and substituting the gas constant (R) for $k\mathcal{N}_a$, ΔS_m reduces to

$$\Delta S_{mix} = -RN_2(\ln \Phi_2 - \frac{u}{2V_1^0 X}N_a\Phi_1) \tag{4.38b}$$

Under these conditions, the free energy of mixing (ΔG_{mix}) is simply equal to $-T\Delta S_{mix}$, a dilute solution being considered as athermic. Hence, the variation of the chemical potential ($\Delta \mu_1$) can be written as

$$\Delta\mu_1 = \Delta G_1 = \frac{\partial \Delta G_m}{\partial N_1} = -RT\left(\frac{\Phi_2}{X} + \frac{u\Phi_2^2}{2V_1^0 X}N_a + \cdots\right) \tag{4.39}$$

Substituting the concentration of polymer for its volume fraction gives

$$\Phi_2 = \frac{C_2 V_1^0 X}{M_2}$$

Thus, $\Delta \mu_1$ becomes

$$\Delta\mu_1 = -C_2 RTV_1^0\left(\frac{1}{M_2} + \frac{uC_2}{2M_2^2}N_a + \cdots\right) \tag{4.39a}$$

The model of excluded volume thus predicts that the second virial coefficient $A_2 = \frac{u}{2M_2^2}N_a$ decreases as the molar mass of the polymer increases, in contrast to the regular Flory–Huggins model.

4.3.2. Flory–Krigbaum Theory: Case of Flexible Polymer Coils

When applying the concept of excluded volume to the case of real polymer coils, it appears that they certainly do not have the same degree of compactness as that of the spheres described in the model of the preceding paragraph. In the case of compact particles, an element of volume is considered excluded or not whether it is occupied or not. For particles that are flexible, such as polymer coils, the degree of exclusion can take any value between 0 and 1. The excluded volume, in this case, is the integral of the degree of exclusion over the entire volume occupied by macromolecular coils. However, the spatial aspect is only one part of the problem.

Insofar as segment–segment interactions are present, they contribute to lower the energy of the system (χ_{12} is positive in the Flory–Huggins theory) and they favor the coil interpenetration; steric exclusion is then counterbalanced by an "interseg-mental" attraction so that flexible coils can even interpenetrate freely (Figure 4.7).

Flory and Krigbaum treated this case and established the expressions of the excluded volume and the second virial coefficient for flexible chains. They considered the interactions existing between pairs of macromolecules whose segment distribution follows a function of radial distribution $\rho(R)$, which starts from the center of gravity of each one of them.

The segment distribution within the envelope of all possible conformations follows a radial Gaussian function of the type

$$\rho(R) = X \left(\frac{\beta^3}{\pi^{1/2}} \right) e^{-\beta^2 R^2} \tag{4.40}$$

where β is a parameter related to the average end-to-end distance $\langle r^2 \rangle^{1/2}$ and hence to the radius of gyration $\langle s^2 \rangle^{1/2}$ and R is the distance separating the point considered from the center of gravity:

$$\beta = \frac{3}{\langle r^2 \rangle^{1/2}} = \frac{3^{1/2}}{2^{1/2} \langle s^2 \rangle^{1/2}} \tag{4.41}$$

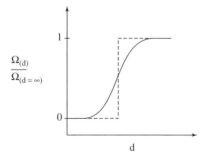

Figure 4.7. Curve describing the variation of the probability of placement Ω as the function of the distance separating the macromolecular barycenters of two coils.

$\rho(R)$ can also be written as a function of the radius of gyration:

$$\rho(r) = X\left(\frac{3}{2\pi s^2}\right)^{3/2} e^{-3R^2/2\langle s^2\rangle} \qquad (4.40a)$$

Describing the segment density in a volume occupied by a macromolecular chain, this function typically takes low average values of about 1%.

To evaluate the excluded volume by flexible chains, Flory and Krigbaum then calculated the increase in free energy that would arise from the fictitious overlap of two isolated coils until their two elementary volumes dV_i and dV_j coincide (Figure 4.8).

The first step consists in writing the expression for the free energy of mixing within dV_i and dV_j volumes using a calculation similar to that of the Flory–Huggins model:

$$d(\Delta G_m)/dn_1 = kT[\ln(1 - \Phi_2) + \chi_{12}\Phi_2] \qquad (4.42)$$

where n_1 is the number of molecules contained in an elementary volume dV and Φ_2 is the volume fraction of polymer. Because the densities in segments $[\rho(r)]$ are small inside the volume occupied by the chain, only the first two terms of the series resulting from the development of $\ln(1 - \Phi_2)$ can be retained so that the preceding expression now becomes

$$d(\Delta G_m)/dn_1 = kT\left[(1 - \chi_{12}) + \frac{\Phi_2}{2}\right]\Phi_2$$

Figure 4.8. Representation of the excluded volume phenomenon in the case of flexible macromolecules.

To express $d(\Delta G_m)$ as a function of $\rho(R)$ of the two entities, one assumes them constant inside the two elementary volumes. The volume fractions of polymer in dV_i and dV_j are $\rho(R_i)v_1$ and $\rho(R_j)v_1$, respectively (v_1 is the volume occupied by a solvent molecule and considered identical to that of a repetitive unit). The numbers of solvent molecules contained in dV_i and dV_j are given by:

$$\frac{dV_i}{v_1}[1 - \rho(R_i)]v_1 \quad \text{and} \quad \frac{dV_j}{v_1}[1 - \rho(R_j)]v_1$$

From these elements the total free energies of mixing in dV_i and dV_j can be written as

$$d(\Delta G_m)_i + d(\Delta G_m)_j = -kT\left[(1 - \chi_{12}) + \frac{\rho(R_i)v_1}{2}\right][1 - \rho(R_i)v_1]\rho(R_i)\,dV$$

$$- kT\left[(1 - \chi_{12}) + \frac{\rho(R_j)v_1}{2}\right][1 - \rho(R_j)v_1]\rho(R_j)\,dV$$

$$(4.43)$$

In the situation corresponding to the perfect coincidence of volumes dV_i and dV_j, the segment volume fraction now has a single value,

$$\Phi_2 = [\rho(R_i) + \rho(R_j)]v_1$$

and the number of solvent molecules contained in this volume is written as

$$dn_1 = \frac{dV}{v_1}\{1 - [\rho(R_i) + \rho(R_j)]v_1\} \tag{4.44}$$

The free energy of mixing inside this volume dV, where the coils overlap until coinciding, is given by

$$d(\Delta G_m)_a = -kT\left[(1 - \chi_{12}) + \frac{\rho(R_i) + \rho(R_j)}{2}v_1\right]$$

$$\times \{1 - [\rho(R_i) + \rho(R_j)]v_1\}[\rho(R_i) + (R_j)]\,dV \tag{4.45}$$

Finally, the increase in free energy due to the pulling of the two coils of Figure 4.8 closer, and whose centers of gravity are at a distant a_1, can be expressed as

$$d(\Delta G_m)^a_\infty = d(\Delta G_m)_a - [d(\Delta G_m)_i + d(\Delta G_m)_j]$$

$$= 2kT\left(\frac{1}{2} - \chi_{12}\right)\rho(R_i)\rho(R_j)v_1\,dV \tag{4.46}$$

Extended to the entire volume V of the solution, this increase in the free energy can be written as

$$(\Delta G_m)^a_\infty = 2kT\left(\frac{1}{2} - \chi_{12}\right)v_1\int_v \rho(R_i)\rho(R_j)\,dV \tag{4.46a}$$

From this relation the expression for the excluded volume can be derived; but prior to that, it is necessary to determine the probability of placement Ω (a) of the two macromolecular coils considered in a dilute solution. This probability obviously tends to decrease as the distance separating the two coils decreases, owing to an unfavorable free energy of overlapping. This decrease can be accounted for by multiplying the probability Ω (∞)—close to 1—by $(\Delta G_m)^a_\infty / kT$:

$$\Omega(a) = \Omega(\infty)e^{-(\Delta G_m)^a_\infty / kT} \qquad (4.47)$$

In the volume $4\pi \, a^2 da$ surrounding the center of gravity (i), the volume really available for another macromolecule corresponds to

$$4\pi a^2 \Omega(a)da = 4\pi a^2 \Omega(\infty)e^{-(\Delta G_m)^a_\infty / kT} \, da$$

The prohibited volume is easily deduced to be

$$4\pi a^2 (1 - e^{-(\Delta G_m)^a_\infty / kT}) \, da$$

and the total excluded volume is then obtained by integration:

$$u = \int_0^\infty \left(1 - e^{-(\Delta G_m)^a_\infty / kT}\right) 4\pi a^2 \, da \qquad (4.48)$$

By substituting in the expression for $(\Delta G_m)^a_\infty$ the distribution functions established by Flory and Krigbaum for ρ_i and ρ_j, one obtains the following expression for the excluded volume (u):

$$u = 2\left(\frac{1}{2} - \chi_{12}\right) X^2 v_1 F(Y) \qquad (4.49)$$

The function $F(Y)$ is a complex integral which has no analytical solution, but which can be evaluated graphically (Figure 4.9).

For low values of Y, $F(Y)$ can be developed into a series of the form

$$F(Y) = 1 - \frac{Y}{2^{3/2}2!} + \frac{Y^2}{3^{3/2}3!} \qquad (4.50)$$

with

$$Y = 2\left(\frac{1}{2} - \chi_{12}\right) X^2 v_1 \left(\frac{3}{4\pi s^2}\right)^{3/2} \qquad (4.51)$$

Introduction of the specific partial volume of the polymer $(\overline{V}_2)$,

$$\overline{V}_2 = X v_1^0 \mathcal{N}_a / M_2$$

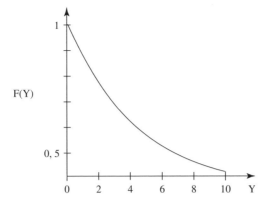

Figure 4.9. Variation of $F(Y)$ versus Y obtained by graphic integration (according to Flory).

into the preceding expressions delivers

$$u = (2/\mathcal{N}_a) \left(\frac{1}{2} - \chi_{12}\right) \frac{\overline{V}_2^2 M_2^2}{V_1^0} F(Y) \tag{4.49a}$$

and

$$Y = [2/\mathcal{N}_a] \left(\frac{1}{2} - \chi_{12}\right) \frac{\overline{V}_2^2 M_2^2}{V_1^0} \left(\frac{3}{4\pi s^2}\right)^{3/2} \tag{4.51a}$$

where V_1^0 is the molar volume of the solvent in the three last expressions. Hence, the second virial coefficient, $A_2 = \frac{u}{2M_2^2}\mathcal{N}_a$, can be written as follows:

$$A_2 = \left(\frac{1}{2} - \chi_{12}\right) \frac{V_2^2}{V_1^0} F(Y) \tag{4.52}$$

The Flory–Krigbaum model leads to an expression for the second virial coefficient (A_2) that differs by the factor $F(Y)$ from relation (4.27), given by the Flory–Huggins model:

$$A_2 = \left(\frac{1}{2} - \chi_{12}\right) \overline{V}_2^2 / V_1^0 \tag{4.53}$$

Because $F(Y)$ is always less than unity, the value of the second virial coefficient predicted by the Flory–Krigbaum theory is necessarily lower than that predicted by the theory of concentrated solutions.

The relation (4.32) relating χ_{12} to θ shows that the excluded volume and Y are equal to 0 and $F(Y)$ is equal to 1 at the temperature θ even in a dilute medium. Thus the chains interpenetrate freely under θ conditions. At temperatures higher than θ, the excluded volume takes positive values; but at temperatures lower than θ, segment–segment attractions prevail and the excluded volume is negative.

4.3.3. Excluded Volume and Expansion Coefficient

In their "unperturbed" dimensions, chains—or rather their repetitive units—develop only short-range interactions. The mean square average end-to-end distance separating the ends of a polymer chain made up of X segments and with a length L is given by

$$\langle r^2 \rangle_0 = CXL^2$$

where C is a parameter that reflects the existence of short-range interactions. The introduction of a good solvent generates long-range interactions that affect units in non-immediate vicinity and belonging to a same chain. These interactions are due to the fact that each of these units tends to maximize its solvation, which is accounted for by the concept of excluded volume. This results in the expansion of the chain, which will occupy a larger volume.

Flory and Fox described the expansion of a macromolecular coil due to the presence of a good solvent by using an approach identical to that of the swelling of a three-dimensional network. They considered all the repetitive units as subjected to the same force field—the mean field—whose effect is to impose an energetic penalty on any placement that would correspond to the the random walk Gaussian statistics. This force field is comprised of two components of opposite nature, one of repulsion—and hence the expansion—and the other one of retraction which is of entropic origin. The expansion of a chain and its conformation are in fact affected by the balance between these two forces; in its perturbed dimensions, the chain adopts a gyration radius $\langle s^2 \rangle^{1/2}$ that is related to the nonperturbed dimension by the expression

$$\langle s^2 \rangle = \alpha^2 \langle s^2 \rangle_0 \tag{4.54}$$

where α, the *expansion coefficient*, is an empirical parameter.

The chain expansion results from forces of osmotic origin exerted by the solvent, which imbibes the macromolecule. The stretching causes the reduction of the number of possible chain conformations, but it is opposed by forces of entropic origin. The free energy of the system becomes

$$\Delta G = \Delta G_{\text{osm}} + \Delta G_{\text{el}}$$

The volume of a spherical shape macromolecule is

$$V_d = \frac{4}{3}\pi s^3 \quad \text{and hence} \quad V_d = \frac{4}{3}\pi \alpha^3 s_0^3$$

the increment of volume (dV_d) associated with an increase in the radius of gyration ds_0 is proportional to

$$dV_d \div \alpha^3 s_0^2 ds_0 \tag{4.55}$$

where $\div$ is a sign of proportionality

Since the extent of swelling depends on the degree of polymerization, the fraction of chain segments present in this element of volume dV_d is

$$\Phi_2 \div X/dV_d \quad \text{and thus} \quad \Phi_2 \div X/\alpha^3 s_0^2 ds_0$$

The variation of free energy corresponding to this swelling and hence to the introduction of dn_1 molecules of solvent is thus

$$d\Delta G = (\mu_1 - \mu_1^\circ)dN_1 \quad \text{with} \quad dN_1 \div dV_d(1 - \Phi_2)/v_1 \tag{4.56}$$

Since the product of a force and a distance ($d\alpha$ being assimilated to the variation in size of the sample) is equivalent to an energy, one can write

$$F_{\text{osm}}\, d\alpha = d\Delta G$$

thus

$$F_{\text{osm}} = \frac{d}{d\alpha}[\mu_1 - \mu_1^0]dN_1 \div \frac{d}{d\alpha}\left[\frac{(\mu_1 - \mu_1^0)\alpha^3(1 - \Phi_2)s_0^2 ds_0}{v_1}\right] \tag{4.57}$$

Knowing that $(\mu_1 - \mu_1^0)$ can also be written as

$$\mu_1 - \mu_1^0 = RT\left[\ln(1 - \Phi_2) + \Phi_2\left(1 - \frac{1}{X}\right) + \chi_{12}\Phi_2^2\right] \tag{4.58}$$

and that $\ln(1 - \Phi_2)$ reduces to

$$\ln(1 - \Phi_2) \cong -\Phi_2 - \tfrac{1}{2}\Phi_2^2$$

with $1/X \to 0$ for $X \to \infty$, one obtains the following for $(\mu_1 - \mu_1^0)$:

$$(\mu_1 - \mu_1^0) \div \left(\tfrac{1}{2} - \chi_{12}\right)\Phi_2 \tag{4.58a}$$

which gives for F_{osm}

$$F_{\text{osm}} \div \frac{\left(\tfrac{1}{2} - \chi_{12}\right)X^{1/2}}{v_1\alpha^4} \tag{4.57a}$$

since s_0^2 is proportional to X.

This force is counterbalanced by an elastic force of entropic origin which can be written as

$$F_{\text{el}} \div \frac{d\Delta S}{dl} \div l_0\frac{d\Delta S}{d\alpha} \tag{4.59}$$

where S denotes the loss of entropy due to stretching, l and l_0 are the lengths of the sample in the stretched and in the relaxed states, and α is the ratio l to l_0.

As swelling occurs in the three dimensions of space, the expression of S will be slightly different from that derived in the case of rubber elasticity, which considers only one direction of stretching:

$$\Delta S \div \tfrac{1}{2}[\alpha_x{}^2 + \alpha_y{}^2 + \alpha_z{}^2 - 3 - \ln(\alpha_x \alpha_y \alpha_z)] \tag{4.60}$$

where α_x, α_y, and α_z denote the expansion in directions x, y, z, thus leading to

$$\Delta S \div \tfrac{1}{2}[(\alpha^2 - 1) - \ln \alpha] \tag{4.60a}$$

then

$$F_{el} \div \frac{d\Delta S}{d\alpha} \div \alpha - \frac{1}{\alpha} \tag{4.61}$$

and as $F_{osm} = F_{el}$, one obtains

$$\frac{\left(\tfrac{1}{2} - \chi_{12}\right) X^{1/2}}{v_1 \alpha^4} \div \alpha - \frac{1}{\alpha}$$

which gives

$$\alpha^5 - \alpha^3 \div \frac{\left(\tfrac{1}{2} - \chi_{12}\right) X^{1/2}}{v_1} \tag{4.62}$$

Taking into consideration the relation (4.34), the preceding expression can be rewritten in the form

$$\alpha^5 - \alpha^3 \div \psi \left[1 - \frac{\theta}{T}\right] \frac{M_2^{1/2}}{v_1} \tag{4.62a}$$

Hence, the Flory theory predicts a dependence of the expansion coefficient α on the amplitude of the interactions between the polymer segments and the solvent [here denoted by the term $(1 - \chi_{12})$] and also on the molar mass M_2 of the sample (or X). Consequently, the excluded volume can be expressed as a function of α from relations (4.49) and (4.62):

$$u \div (\alpha^5 - \alpha^3) \left[\frac{4\pi \langle s^2 \rangle_0}{3}\right]^{3/2} F(Y) \tag{4.63}$$

with

$$Y = 2(\alpha^2 - 1) \tag{4.64}$$

The expressions (4.62) and (4.53) show that the radius of gyration of a chain subjected to the phenomenon of excluded volume is proportional to $X^{3/5}$, whereas that of an unperturbed chain is proportional to $X^{1/2}$. For example, the average dimension of a chain with a degree of polymerization of $\overline{X}_n = 1000$ is expected to grow by a factor of 2, and its volume under unperturbed conditions is expected to increase by a factor of $2^3 = 8$.

4.3.4. Perturbations Theory: Other Expressions of the Expansion Coefficient

In the Flory–Krigbaum theory, the chains are regarded and treated as isolated species whose volume would be eight times larger than their own volume. In the interior of the sphere, the segments are assumed to adopt a Gaussian distribution, a statement that has been questioned by various authors. According to them, the excluded volume phenomenon occurs intramolecularly at the level of each monomer unit of the macromolecule and not only intermolecularly between macromolecules. In the improvements introduced in the Flory–Krigbaum theory, the elementary excluded volume β corresponds to the level of each polymer segment and is related to the exclusion volume u by the expression

$$\beta = \frac{u}{X^2} \tag{4.65}$$

Depending on whether these models consider the contacts between segments as involving just pairs of them or more than two of these segments, the expressions relating α to z—the parameter of excluded volume—differ:

$$z = \left(\frac{1}{4\pi s^2}\right)^{\frac{3}{2}\beta X^2} \tag{4.66}$$

In the Zimm theory of perturbations known as "first-order perturbations," where only the interactions between segments are taken into consideration, the relation between α and z is written as

$$\alpha_r^2 = 1 + \frac{4}{3}z - \cdots \qquad \text{with} \qquad \alpha_r^2 = \frac{r^2}{r_0^2} \tag{4.67}$$

$$\alpha_s^2 = 1 + \frac{134}{105}z - \cdots \qquad \text{with} \qquad \alpha_s^2 = \frac{s^2}{s_0^2} \tag{4.68}$$

Using the Flory formalism, one obtains

$$\alpha_r^5 - \alpha_r^3 = \frac{4}{3}z \tag{4.69}$$

and

$$\alpha_s^5 - \alpha_s^3 = \frac{134}{105}z - \cdots \tag{4.70}$$

In the case of the refined Yamakawa model, this gives

$$a_r^2 = 1 + 1.33z - 2.07z^2 + 6.459z^3 \tag{4.71}$$

$$\alpha_s^2 = 1 + 1.276z - 2.082z^2 \tag{4.72}$$

Using again the Flory formalism, one obtains $\alpha^5 - \alpha^3 = 2.60z$. All the improvements introduced in the Flory–Krigbaum theory lie within the traditional framework of "mean field" theories.

4.4. SEMI-DILUTE MACROMOLECULAR SOLUTIONS

If the Flory theory is indisputably a reference for the thermodynamics of polymer solutions, it suffers from a lack of accuracy in its description of dilute polymer solutions as previously mentioned. Well suited to the case of concentrated solutions, this theory depicts the behavior of dilute solutions and describes the forces due to excluded volume as the result of a perturbation to "random walk statistics"; for example, it does not account for the significant variations experienced by the density of segments in dilute media. Indeed, the replacement of the radial variation of this function (which describes the density of interaction in the medium) by an average value is not satisfactory.

In particular, the transition from a dilute regime ($s \approx X^{3/5}$) to a concentrated one ($s \approx X^{0.5}$) that is accompanied by a progressive contraction of the chain as the concentration increases is ill-explained by a mean field theory such as that of Flory. The mean field approach predicts a result close to reality for the relation between the dimensions of a chain in a dilute medium and its degree of polymerization ($s \sim X^{3/5}$). This result is, however, fortuitous according to de Gennes. It results from the cancellations of the errors introduced into the calculation of ΔS_m and ΔH_m.

De Gennes tackled the problem of polymer solutions with another point of view and treated it as a second-order transition, a process characterized by a continuous variation of thermodynamic potentials and by the divergence of some of their second derivatives. According to de Gennes, the concentration C^* at which the chains overlap—defined as the start of the semi-dilute concentration region—is a second-order transition that can be described by the renormalization of certain variables, using tools developed by Wilson. In the language of modern thermodynamics, critical points designate points that are subject to a second-order transition; in the vicinity of such critical points, the physical behavior of a system can be described in the form of scaling laws that contain critical exponents (see Appendix).

Even if nothing peculiar occurs at this critical concentration (C^*) with respect to the solution properties, the medium is the subject of fluctuations of concentration while passing from the dilute regime to the semi-dilute one (Figure 4.10).

According to de Gennes, the fluctuations in this order parameter (the concentration) are reflected in the correlation length (ξ)—characterizing their amplitude in space—and in the correlation function $g(a)$ between pairs of repetitive units.

The existence of a critical point is observed when the correlation length associated with the correlation function of the order parameter diverges. By definition, a mean field theory ignores the fluctuations of the order parameter and affords satisfactory results only far away from the critical point; in addition to the calculation of the entropy and enthalpy of mixing, de Gennes criticized this main point in the classical theories and observed that the variation of $g(a)$ with the concentration and the distance (a) considered cannot be overlooked or neglected:

$$g(a) = \frac{1}{2}[C(0)C(a) - \overline{C}^2] \tag{4.73}$$

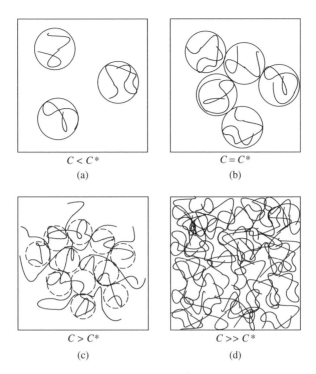

Figure 4.10. Chain overlapping; (a) dilute solution; (b) overlap concentration; (c) semi-dilute solution; (d) concentrated solution.

where $C(a)$ denotes the local concentrations at the reference point (0) and at the point (a) under consideration, whereas $\overline{C}$ is the average concentration in the medium.

As for the correlation length (ξ), it represents the average distance that separates the segments of various chains which come into contact above the critical concentration of overlapping C^*. $C^*(\equiv X/r^3)$ is attained when the concentration of the solution is roughly equal to the average concentration in segments inside a macromolecular coil. Hence, the macromolecular solutions can be classified in three categories depending upon their concentration domains:

$$C < C^* \qquad \text{dilute solution}$$

$$C \geq C^* \qquad \text{semi-dilute solution}$$

$$C \gg C^* \qquad \text{concentrated solution}$$

In the dilute regime, polymer chains behave like isolated spheres called "blobs"; the radius of gyration or the end-to-end distance follows a variation in $r \div X^\gamma$ with $\gamma = 0.586 \equiv 0.6$ and $\Phi_2(\leq \Phi_2^* \equiv XL^3/r^3)$, the volume fraction of polymer, follows a variation in $\Phi_2 \div X/(X^{3/5})^3$, which gives $X^{-4/5}$. This is in agreement with the predictions of the mean field theories. The correlation length ξ is close to r and $g(a) = 0$.

As the average concentration increases, polymeric chains overlap and establish interactions. They can be depicted as a collection of "blobs" of a diameter ξ whose periphery witnesses intersegmental interactions (Figure 4.11). The latter contributes to create a sort of network whose mesh size corresponds precisely to the correlation length. Above C^* the end-to-end distance is not relevant as characteristic length controlling the solution properties. r has to be replaced by ξ, which then decreases from $\xi = r$ for $C = C^*$ to $\xi = L$ as the concentration increases. Inside the "blob" $(a < \xi)$ the subchain is subjected to the excluded volume effect. As for the excluded volume forces generated by segments located outside the "blob," they are screened by the presence and the interactions of immediatly adjacent chains according to a model proposed by Edwards: beyond ξ, chains adopt random walk Gaussian conformations whereas the correlation length decreases exponentially with the volume fraction—or the concentration—above C^*:

$$\xi \div r[\Phi_2^*/\Phi_2]^y \qquad (4.74)$$

In such a representation of the medium above C^*, ξ is independent of X. In other words, when macromolecules overlap, the characteristic length ξ is not determined by the size of the chains but by their concentration. y, the exponent in expression (4.74), must take a value that yields an independent ξ with respect to the molar mass of the chain:

$$\xi/r = \Phi^{-3/4} \qquad (4.75)$$

Thus, for concentrations higher than C^* beyond ξ, a macromolecule can be viewed as Gaussian chain made up of $X_e(\Phi)$ segments of size ξ and therefore $r(\Phi)$ can be expressed as

$$r(\phi) = X_e(\Phi)\xi^2(\Phi) \qquad (4.76)$$

Figure 4.11. Representation of a chain in a semi-dilute solution.

As for ξ, it is equal to

$$\xi = Lg(a)^{3/5} \tag{4.77}$$

where $g(a)$, the correlation function between pair of repetitive units, also represents their number inside the "blob" (good solvent and effect of excluded volume).

Because we have

$$X_e(\Phi) \equiv X/g(a) \tag{4.78}$$

one can deduce

$$r(\Phi)/r(\Phi^*) \div \Phi^{-1/4} \tag{4.79}$$

For distances lower than ξ, $g(a)$ corresponds to

$$g(a) \approx 1/a^{4/3} L^{5/3} \tag{4.80}$$

whereas beyond ξ $(a > \xi)$

$$g(a) \approx C\xi/a \tag{4.81}$$

and its variation is represented in Figure 4.12.

As the concentration increases, g tends to 1 $(\xi = L)$ due to the "screening" of the excluded volume phenomenon: in the high concentration regimes the de Gennes approach delivers the same traditional well-known expression for the unperturbed dimensions of a chain as the Flory model, but with its renormalization methods it better accounts for the shrinkage undergone by the chain as the polymer concentration increases.

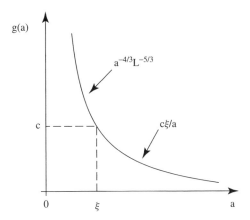

Figure 4.12. Variation of the correlation function $g(a)$ as the function of the distance a (according to de Gennes).

4.5. POLYMER–POLYMER BLENDS

When two polymers of different chemical nature are mixed, a macroscopic phase separation is generally observed, due to the fact that they tend to minimize their surface of contact. Their macromolecular nature is responsible for a low entropy of mixing, which, associated with an endothermic enthalpy of mixing, explains their immiscibility. For two polymers, the higher their molar mass the more difficult their mixing.

Until the mid-1970s, polymer chemists used to think that attempts at mixing two polymers would necessarily wind up in immiscible blends so that very few examples of homogeneous blends were known. The polystyrene/poly(2,6-dimethyl-1,4-oxyphenylene) blend (sold under the trademark Noryl®) was an exception. This field witnessed a real development only after it was realized that most homogeneous blends phase separate upon raising the temperature (LCST) (in general the researchers tend to increase the temperature in order to improve the miscibility of polymers).

From this observation, studies were carried out to identify the window of miscibility by varying the temperature or by introducing specific interactions between polymers. Thus, miscible blends could be obtained through the use of statistical copolymers; the most well known example is the blend of poly(vinyl chloride) (PVC) with poly[ethylene-co-(vinyl acetate)], which is homogeneous although PVC is not miscible with either polyethylene or poly(vinyl acetate). It could be shown that specific attractive interactions develop in such miscible blends between the PVC repeating units and the copolymer. These interactions counterbalance the repulsive ones opposing the formation of a homogeneous blend. ΔG is, of course, negative in the case of homogeneous blend formation. Expression (4.14) for ΔG using the Flory–Huggins model is

$$\Delta G_{mix} = RT(N_1 \ln \Phi_1 + N_2 \ln \Phi_2 + \chi_{mix} N_1 \Phi_2) \qquad (4.82)$$

where χ_{mix} is now equal to

$$\chi_{mix} = \Phi_1 \chi_{13} + \Phi_2 \chi_{23} - \chi_{12} \Phi_1 \Phi_2 \qquad (4.83)$$

where Φ_1 and Φ_2 are the volume fractions of the two comonomers in the copolymer.

4.5.1. Theories Based on Free Volume and on the Equation of State

Although it is very useful in the description of phenomena such as those mentioned above in blends involving statistical copolymers, the Flory–Huggins theory is unable to account for the variation of the properties of polymer–polymer blends as functions of the temperature. Indeed, the coexistence of both a lower critical solution temperature and an upper critical solution temperature of mixing, which is a rather common phenomenon in polymer blends, could not be described by the Flory–Huggins theory.

The assumptions that the volume of the solution remains constant upon mixing ($\Delta V_{\mathrm{mix}} = 0$) and thus it is incompressible are the reasons generally put forward to explain the deficiencies of this theory. These assumptions indeed imply no variation of a variable such as the pressure and more generally no variation of the equation of state. For the calculation of the enthalpy and the entropy of mixing, a theory that would include relations between the temperature, the pressure, and the volume of the system considered is thus a necessity.

From the observation that free volumes—due to the difference between specific volumes at a given temperature T and at $0°K$—can be very different, depending upon the nature of the components, theories based on the concept of free volume and including the equations of state were thus proposed. Polymer connectivity implies the existence of free volume and a lower coefficient of thermal expansion than that of a solvent, and under these conditions an increase of temperature necessarily induces differentiation in the densities of the components, which cannot adjust one with another to make a homogeneous solution. As a result of an increase in temperature, the solution demixes and this phase separation can occur at an even lower temperature if the polymer molar mass is higher.

To overcome the limitations due to the hypothesis of volume incompressibility, Lacombe and Sanchez proposed to introduce vacant cells in the lattice model as a means to accommodate free volume (V_0). A system of m constituents would thus be mixed with n_0 vacant cells and V_0/V would be the free volume fraction. Under these conditions the entropy of mixing can be shown to be

$$\frac{\Delta S_{\mathrm{mix}}}{k} = -n_o \ln\left(\frac{V_0}{V}\right) - \sum_i n_i \ln\left(\frac{V_i}{V}\right) \tag{4.84}$$

Each component occupies an incompressible volume

$$V_i^* = n_i v_i^* \equiv n_i M_i / \rho_I^* \tag{4.85}$$

and $\rho_i^* (\equiv M_i / v_i^*)$ is the incompressible density of component i.

The total incompressible volume is written as

$$V^* = \sum_i V_i^* \tag{4.86}$$

These incompressible or characteristic volumes and densities can be identified with those prevailing at $0\,\mathrm{K}$.

The Lacombe–Sanchez model then defines reduced values $\tilde{v}$ and $\tilde{\rho}$ (reduced volume and reduced density) as being equal to

$$\tilde{v} = V/V^* \geq 1 \quad \text{and} \quad \tilde{\rho} = \frac{1}{\tilde{v}} \tag{4.87}$$

The free volume fraction is easily deduced as

$$V_0/V = (V - V^*)/V = 1 - \tilde{\rho}$$

The volume fraction of the component i considered in the incompressible state corresponds to

$$\Phi_i = \frac{V_i^*}{V^*} = \frac{n_i M_i / \rho_i^*}{\sum_i (n_i M_i / \rho_i^*)} \tag{4.88}$$

Knowing that V_i/V can be written as $\tilde{\rho}\tilde{v}_i\Phi_i$, the entropy of mixing is given by

$$\frac{\Delta S_{\text{mel}}}{kV^*} = -\tilde{v}\left[\frac{(1-\tilde{\rho})\ln(1-\tilde{\rho})}{V_0} + \frac{\tilde{\rho}\ln\tilde{\rho}}{V^*}\right] + \sum_i \frac{\Phi_i \ln \Phi_i}{V_i^*} + h(\tilde{v}_i) \tag{4.84a}$$

where $h(\tilde{v}_i)$ is a function of the reduced volume of the pure components and where $\sum_i \Phi_i / V_i^*$ is equivalent to $1/V^*$.

The term $h(\tilde{v})$ is a positive contribution to the entropy which results from the expansion of the system from a volume V^* to volume V.

V_0, which denotes the volume of a vacant cell or a "hole," is supposed to vary with the composition of the solution or the blend considered. The potential energy of the system is given by the expression

$$\frac{E}{V^*} = -\tilde{\rho}P^* \equiv \tilde{\rho}\left[\sum_i \Phi_i P_i^* - kT \sum_i \sum_{<j} \Phi_i \Phi_j \chi_{ij}\right] \tag{4.89}$$

where

$$P_i^* = (\delta_I^*)^2 \tag{4.90}$$

denotes the characteristic pressure or the cohesive energy density.

Consequently, the expression for the free energy density can be written as

$$G = E + PV - TS/V^* = G_{\text{nco}} + G_{\text{co}} \tag{4.91}$$

with

$$G_{\text{nco}} = -\tilde{\rho}P^* + P\tilde{v} + kT\tilde{v}\left[\frac{(1-\tilde{\rho})\ln(1-\tilde{\rho})}{V_0} + \frac{\tilde{\rho}\ln\tilde{\rho}}{V^*}\right] \tag{4.92}$$

and

$$G_{\text{co}} = kT \sum_i \Phi_i \frac{\ln \Phi_i}{V_i^*} \tag{4.93}$$

G_{nco} and G_{co} are the nonconformational and conformational contributions, respectively.

From the expression for the free energy, the expression for the equation of state for a mixture of m constituents can be established from the condition that at equilibrium the system occupies a given volume $(\partial G/\partial \tilde{v} = 0)$.

$$\tilde{\rho}^2 + \tilde{P} + \tilde{T}[\ln(1 - \tilde{\rho}) + (1 - 1/q)\tilde{\rho}] = 0 \qquad (4.94)$$

where q is the ratio of volumes V^*/V_0 and $\tilde{P}(= P/P^*)$ and $\tilde{T}(= T/T^* = kT/P^*V_0)$ are the reduced pressures and temperatures of this system.

Each component considered separately complies with the same equation of state, but with its own characteristic (P_i^*, T_i^*, q_i).

For a polymer of high molar mass, $q \to \infty$, and the expression reduces to

$$\tilde{\rho}^2 + \tilde{P} + \tilde{T}[\ln(1 - \tilde{\rho}) + \tilde{\rho}] = 0 \qquad (4.94a)$$

A component can be completely defined through the knowledge of T_i^*, P_i^* and ρ_i^*; the number of sites occupied, characterized by q, is related to the parameters of the equation of state by

$$q = V^*/V_0 = P^*V^*/kT^* \qquad (4.95)$$

Practically, T_i^* P_i^*, and ρ_i^* can be experimentally determined from the measurements of their cohesive energy density, thermal expansion, and density, extrapolated to $0°K$. For blends comprised of m constituents, one resorts to the following procedure: from the relation

$$\frac{1}{V_0} = \sum_i \left(\frac{P_i^*}{kT_i^*} \right) \qquad (4.96)$$

V_0 can be deduced.

The characteristic pressure of the mixture P^* can be calculated from relation (4.89):

$$P^* = \sum_i \Phi_i P_i^* - kT \sum_{i<j} \sum \Phi_i \Phi_j \chi_{ij}$$

Knowing V_0 and P^*, the characteristic temperature of the mixture (T^*) can be deduced using the expression $kT^* = P^*V_0$. Table 4.1 gives the characteristic values for some polymers that can be used to determine the miscibility of any two of them. Practically, two polymers are miscible when their temperatures T_i^*, and hence their products $P_i^*V_0$, are close. There is miscibility if T_1^* is higher than T_2^*, P_1^* is higher than P_2^*, and, thus, the two polymers have similar expansion coefficients.

Simple in its use, this model describes appropriately the behavior of polymer blends and of concentrated macromolecular solutions. It predicts the change in volume resulting from the mixing and also the existence of lower and higher critical temperatures of phase separation when there is a great disparity in the size of components 1 and 2. At high pressure and/or low temperature ($\tilde{\rho} \to 1$), this model is close to that of Flory–Huggins.

Table 4.1. Characteristic values of the most common polymers (according to Lacombe and Sanchez)

Polymer	T^* (K)	P^* (MPa)	$\rho^*(\text{g}\cdot\text{cm}^{-3})$
Poly(dimethylsiloxane), PDMS	476	0.30	1.104
Poly(vinyl acetate), PVAc	590	0.51	1.283
Poly(butyl methacrylate), PBuMA	627	0.43	1.125
Polyisobutene, PIB	643	0.35	0.974
Polyethylene (low density), LDPE	673	0.43	0.887
Poly(methyl methacrylate), PMMA	696	0.50	1.269
Poly(cyclohexyl methacrylate), PCHMA	697	0.43	1.178
Polystyrene (atactic), PS	735	0.36	1.105
Poly(2,6-dimethyl-1,4-oxyphenylene), PPO	739	0.52	1.161
Poly(α-methylstyrene), PαMS	768	0.38	1.079

4.5.2. Case of Block Copolymers

The case of block copolymers is peculiar and deserves a specific development. In such a structure, the A and B blocks are connected to one another by a covalent bond, and their respective molar mass and composition can be varied independently. Being incompatible, A and B blocks tend to minimize their surface of contact but, contrary to the mere blends of two polymers they cannot phase separate to a macroscopic scale due to the bond which links them. Classical composition-temperature phase diagrams cannot be constructed for block copolymers as for the corresponding blends. Indeed the A and B blocks are forced to self-organize in domains of more reduced nano- or mesoscopic size. The transition from a homogeneous blend to a system composed of ordered phases as well as the size and the morphology of these ordered phases depend on two elements; the product $\chi_{AB}X$ ($X = $ total degree of polymerization) and the dissymmetry in size of the two blocks.

Thus, Leibler predicted that copolymers with two symmetrical blocks phase-segregate into mesodomains only for values of $\chi_{AB}X > 10.5$, which has been experimentally confirmed. Beyond that value, the repulsive interactions of enthalpic origin dominate and force the system to phase-segregate. The term "order–disorder transition" (ODT) is used to describe this type of phase separation. The size of the domains corresponding to the A and B blocks then grow, which results in the concomitant reduction of the surface of contact between phases. To form the thickest possible domains, the chains tend to stretch themselves, provoking a loss of conformational entropy that opposes the phase growth. In addition, the confinement of the links between the blocks in an increasingly smaller interface also contributes to the loss of entropy. The structure of the domain which is formed at equilibrium is thus governed by factors of opposite nature, of enthalpic and entropic origins. In the case of blocks of symmetrical size, the copolymers self-organize in a lamellar structure.

In the case of asymmetric blocks, the system is unable to accommodate their lamellar organization. Indeed, a lamellar structure with a thickness proportional to the length of the blocks would imply an excessive stretching of the short block; the system actually eases this constraint by curving its interface and adopting other morphologies. This results in an increase of the contact area between A and B phases and in the formation of cylindrical and spherical morphologies. The complexity of the organization increases with the block dissymmetry: the unidimensional lamellae order gives way to the two-dimensional network of cylinders and then to the three-dimensional macrocrystal of close-packed spheres (see Figure 5.35). Between lamellar and cylindrical morphologies, the so-called ordered *double diamond* morphology was also discovered, which corresponds to the overlap of two bicontinuous phases.

One of the main applications of diblock copolymers is their use in very small amounts—typically a few percent—in polymer blends of the same chemical nature as that of their constituting blocks. The diblock copolymer moves spontaneously to the interface—when used in suitable quantities—so as to ease the interfacial tension and increase the interface area. With an enlarged interface, the system self-orders in smaller domains, facilitating the dispersion of one phase in the other one. Moreover, the presence of the diblock copolymer at the interface improves the interphase adhesion, and in that respect it can improve the mechanical properties of polymer blends (impact strength, etc.) (see Section 5.6.5).

In addition to diblock copolymers, copolymers with multiple blocks—in particular with three blocks—have also been prepared and shown to exhibit original morphologies and interesting properties. Triblock copolymers of the poly(styrene-*b*-butadiene-*b*-styrene) (SBS) type and poly(styrene-*b*-isoprene-*b*-styrene) (SIS) are widely used in the field of *thermoplastic elastomers* (see Section 5.5.5). Recently, ABC triblock copolymers including three different blocks were shown to afford a host of exotic phases.

APPENDIX

Scaling Theory and Semi-dilute Solutions

The scaling theory has been developed to explain critical phenomena and to account for the dramatic changes undergone by a system when approaching the order-disorder transition point. The phase transition in liquid crystals and the demixing transition of a two-liquid mixture near LCST or UCST are two well-known examples of such order–disorder transition.

As T approaches T_c, the critical temperature, fluctuations grow in the local alignment of liquid crystalline molecules or in the local composition of two-component mixtures so that increasingly larger domains develop. Near T_c the domain size grows in a power of the temperature difference $|T - T_c|$ and for $T = T_c$ the domain size becomes infinite, the system segregating into macroscopically ordered phases.

By analogy between $|T - T_c| \to 0$ and $X^{-1} \to 0$ or $C/C^* \gg 1$, the scaling theory has also been applied to polymer solutions. In a solution of overlapping polymer chains, a number of thermodynamic parameters, including the correlation length (ξ) of monomer density fluctuations, the osmotic pressure (Π) (see Section 6.1.2.1) or the excess of scattering intensity (I_{ex}) depend on C/C^*, and not on X.

At low concentration, the correlation length can be assimilated to r ($\xi \cong r$), but as the chains overlap in the semi-dilute regime, this correlation between the density fluctuation and r is lost and ξ becomes shorter.

A scaling function (also called *universal function* because it is independent of X and of the system considered), $f\xi(C/C^*)$, is to be defined which can be expressed as

$$\xi = rf_\xi(C/C^*) \quad \text{with} \quad f_\xi(C/C^*) \begin{cases} = 1 & \text{for } (C/C^*) \to 0 \\ = (C/C^*)^u & \text{for } (C/C^*) \gg 1 \end{cases}$$

The scaling theory does not provide the numerical value taken by u—also called scaling exponent—but the latter can be obtained by different means (experiments, computer simulation, or renormalization group theory).

As demonstrated in Section 4.4, ξ decreases above C^* in a power law with an exponent $u = -3/4$.

LITERATURE

P. Flory, *Principles of Polymer Chemistry*, Chapters XII and XIII, Cornell University Press, Ithaca, New York, 1971.

P. G. De Gennes, *Scaling Concepts in Polymer Physics*, Cornell University Press, Ithaca, NY, 1979.

I. Sanchez, in *Encyclopedia of Physical Science and Technology*, Chapter 1, Vol. 11: *Polymer Phase Separation*, Academic Press I., Orlando, FL, 1987.

I. Teraoka, *Polymer Solutions: An Introduction to Physical Properties*, Wiley, New York, 2002.

5

CONFORMATIONAL STRUCTURES AND MORPHOLOGIES

Besides the molecular structure reflecting the position of atoms within monomer units and the distribution (or order of placement) of the latter along the chain, there are higher forms of organization that may concern the whole chain and even an assembly of chains. Because the latter can self-organize into morphologies that in turn may affect the physicochemical and the mechanical characteristics of the corresponding polymeric material, it is essential to describe the structures formed in all their complexity.

The polymer chain will first be described individually in the disordered state and in the ordered state, and then it will be described as an integral part of an assembly.

5.1. DESCRIPTION OF A RANDOM POLYMER CHAIN

Since its origin, polymer science has been confronted with a same duality that is, on the one hand, the measurement of the average molar mass of macromolecules and, on the other, the determination of their size and their specific shape. The average molar mass of a sample and its dispersity (DM) depend mainly on the experimental conditions of synthesis. As for the size and the shape of the macromolecules, they are closely controlled by the topological placement of their repetitive units (linear, branched...) and also by the interactions that develop between each one of these units and the surrounding medium (solvent, another repetitive unit, stretched or chains at rest).

Organic and Physical Chemistry of Polymers, by Yves Gnanou and Michel Fontanille
Copyright © 2008 John Wiley & Sons, Inc.

A solution containing macromolecules differs from a solution of simple molecules, first of all by the range of interactions that build up intra- or inter-molecularly. Indeed, the large size and the connectivity of macromolecules imply that repetitive units can interact with each other at distances of several tens—even hundreds—of nanometers and not only locally in a few atom scale. A very clear distinction prevails in the field of polymers between long-range and short-range interactions that refer to the distance separating them along the backbone.

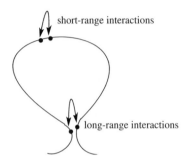

Short-range interactions refer to the restrictions to the free rotation of chemical bonds and their valence angles and they are always present; on the contrary, long-range interactions manifest themselves only when two monomer units come close in the presence of a good solvent; they are responsible for the so-called excluded volume effect, which reflects the exclusion by a monomer unit of any alike one from the volume it occupies.

In addition to the multiple interactions (either between repetitive units within short and long range or with the solvent) that characterize them, polymers may also lend themselves to cooperative and amplified effects: a small interaction between polymer segments and other molecules repeated many times may indeed result in a phenomenon perceptible at the macroscopic level. Before describing the behavior of polymers in solution, it is necessary to consider the various factors that condition the conformation of a chain in the absence of any solvent and, in first instance, short-range interactions.

The term **conformation** is used in organic chemistry to describe all the positions spatially occupied by atoms and groups of atoms by rotation around a σ chemical bond. In the field of polymers, the term *local conformation* is used to describe the position of a limited number of atoms, and molecular conformation or *macro-conformation* refers to a sequence of *microconformations* along the chain. Local conformations are mainly determined by the nature and the configuration of the repeating units, but they also depend on the temperature and on the interactions of these units with the surrounding medium.

Starting from the conformations of a molecule such as butane, the conformations of a polymethylene chain $-(CH_2)_n-$ can be calculated by extension to the entire chain. In butane, the conformations are determined by three parameters: the carbon–carbon bond length, the values taken by the valence angle θ, and the torsional angle (dihedral) ϕ corresponding to the rotation around the carbon–carbon bonds (Figure 5.1).

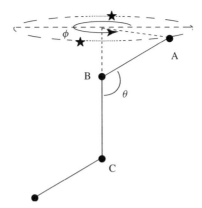

Figure 5.1. Rotation around bonds. Atom A turns around the axis B-C with a constant angle θ; it is represented in a transoidal conformation (•). (★) corresponds to the positions of A for gauche conformations.

Thus, it can be easily demonstrated that the most energetically favored conformations are transoidal (T) *(antiperiplanar)* and gauche conformations *(synclinal)*; cisoidal conformations (C) *(synperiplanar)* and the two anti conformations *(anticlinal)* are less probable (Figure 5.2).

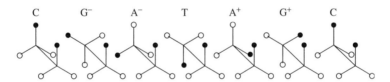

Figure 5.2. Local conformations of the central bond in a butane molecule; the methyl groups are denoted by (•).

By a similar approach and using the data obtained from butane, one could thus calculate the most favorable conformations for a macromolecular chain, but obviously the problem becomes more complex as the number of bonds increases. If one observes that only three local conformations are energetically favorable in a sequence of three carbon–carbon bonds, a chain of degree of polymerization equal to 102 can exhibit up to 3^{100} possible molecular macroconformations.

A chain whose local conformations would all be transoidal and whose valence angles would all be equal to $109° \ 28'$ corresponds to the case of a completely stretched hydrocarbon chain (polyethylene chain) as shown in the diagram of Figure 5.3. The distance r separating the two chain ends then denotes the maximum length that a macromolecular chain can exhibit; this distance is also called *contour length*:

$$r_{\text{cont}} = nL \, \sin\frac{\theta}{2} \qquad \text{with } L = 0.154 \text{ nm}$$

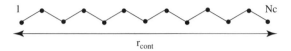

Figure 5.3. Representation of a hydrocarbon chain in total extension.

The case of a completely stretched chain can be observed in a crystalline state. In solution, none of the 3^{100} conformations exists beyond a negligible fraction of time; to obtain information such as the size or the overall shape of a macromolecule, it is necessary to calculate the average over all possible conformations. By effecting this, one realizes that macromolecular chains do not exhibit a Euclidean shape of either "sphere" or "rod" type but behaves instead like a statistical coil. Confronted with the tiresome character of such a conformational calculation, one finds it more advantageous to use statistical mechanics and probabilistic reasoning to determine the average size of the chains.

5.1.1. Freely Jointed Chains

5.1.1.1. Calculation of the End-to-End Distance. The characteristic dimensions of a chain with perfectly flexible joints can be described by means of the *random walk* model. In this case, the N-step trajectory of a random walker—who can freely return to a site previously visited—is assimilated to a chain of length NL: the only element taken into consideration is thus the bond length (L). The valence (θ) and rotation (φ) angles are subjected to no particular constraint, being free to take any value between $0°$ and $360°$. Such chains behave as if they were *"phantom"* objects, which can freely intersect and occupy the same element of volume. One convenient means to evaluate the dimensions of such a macromolecular chain is to calculate the end-to-end distance ($\mathbf{r}$) connecting its two chain ends by vector analysis.

End-to-end vector $\mathbf{r}$ can be expressed as the sum of vectors $\mathbf{r}_i$ corresponding to each bond (Figure 5.4):

$$\mathbf{r} = \sum_i \mathbf{r}_i$$

For polymers in the molten state, this end-to-end distance is subject to constant fluctuation and experimental techniques permitting the determination of the instantaneous values of $\mathbf{r}$ do not exist. In such a situation one has no option but to deal with average values; as $\mathbf{r}$ is likely to be continuously reoriented due to Brownian motion its average value tends to 0. Thus, for the determination of chain dimensions it is more meaningful to calculate the average of the product (r^2), which squares positive and negative components before averaging, affording a more realistic route to characterize a polymer chain:

$$r^2 = \mathbf{rr} = \sum_{i=1}^{n} \mathbf{r}_i \sum_{j=1}^{n} \mathbf{r}_j = \sum_{i=1}^{n} \mathbf{r}_i^{\,2} + 2 \sum_{i=1}^{n-1} \sum_{j=i+1}^{n} \mathbf{r}_i \mathbf{r}_j \qquad (5.1)$$

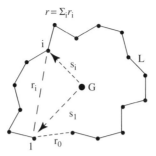

$r = \Sigma_i r_i$

Figure 5.4. End-to-end distance (r_0) in the case of a freely jointed chain.

After averaging over an assembly of the M chains present in the medium, expression (5.1) becomes

$$\langle r^2 \rangle_0 = \frac{1}{M} \sum_{k=1}^{M} r_k^2 \sum_{i=1}^{n} \langle r_i^2 \rangle + 2 \sum_{i=1}^{n} \sum_{j=i+1}^{n} \langle r_i r_j \rangle \tag{5.1a}$$

where $\langle r^2 \rangle_0$ denotes the *mean square end-to-end distance*.

To calculate this double sum, it is practical to represent it in the form of a matrix of n lines and n columns:

$$\langle r^2 \rangle_0 = \begin{vmatrix} \langle r_1 r_1 \rangle + \langle r_1 r_2 \rangle + \cdots + \langle r_1 r_n \rangle + \\ \langle r_2 r_1 \rangle + \langle r_2 r_2 \rangle + \cdots \\ \langle r_n r_1 \rangle + \langle r_n r_2 \rangle + \cdots + \langle r_n r_n \rangle \end{vmatrix}$$

The first term, $\Sigma_i \langle r_i^2 \rangle$, corresponds to the sum of elements appearing in the diagonal, whereas the $\Sigma_i \Sigma_j \langle r_i r_j \rangle$ term denotes those located above this diagonal. The scalar product of two vectors r_i and r_j is written as

$$\langle r_i r_j \rangle = L^2 \langle \cos \tau_{ij} \rangle \tag{5.2}$$

where τ is the angle between vectors r_i and r_j. Expression (5.1a) thus reduces to

$$\langle r^2 \rangle_0 = nL^2 + 2L^2 \sum_{i=1}^{n-1} \sum_{j=i+1}^{n} \langle \cos \tau_{ij} \rangle \tag{5.3}$$

As the orientation of the various bonds is completely uncorrelated, there is no privileged direction so that

$$\langle \cos \tau_{ij} \rangle = 0 \qquad \text{for } i \neq j$$

which gives for $\langle r^2 \rangle$

$$\langle r^2 \rangle_0 = nL^2 \tag{5.4}$$

5.1.1.2. Radius of Gyration. The mean square end-to-end distance is a value that has a meaning only for perfectly linear polymers having two ends. Macromolecules, which do not have any free end as in the case of macrocycles or have several free ends as in the case of star polymers, cannot be described by an end-to-end distance. It is more appropriate to use the *radius of gyration* or *root-mean-square radius of gyration* as a measure of the chain dimensions. Like any material object, a macromolecule can be identified with i entities (here atoms) of mass m_i, located at distances s_i from its center of mass S; by definition, the square of the radius of gyration is the second moment around the center of mass and is defined as the square of the distances between the various entities (i) and the center of mass (Figure 5.4):

$$\langle s^2 \rangle = \frac{\langle \sum_i m_i \, s_i^2 \rangle}{\sum_i m_i} \tag{5.5}$$

To find how this expression is established, one has to remember that the moment of inertia of an object for rotational motion around its center of mass is defined as

$$I = \sum_i m_i s_i^2 \tag{5.6}$$

The radius of gyration can also be defined as the distance that separates the rotational axis of an object from the fictitious point where all its mass would be concentrated:

$$I = s^2 \sum_i m_i = \sum_i m_i s_i^2$$

which means writing expression (5.5). In other words, the radius of gyration is equal to the radius of a sphere of uniform surface density whose moment of inertia is equal to ms^2, where m represents the mass of that sphere.

Let us consider a chain made up of n repetitive units that are randomly selected. The vector that connects the reference unit to an unspecified unit i is called $\mathbf{r}_i$; as for $\mathbf{s}_i$, it refers to the vector connecting the center of mass to this unit. The reference unit and the center of mass are connected by the vector $\mathbf{z}$:

$$\mathbf{s}_i = \mathbf{r}_i - \mathbf{z} \tag{5.7}$$

By definition of the center of mass, we have $\sum \mathbf{s}_i = 0$, which gives

$$\mathbf{z} = \frac{1}{n} \sum \mathbf{r}_i \quad \text{and} \quad \mathbf{z}^2 = \left(\frac{1}{n^2} \right) \sum_i \sum_j \mathbf{r}_i \, \mathbf{r}_j \tag{5.8}$$

To obtain the relation between the root-mean-square radius of gyration $\langle s^2 \rangle^{1/2}$ and the root-mean-square end-to-end distance $\langle r^2 \rangle^{1/2}$, one has to calculate the sum $\sum_i s_i^2$:

$$\sum_i s_i^2 = \sum_i \mathbf{s}_i \mathbf{s}_i = \sum_i (\mathbf{r}_i - \mathbf{z})(\mathbf{r}_i - \mathbf{z}) \tag{5.9}$$

$2\mathbf{z}\sum_i \mathbf{r}_i$ can also be written as $2n\mathbf{z}\mathbf{z} = 2nz^2$, which gives

$$\sum_i s_i^{\,2} = \sum_i r_i^{\,2} - nz^2$$

$$= \sum_i r_i^{\,2} - \frac{1}{n}\sum_{i=1}^{n-1}\sum_{j=i+1}^{n} \mathbf{r}_i\mathbf{r}_j \qquad (5.10)$$

The product $\mathbf{r}_i\mathbf{r}_j$ is nothing but the projection of a vector on another one—that is, the product of their respective scalars by $\cos\theta$, the angle formed by the two vectors:

$$\mathbf{r}_i\mathbf{r}_j = r_i r_j \cos\omega = (r_i^2 + r_j^2 - r_{ij}^2)/2 \qquad (5.11)$$

where r_{ij} represents the length of the vector connecting the units i and j. Introducing this expression into $\sum s_i^2$ gives

$$\sum_i s_i^{\,2} = \frac{1}{2n}\sum_{i=1}^{n-1}\sum_{j=i+1}^{n} r_{ij}^2 \qquad (5.10a)$$

which leads to

$$\frac{1}{n}\sum_i s_i^{\,2} = \frac{1}{2n^2}\sum_i\sum_j \langle r_{ij}^2\rangle \qquad (5.10b)$$

This represents the expression for the mean square distance from the unit i to the center of mass for a given conformation. To obtain the average $\langle s^2\rangle$, it is necessary to consider all possible conformations:

$$\langle s^2\rangle = \frac{1}{n}\sum_i s_i^2 = \frac{1}{2n^2}\sum_i\sum_j \langle r_{ij}^2\rangle \qquad (5.12)$$

$\langle r_{ij}^2\rangle$ is the mean square end-to-end distance for a fictitious chain made up of $(j-i)$ units of length L.

For a Gaussian chain, such a distance depends only on the path to cover along the chain to go from one end to another:

$$\langle r_{ij}^2\rangle = |j-i|L^2 \qquad (5.13)$$

which gives

$$\langle s^2\rangle = \frac{L^2}{2n^2}\sum_i\sum_j |j-i|$$

$$\sum_{j=1}^{j=n} |i-j| = \sum_{j=1}^{i}(i-j) + \sum_{j=i+1}^{n}(j-i) = i^2 - \tfrac{1}{2}i(i+1) + \tfrac{1}{2}(n-i)(n+i+1)$$

$$\langle s^2\rangle = i^2 - in + \tfrac{1}{2}n^2 + \tfrac{1}{2}n - i \qquad (5.14)$$

In addition, the sum of the i squares is equal to

$$\sum_i i^2 = 1^2 + 2^2 + \cdots + n^2 = \frac{n(n+1)(2n+1)}{6} \tag{5.15}$$

$$\sum_{i=1}^{i=n} \sum_{j=1}^{j=n} |j - i| = \frac{n^3 - n}{3} \simeq \frac{n^3}{3} \tag{5.16}$$

from which the expression of the mean square of the radius of gyration for a chain with free rotations can be written as follows:

$$\langle s^2 \rangle_0 = \frac{L^2}{2n^2} \frac{n^3}{3} = \frac{nL^2}{6} \tag{5.17}$$

5.1.1.3. Distribution of the End-to-End Distances of the Chain

Analysis of the "Random Walk" Model. If the average dimensions of freely jointed coils can be determined using the previously described calculation, the latter cannot be used to evaluate the distributions corresponding to these average values. Such an information can be obtained by identifying a freely jointed chain with the random walk of a particle.

The average trajectory of such a random walk particle and the distribution of displacements for a given number of steps were calculated by Chandrashekar. He initially dealt with the problem in one dimension, by considering the motion of a particle which would make n steps of length L along an axis x, either to the right (n_+) or to the left (n_-) with an equal probability. The number of steps that separate the origin of axis x (which is also the starting point of the particle) from the final point after n steps (Figure 5.5) is

$$u = n_+ - n_- \tag{5.18}$$

$$n = n_+ + n_- \tag{5.19}$$

where $n = n_+ + n_-$ is the total number of steps.

Thus, the number of steps in the positive direction is

$$n_+ = \frac{n + u}{2} \tag{5.20}$$

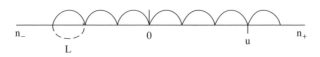

Figure 5.5. Distance covered by a particle in a direction of space, according to the random walk model.

and, in the negative direction,

$$n_- = \frac{n - u}{2} \tag{5.21}$$

The probability that the origin and the final point are separated by the distance u after n steps is the product of the probability of having a sequence of n_+ positive moves and n_- negative moves, each one of probability 1/2 times the number of ways of executing n_+ positive steps among n.

$$w(n, m) = \left(\frac{1}{2}\right)^n \frac{n!}{[\frac{1}{2}(n + u)]![\frac{1}{2}(n - u)]!} \tag{5.22}$$

Using the Sterling approximation

$$\ln n! = \left(\frac{n + 1}{2}\right) \ln n - n + \frac{1}{2} \ln 2\pi \tag{5.23}$$

this expression can be simplified:

$$\begin{aligned}
w(n, u) = {}& n \ln 2 + \left(\frac{n + 1}{2}\right) \ln n - n + \frac{1}{2} \ln 2\pi \\
& - \frac{1}{2}(n + u + 1) \ln \left[\frac{n}{2}\left(1 + \frac{u}{n}\right)\right] + \frac{n + u}{2} - \frac{1}{2} \ln 2\pi \\
& - \frac{1}{2}(n + u - 1) \ln \left[\frac{n}{2}\left(1 - \frac{u}{n}\right)\right] + \frac{n - u}{2} - \frac{1}{2} \ln 2\pi
\end{aligned} \tag{5.24}$$

as $u < n$, expression $\left(1 + \frac{u}{n}\right)$ can be also written as

$$\ln \left(1 + \frac{u}{n}\right) = \frac{u}{n} - \frac{u^2}{2n^2} + \frac{u^3}{3n^3} - \cdots \tag{5.25}$$

which gives

$$\begin{aligned}
\ln w(n, u) = {}& n \ln 2 + \left(\frac{n + 1}{2}\right) \ln n - \frac{1}{2} \ln 2\pi \\
& - \frac{1}{2}(n + u + 1) \left(\ln n - \ln 2 + \frac{u}{n} - \frac{u^2}{2n^2}\right) \\
& - \frac{1}{2}(n - u + 1) \left(\ln n - \ln 2 + \frac{u}{n} - \frac{u^2}{2n^2}\right)
\end{aligned} \tag{5.26}$$

Finally,

$$\ln w(n, u) = -\frac{1}{2} \ln n + \ln 2 - \frac{1}{2} \ln 2\pi - \frac{1}{2} \frac{u^2}{n} \tag{5.27}$$

The exponential function of this expression gives

$$w(n, u) = \left(\frac{2}{\pi n}\right)^{1/2} \exp\left(\frac{-u^2}{2n}\right) \tag{5.28}$$

In the case of polymer systems, it is more appropriate to express the random walk distribution as a function of the net displacement x rather than the number of steps u. Consequently, it is necessary to carry out a change of variable

$$x = uL \tag{5.29}$$

where L is the step length.

The probability $w(n, x)$ that the random walk ends between x and $(x + \Delta x)$ (Δx being the characteristic interval along the x axis) as the final position can be calculated using the following expressions:

$$w(n, x)\Delta x = w(n, u)\left(\frac{\Delta x}{2L}\right)$$

$$w(n, x) = (2\pi n L^2)^{-1/2} \exp\left(\frac{-x^2}{2nL^2}\right) \tag{5.28a}$$

These relations show that the random walk and thus the dimensions of a macromolecular chain are given by a distribution function of Gaussian nature.

This one-dimensional random walk calculation can then be generalized to three dimensions because macromolecular chains are not confined in only one direction. Assuming that the three directions of space are equiprobable—which is the case for flexible chains—the probability for the occurrence of n steps in a given direction of space has to be calculated with a number of steps equal to $n/3$ for each of the three directions of space x, y, z:

$$P(n, x, y, z)dx\,dy\,dz = P\left(\frac{n}{3}, x\right)P\left(\frac{n}{3}, y\right)P\left(\frac{n}{3}, z\right) \tag{5.30}$$

$P(n, x, y, z)dxdydz$ is the probability to find the end of the macromolecular chain considered in the element of volume of coordinates x, y, z. This probability is written as

$$P(n, x, y, z)dx\,dy\,dz = \left(2\pi\frac{n}{3}L^2\right)^{-3/2} \exp\left(-\frac{3(x^2 + y^2 + z^2)}{2nL^2}\right)dxdydz \tag{5.30a}$$

For the sake of simplicity, it is more practical to use spherical coordinates:

$$x^2 + y^2 + z^2 = r^2 \quad \text{and} \quad dx\,dy\,dz \rightarrow 4\pi r^2 dr$$

Therefore, this probability does not apply to a cubic element of volume $dx \cdot dy \cdot dz$, but instead applies to a spherical envelope of radius r and thickness dr. The probability to find the chain end inside this spherical envelope is written as

$$P(n,\ r)dr = 4\pi \left(2\pi \frac{n}{3}L^2\right)^{-3/2} r^2 \exp\left(\frac{-3r^2}{2nL^2}\right)dr \qquad (5.30\text{b})$$

From this expression the mean square end-to-end distance can be deduced:

$$\langle r^2 \rangle_0 = \int_0^\infty P(n,\ r)r^2 dr \qquad (5.31)$$

$$\langle r^2 \rangle_0 = 4\pi \left(\frac{2}{3}\pi n L^2\right)^{-3/2} \int_0^\infty r^4 \exp\left(\frac{-3r^2}{2nL^2}\right)dr \qquad (5.31\text{a})$$

For this calculation, one has to take into consideration the fact that $P(n,\ r)$, the probability function, is normalized and satisfies the criterion of an integral equal to 1. Thus, the properties of the Gauss integrals apply to this case, which for $P(n,\ r)$ gives

$$P(n,\ r)dr = 4\pi A \int e^{-B^2 r^2} r^2 dr = \pi^{3/2} A B^{-3} \qquad (5.32)$$

with

$$A = \left(\frac{2\pi}{3}n L^2\right)^{-3/2} \quad \text{and} \quad B^2 = \left(\frac{3}{2nL^2}\right)$$

and for the mean square $\langle r \rangle_0^2$:

$$4\pi A \int e^{-B^2 r^2} r^4 dr = \frac{3}{2} \pi^{2/3} A B^{-5}$$

giving the same result as the one established in expression (5.4):

$$\langle r^2 \rangle_0 = n L^2 \qquad (5.33)$$

5.1.2. Freely Rotating Chains

Freely jointed chains in which short-range interactions are not taken into consideration do not reflect the reality. Indeed, in a real chain, restrictions such as the effective valence angles and the substitution of valence angles of $109°\ 28'$ for arbitrary values result in an increase of the end-to-end distance $\langle r \rangle_{0f}^{1/2}$.

In the expression of $\langle r^2 \rangle_0$ for freely jointed chains the term $\langle \cos \tau_{ij} \rangle$ was considered equal to 0 for $i \neq j$, which cannot be the case for freely rotating chains since their valence angles θ are imposed. In this case, $\langle \mathbf{r}_i \mathbf{r}_{i+1} \rangle$ can be written as

$$\langle \mathbf{r}_i \mathbf{r}_{i+1} \rangle = L^2 \cos(180 - \theta)$$
$$\langle \mathbf{r}_i \mathbf{r}_{i+2} \rangle = L^2 \cos(180 - \theta)$$

and thus

$$\langle \mathbf{r}_i \mathbf{r}_j \rangle = L^2 [\cos(180 - \theta)]^{j-i} \tag{5.34}$$

The expression for $\langle r^2 \rangle$ is now written as

$$\langle r^2 \rangle_{0f} = nL^2 + 2L^2 \sum_{i=1}^{n-1} \sum_{j=i+1}^{n} [\cos(180 - \theta)]^{j-i} \tag{5.35}$$

If one replaces $(j - i)$ by a single variable k, $\langle r^2 \rangle$ establishes as

$$\langle r^2 \rangle_{0f} = nL^2 \left\{ 1 + \frac{2}{n} \sum_{k=1}^{n=1} (n - k)[\cos(180 - \theta)^k] \right\} \tag{5.35a}$$

which can be rewritten in the form

$$\langle r^2 \rangle_{0f} = nL^2 \left\{ 1 + 2\sum_{k=1}^{n=1} [\cos(180 - \theta)]^k - \frac{2}{n} \sum_{k=1}^{n=1} k[\cos(180 - \theta)^k] \right\}$$

$$\langle r^2 \rangle_{0f} = nL^2 \left[1 + \frac{2\cos(180 - \theta)}{1 - \cos(180 - \theta)} - \frac{2\cos(180 - \theta)}{n} \frac{[1 - \cos(180 - \theta)]^n}{[1 - \cos(180 - \theta)]^2} \right]$$

$$\tag{5.35b}$$

For long chains, $n = \infty$:

$$\langle r^2 \rangle_{0f} = nL^2 \left[1 + \frac{2\cos(180 - \theta)}{1 - \cos(180 - \theta)} \right] = nL^2 \frac{1 + \cos(180 - \theta)}{1 - \cos(180 - \theta)}$$

$$\langle r^2 \rangle_{0f} = nL^2 \left[\frac{1 - \cos\theta}{1 + \cos\theta} \right] \tag{5.36}$$

which gives for a polyethylene chain with $\theta = 109° \, 28'$,

$$\langle r^2 \rangle_{0f} \cong 2nL^2 \tag{5.37}$$

5.1.3. Chain with Restricted Rotations

The preceding result, which is based on the assumption of a free rotation of repeating units (unrestricted dihedral angle), represents a good approximation at high temperature when the variations of energy between transoidal and gauche conformations are negligible if compared to the product RT. At lower temperatures, conformations of lower energy dominate—in particular, anti conformations—so that it is necessary to take into account the restriction to the free rotation of bonds.

In view of the complexity of such calculation, it is useful to reconsider the case of simple molecules. The potential energy of a molecule such as ethane, which can be taken as a model of polyethylene, is represented in Figure 5.6: the potential

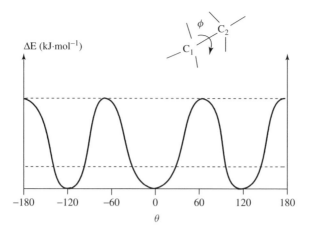

Figure 5.6. Variation of the potential energy of the ethane molecule as a function of the internal rotation angles.

energy of this molecule varies as a function of ϕ due to the electronic repulsion between the atoms carried by carbons 1 and 2. The maxima correspond to the alignment of the hydrogen atoms carried by carbon atoms and the minima to their rotation out-of-phase. The barrier of energy is about $12\,\text{kJ·mol}^{-1}$.

In the case of polyethylene, the potential energy has, in addition to the characteristic minimum for $\phi = 0$ (transoidal conformation T), two secondary minima (rotation of $\pm 120°$) which correspond to the conformations G^{+} and G^{-}. This time, the presence of chain segments on both sides of the bond slightly modifies the situation as compared to the case of ethane. The difference ΔE in energy between transoidal and gauche conformations is about $4\,\text{kJ·mol}^{-1}$ (Figure 5.7). Due to the variation of the potential energy with the rotation angle, the probability of an angular position on the cone of rotation depends on the potential energy at this site, E_{ϕ}.

After taking into account these rotational constraints, one can write

$$\langle r^2 \rangle_{0g} = nL^2 \left[\frac{1 - \cos\theta}{1 + \cos\theta} \right] \left[\frac{1 + \langle \cos\phi \rangle}{1 - \langle \cos\phi \rangle} \right] \tag{5.38}$$

where $\langle \cos\phi \rangle$ is the average value of $\cos\phi$ and is determined using the constant of Boltzmann $[\exp(-E_{\phi}/RT)]$. Because $[\exp(-E_{\phi}/RT)]$ depends on the temperature, this expression can account for the variation of $\langle r^2 \rangle_{0g}$ with the temperature.

The restrictions introduced, which reflect short-range interactions, cause the stretching of the chain. In a Markovian approach a given local conformation is conditioned by the preceding one; in such case, a new multiplicative factor should be added to the preceding expression. It is frequently done to gather all local constraints under a same factor C:

$$\langle r^2 \rangle_0 = CnL^2 \tag{5.39}$$

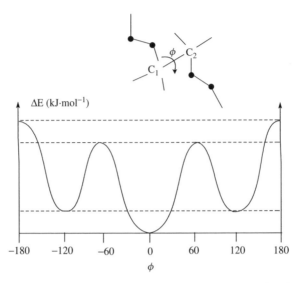

Figure 5.7. Variation of the potential energy of a polyethylene chain as a function of the internal rotation angles.

C can take values ranging between 4 and 10, depending on the type of polymer; expression (5.39) thus corresponds to the "unperturbed" dimensions of a macromolecular chain. The term "unperturbed" conveys the idea that the chain dimension is not perturbed by long-range interactions, as in the presence of a good solvent which would cause the chain to expand further. "Unperturbed" dimensions are characteristic of the so-called "θ conditions" and are observed at temperature θ.

5.1.4. Kuhn Chain Equivalent

Factors such as the valence angles or the hindrance to the free rotation contribute to the stretching of the chain which can thus be represented as n_k units of length L_k linked together (Figure 5.8). Obviously, n_k and L_k (or *Kuhn's length*) do not

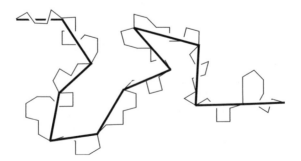

Figure 5.8. Representation of the Kuhn chain equivalent.

actually exist; L_k can be calculated starting from the ratio

$$L_k = \langle r^2 \rangle_0 / r_{\text{cont}}$$

and is defined as follows:

$$r_{\text{cont}} = n_k L_k \tag{5.40}$$

Thus

$$\langle r^2 \rangle_0 = n_k L_k^2 \tag{5.41}$$

Such representation gives an idea of the rigidity of the chain. The latter is more flexible if L_k is small and close to L. Macromolecules whose Kuhn's length is large (and n_k small compared to n) cannot be viewed as random coils but rather as *worm-like* chains.

5.1.5. Chain with Smooth Curvatures or *Worm-like* Chains

Kratky and Porod proposed a model of chain with smooth curvatures to interpret the results of their work on biological macromolecules. Proteins and DNA are rigid macromolecules in which local interactions (hydrogen bonds) exist that were ignored by the previously described models (see Figure 5.13).

The Kratky and Porod model is an extension of the freely rotating chain model in which the valence angles between two units are close to $180°$. This confers to such chains a certain persistence which prevents them from behaving like a random coil; such chains are *semi-rigid* and are also called worm-like. The persistence length (a) is defined as

$$a = L/(1 + \cos \theta) \tag{5.42}$$

The mean square end-to-end distance of a worm-like chain can be calculated as a particular case of a chain with free rotations whose bond angles are close to $180°$, bond lengths are short ($\rightarrow 0$) and number of bonds tend to infinity. The length of such a chain is

$$L_{\text{worm}} = nL = n_a a$$

with

$$n_a = nL/a$$

The mean square end-to-end distance $\langle r^2 \rangle_{\text{worm}}$ is thus established as follows:

$$\langle r^2 \rangle_{\text{worm}} = 2n_a a^2 \left[1 - \left(\frac{1}{n_a} \right) + \left(\frac{1}{n_a} \right) \exp(-na) \right]$$
$$= 2aL_{\text{worm}} - 2a^2 [1 - \exp(-L_{\text{worm}}/a)] \tag{5.43}$$

When the chains are flexible, the persistence length (a) is smaller than nL; therefore $nL/a = n_a \gg 1$ which implies that the term $\exp(-n_a)$ tends to 0. Equation 5.43 can be written as follows:

$$\langle r^2 \rangle_{0f} = 2n_a \, a^2 \tag{5.44}$$

In addition, because $n_a = n_k L_k / a$ and $\langle r^2 \rangle = n_k L_k^2$, one obtains $L_k = 2a$. The Kuhn length is thus twice the persistence length.

When the chains are completely rigid, the persistence length is much larger than the product nL. The term $\exp(-n_a)$ can be expanded for $a/nL \ll 1$:

$$\exp(-n_a) \text{ can be written as } 1 - n_a + n_a{}^2/2 - \cdots$$

$$\langle r^2 \rangle_{\text{worm}} = n_a{}^2 a^2$$

and thus

$$\langle r^2 \rangle_{\text{worm}}^{1/2} = n_a a \tag{5.45}$$

When the end-to-end distance of the chain corresponds to its contour length, this macromolecule can be identified with a rod-like chain. Thus, the Kratky–Porod model is appropriate to describe the transition from a rigid rod to a flexible chain.

5.1.6. Relation Between the Radius of Gyration and the Molar Mass of a Chain in the Swollen State

In the unperturbed state, the radius of gyration s varies as

$$\langle s^2 \rangle = \frac{nL^2}{6}$$

that is, it is proportional to $M^{1/2}$ ($M = $ molar mass of the chain). In the swollen state a different variation is observed due to the excluded volume effect. Using the expression of the free energy of the system as established in the preceding chapter, we have

$$\Delta G = \Delta G_{\text{osm}} + \Delta G_{\text{el}}$$

ΔG_{osm} is written according to the Flory–Huggins model; expressing ΔG_{osm} as a function of Φ_2, one can write

$$\Delta G_{\text{osm}} \div K \Phi_2 \tag{5.46}$$

since

$$\Phi_2 \div \frac{n}{V_{\text{macr}}} = \frac{n}{\frac{4}{3}\pi \langle s \rangle^3} \tag{5.47}$$

In addition, the terms N_1 and N_2 (molar fractions) are also proportional to X and thus we have

$$\Delta G_{\text{osm}} \div K_1 n^2 \langle s \rangle^{-3} \tag{5.46a}$$

As for $\Delta G_{\mathrm{el}} \equiv \Delta S_{\mathrm{el}}$, its expression was established as follows:

$$\Delta G_{\mathrm{el}} \doteq \tfrac{1}{2}[3\alpha^2 - 3]$$

α is the expansion coefficient of the chain and denotes the swelling of the latter in the presence of a solvent (Figure 5.9):

$$\alpha^2 = \langle s^2 \rangle / \langle s^2 \rangle_0$$

where $\langle s^2 \rangle^{1/2}$ and $\langle s^2 \rangle_0^{1/2}$ are the radii of gyration in the presence of a solvent and in the unperturbed state, respectively. Because $\langle s^2 \rangle_0$ can be also written in the form

$$\langle s^2 \rangle_0 = n \frac{L^2}{6} \tag{5.48}$$

one deduces that

$$\Delta G_{\mathrm{el}} \doteq K^2 n^{-1} \langle s^2 \rangle. \tag{5.49}$$

At equilibrium, the derivative of the free energy of the system (ΔG) with respect to the radius of gyration should be equal to 0. After dividing the terms by s and taking into consideration the fact that $X \doteq M$, one can write $\langle s \rangle^2$ as

$$(\langle s^2 \rangle)^{1/2} \doteq M^{3/5} \tag{5.50}$$

or

$$\langle s^2 \rangle^{1/2} \equiv s = KM^{\gamma} = KM^{3/5} \tag{5.50a}$$

Thus, in a good solvent and dilute medium, the radius of gyration increases with a 3/5 power as a function of the molar mass of the polymer (Figure 5.10).

Except for some exceptions, the Flory mean field theory accounts well for the observed phenomena. Experimentally measured exponents γ vary between 0.59 and

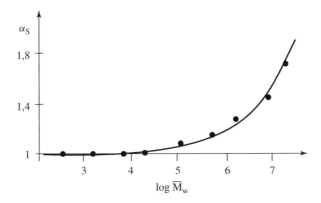

Figure 5.9. Variation of the expansion coefficient α_s as a function of the mass-average molar mass for polystyrene chains in toluene solution at 15°C.

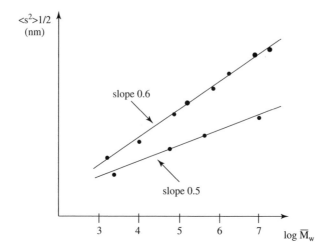

Figure 5.10. Variation of the mean square radius of gyration for polystyrene chains in toluene solution at 15°C.

0.60, which is in agreement with the predictions of this theory. In θ conditions the γ value is equal to 0.50. Compared to the unperturbed state, one observes that s changes from a variation in $M^{1/2}$ to a variation in $M^{3/5}$, which corresponds to an increment of $M^{1/10}$. For a polymer of $\overline{DP}_n = 1000$, the radius of gyration should increase by a factor 2 from the unperturbed state to the perturbed state and the volume by a factor 8.

APPENDIX: CALCULATION OF THE RADIUS OF GYRATION FOR A STAR POLYMER

Like any branched macromolecule, star polymers occupy smaller volumes than their linear equivalents due to the presence of branch points. Their radius of gyration is in turn smaller, and this can be expressed by the factor g:

$$g = \langle s^2 \rangle_{st} / \langle s^2 \rangle_{lin} \tag{5.51}$$

The calculation of the radius of gyration of a star polymer carrying uniform branches and obeying Gaussian statistics can be carried out on the same bases as described previously for a linear polymer:

$$\mathbf{s}_{i,st} = \mathbf{r}_i - \mathbf{z} \tag{5.52}$$

By definition of the center of mass, since $\sum_i \mathbf{s}_{i,st} = 0$, we have

$$\mathbf{z} = \frac{1}{nf} \sum_i \mathbf{r}_i \tag{5.53}$$

In this case, the reference unit will not be randomly selected but will be specifically the core of the star.

The mean square end-to-end distance $\langle s^2 \rangle_{st}$ is expressed logically as the difference between the mean square distance $\langle y^2{}_{st} \rangle$ separating the reference unit from the rest of the chain and $\langle z^2 \rangle_{st}$, which is the mean square distance from the center of mass to this reference unit.

If for a branch $y_i{}^2$ is the mean square distance for the unit of reference to the unit i, then we obtain

$$y_i{}^2 = L^2 i \tag{5.54}$$

$$\langle y^2 \rangle_{br} = \frac{1}{n_f} \sum_i r_i{}^2 \tag{5.55}$$

$$\langle y^2 \rangle_{br} = \frac{L^2}{n_f} \sum_i r_i{}^2$$

as

$$\sum_{i=1}^{n} i = n_f \left(\frac{n_f + 1}{2} \right)$$

$$\langle y^2 \rangle_{br} = \frac{n_f L^2}{2} \tag{5.55a}$$

As for the mean square radius of gyration $\langle s^2 \rangle$, it was previously established:

$$\langle s^2 \rangle_{br} = \frac{n_f L^2}{6} \tag{5.56}$$

$z^2{}_{br}$ is written as

$$z_{br}^2 = \frac{n L^2}{3} \tag{5.57}$$

For a star containing f branches, we have

$$\sum_{1}^{f} n_f = n$$

and

$$\langle y^2 \rangle_{st} = \frac{1}{n} \sum n_f \langle z^2{}_{br} \rangle \tag{5.58}$$

$$\bar{z}_{st}^2 = \frac{1}{n^2} \sum n_f^2 \langle z_{br}^2 \rangle \tag{5.59}$$

For a star, one obtains

$$\langle y^2 \rangle_{st} = \frac{L^2}{2n} \sum_{1}^{f} n_f^2 \tag{5.58a}$$

$$\langle z^2 \rangle_{\text{st}} = \frac{L^2}{3n^2} \sum_1^f n_f^3 \tag{5.59a}$$

and one deduces

$$\langle s^2 \rangle_{\text{st}} = \frac{L^2}{n} \sum_1^f \left[\frac{n_f^2}{2} - \frac{n_f^3}{3n} \right] \tag{5.60}$$

$$g = \langle s^2 \rangle_{\text{st}} / \langle s^2 \rangle_{\text{lin}} = \sum_1^f \left[\frac{3n_f^2}{n^2} - \frac{2n_f^3}{n^3} \right] = f \left[\frac{3}{f^2} - \frac{2}{f^3} \right] \tag{5.61}$$

$$g = \frac{3f - 2}{f^2} \tag{5.61a}$$

5.2. POLYMER CHAINS WITH REGULAR CONFORMATIONS

When kinetic (appropriate rate of cooling in bulk or in solution, rate of evaporation of the solvent) and/or thermodynamic (temperature of the system) conditions are favorable, polymers exhibiting high structural regularity—due to positional isomerism and tacticity—tend to crystallize. The chains crystallize upon concentrating their solutions or, more commonly, upon cooling from the molten state. A crystallization phenomenon corresponds to a decrease of the potential energy of the system, with the loss of entropy being largely compensated by the increase in enthalpy. Thus, crystallization is mainly the result of the development of molecular interactions, and, in most cases, intramolecular interactions play the major role.

Due to its complexity, the calculation of regular conformational structures could be carried out only in a limited number of cases for short sequences of those polymers exhibiting a simple molecular structure. Purposely, the structures considered will be intentionally restricted to their description, and their classification will be accomplished by means of simple rules established from experimental observations.

Because the architecture of individual chains is significantly affected by the symmetry of the constitutive units, this parameter will be used to classify the various regular conformational structures.

5.2.1. Chains with Symmetrically Perturbed Rotations

When each of the molecular groups carried by a chain appear along the backbone according to a plane of symmetry and insofar as steric effects or intramolecular interactions are negligible, the conformations of such chains are transplanar (succession of transoidal conformations). The most known and important example of such a conformation is that taken by polyethylene whose carbon atoms of the chain appear in the plane, being flanked laterally by hydrogen atoms (Bunn model):

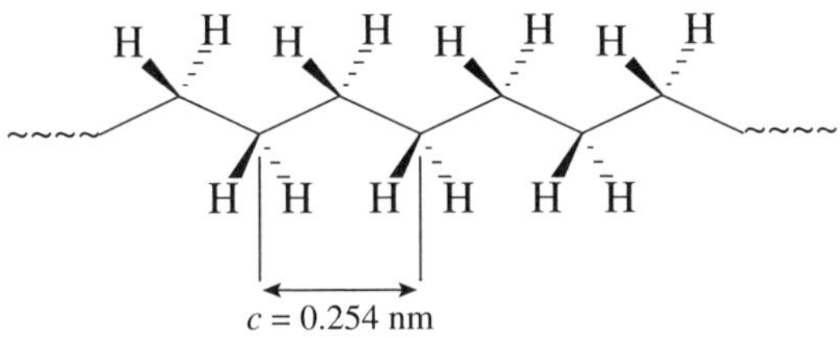

$c = 0.254$ nm

The repeating distance along the chain axis is called the *fiber period* and is denoted c (refer to Section 6.5); here it corresponds to a monomer unit ($-CH_2-CH_2-$) and two constitutive repeating units ($-CH_2-$). Such a chain can adopt a macroconformation of a "flat" helix in which the number of constitutional repeating units per helix turn is equal to two. Such conformation is denoted 2_1.

One observes exactly the same structures in polymers such as:

- Polyamides—in this case polyamide-6 (PA-6 or polycaprolactam)—where the planar transoidal conformation ("helix 2_1") shown below is stabilized by interchain hydrogen bonds

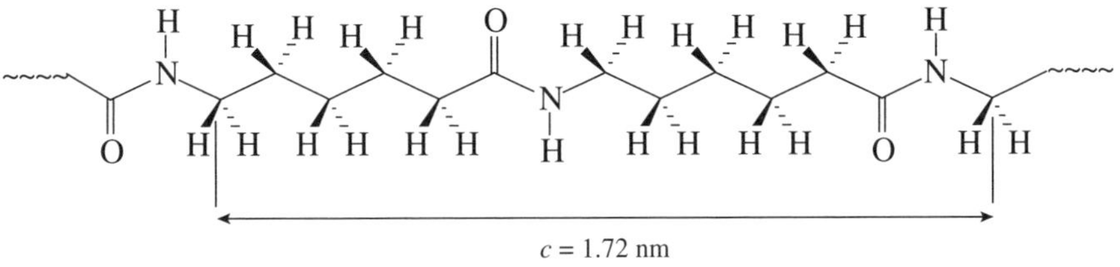

$c = 1.72$ nm

- Vinyl and related polymers obtained from symmetrically disubstituted ethylene monomers; represented below, poly(vinylidene fluoride)

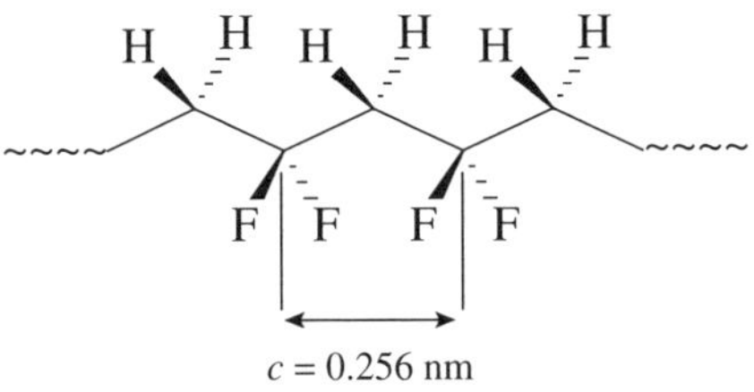

(1_1 helix)

$c = 0.256$ nm

- Polydienes such as 1,4-polybutadienes

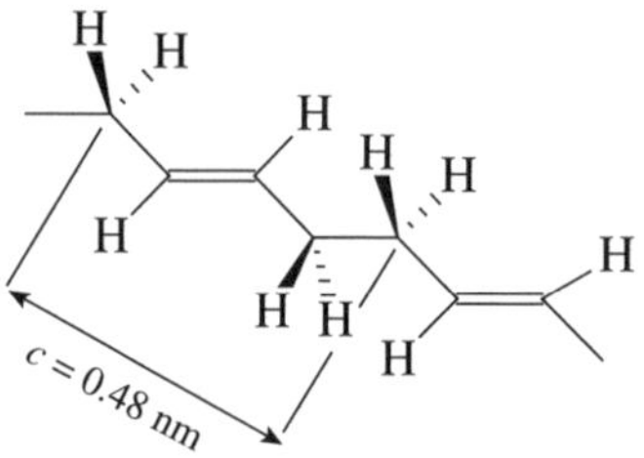

trans (1_1 helix)

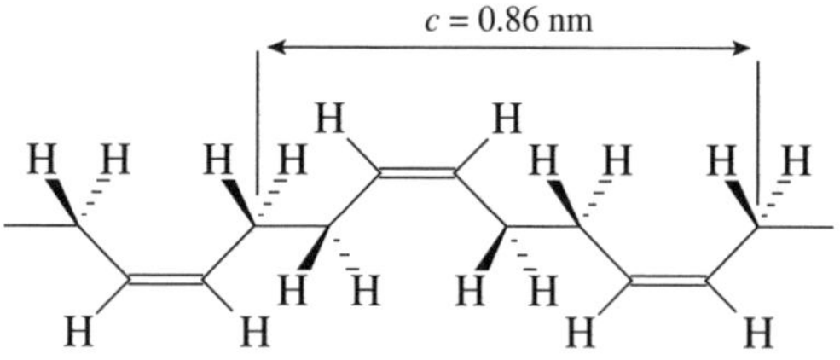

cis (2_1 helix)

In this cis polymer, the reason for the formation of 2_1 helix is the minimization of the repulsion between π electrons of the double bonds.

- Those whose substituent is too small to perturb the planar transoidal conformation found, for instance, in PE; it is the case of poly(vinyl alcohol) ($-CH_2-CHOH-)_n$ and poly(vinyl fluoride) ($-CH_2-CHF-)_n$, whatever their tacticity

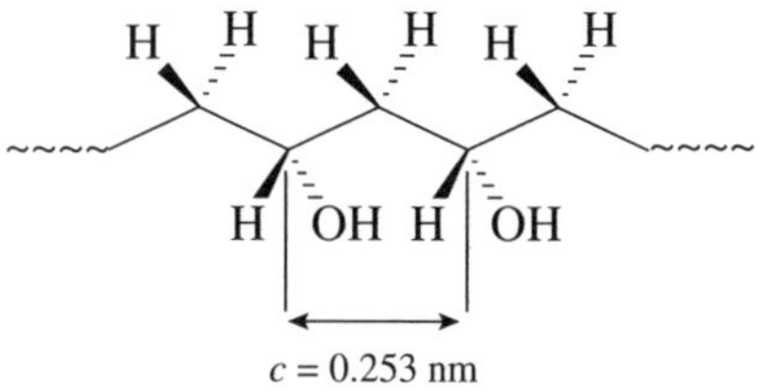

1_1 structure of poly (vinyl alcohol)

For the latter polymer—and also for crystalline atactic polymers—an analogy can be made with the *syncristallization* of small molecules since all unit cells do not consist of molecular groups with a same configuration.

However, there are exceptions among macromolecules exhibiting a symmetry of their molecular structure. Indeed, a number of them adopt a nonplanar helical conformation in spite of this symmetry. The reasons for such exceptions are as follows:

1. They can be due to *steric hindrance*, as in the case of polytetrafluoroethylene ($-CF_2-CF_2-)_n$ (PTFE), whose four fluorine atoms and their size ($r_F = 0.061$ nm) prohibits purely transoidal conformation unlike the four hydrogen atoms ($r_H = 0.032$ nm) in PE; however, gauche conformations are not allowed because

they would require an interpenetration of repulsive electronic clouds of the other fluorine atoms. Accordingly, the chain crystallizes in a helical conformation with 13 constitutive units per turn (equivalent to 6.5 monomeric units), which corresponds to the best compromise between the stability of the transoidal conformations and the minimal repulsion between fluorine atoms.

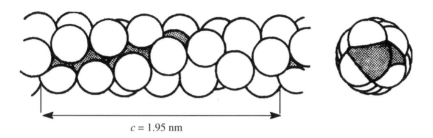

$$c = 1.95 \text{ nm}$$

Such geometry brings about an extreme rigidity of the chain which results in particular thermal, mechanical, and rheological characteristics.

Another interesting exception due to steric hindrance is provided by the example of polyisobutene $[-CH_2-C(CH_3)_2-]_n$ (PIB); although regular and symmetrical, this polymer does not develop sufficiently strong molecular interactions to stabilize a regular conformation. Under the effect of thermal agitation, the chains thus adopt spontaneously a disordered conformation. However, if a mechanical constraint is applied forcing the chains to stretch, they tend to align, and as they come closer, they develop stronger molecular interactions and crystallize. For steric reasons, this crystallization occurs in the form of 8_5 helix ($c = 1.863$ nm).

2. Other exceptions include polymers with restricted symmetry which can, however, crystallize in a helical macroconformation because of *electrostatic interactions* between molecular groups of the main chain ("intrachain" interactions). However, for such "helicity" to occur, the chain should exhibit a great mobility, which is the case, for example, in the family of polyethers; in the latter case, the dipole attraction due to $-C^{\delta+}O^{\delta-}$ groups is responsible for the stability of the crystalline state; poly(ethylene oxide) $(-CH_2-CH_2-O-)_n$ crystallizes in a 7_2 helix ($c = 1.94$ nm), whereas poly(oxymethylene) $(-CH_2-O-)_n$ (also called polyformaldehyde or "polyacetal") does in a 9_5 helix ($c = 1.72$ nm). In the latter case, multiple dipole interactions contribute to stiffen the chains and enhance mechanical properties of the corresponding materials.

When dipolar groups in these polyether chains are located at sufficiently large distance from each other, the interactions among them weaken and the chain again adopts a "normal" planar transoidal conformation as observed in polytetrahydrofuran [poly(oxybutylene)] $[-O-(CH_2)_4-]_n$.

Many other exceptions could be mentioned; depending upon the intensity of the interactions or the repulsions which cause such conformations, more or less pronounced deviations from transoidal–planar conformation are observed.

5.2.2. Chains with Unsymmetrically Perturbed Rotations

As a matter of fact, these perturbations should occur regularly to allow chains to crystallize. In the family of **vinyl and related polymers** with the general formula $(-CH_2-CAB-)_n$ two molecular regularities are to be considered:

- Those found in **syndiotactic polymers** that perturb only slightly the planar zigzag typical of polyethylene (transoidal conformation). For instance, in syndiotactic 1,2-polybutadiene, the $-CH=CH_2$ substituent carried by carbon atoms (whose configuration changes for each monomer unit) induces a slight *gauche* effect on the main carbon chain. The crystallographic repeating unit corresponds to the chemical repeating unit, itself comprised of two monomer units differing at the level of their configuration; the crystallized chain is formed as a result of a simple translation of this dyad but nevertheless corresponds to a 1_1 helix. For most of vinyl and related polymers, which adopt an "extended" conformation, the value taken by the fiber period is characteristic of this type of tacticity; it lies between 0.500 and 0.515 nm, depending upon the nature of the substituents.

Remarks

1. Syndiotactic polypropene crystallizes mainly in a conformation $(TTGG)_2$ which corresponds to a 2_1 helix.
2. Depending upon the conditions of crystallization, macromolecular chains can adopt different conformations; the conformational polymorphism is a common phenomenon among polymers.

- Those found in **isotactic** polymers whose regular perturbation along the chain favors the formation of successive transoïdal and gauche conformations. When no strong steric compression is exerted, a helix with three monomer units per turn (3_1 helix) is formed which can be denoted $(TG)_3$ to indicate that, in one helix turn, three successive transoidal and gauche conformations (TG) occur:

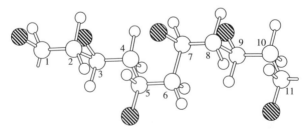

In the above representation of isotactic polypropylene, carbon atom 7 is the translational repetition of carbon 1 along the fiber axis, which corresponds to a fiber period of $c = 0.650$ nm. Figure 5.11 shows the results of the calculation of the potential energy as a function of the angles of internal rotation, in the case of isotactic polypropene; a good agreement is observed between the experimental results and predictive calculation.

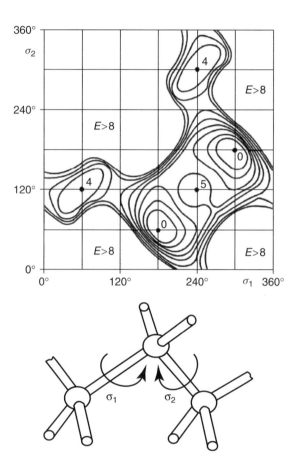

Figure 5.11. Variation of the potential energy of isotactic polypropylene as a function of angles of internal rotation σ_1 and σ_2. Energies are given in kJ/mole of monomer units, and reference level corresponds to gauche conformation.

Isotactic polystyrene also crystallizes and forms a 3_1 helix ($c = 0.665$ nm), with its phenyl rings (also hindering in two dimensions) ordering themselves into a helical macroconformation of ternary symmetry like isotactic polypropene.

When the steric hindrance of the substituents further augments, the chain attains the minimum of its potential energy for 7 monomeric units in two helix turns (helix 7_2); it is also the case for polyhex-1-ene and poly(4-methylpent-1-ene), which are characterized by a fiber period (c) of 1.38 nm.

If steric hindrance still increases due to the size of the substituent (*tert*-butyl, cyclohexyl, etc.), the number of monomeric units per helix turn augments accordingly and macroconformations of 4_1 type are eventually obtained as indicated in the following examples:

Isotactic poly(m-methylstyrene) 11_3 helix ($c = 2.17$ nm)

CH_3

Poly(vinyl cyclohexane) 4_1 helix ($c = 0.645$ nm)

Poly(vinyl naphtalene) 4_1 helix ($c = 0.810$ nm)

Some of the most typical helical macroconformations with different symmetry are represented in Figure 5.12.

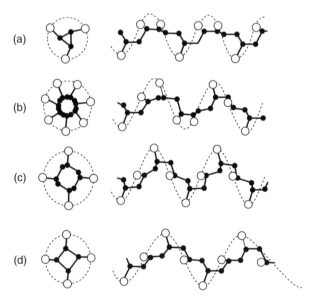

Figure 5.12. Representation of 3_1, 7_2, and 4_1 helices of isotactic polyolefins. The substituent **R** of the carbon chain appears as a large circle. (a) **R** = methyl, ethyl, vinyl, phenyl, etc. (b) **R** = isobutyl, isopentyl, etc. (c) **R** = isopropyl, etc. (d) **R** = *tert*-butyl, etc.

Among helical conformational structures, those taken by polypeptides (-NH-CH**R**-CO-)$_n$ occupy an unique place since some of the most important properties of proteins result from them.

Chains adopt a helical conformation of 18$_5$ type (α-helix) with a direction of rotation (right- or left-handed helix) determined by the absolute configuration of monomer units (called "residues") to maximize the intrachain hydrogen bonding due to amide -CO- and -NH- functional groups; in general, one observes right-handed helices for residues with L configuration. Such α-helix is represented in Figure 5.13.

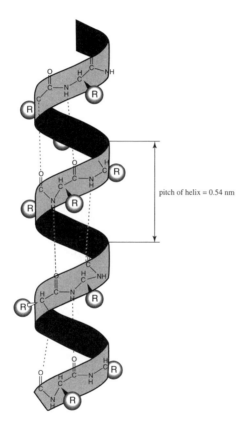

pitch of helix = 0.54 nm

Figure 5.13. Polypeptide chain in its α-conformation.

Spontaneously (as for silk) or under mechanical stress, chains can also take a planar transoidal conformation (β-conformation) stabilized by interchain hydrogen bonding (Figure 5.14) and not intrachain ones as in the case of α-helix. The type of structure formed can be identified by means of infrared spectrometry, with the carbonyl groups vibrating in the elongation mode at $1660 \, \text{cm}^{-1}$ for α-conformation and at $1430 \, \text{cm}^{-1}$ for β-conformation.

Natural proteins can be viewed as copolycondensates built from 22 amino acids differing by the nature of their lateral substituent **R**.

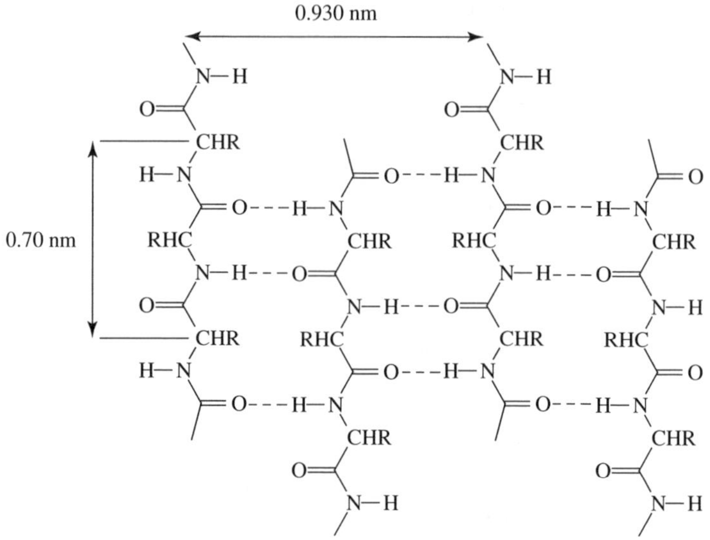

Figure 5.14. Structure β (in layers) of polypeptides, revealing the prevalence of interchain interactions.

Nucleic acids such as ribonucleic acid or RNA and such as deoxyribonucleic acid or DNA are linear copolyesters of phosphoric acid and substituted ribose (RNA) or substituted Z'-deoxyribose (DNA). Below, the repeating unit of DNA, also called deoxyribose polyphosphate, is represented by

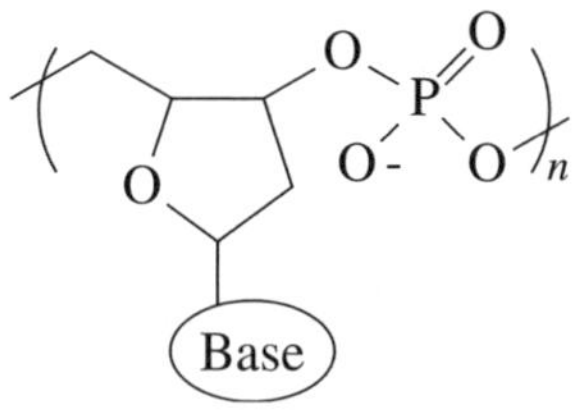

whose saccharide cycle is substituted by one of the four following bases:

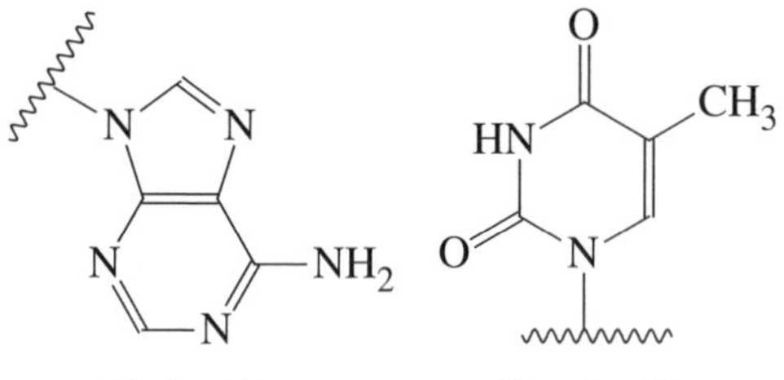

Adenine (A)   Thymine (T)

Cytosine(C)   Guanine(G)

Chain moieties comprising these saccharide cycles and the bases are called nucleosides (adenosine, guanosine, cytidine, and thymine in the case of DNA), and the phosphoric esters of these nucleosides are the nucleotides (adenylic acid, guanylic acid, cytidylic acid, and thymidylic acid).

Hence, DNA can be regarded as a statistical copolymer between four different comonomers (hence it is called *quaterpolymer*):

Adenine-thymine   Guanine-cytosine

Unlike ribonucleic acids, which are single-strand polymers, DNA form double-strand helices, a sort of twisted ladder, consisting of two complementary chains. This complementarity occurs through intermolecular hydrogen bonding between two pairs of bases, between the adenosine of one strand and the thymidine of another, and likewise between cytidine and guanosine.

This double helix was identified by Crick and Watson in 1954; the rungs of the twisted ladder correspond to the pair of bases. DNA is always formed by replication/duplication upon separation of the double strands; the intermediate single strands are the matrices for the generation of new DNA chains. After replication, each double helix includes one old strand and a new one. Three successive nucleotides of DNA provide the code for one amino acid, and the genetic code is determined by the sequence of these triplets.

Each nucleus in a living cell contains long, thread-like structures called chromosomes, which carry bits of genes. Both chromosomes and genes are made of DNA, which is often called the blueprint for life; every living cell contains indeed a copy of the blueprint.

Figure 5.15 shows the complexity of such a structure, underscoring the progress made by biology to unveil it.

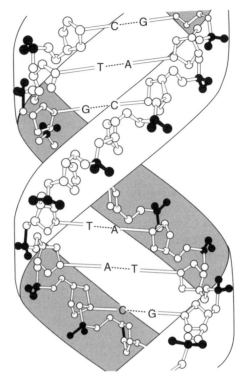

Figure 5.15. Representation of the DNA double-helix organization showing A-T and C-G links.

5.3. CHAIN PACKING

5.3.1. Assembly of Random Coils

Taken individually, random coils commonly exhibit a Gaussian distribution of their constitutive units if one considers the distance of the latter to the center of mass of the macromolecule under consideration. The chains constituting a sample add their distributions and give an apparently homogeneous material down to the nanometer level. Figure 5.16 schematizes the situation of an assembly of chains of different size, showing the interpenetration of random coils in the condensed state. Such an interpenetration leads to interchain entanglements and enhances the cohesion of the corresponding material.

For reasons that are related to the rigidity of constitutive units, atoms in certain polymers do not occupy entirely the space available to them in spite of the apparent homogeneity of the latter as illustrated by the horizontal straight line of the $P = f(r)$ diagram (line corresponding to the addition of the probabilities of presence of monomer units belonging to different chains). To account for this unfilled space,

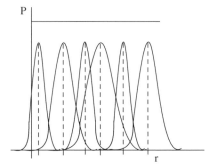

Figure 5.16. Diagram showing the variation of the probability P of presence of monomeric units belonging to an assembly of polymeric chains as a function of their distance R to a reference point.

the concept of *free volume* was introduced. The free volume (see Chapter 11) plays an important role with respect to thermomechanical properties (glass transition) and transfer properties (permeability, etc.).

5.3.2. Packing of Sequences of Regular Chains

Due to the possible existence of defects in the molecular structure of monomeric units and in their placement, an assembly of regular chains can be described only for short sequences whose length is closely related to the extent of their regularity. In that respect, only linear and stereoregular sequences can be taken into account since branching points, junction points in networks, chain ends, and configurational irregularities are structural defects that oppose the regular chain packing in their totality. Given the difficulty for chains to organize on a large scale due to the macromolecular state, only assemblies made up of a limited number of constitutive units will be described.

Three categories of assemblies can be arbitrarily distinguished whose geometry is determined by the molecular structure of the constitutive unit, the relative size and bulkiness of side groups, and the conformation of the isolated chain. This geometry is governed by the tendency of these assemblies to minimize their potential energy and maximize their molecular interactions (intra- and interchain).

The **first category** of assemblies is that of chains which exhibit a cylindrical overall shape and can be viewed as screws with small "threads." For the maximum development of molecular interactions, the chains tend to minimize the distance between them (which corresponds to the maximum density), and it is the hexagonal packing which complies best with this criterion as shown in Figure 5.17.

A typical example of such a packing is that of polytetrafluoroethylene which crystallizes in a hexagonal system with $a = 0.554$ nm and $b = 1.680$ nm and whose regular conformation was previously described (see page 111).

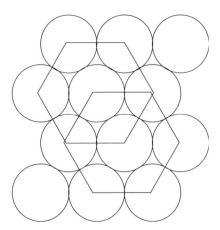

Figure 5.17. Hexagonal packing of chains similar to cylinders.

The **second group** is that of chains with helical conformation, which are different from cylinders: their "threads" are indeed more prominent, and the number of constitutive units per helix turn is fractional. The chain packing takes a tetragonal symmetry with an interpenetration of the "threads" of a left-handed helix with those of the four right-handed helices which surround it and vice versa. Figure 5.18 represents an arrangement of such chains in which the size of the thread is measured by the ratio R/r. It also shows the way chains assemble and the relative direction of the interpenetrated helices. In this group, isotactic poly(4-methylpent-1-ene) is found whose regular conformation is a 7_2 helix. This means that the period of identity along the fiber axis of the chain is 7 repetitive units regularly placed on 2 helix turns; the parameters of the corresponding tetragonal cell are: $a = b = 1.86$ nm and $c = 13.7$ nm.

$$\begin{array}{c} \left(CH_2-\underset{\displaystyle \underset{\displaystyle \underset{\displaystyle CH_3}{|}}{\underset{\displaystyle \underset{\displaystyle CH-CH_3}{|}}{\underset{\displaystyle \underset{\displaystyle CH_2}{|}}{CH}}}\right)_n \end{array}$$

The **third group** includes chains similar to the preceding ones, whose number of constitutive units per helix turn is a nonfractional number. In this case, the symmetry of the assembly reflects that of the individual chain: ternary symmetry for an assembly of chains with a ternary symmetry, and so on. Figures 5.19 and 5.20 show such an assembly with ternary and quaternary symmetries, respectively.

It is worth stressing that the criteria that distinguish the above groups tolerate a number of exceptions which can be found even for an usual polymer with a simple structure.

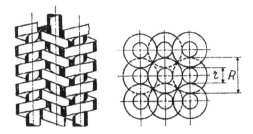

Figure 5.18. Diagram of the packing of helical chains exhibiting a tetragonal symmetry.

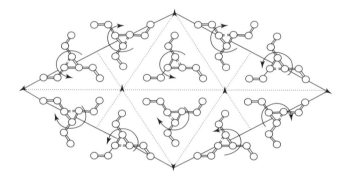

Figure 5.19. Packing of ternary symmetry chains: isotactic polybut-1-ene (conformation 3_1).

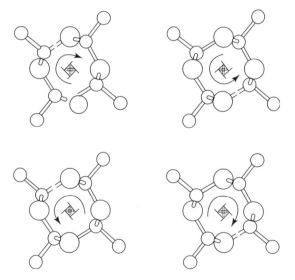

Figure 5.20. Packing of quaternary symmetry chains: isotactic polyacetaldehyde (conformation 4_1).

5.4. MORPHOLOGY OF MACROMOLECULAR SYSTEMS

The term *morphology* corresponds to the structure taken by polymers at the micro-scopic level. The morphology of a macromolecular system is primarily determined by the molecular regularity (placement of the constitutive units and configurational regularity) and by the treatment undergone by the sample prior to or being in the solid state. All situations can exist between the amorphous state corresponding to the maximum entropy for a macromolecular system and the single crystal whose only imperfections are due to chain folding and the molecular irregularity of their ends. The various situations will be successively examined.

5.4.1. Homogeneous Amorphous State

The amorphous state can be depicted as a multitude of random coils being thoroughly entangled. At the microscopic level, this brings about an apparent homogeneity, which is responsible, in particular, for the transparency of these systems to the visible light; such polymers are often called organic glasses.

The amorphous state results from the impossibility of chains to crystallize due to the existence of defects at the molecular level or difficulty for the chains to disentangle when cooling from the molten state. In the latter case, a fast cooling quenches the disordered molten state.

Poly(methyl methacrylate) (PMMA) and polystyrene (PS) obtained by free radical polymerization are amorphous due to their atacticity. Poly(ethylene terephthalate) (PET) is also amorphous when quenched from the molten state but is potentially crystallizable; the rigidity of the chains prevents them from disentangling rapidly enough so that they remain as in the molten state—that is, in a disordered state.

5.4.2. Extended Chain Polymers

Due to the molecular agitation at the time of the transition, chains can hardly crystallize in an extended form and without folding. However, chains that are highly rigid such as aromatic polyamides—for example, poly(p-phenyleneterephthalamide)

with their rigid phenylene moieties and interchain hydrogen bonds between amide functional groups (-CO-NH-)—crystallize almost unfolded. Application of an external stress can also prevent chain folding. For instance, poly(oxymethylene) (POM) [-(CH$_2$-O)$_n$-] obtained by solid-state radiation polymerization of cyclic trioxane (CH$_2$-O)$_3$ form extended crystalline chains. In this case, the monomer is polymerized in its crystal form which affords directly stretched chains; the length of the

Figure 5.21. Electron micrography of a fracture of a PE sample revealing zones comprising extended chains.

extended crystalline part corresponds to the molar mass of the chain (chains in total extension).

Extended chains whose folding occurs only beyond $\sim$100 nm are also considered. Such length corresponds to degrees of polymerization (for common vinyl polymers) higher than 500. Such structures are observed in "nascent" polytetrafluoroethylene (PTFE), which forms partially extended highly rigid helical chains. In another example, when polyethylene is crystallized under strong pressure (about 100 MPa), it can also give rise to extended chains of the type shown in Figure 5.21.

Chains extended under stressed conditions do not exhibit the same attractiveness applicationwise as those oriented monodimensionally (fibers and films) or two-dimensionally (films) stretched. Section 5.5 will be devoted to the description of orientated polymers.

5.4.3. Single Crystals

Upon cooling slowly dilute solutions of a polymer of great molecular regularity, it is possible to obtain single crystals with a morphology close to that of simple molecules as shown in Figure 5.22a (electron diffraction).

These single crystals exhibit the most regular arrangement possibly formed in a polymer; they form lamellae (Figure 5.23) whose thickness (a few tens of nanometers) is determined by the nature of the polymer and the thermodynamic conditions of crystallization.

These lamellae can also pile up by means of screw dislocations (see Figure 5.28) and afford more complex structures such as those shown in Figure 5.22b. Their dimensions (about a few tens of micrometers) are such that optical or electron microscopies are essential techniques to visualize them. By analysis of the electron diffraction patterns of these single crystals, it could be established that they

Figure 5.22. Electron diffraction pattern (a) and transmission electron micrography (b) of a single crystal of polyethylene of pyramidal shape. [Courtesy of J. C. Wittmann, ICS, CNRS Strasbourg (France).]

Figure 5.23. Transmission electron micrography of monolamellar single crystals of polyethylene.

comprise regularly folded linear chains, as schematically shown in Figure 5.24, and that the chain axes are perpendicular to the surface of the lamellae. The existence of such folding is proved by the fact that the thickness of the lamellae is generally much lower than the length of totally stretched chains. Due to this folding, irregularities related to the dispersion in the molar masses of the chains forming a same single crystal can be somehow compensated.

From a thermodynamic point of view, the regular folding of the chains is the result of a compromise between the increase of the free energy of the system related to torsional and longitudinal oscillations of extended chains under the effect of molecular agitation and the tendency of the crystal to exhibit a minimum surface free energy. The existence of such a compromise indicates that the dimension of the extended segments (thickness of the lamellae) is likely affected by the temperature of crystallization.

This is what is actually observed experimentally. It is even possible to augment the thickness of a preexisting single crystal by means of a thermal treatment

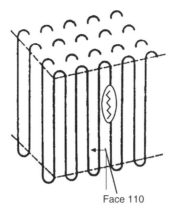

Face 110

Figure 5.24. Schematic representation of an ideal polyethylene single crystal resulting from the folding of chains planar zigzag conformation.

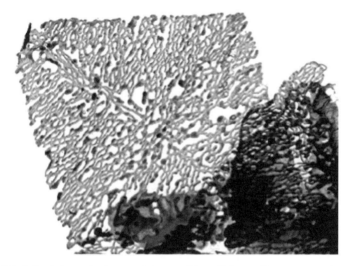

Figure 5.25. Polyethylene single crystal "reconditioned" in a different thermodynamic environment from that of the initial crystallization and resulting in the formation of "holes."

(annealing); as the other dimensions remain constant, "holes" appear in the crystal to compensate the increase of the lamellae thickness (Figure 5.25).

5.4.4. Semi-crystalline State

It corresponds to a state intermediate between the amorphous state and a strongly ordered one such as that of a single crystal. All polymers that exhibit a sufficiently high molecular regularity to generate crystalline zones organize in a semicrystalline state when subjected to favorable thermal and kinetic conditions (see Section 12.3). Before describing various morphologies referring to this physical state, it is necessary to define the *degree of crystallinity* (X) of semicrystalline polymers, which is simply the proportion of crystalline matter; depending on whether this proportion is expressed in mass or in volume, slightly different values will result.

The *degree of crystallinity* (X_v) *in volume* is defined as

$$X_v = V_c/(V_a + V_c) = V_c/V$$

a relation in which V_a, V_c, and V denote the respective volumes of the amorphous and crystalline phases and the total volume phases of the sample studied.

In the same way, the *degree of crystallinity* (X_m) *in mass* can be defined as

$$X_m = m_c/(m_a + m_c) = m_c/m$$

a relation in which m_a, m_c, and m denote the mass of the amorphous and crystalline phases and the total mass phases, respectively.

If $\overline{V}_c$, $\overline{V}$, $\overline{\rho}_c$, and $\overline{\rho}$ are the bulk volumes and the densities of the crystalline phase and of the entire sample, one can write

$$X_v = \frac{\overline{V}_c m_c}{\overline{V} m} = \frac{\overline{V}_a}{\overline{V}} X_m = \frac{\overline{\rho}}{\overline{\rho}_c} X_m$$

For the majority of polymers, both mass and volume degrees of crystallinity are not very different; both are thus used indifferently, depending upon the method utilized to carry out the measurement.

It is worth emphasizing that the physical, chemical, mechanical, and so on, properties of amorphous and crystalline phases are very different. In most cases, there is a proportional additivity of the specific properties. If $\overline{P}_a$, $\overline{P}_c$, and $\overline{P}$ represent, respectively, the specific property of the amorphous phases, the crystalline phases, and all the phases, one can write

$$\overline{P} = X\overline{P}_c + (1 - X)\overline{P}_a$$

and this relation is used for the measurement of the degree of crystallinity (see Section 6.5).

Depending upon the degree of crystallinity of a polymer, regardless of whether it is low or high, two different types of morphology for semicrystalline systems can be distinguished.

For **low degrees of crystallinity**, the morphology can be described by the *fringed micelle* model, with small-size *crystallites* being dispersed in an amorphous polymer matrix. The size and the degree of perfection of these crystallites are closely related to the length of the regular—and thus crystallizable—sequences in the chains constituting the sample. Figure 5.26 schematically represents a polymer exhibiting such a morphology. In this representation the crystallites result from the packing of more or less long sequences belonging to different chains. In addition, the same chain can be involved in the formation of several crystallites; it

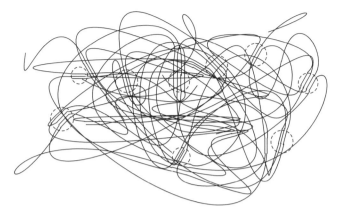

Figure 5.26. Diagram showing the morphology of a semicrystalline polymer with a low degree of crystallinity (crystallites are ringed).

is thus impossible to physically separate crystalline domains from the amorphous phase. Crystallites are zones of high density of cohesive energy; they play the role of physical cross-links and, even in small proportion, can significantly affect the mechanical properties of polymers in the solid state. For instance, the difficulties encountered in the processing of PVC are of rheological origin and due to the crystallization of short syndiotactic sequences.

For **high degrees of crystallinity**, the crystalline zones give rise to an organization of higher order. They represent the majority of the sample and self-organize in lamellae made of folded chains as described in the case of monocrystals. Both impurities present in the medium and noncrystallizable sequences or sequences that could not crystallize due to kinetic reasons are rejected into interlamellar zones and form the amorphous phase (Figure 5.27). Because the matter in such amorphous zones is less cohesive than that in the crystalline layers, its proportion and its thermomechanical characteristics can considerably affect the overall mechanical properties of the whole sample.

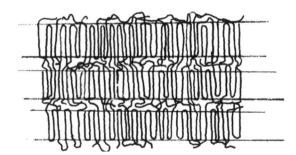

Figure 5.27. Diagram showing the detail of the lamellar structure of a spherulite in a polymer with a high degree of crystallinity.

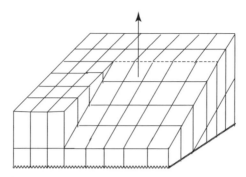

Figure 5.28. Representation of a screw dislocation: the arrow indicates the direction of growth of the dislocation.

The examination of this structure shows that a same chain can be involved in the formation of different lamellae, thus bringing about the cohesion between the various layers. The three-dimensional filling of space by crystallized matter occurs by means of dislocations (Figure 5.28) and various lamellae stuctures resulting from crystallographic defects.

At the microscopic level a structure with an apparent spherical symmetry is obtained, which is referred to as *spherulite*. Figure 5.29 explains how the growth of such lamellae (direction of the arrows) and the space filling by the crystallized matter in the perpendicular direction to the orientation of the chains occur.

The electron microscopy image of Figure 5.30 clearly shows that the crystallization starts from a central nucleus and grows through the formation of layers whose orientation corresponds to the representation of the Figure 5.29.

In addition, the examination of spherulites in polarized light reveals textures which can be related to the orientation of lamellae and thus to that of the chains.

Figure 5.29. Representation of the directions of the growth of the lamellae starting from a microcrystalline nucleus (crystallization from the molten state).

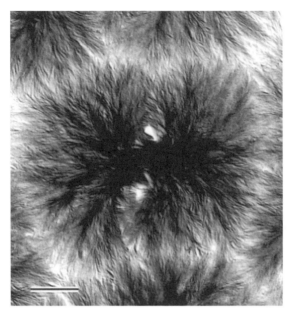

Figure 5.30. Electron micrography of a polyethylene spherulite at the beginning of growth. [Courtesy of B. Lotz, ICS-CNRS, Strasbourg (France).]

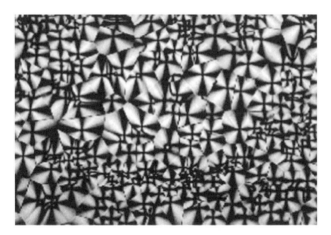

Figure 5.31. Microscopic texture of a polyethylene spherulite with radial lamellae. Observation between crossed nicols. [Courtesy of B. Lotz, ICS-CNRS, Strasbourg (France).]

Figure 5.31 is characteristic of a highly crystalline structure in which the lamellae are oriented along the radius of the spherulite; in Figure 5.32, the existence of concentric extinction lines in the texture shows that the lamellae are twisted as schematically represented in Figure 5.33.

Crystallization from the molten state is an important phenomenon whose mechanism, thermodynamic aspects and kinetics will be described in Chapter 12.

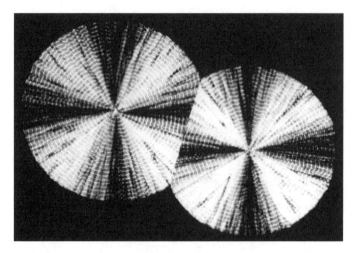

Figure 5.32. Microscopic texture of a spherulite of poly(ethylene adipate) with twisted lamellae. Observation between crossed nicols. [Courtesy of B. Lotz, ICS-CNRS, Strasbourg (France).]

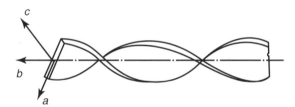

Figure 5.33. Schematical representation of a twisted lamella corresponding to a ringed texture as shown in Figure 5.32.

5.4.5. Morphology of Phase-Separated Polymer Systems

Even if the heterogeneity of their structure is well established, semicrystalline homopolymers are generally not included in the category of heterogeneous systems. This term is reserved to polymers subject to a clear phase separation in which the nonmiscibility is due to the presence of different molecular structures. The thermodynamic aspect of the nonmiscibility of polymers is discussed in Section 4.4, where it is shown that this phenomenon is practically general, with the only exception of polymers with strong specific molecular interactions. Upon mixing two nonmiscible polymers molecular interactions develop within each phase but interphase interactions remain weak. It results in poor mechanical characteristics for the corresponding materials. The electron micrography of Figure 5.34 clearly shows the lack of interphase cohesion between two nonmiscible homopolymers.

The situation is different when a covalent link can be established between the phases. In block copolymers the various blocks are also nonmiscible, but the dyad linking the two blocks ensures the cohesion of the system and the phase dispersion.

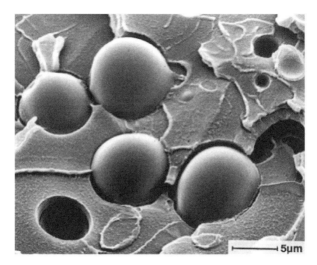

Figure 5.34. Scanning electron micrography of a polystyrene–polybutadiene blend in absence of a compatibilizing agent. [Courtesy of BASF Cy (Ludwigshaffen Germany).]

5.4.5.1. Morphology of Block Copolymers — "Self-Organized Polymers".

Two situations have to be considered in the case of block copolymers, depending upon their situation with respect to the limit of segregation. If the system is far from this limit, the segregation is clear and the morphology is determined only by the molecular composition of the block copolymer. In the second case, an increase (in absolute value) of the entropy of mixing with the temperature can result in a certain compatibility at high temperature (see Section 4.4). For the sake of simplicity, only systems that are far from the limit of segregation will be described.

When block copolymers exhibit a relatively well-defined molecular structure (uniform composition and molar mass), four types of morphologies are observed, depending upon their composition. Such systems are referred to as *self-organized polymers*.

For a poly(A-*b*-B) copolymer whose mass ratio [A]/[B] (or [B]/[A]) is lower than approximately 20%, the minority phase forms spherical domains that are regularly dispersed in the matrix based on the majority block. The system self-organizes in a centered cubic symmetry (Figures 5.35a and 5.35e). With the increase of the proportion of the minority phase and for compositions [A]/[B] ranging between 20% and 35%, the spheres self-assemble into cylinders exhibiting a hexagonal symmetry (Figures 5.35b and 5.35d). In the case of block copolymers with a balanced composition ([A]/[B] from 40% to 60%), the cylinders self-assemble in lamellae whose thickness is determined by the composition (Figure 5.35c). For intermediate compositions [A]/[B] (or [B]/[A]) ranging from 35% to 40%, structures such as *bicontinuous phases* are observed. These biphasic morphologies (not represented in the Figure 5.35) are intermediate between cylinder ones and lamellae ones. The minor phase can form a sort of "network" (or two interpenetrated

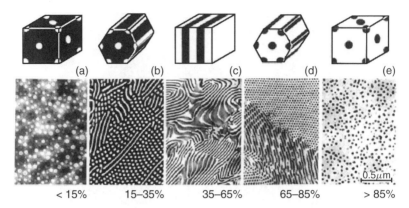

Figure 5.35. Morphologies and micrographies of poly (styrene-*b*-butadiene) organized polymers: (a, e) Spheres; (b, d) cylinders; (c) lamellae. (Percentages are those of styrene).

"networks") regularly distributed in the matrix constituted by the major phase. Star block copolymers also give rise to this type of structure.

Morphologies of self-organized copolymers are observed not only in diblock copolymers but also in copolymers with a higher number of blocks. However, as the number of blocks increases and their individual length decreases (*segmented* polymers), it is more difficult to obtain a clear phase separation.

Triblock copolymers comprising a central block of polybutadiene (BR) and two external blocks of polystyrene (PS) are commonly used as *thermoplastic elastomers*.

Their composition induces the formation of a morphology in which polystyrene spheres are distributed with a cubic symmetry in an elastomeric matrix of polybutadiene. At the service temperature the spherical polystyrene nodules are in the glassy state, whereas the polybutadiene chains connecting them are in the elastomeric state. Figure 5.36 shows the rigid nodules of PS in their role of physical cross-links, which are responsible for the reversibility of the deformations undergone by the sample.

When the temperature of the system is increased beyond the glass transition temperature of polystyrene, the material becomes plastic and can be processed as a viscous liquid. Upon lowering the temperature, the PS nodules become glassy again and the elastomeric character of the material is restored.

5.4.5.2. "Compatibilization" of Polymer Mixtures — High Impact Polymers.

In block copolymers that are ill-defined (heterogeneity in composition, dispersity of molar masses, etc.) phase separation occurs with no defined order. This is also observed in graft copolymers and polymer blends that are *"compatibilized"* by means of a block (or graft) copolymer consisting of the same monomeric units as those of the homopolymers to be mixed.

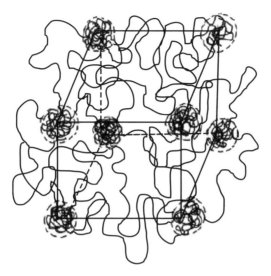

Figure 5.36. Diagram of the morphology of relaxed S-B-S thermoplastic elastomers revealing the physical cross-linking of the system by the glassy nodules of polystyrene.

Block or graft copolymers used in the latter case act as *compatibilizers* (surfactants) of the polymer blend so as to "emulsify" the two homopolymers (Figure 5.38) and thus improve its mechanical characteristics.

> **Remark.** It is important to distinguish between *miscibility* and *compatibility*: the *miscibility* is a thermodynamic characteristic, whereas the *compatibility* is a phenomenon affecting a service property.

Upon thoroughly blending two homopolymers, such compatibilizers can sometimes be generated. Indeed, such a mechanical mixing can cause the homolytic breaking of σ bonds and generate macromolecular free radicals that can give rise to block copolymers by random recombination.

A particularly interesting case of polymer systems with heterogeneous morphology is that of polymers with high-impact strength. These polymers exhibit a strongly improved impact resistance as compared to that of common polymers as a result of the dispersion of micron-size nodules of elastomers in a rigid phase. High-impact polystyrene (HIPS) and ABS (acrylonitrile, butadiene, and styrene copolymers) are the best-known examples.

HIPS is obtained by free radical polymerization of styrene in the presence of polybutadiene. The labile character of the allylic hydrogen atom of polybutadiene favors radical transfer from the growing polystyrene chains (see section 8.5.6.4), which generates graft copolymers along with homopolystyrene:

R$^{\bullet}$ + Styrene ⟶ Polystyrene

R$^{\bullet}$ + ⌇⌇CH_2-CH=CH-CH_2⌇⌇ ⟶ RH + ⌇⌇$\overset{\bullet}{C}$H-CH=CH-CH_2⌇⌇
         Polybutadiene

⌇⌇$\overset{\bullet}{C}$H-CH=CH-CH_2⌇⌇ + Styrene ⟶ ⌇⌇$\underset{}{C}$H-CH=CH-CH_2⌇⌇ ⎧Polystyrene$^{\bullet}$

How does the morphology of the resulting polymer material occur and build up?

Polybutadiene is first solubilized in styrene, which plays the role of a solvent. Upon polymerizing, the polystyrene formed is still in a minor proportion—compared to the polybutadiene present—and is thus dispersed as nodules in the polybutadiene solution. As the yield of styrene increases, the polystyrene phase becomes predominant, provoking an inversion of phases and in turn the dispersion of the polybutadiene phase in the PS matrix. Figure 5.37 shows the various steps of such a process.

The covalent links between the elastomeric nodules and the matrix are responsible for the cohesion of the whole material and the dispersed soft phase for stopping the propagation of cracks resulting from an impact (Figure 5.38).

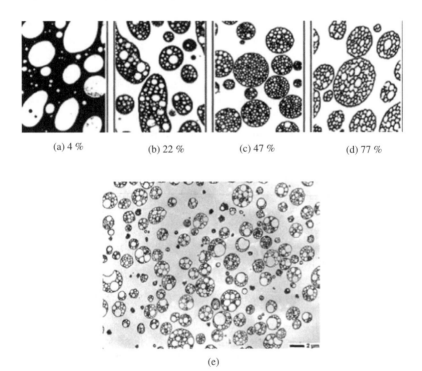

(a) 4 %   (b) 22 %   (c) 47 %   (d) 77 %

(e)

Figure 5.37. Scanning electron micrographies carried out at various steps of the formation of a HIPS. (a–d) The given percentages correspond to the yield in styrene. (e) Final state; dark parts consist of polybutadiene phase. [Courtesy of BASF Cy (Ludwigshaffen Germany).]

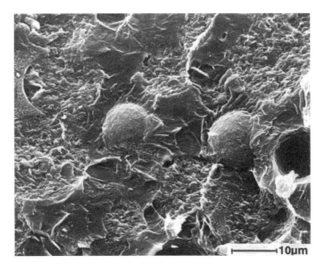

Figure 5.38. Scanning electron micrography of a high-impact polystyrene showing the cohesion between the polybutadiene nodules and the polystyrene matrix. [Courtesy of BASF Cy (Ludwigshaffen Germany).]

Terpolymers referred to as ABS are obtained in the same way as HIPS, a styrene/acrylonitrile mixture replacing styrene. The copolymerization of acrylonitrile with styrene gives rise to a material with better cohesive properties.

It is worth mentioning that in heterogeneous multiphase systems, each phase retains its characteristics. The best method to check the presence of a phase separation in a system consists of the observation of two glass transition temperatures. Techniques of microscopy are also widely used to characterize heterogeneous systems (Figures 5.34 and 5.38) because they afford extremely precise information with respect to the phase dispersion and the structure of the interphase zones.

5.5. ORIENTED POLYMERS

5.5.1. Intrinsic and Shape Anisotropy of Polymers

A system consisting of oriented molecules generates anisotropic properties. This anisotropy can take various facets—for example, an anisotropy of the refractive index (i.e., birefringence).

The anisotropy of a material depends on the degree of orientation of its constitutive molecules and on the molecular anisotropy of the latter. Both a *shape* anisotropy and the molecule *intrinsic* anisotropy contribute to this molecular anisotropy.

The shape anisotropy results from the molecular asymmetry: when placed in an external electric field, an object of refractive index $\tilde{n}_0$ modifies it in a nonisotropic way due to an asymmetrical polarization of its charges. This shape anisotropy, which is proportional to $(\tilde{n} - \tilde{n}_0)^2$, can be only positive. Due to their asymmetry, macromolecular chains are characterized by a shape anisotropy that can only increase with the deformation applied.

As for the intrinsic anisotropy of molecules, it depends on their chemical structure and more particularly on their polarizability. Double bonds and more particularly conjugated ones—mainly in aromatic cycles—contribute to the anisotropy of molecules. For a macromolecule to exhibit an intrinsic anisotropy, it has to consist of anisotropic groups organized in a very regular manner along the chain. Indeed, an assembly of anisotropic molecules, which would be randomly oriented, would generate a macroscopically isotropic system. Polymers forming double helixes (DNA) or which are rigid [poly(benzyl glutamate)] are among the best-known examples of polymers with strong intrinsic anisotropy. Macromolecular coils exhibit an intrinsic anisotropy that is less pronounced than that of highly organized polymers; it depends essentially on differences in the polarizabilities of the backbone and of the side-chain substituents. This difference is minimum in the case of the poly(methyl methacrylate) whose main chain anisotropy is almost completely counterbalanced by that of the side groups.

In the case of polystyrene, the contribution of the aromatic moieties perpendicular to the chain results in a negative intrinsic anisotropy of its segments. As observed for the shape anisotropy, the intrinsic anisotropy of macromolecules is also expected to strongly increase upon orientation of each of their anisotropic segments. In the solid state, this orientation can be obtained by a mechanical solicitation (stretching); in the liquid state, it can be obtained by (a) an elongational flow (extrusion, spinning) or (b) an electric (Kerr effect) or magnetic effect (Colton–Sheep effect) when the chemical structure of the macromolecules is appropriate.

All macromolecules are not prone to undergo an orientation by one of the means previously mentioned. The structural criteria determining the "orientability" of polymer chains are almost the same as those required for manufacturing fibers. Highly symmetrical and stereoregular chains that possess groups with a strong energy of interaction [poly(vinyl chloride) (PVC), polyacrylonitrile (PAN), etc.] are the best suited to retain an orientation upon drawing or in an elongational flow.

5.5.2. Orientation of Polymers

5.5.2.1. Uniaxial Drawing in a Solid State. Upon drawing, initially disordered chains in an amorphous polymer undergo a phenomenon of orientation. The fact that macromolecular coils are oriented upon application of an elongational stress is materialized by the phenomenon of *necking* (point B of Figure 5.39); in the subsequent step (B-C), the stretched chains self-organize in fibrillae, which reduces the intermolecular distances and contributes to reinforce the interactions between chains and the density of the cohesive energy of the system. PVC is an example of an amorphous polymer whose mechanical properties can be improved to give textile fibers after undergoing such a drawing.

Drawing also orientates the crystalline zones (when they exist) inducing the transformation of spherulites into lamellar structures (see Figure 5.40). However,

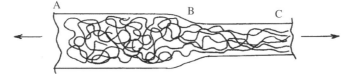

Figure 5.39. Schematic presentation of the necking phenomenon (B) obtained by the drawing of an amorphous polymer sample.

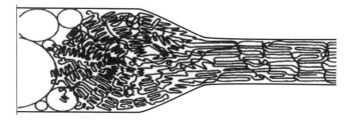

Figure 5.40. Deformation under stretching of spherulites in a semicrystalline polymer.

more than the latter phenomenon, it is the orientation of the amorphous parts which contributes to the increase of the Young modulus of such a stretched material in the direction of drawing and to the increase of its fracture strength. Indeed, an initially stretched sample that is subjected to a new drawing test exhibits a higher Young modulus in the direction of stretching and a lower one in the perpendicular direction. Such a drawing should be imperatively carried out at a temperature close to, but lower than, the melting point (T_m) of the crystalline zones to favor the chain rearrangement and prevent their immediate relaxation as soon as the stress is suppressed. When treated under such conditions, semicrystalline polymers such as aromatic polyamides can exhibit a Young modulus equal to 130 GPa and a fracture strength of 2.8 GPa.

5.5.2.2. Orientation by Elongational Flow. Crystallizable polymers can undergo an orientation of their chains in dilute solutions. When subjected to an elongational flow, these chains are forced to be oriented in the parallel direction and thus form fibrillae; this phenomenon is referred to as *fibrous crystallization*. Inside these fibrous parts which behave as nuclei for the epitaxial growth of remaining macromolecules, shear stresses are low. This growth results in crystalline lamellae that are perpendicular to the fiber axis. As for the axis of the chains folded in these lamellae, it is parallel to that of the fibers: one speaks of "shish-kebab" (skewer structures) as shown in Figure 5.41.

Dilute solutions of either polyethylene in toluene or isotactic polypropene in chloronaphthalene afford such "shish-kebab" structures when crystallized in an elongational flow. In the case of polyethylene of high molar mass spun from a viscous solution, fibers of 3-GPa fracture strength and of 90-GPa Young modulus could be obtained. These are very high values if one considers that only London-type molecular interactions are responsible for the cohesion of the material.

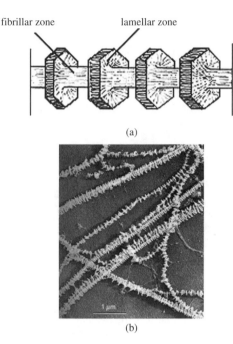

(a)

(b)

Figure 5.41. (a) Representation and (b) scanning electron microscopy of a "shish-kebab" structure obtained in an elongating flow.

Whatever the method used to obtain the orientation of the macromolecular systems, an orientation function (referred to as *"Hermans function"* and indicated by F_{her}) can be defined to characterize the chain alignment with the reference direction (in general the fiber axis in the case of an uniaxial drawing):

$$F_{her} = \tfrac{1}{2}(3\langle\cos^2\Phi\rangle - 1)$$

where Φ is the angle between the direction of drawing and that of the chains axis. If all the chains are completely oriented, then $\Phi = 0$ and $F_{her} = 1$. The orientation function is equal to 0.5 for a perpendicular orientation of the chains and 0 for a random orientation.

5.5.2.3. Effect of Biaxial Drawing. Such a biaxial drawing is observed in film-forming processes. It can be carried out on either amorphous or semicrystalline polymers. The two-dimensional orientation can be described using the angles Φ_X and Φ_Y between the chain direction and those (X and Y) of drawing (generally orthogonal). The Hermans functions relative to each reference direction are given by

$$F_{herX} = 2\langle\cos^2\Phi_X\rangle + \langle\cos^2\Phi_Y\rangle - 1$$
$$F_{herY} = 2\langle\cos^2\Phi_Y\rangle + \langle\cos^2\Phi_X\rangle - 1$$

5.6. LIQUID CRYSTALLINE POLYMERS

5.6.1. Molecular Liquid Crystals

Between the crystalline state (characterized by a long-range three-dimensional order), and the amorphous isotropic state, there is an intermediate state of matter referred to as liquid crystal. It is specific to certain molecules, which simultaneously exhibit order like crystals and flow like fluids. Reinitzer, who observed that cholesteric esters form opaque liquids that become transparent upon raising the temperature, is considered as the precursor of this field. In addition to the term liquid crystal, *mesomorphic* or *mesophase* can also be used (from Greek mesos meaning *"median"*) to name this intermediate state between an isotropic liquid state and the three-dimensional crystalline order as first proposed by Friedel. Molecules that adopt a preferential orientation and result in an anisotropy are called *mesogens*.

Liquid crystalline molecules are classified in two families referred to as *thermotropic* and *lyotropic*, respectively. In thermotropic liquid crystals, the formation of *mesophases* is temperature-dependent; as for lyotropic liquid crystals, they necessitate the use of a solvent for forming mesophases. Liquid crystals are also sensitive to other stimuli such as magnetic or electric fields, pressure, and so on.

Molecules that are prone to generate mesophases exhibit either "rod-like" or "disk-like" rigid structures:

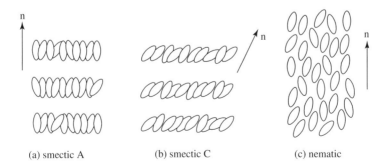

Mesophases are characterized by a long-range orientational order which results from the longitudinal alignment of the mesogenic groups along a directing axis.

Two categories of mesophases can be distinguished, depending upon the dimensions and the degree of order that is attained (Figure 5.42). When mesogens are organized in two-dimensional layers of regular size, the corresponding mesophases

Figure 5.42. Representation of the organization of mesogens in smectic phases (S_a, S_c) and nematic one (N).

are called *smectic* (S). As lateral forces between the molecules of a mesophase are quite larger than those between the layers, the fluid character results from the relative slipping of the layers.

At least eight different smectic phases are known; they can be distinguished from one another by adding alphabetical index to the letter S. The phase corresponding to the highest degree of order and to a hexagonal ordering of mesogens within a layer is referred to as smectic B (S_B). In addition to S_B phases, which exhibit a three-dimensional arrangement of the mesogen groups, smectic mesophases E, G, and H are also characterized by a similar degree of order, but in S_G and S_H the direction of alignment of mesogens is tilted with respect to the axis perpendicular to the layer plane. S_A mesophases are the least-ordered smectic structures that are characterized by a random lateral distribution of mesogens even if their longitudinal axis is perpendicular to the layer plane. S_C mesophases possess the same characteristics as S_A ones, but in the latter case the mesogens are tilted by an angle θ with respect to the axis perpendicular to the layer plane. Mesophases S_F and S_I correspond to an intermediate degree of order between S_B and S_A mesophases.

Nematic mesophases (N) are less ordered than smectic ones because they exhibit only a monodimensional order. In this case, even if the orientation order is retained with respect to a directing axis—which can thus be regarded as the main directional axis of the molecule—the centers of mass are not necessarily within a layer but can be distributed in a random way. Within these domains, whose size is in general in the micron range, the average degree of alignment with respect to a preferential axis is described by the Hermans orientation factor (F_{her}) (order parameter). The closer this factor to 1, the higher the degree of order of the phase. Nematic mesophases are more fluid than their smectic homologs.

The family of *cholesteric* mesophases is also part of the family of nematic mesophases (N*) (Figure 5.43). Only mesogens carrying a chiral center (denoted by the presence of * next to the letter N) and ordering themselves in nematic phases can generate such phases. The presence of this chiral center forces mesogen groups to adopt a screw-type structure corresponding to a helical variation of the nematic directing axis. Smectic structures S_C and S_A can also afford chiral phases insofar as the mesogenic group carries a chiral center. Chiral smectic phases C (S_{C*}) are known for their ferro- and piezoelectric properties.

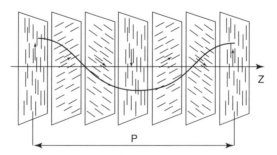

Figure 5.43. Representation of a chiral nematic phase known as "cholesteric."

Finally, it is worth mentioning that a molecule can undergo several transitions and experience successively the highly ordered state of a crystal (the nematic and then smectic) and finally go to the isotropic states upon raising the temperature $(S_B \rightarrow S_C \rightarrow S_A \rightarrow N \rightarrow I)$.

5.6.2. Liquid Crystalline (Mesomorphic) Polymers (LCP)

The association of simple molecular groups exhibiting mesomorphic properties with polymers was considered soon after Flory predicted in 1956 that concentrated solutions of "rod"-type rigid polymers could form ordered structures. Investigations of the behavior of polymers with helical conformation such as those of poly(methyl *and/or* benzyl glutamate) type showed that they self-align in a given direction, thus corroborating this prediction. By associating polymers and mesogenic groups within the same structure, one can design materials exhibiting simultaneously the anisotropic characteristics of liquid crystals and the thermoplastic behavior of certain liquids. These mesogenic groups can be incorporated either in the main chain or as side chains—that is, laterally grafted onto the backbone (Figure 5.44).

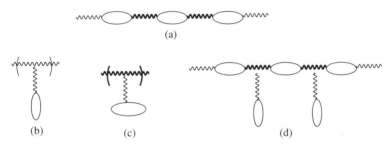

Figure 5.44. Representation of liquid crystalline polymers with main-chain (a) and side-chain (b, c) mesogenic groups and combination of both types of chains (d).

Certain liquid crystalline polymers comprise both main-chain and side-chain mesogenic groups. Polymers that include mesogenic groups in the main chain are generally obtained by step-growth polymerization (see Chapter 8). Depending upon the nature and the size of the links which connect the mesogenic groups together, main-chain mesomorphic polymers form either very rigid or semiflexible structures.

Polymers carrying side-chain mesogenic groups can be prepared in various ways:

- By chemical modification of a flexible polymeric backbone as in the case of polysiloxanes
- By chain polymerization of a vinyl or related monomer carrying a mesogenic group
- By step-growth polymerization of mesomorphic monomers.

In the latter case, the polymeric backbone and the mesogens are separated by a bivalent flexible molecular group called *spacer*.

5.6.2.1. Main-Chain Liquid Crystalline Polymers.   Of the two types of main-chain liquid crystalline polymers, it is those containing mesogenic groups in the main-chain that have generated most of the applications. Indeed, such structures exhibit exceptional mechanical properties. When the mesogenic units forming the backbone are connected to one another by small-size molecular species, the resulting materials exhibit strong rigidity and decompose before melting and expressing their liquid crystalline properties.

For lyotropic polymers, the factor Λ *(anisotropy ratio)* which denotes the ratio of the length L of the mesogen to its diameter is high ($\Lambda > 6$). The attractive forces generated by $\pi-\pi$-type interactions between the mesogens aligned in a parallel way can consequently develop and contribute to the rigidity of the material. Only the addition of a highly polar solvent (dimethylsulfoxide, dimethylformamide, etc.) can weaken these intermolecular interactions and reveal the liquid crystalline behavior of such materials. Such birefringent solutions are referred to as *lyotropic*. The best-known and commercially available example is that of "Kevlar®," which is an aromatic polyamide, poly(*p*-phenylene terephtalamide):

Lyotropic polymer solutions exhibit a peculiar behavior; above a critical concentration, they separate in two phases: the highly concentrated phase is liquid crystalline, whereas the least concentrated phase is isotropic. The higher the factor Λ, the lower the critical volume fraction at which phase separation occurs.

At the critical concentration, the viscosity of the medium increases dramatically. Along with aromatic polyamides, some cellulosic derivatives (hydroxypropylcellulose, etc.) and polypeptides belong to the category of lyotropic polymers. Fibers can be produced from lyotropic solutions of these polymers at a concentration lower than their critical concentration.

To reduce the melting point of rigid structures corresponding to main-chain mesogen-containing polymers, it is necessary to break the symmetry of these assemblies by introducing either bulky lateral substituents or semiflexible connections. By this means it is possible to increase the molecular mobility and attain the properties of liquid crystals without the contribution of a solvent. In other words, the liquid crystals behavior becomes observable at temperatures lower than that of degradation; such structures are known as *thermotropic*. This strategy was successfully utilized, in particular, in the case of aromatic polyesters. Many thermotropic polymers have been obtained by "altering" the chain symmetry of polyesters. Polyesters such as form nematic mesophases for odd values of m and smectic mesophases for even values of m.

5.6.2.2. Side-Chain Liquid Crystalline Polymers.

Contrary to main-chain LCP, those carrying side-chain mesogens did not find significant applications until recently, although many studies were devoted to them. In 1978 the first polymers with side-chain mesogens were prepared by Finkelman et al. To reveal the liquid crystalline behavior of such structures, these authors introduced a spacer between the backbone and the side-chain mesogens. They could uncouple in this way the movements of the main chain from those of the mesogenic groups. In addition to the nature of the mesogen, the type of polymer chosen as the backbone and the length of the spacer play a vital role in the formation of mesophases.

The more flexible the main chain, the broader the range of thermal stability of such mesophases. As for the size of the spacer, it determines to a large extent the type of resulting mesophases, with their degree of order increasing with the length of the spacer.

Liquid crystalline polymers with side-chain mesogenic groups are also interesting materials, but for totally different reasons than those mentioned for main-chain mesogenic structures. Because the mesogen groups and the backbone are uncoupled, these structures exhibit a behavior close to that of simple mesogenic molecules (Figure 5.45). Attempts have been made to prepare by this means materials with the characteristics of processability and mechanical resistance characteristic of polymers and the same sensitivity to stimuli (magnetic or electric field) as that of simple mesogenic molecules. Such properties are of considerable interest in electro-optical technologies.

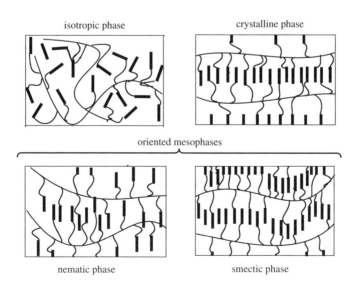

Figure 5.45. Mesophases in LCP side chain.

Molecules that self-organize in chiral nematic or smectic mesophases and comprise a permanent dipole exhibit ferroelectric properties. This means that they are spontaneously polarized and are able to orient themselves under the effect of an electric field. Polymers with chiral mesogens could be used for the manufacture of large-size video screens and more generally for displays. To this end, they require a response time (i.e., the time required for mesogens to flip from one position to another) that is comparable to that of molecular liquid crystals. This response time t depends on the electric field (E), the polarization (P), and the viscosity (η):

$$t = \eta/PE$$

These materials are the subject of very active studies.

Polymers with side-chain mesogens are also useful for applications in the field of optical information storage (Figure 5.46). Using a mesomorphic polymer whose mesogenic groups would be aligned in a homeotropic way by application of a field and upon retaining this orientation by cooling, it is possible to produce localized isotropic domains with a laser beam. These domains scatter the light and lose their transparency; thus, information can be "written" on a film and subsequently erased by simple increase of the temperature.

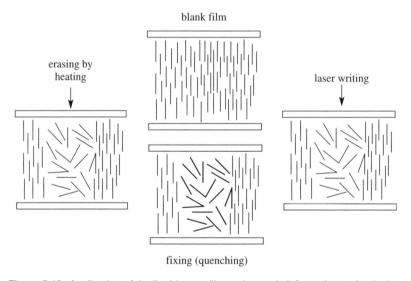

Figure 5.46. Application of the liquid crystalline polymers in information technologies.

To identify mesophases, the experimenter can use a polarizing microscope to analyze the characteristic textures of the various mesophases; he can also make use of differential scanning calorimetry and of X-ray diffraction.

LITERATURE

P. Flory, *Principles of Polymer Chemistry*, Cornell University Press, Ithaca, NY, 1971.

P-G. De Gennes, *Introduction to Polymer Dynamics*, Cambridge University Press, Cambridge, 1990.

W. Mattice, and U. Sutter, *Conformational Theory of Large Molecules—The Rotational Isomeric State Model in Macromolecular Systems*, Wiley Interscience, New York, 1994.

L. Sperling, *Introduction to Physical Polymer Science*, 4th edition, Wiley Interscience, New York, 2006.

C. Booth and C. Price (Eds.), *Comprehensive Polymer Science*, Vol. 2: *Polymer Properties*, Pergamon Press, Oxford, 1989.

B. Wunderlich, *Macromolecular Physics,* Vol. 1 (1973), Vol. 2 (1976), Vol. 3 (1980), Academic Press, New York.

F. A., Bovey and L. W. Jelinski, *Chain Structure and Conformation of Macromolecules*, Academic Press, New York, 1982.

6

DETERMINATION OF MOLAR MASSES AND STUDY OF CONFORMATIONS AND MORPHOLOGIES BY PHYSICAL METHODS

After the polymer synthesis step (see Chapters 7–10) comes that of their structural characterization. The methods available for this can be grouped into two categories:

- The first category includes methods used for the identification of any organic (simple or macromolecular) molecule, primarily spectroscopic ones; the study of polymer tacticity by NMR spectroscopy described in Chapter 3 is an example.
- The second category includes methods that correlate the variation of a property characteristic of the macromolecular state with structural parameters; they are described in this chapter.

6.1. DETERMINATION OF MOLAR MASSES BY COLLIGATIVE METHODS

Colligative methods are those that involve the determination of the number of macromolecules present in a polymer sample of given mass; it is then easy to deduce the number average molar mass $(\overline{M}_n)$ of the sample analyzed.

6.1.1. End-Group Titration

Due to the low concentration of the chain ends and the lack of precision of titration, this method is well-suited for polymers of relatively low molar mass. It involves the identification and titration of the functional groups located at one or each of the two

Organic and Physical Chemistry of Polymers, by Yves Gnanou and Michel Fontanille
Copyright © 2008 John Wiley & Sons, Inc.

ends of strictly linear polymers. Chemical titration requires only simple equipment, which is why it is still frequently used for the characterization of condensation polymers (see Chapter 7). It is also used when the polymers analyzed are obtained by "living" polymerization and functionalized at their terminal positions (that is, either through the initiator or through a functional deactivator) by a reactive group easy to titrate using spectrophotometric techniques.

An example is the characterization of a polyamide sample whose solubility is limited to highly polar solvents, preventing it from being characterized by other methods for the measurement of molar masses. The titration of a primary amine in a polyamide can be carried out in *m*-cresol solution using perchloric acid as the reagent and can be monitored by potentiometry. For the titration of carboxylic acids, it is preferable to use a solution in benzyl alcohol and operate at high temperature with sodium hydroxide as the base and an indicator. For polyamides grown from α-amino,ω-carboxylic acids, only one titration is required but a double titration can serve to confirm the values obtained in the first instance. The accuracy of the titration is approximately 3×10^{-6} moles of titrated functional groups per gram of polyamide.

A second example is the titration of end groups by NMR spectroscopy. The number average degree of polymerization of an α,ω-heterodifunctional acetal-terminated polystyrene can be easily deduced from the ratio of intensities of the signals corresponding to the protons of the terminal functional groups and to those of aromatic protons (polystyrene) (Figure 6.1). Moreover, since one of the two terminal functional groups is necessarily carried by the chain—the one that is a fragment of the initiator—the ratio of the intensities of their signals corresponds to the polymer

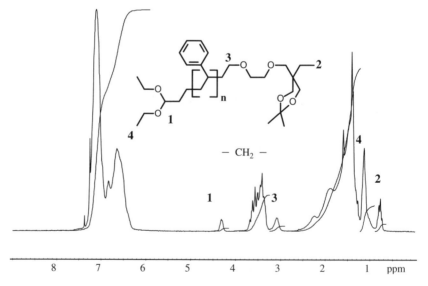

Figure 6.1. ^{1}H NMR spectrum of a polystyrene (with $\overline{M}_n = 1800 \, \text{g·mol}^{-1}$) activated by two different acetal functional groups.

functionality. The accuracy of NMR spestroscopy for end-group titration is satisfactory for values of the degree of polymerization within the range of a few tens. This example illustrates the limits of end-group titration as a method of determination of molar masses. Indeed, this method requires the molecular structure of the polymer analyzed to be perfectly known: branching and/or the presence of cyclic products in the sample can lead to completely erroneous values of the number average molar mass.

6.1.2. Osmometry

Colligative properties of dilute solutions—polymer solutions particularly—directly result from the variation of the chemical potential of the solvent into which a solute is added. Such properties can be assessed by measuring the osmotic pressure (membrane osmometry), the decrease of the vapor pressure (vapor phase osmometry) or of the freezing point (cryometry). Contrary to the titration of the terminal functional groups, colligative methods do not require a prior knowledge of the polymer structure and depend exclusively on the number of solute molecules.

6.1.2.1. Origin of the Osmotic Pressure.
The differential of the free energy corresponds to the variation of the free energy G_i of the various molecular species present in the solution with pressure, temperature, and mole fraction f_i:

$$dG_i = (\partial G_i/\partial p)_{T,\text{fi}}dp + (\partial G_i/\partial T)_{p,\text{fi}}dT + (\partial G_i/\partial f_i)_{p,T,\text{fj}}df_i \qquad (6.1)$$

The term $(\partial G_i/\partial f_i)_{p,T,\text{fj}}$ represents the chemical potential μ_i of the species i. The third term of expression (1) can then be written in terms of $d\Delta\mu_i$, the difference in chemical potential between the solvent in solution and the pure solvent. This difference may then be expressed in terms of the activity of the solvent:

$$\Delta\mu = (\mu_i - \mu_i^0) = RT\ln a_i$$

The terms $(\partial G_i/\partial p)_{T,\text{fi}}$ and $(\partial G_i/\partial T)_{P,\text{fi}}$ represent the molar volume V_i^0 (assuming the solution is dilute enough that the molar volume of the pure solvent is approximately equal to the partial molar volume of the solvent in solution) of the species i and its partial molar entropy $\overline{S}_i$, respectively; expression (6.1) can thus be written as

$$dG_i = V_i^0 dp + \overline{S}_i dT + d\Delta\mu_i \qquad (6.2)$$

This expression clearly shows that in an isothermal system ($dT = 0$), where a solute is added to pure solvent, the only means of equalizing the solvent activity in the solution and that of the pure solvent is to exert an additional pressure (dp) on the solution. At equilibrium ($dG_i = 0$), this additional pressure is defined as the osmotic pressure. Assuming that the solution is incompressible, expression (6.2) may be integrated to give

$$V_1^0 \Pi = -RT\ln a_1 = -\Delta\mu_1 \qquad (6.3)$$

For ideal solutions, the activity and the mole fraction (f_2) of a species (6.2) are identical in the entire range of concentration; ln a_1 is thus equal to ln f_1, which itself is equivalent to $\ln(1 - f_2)$, where f_2 refers to the solute mole fraction. Since the solution is dilute, the latter term is approximately equal to $-f_2$. Equation (6.3) can now be written as

$$V_1^0 \Pi = -RT f_2 \tag{6.4}$$

The mole fraction (f_2) and the mass concentration of the solute (C_2) can be related:

$$C_2 = m_2/(V_1 + V_2) \sim m_2/V_1$$

which can be also expressed as

$$C_2 = N_2 M_2/V_1 = N_2 M_2/N_1 V_1^0 \sim f_2 M_2/V_1^0$$

where m_2 represents the mass of the solute, V_1 and V_2 are the volumes occupied by the solvent and the solute, N_1 and N_2 are the number of molecules of solvent and solute, and M_2 is the molar mass of the solute.

One thus obtains the Van't Hoff law, which shows that at infinite dilution the osmotic pressure (Π) is inversely proportional to the molar mass of species (6.2) under consideration:

$$\lim_{C_2 \to 0} \frac{\Pi}{C_2} = \frac{RT}{M_2} \tag{6.5}$$

A polymer sample consists of many macromolecular species. Equation (6.5) can be easily generalized in this case:

$$\Pi = RT \sum_i \left(\frac{C_i}{M_i} \right)$$

Since by definition $C_i = N_i M_i$, one obtains

$$\Pi = RT \frac{\sum_i N_i M_i}{\sum_i M_i}$$

Recalling that $\overline{M}_n = \sum_i N_i M_i / \sum_i N_i$, the relation between the osmotic pressure and the sample molar mass reduces to

$$\frac{\Pi}{C_2} = RT \frac{\sum_i N_i}{\sum_i N_i M_i} = \frac{RT}{\overline{M}_n} \tag{6.6}$$

Thus, the molar mass accessible by osmometry is the number average one.

Actually, equation (6.6) applies only in the limit of infinite dilutions. The assumption that ln $a_1 = -f_2$ becomes invalid at finite concentrations (athermic or regular solutions). Indeed, the osmotic pressure depends on the concentration

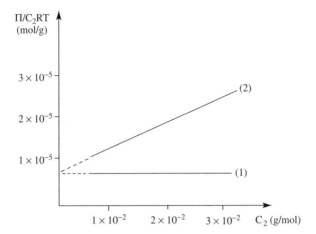

Figure 6.2. Variation of the reduced osmotic pressure ($\Pi/C_2RT \equiv 1/\overline{M}_n$) with the concentration: case of poly(methyl methacrylate) in m-xylene (1) at 20°C (under θ conditions, $A_2 = 0$) and in dioxane (2) (good solvent).

of the solution studied except for very dilute solutions (Figure 6.2). For all other solutions, the solvent activity can be expressed as a power series with respect to solute mole fraction:

$$-\ln a_1 = f_2 + B_2 f_2^2 + B_3 f_2^3 + \cdots$$

and since

$$f_2 = V_1^0 C_2 / M_2$$

relation (6.6) can be written as

$$\frac{\Pi}{C_2} = RT \left[\frac{1}{M_2} + B_2 \frac{V_1^0}{M_2^2} C_2 + \frac{B_3 V_1^{0^2}}{M_2^3} C_2^2 + \cdots \right]$$

which gives

$$\frac{\Pi}{C_2} = RT \left[\frac{1}{M_2} + A_2 C_2 + A_3 C_2^2 + \cdots \right] \tag{6.7}$$

A_2 and A_3 are the second and third virial coefficients, and $1/M_2$ corresponds to the first virial coefficient.

A_2 is an indicator of the polymer–solvent interactions and is useful for the determination of the solvent–polymer interaction parameter $[A_2 \div (1 - 2\chi)]$; a solution with $A_2 = 0$ is called a *theta solution*, and the polymer is said to be in the *theta state*. The temperature at which $A_2 = 0$ is the *theta temperature* (θ), and the solvent becomes a *theta solvent* at this precise temperature. These are called *theta conditions*.

In the same way, the third virial coefficient is an indicator of the interactions between three entities and takes positive values as the concentration of polymer increases in the solution. Figure 6.2 shows the dependence of Π/C_2 on the concentration for two different examples.

As shown in Chapter 4, the second virial coefficient (A_2) is proportional to the excluded volume. However, de Gennes has shown that above the overlap concentration (C^*), the effect of excluded volume is "screened" beyond the correlation length (ξ) or the distance between two contact points. In the semi-dilute region, chains shrink [equation. (79), Chapter. 4] in the proportion $r(\Phi)/r(\Phi^*) \div \Phi^{-1/4}$ and, likewise, the osmotic pressure decreases by a factor $\Phi^{-1/4}$ with respect to the value predicted by equation (6.7).

6.1.2.2. Measurement by Membrane Osmometry. Membrane osmometry

is, in principle, a very simple technique (Figure 6.3). A semipermeable cellulose-based membrane for organic solvents or a cellulose acetate-based one for water separates the solution chamber containing the solute from that of pure solvent. The membrane is permeable only to the solvent. At the onset of an osmometry experiment, the levels of the solution and pure solvent are equal; the solvent then flows into the solution chamber and after a time that can be very long, the osmotic pressure (Π) on the solution is attained and the system is at equilibrium.

Modern osmometers are equipped with a servo-control, which compensates the increase of the osmotic pressure by a hydrostatic backpressure. With such equipment, equilibrium is attained in a few minutes instead of several hours.

Since the osmotic pressure is inversely proportional to the molar mass of the sample, measurement of the osmotic pressure becomes less precise with the increase of the molar mass of the macromolecules analyzed; the maximum molar mass, which can be measured by this method, is in the range of 5×10^5 g·mol^{-1}.

However, the problem of the permeation of species of low molar mass through the membrane is a real one, especially for samples under the molar mass threshold of 5×10^3 to 10^4 g·mol^{-1}; diffusion toward the solvent chamber can indeed

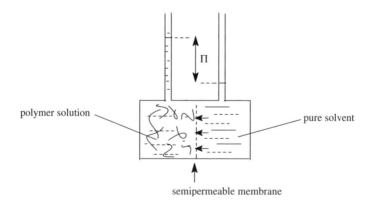

Figure 6.3. Diagrammatic presentation of the osmometer operation.

dramatically affect the value of the osmotic pressure measured. But with modern osmometers, the risk of permeation of small molecules is minimized (since the measurement time is short).

6.1.3. Vapor-Phase Osmometry (VPO)

Vapor-phase osmometry is also a colligative method and is based on the decrease of the vapor pressure, due to the addition of a solute into a pure solvent (Figure 6.4). Like osmotic pressure this property depends exclusively on the number of molecules of solute introduced and thus gives access to the number average molar mass. The disadvantage of VPO is its lack of sensitivity, due to the very small variations of the vapor pressure exhibited by dilute polymer solutions. It is, in fact, well suited to the analysis of low molar mass samples ($\leq 2 \times 10^4$ g·mol^{-1}) and is thus complementary to membrane osmometry. Because it is easier to measure small variations in temperature than those in vapor pressure, a thermoelectric device is used to transform the increase of vapor pressure in a VPO experiment into a variation of temperature. Thus, two thermistors are placed in a closed chamber containing a pure solvent at a given temperature. If a pure solvent drop is placed on each of the two thermistors, they will indicate the same temperature. On the other hand, if a drop of a dilute polymer solution is placed on one of the two thermistors, a variation in temperature will result, caused by the condensation of pure solvent on this thermistor due to the difference in chemical potential between the drops.

As long as the vapor pressure is not the same as that of pure solvent, the temperature will rise:

$$\Delta T \approx -\frac{RT^2}{\Delta H_v}\Delta\mu_1 \tag{6.8}$$

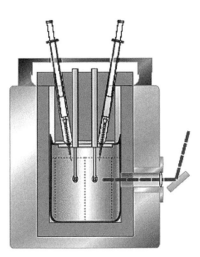

Figure 6.4. Representation of a vapor phase osmometer (Gonotec scheme).

where ΔH_v is the enthalpy of vaporization. Similarly to the relation for osmotic pressure (6.5), one obtains here

$$\lim_{C_2 \to 0} K \Delta T = \frac{RT C_2}{\overline{M}_n} \quad \text{with} \quad K = \rho \frac{\Delta H_{v,m}}{T M_1} \quad (6.9)$$

where ρ is the density of the solution and M_1 is the molar mass of the solvent. In practice, one measures differences in resistance, with the first member of the expression (6.9), $K \Delta T$, being replaced by $K' \Delta \Omega$.

6.2. MACROMOLECULAR MASSES AND SIZES BY SCATTERING TECHNIQUES

In addition to the molar mass, many parameters contribute to "shape" a polymer molecule in a given form: the nature of repetitive units, their configuration, the presence (or absence) of branches, the existence of intra- or intermolecular combinations, the nature of the solvent in which the polymer is dispersed, the temperature, and so on; all these factors are to be taken into consideration to account for the fact that the macromolecule may take the shape of a random coil, a sphere, or a rod. Thanks to the existence of techniques such as electron microscopy or atomic force microscopy, it is now possible to visualize the shape of big size macromolecules. These images, however, do not necessarily reflect the real structure of the chains—in particular, the three-dimensional architecture. Indeed, the absence of solvent can cause a collapse of the internal structure and/or a two-dimensional spreading of the macromolecule. Cryomicroscopy is a promising technique that prevents this drawback since macromolecules are quenched under the real solution conformation at $-180°C$. However, the contrast between the various elements has to be significant enough, or coloring techniques may be required to visualize the structure.

With characterizations based on the interaction between matter and an electromagnetic wave (X-rays, light) or the so-called "matter waves" (neutrons, electrons), the risk of degradation and/or of a bias induced by such techniques is normally not to be feared; the measurement is direct and absolute and is not a comparative or relative one as in the case of the evaluation of hydrodynamic properties (viscosity) or with chromatographic techniques.

Experimentally, when an electromagnetic radiation impinges on matter, the radiation is in any case modified and may take one of several paths, depending on the material analyzed and its wavelength.

Matter may absorb electromagnetic waves (UV, visible, infrared), and according to quantum mechanics, the absorbed energy may be transferred to other modes of motion, perhaps emitted as a photon or/and subsequently dissipated as heat. Depending upon the type of excited state (electronic or higher level), either fluorescence, phosphorescence, or X-ray emission may be observed.

The radiation may also be reemitted as scattered radiation in all directions with the same wavelength as the incident beam.* In the case of periodic structures in which atoms, molecules, or particles are organized in regular arrays, scattering will be observed only at specified angles because of destructive interference by the radiation scattered by different parts of the array at all other angles. In this case the radiation is said to be diffracted. X-ray diffraction by a crystal is an important example.

In addition to elastic scattering, there is often a less intense inelastic scattering resulting from the loss or gain of a quantum of energy by the transmitted photon; under these conditions, the spectral analysis of the transmitted radiation shows a distribution of wavelengths around λ_0. Information obtained from elastic and inelastic scatterings are often complementary. For instance, the first allows the determination of the molar mass and the radius of gyration of macromolecules while the second, in the case of quasi-elastic scattering, allows the determination of the diffusion coefficient and the hydrodynamic diameter.

The scattering phenomenon results from the interaction of an electromagnetic wave with the electron cloud of an atom or from the interaction of a neutron beam with its nucleus (neutron scattering). While traversing a material, radiation and the electrical field associated with it causes a distortion of the distribution of charges (the electrons) in the element of matter impinged. This induces the formation of a dipole that, since it is oscillating, becomes a secondary source of radiation. These oscillating dipoles generate the scattered elastic wave. In the interaction between the incident electric field and the electrons of the element concerned, only the outer electronic layers are involved; indeed, even a wave of low energy is capable of inducing their distorsion. On the other hand, it is necessary to use radiation of higher energy (X-ray) to produce vibrations of electrons in more closely held layers that are more strongly bound to atomic nuclei.

Because the structural information that can be obtained by scattering is in the same range as the wavelength of the incident radiation, either visible light ($\lambda = 300$ to $700\,nm$) or neutrons ($\lambda = 0.2$ to $2\,nm$) or X-rays ($\lambda = 0.1\,nm$) can be used to probe different length scales. Independently of the wavelength of the incident beam, the same types of structural and dynamic properties of the macromolecular medium are responsible for the scattering. Depending upon the radiation used, the contrast will depend on the variation of the refractive index (light), on the difference of scattering length between hydrogenated and deuterated elements (neutrons), or on the difference of electron density (X-rays).

6.2.1. Light Scattering

Light scattering by particles of small size compared to the wavelength of the incident beam can be described using both electromagnetic particulate and thermo-dynamic fluctuation approaches. Both are important since the first can be applied

*Radiation is said to be scattered when the interacting object is the same size as or smaller than the radiation wavelength; this radiation is said to be reflected when the object is much larger than the incident beam wavelength.

to both isotropic and anisotropic particles, and the second can more easily treat the existence of interactions in the systems studied.

6.2.1.1. Refraction.
As previously mentioned, the interaction between matter and an electric field results in the formation of induced dipole moments from elements having a polarizability α. These dipole moments slow down the propagation of the incident beam from its velocity in vacuum ($\tilde{c}$) to a lower velocity in the material traversed ($\tilde{c}_m$). A consequence of the medium polarizability is the refraction of the incident beam. The refractive index of the material ($\tilde{n}$) is defined as the ratio of the velocity of the electromagnetic radiation in vacuum to that in the material:

$$\tilde{n} = \frac{\tilde{c}}{\tilde{c}_m}$$

The frequency of propagation (ν) is the same in the vacuum and in the medium traversed; and since $\tilde{c} = \nu\lambda_0$, the previous equation can be written as

$$\tilde{n} = \frac{\tilde{c}}{\tilde{c}_m} = \nu\lambda_0/\nu\lambda = \lambda_0/\lambda$$

where λ_0 represents the wavelength in vacuum and λ represents that in the material. The refractive index of polymers for visible light is generally greater than one. For X-rays, the refractive index is close to one because they are only slightly retarded by the medium. Neutron radiation is also amenable to refraction, although its interaction occurs with the atom and not with the electron cloud as with light or X-rays.

Since electromagnetic radiation undergoes a retardation of its velocity upon traversing a polarizable medium and is refracted, the changes of direction of the incident beam at the air/medium interface can be used to measure $\tilde{n}$.

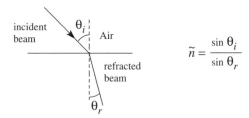

$$\tilde{n} = \frac{\sin\theta_i}{\sin\theta_r}$$

The polarizability of an element of matter is not directly accessible to experiment; it can, however, be determined from $\tilde{n}$. The Clausius–Mosotti equation relates the macroscopic dielectric constant (ε) to the molecular polarizability α:

$$\frac{\varepsilon - \varepsilon_0}{\varepsilon + 2\varepsilon_0} = \frac{4}{3}\pi N_t \alpha$$

where N_t is the number of dipoles of moment μ (see Section 6.2.1.2) and ε_0 is the dielectric constant of vacuum. Noting that ε is only slightly greater than ε_0, we may rearrange the above expression to obtain

$$\frac{\varepsilon - \varepsilon_0}{3\varepsilon_0} = \frac{4}{3}\pi N_t \alpha$$

This equation may be written in terms of the relative dielectric constant $\varepsilon_r = \varepsilon/\varepsilon_0$:

$$\varepsilon_r - 1 = 4\pi N_t \alpha$$

Since relative dielectric constants and refractive indices are related according to the Maxwell theory, $\tilde{n}^2 = \varepsilon_r$, polarizability and refractive index can be expressed one as a function of the other by

$$\tilde{n}^2 - 1 = 4\pi N_t \alpha$$

6.2.1.2. Scattering by an Isotropic Particle – Extension to a Collection of Isotropic Independent Particles (Electromagnetic Approach).

An electromagnetic wave interacting with matter and generating an electric field (E) can be expressed as

$$E = E_0 \sin 2\pi \nu t \tag{6.10}$$

where E_0 is the maximum amplitude of the incident field and ν the frequency of the field.

$$\tilde{c} = \nu \lambda_0$$

Depending upon the polarizability (α) of the particle interacting with the incident beam, a dipole moment (μ) will be induced ($\mu = \alpha E$) and this dipole will vibrate in phase with the incident field. According to the Maxwell theory, this oscillating dipole will be a secondary source of radiation of amplitude $\mathbf{E}_s$.

$\mathbf{E}_s$ will depend on the position of the observer relative to this dipole. If the vector ($\mathbf{r}$), separating this observer from this dipole, is perpendicular to μ (point A) (Figure 6.5), $\mathbf{E}_s$ is then parallel to μ and can be written as

$$\mathbf{E}_s = \frac{d^2\mu/dt^2}{\tilde{c}^2 r}$$

$\mathbf{E}_S$ is equal to 0 in the direction parallel to the dipole ($\mathbf{r}$ parallel to μ). For a random position in the space, only the perpendicular component $\mu_\perp$ has to be taken into consideration and thus the amplitude of the scattered field at this point is

$$E_s = \frac{d^2\mu_\perp/dt^2}{\tilde{c}^2 r} = \frac{d^2\mu \sin \nu}{dt^2 \tilde{c}^2 r}$$

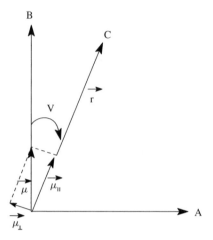

Figure 6.5. Electric field generated by an oscillating dipole and its variation according to the orientation of the dipole and the position of the observer.

In this case, E_s is parallel to $\mu_\perp$ and ν represents the angle between μ and $\mathbf{r}$.

Considering the two preceding equations, the amplitude (E_s) can also be expressed as

$$E_s = E_0\alpha^4 \frac{\sin\nu \sin 2\pi\nu t}{\tilde{c}^2 r}$$

In scattering experiments, the intensity of the scattered light is more interesting than the amplitude of the electric field. Using the Maxwell theory, we can express the intensity of light scattered by an isolated molecule or a particle in terms of the mean-square amplitude:

$$I = \frac{\tilde{c}\tilde{n}|E^2|}{4\pi}$$

Taking into consideration the preceding relations and for a harmonic field $|E^2| = E_0^2/2$, the expression of the scattered intensity is given by

$$I_d = I_0 16\pi^4\nu^4\alpha^2 \frac{\sin^2\nu}{\tilde{c}^4 r^2} = I_0 16\pi^4\alpha^2 \frac{\sin^2\nu}{\lambda_0^4 r^2} \qquad (6.11)$$

The ratio of the intensity of the scattered wave (I_d) to that of the incident wave $(I_0 = \tilde{c}\tilde{n}/4n|E_0^2|)$ multiplied by r^2 is called the *Rayleigh ratio*, after Lord Rayleigh, the "father" of light scattering theory:

$$\mathcal{R} = \frac{I_d r^2}{I_0}$$

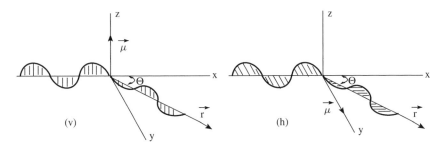

Figure 6.6. Orientation of both oscillating dipole μ and r establishing the position of the observer for a vertically (v) and a horizontally (h) polarized incident beam.

When the incident beam is vertically polarized (μ parallel to z axis) and if the observer (or the detector) is placed in the horizontal plane and at a position forming an angle θ with the axis of motion of the incident wave, the Rayleigh ratio becomes (Figure 6.6v):

$$\mathcal{R} = \frac{16\pi^4\alpha^2}{\lambda_0^4}$$

since $\sin v$ is equal to 1 whatever the value of θ.

In the case of a horizontal polarization of the beam (Figure 6.6h), the angles v and θ are complementary so that $\sin v = \cos \theta$; the Rayleigh ratio reduces to

$$\mathcal{R} = 16\pi^4\alpha^2 \frac{\cos^2\theta}{\lambda_0^4}$$

In this case, the scattered wave is also horizontally polarized and no signal can be detected for $\theta = 90°$.

When the incident beam is not polarized and if the scattered wave is simply the superposition of two vibrations moving along Ox and Oz axes, the Rayleigh ratio is defined as follows:

$$\mathcal{R} = 8\pi^4\alpha^2(\sin^2 v_x + \sin^2 v_z)/\lambda_0^4$$

which corresponds to

$$\mathcal{R} = 8\pi^4\alpha^2(1 + \cos^2\theta)/\lambda_0^4$$

In the latter case, the light scattered by an isotropic particle shows a partial or total polarization, depending on the direction under consideration. At $90°$, the scattered light is also vertically polarized.

Except for the vertically polarized incident beams that induce scattered intensities with no angular variation, an either horizontally polarized or nonpolarized

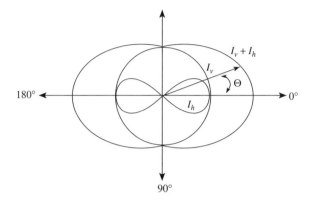

Figure 6.7. Pattern of the intensity of the beam scattered by a small isotropic particle in case of vertically (I_v), horizontally (I_h) polarized and nonpolarized (I_{v+h}) light.

radiation gives rise to an angular distribution of the scattered intensity. The intensity profile of the beam scattered by a small isotropic particle (i.e., its variation with the observation angle) is shown in Figure 6.7 for these three cases. For an assembly of N_t independent isotropic particles, the total intensity of the scattered beam is simply the sum of the N_t contributions and its Rayleigh ratio is expressed as

$$\mathcal{R} = 8\pi^4 N_t \alpha^2 (1 + \cos^2\theta)/\lambda_0^4 \tag{6.12}$$

Currently, lasers are used as source of radiation. These lasers are normally vertically polarized and the scattered intensity is detected in the plane perpendicular to the polarization direction. Under these conditions the Rayleigh ratio is equal to

$$\mathcal{R} = \frac{16\pi^4 \alpha^2}{\lambda_0^4} \tag{6.12a}$$

As the polarizability of a particle is usually not directly experimentally accessible, it is common practice to relate the Rayleigh ratio to the refractive index.

Since the refractive index of a gas of particles is close to 1 and is also concentration (c)-dependent, it can be expressed as a power series with respect to c in the form

$$\tilde{n} = 1 + \left(\frac{d\tilde{n}}{dc}\right)c$$

where

$$c = N_t M/\mathcal{N}_a,$$

and ($d\tilde{n}/dc$) is the refractive index increment, with $\tilde{n}$ and α being related by the expression

$$\tilde{n}^2 - 1 = 4\pi N_t \alpha$$

The Rayleigh ratio now becomes

$$\mathcal{R} = \frac{2\pi^2}{\lambda_0^4} \left(\frac{d\tilde{n}}{dc}\right)^2 c \frac{M}{\mathcal{N}_a} \tag{6.13}$$

Three major conclusions can be drawn from this expression:

- Light scattering and polarizability are actually related to the variations/fluctuations of the refractive index associated with the outer electrons. X-ray scattering by contrast arises from the fluctuations of the electron density, and neutron scattering is related to fluctuations in neutron scattering wavelength that reflect the fluctuations in the correlation of isotopic species.
- The scattering intensity is proportional to both the molar mass of the particle and the square of the refractive index increment. Thus, the experimenter can determine the molar mass of a molecular object by a simple scattering experiment. The scattering profile by a small isotropic particle is shown above in the three cases of vertically polarized, horizontally polarized, and nonpolarized light.
- The intensity of the scattered beam is inversely proportional to the fourth power of wavelength; that is, the shorter the wavelength of the incident source, the larger the scattering.

6.2.1.3. Scattering by an Anisotropic Particle.

Molecules are generally anisotropic. Under the effect of an electric field, the dipole induced on an anisotropic particle will be characterized by a molecular polarizability that can be described only by a third-order tensor.

The intensity of the wave scattered by such a particle is actually the average intensity of all scattered polarizations, taking into account the polarizability tensor of the particle. By splitting up the Rayleigh ratio into contributions of the vertically (V) and horizontally (H) polarized scattered light, one obtains

$$\mathcal{R}_t = \frac{V_n + H_n}{2} = \frac{V_v + V_h + H_v + H_h}{2}$$

where the indices refer to the nature of incident wave (n, nonpolarized; v, vertically polarized; h, horizontally polarized); V_n and H_n thus refer to vertically and horizontally polarized scattered light, resulting from a nonpolarized incident beam. For isotropic particles, $V_h = H_v = H_h = 0$; for anisotropic particles V_v and $V_h = H_v = H_h$ can be written as

$$V_v = \frac{16\pi^4}{\lambda_0^4} \left(\bar{\alpha}^2 + \frac{4\gamma^2}{45}\right)$$

$$V_h = H_v = H_h = \frac{16\pi^4}{\lambda_0^4} \frac{3\gamma^2}{45}$$

The average polarizability ($\bar{\alpha}$) of the particle and its anisotropy γ^2 is defined relative to the eigenvalues of the polarizability tensor (α_1, α_2, α_3),

$$\bar{\alpha} \simeq \frac{\alpha_1 + \alpha_2 + \alpha_3}{3}$$

$$\gamma^2 = \alpha_1^2 + \alpha_2^2 + \alpha_3^2 - \alpha_1\alpha_2 - \alpha_1\alpha_3 - \alpha_2\alpha_3$$

The Rayleigh ratio corresponding to the total intensity of scattered wave becomes

$$\mathcal{R}_t = \frac{8\pi^4}{\lambda_0^4}\left(\alpha^2 + \frac{13\gamma^2}{45}\right)$$

The depolarization resulting from the anisotropy of these particles can be quantified through the depolarization factor

$$\delta = \frac{H}{V} = \frac{H_v + H_h}{V_v + V_h} = \frac{6\gamma^2/45}{\alpha^2 + 7\gamma^2/45}$$

Both isotropic ($\mathcal{R}_{is}$) and anisotropic ($\mathcal{R}_t = \mathcal{R}_{is} + \mathcal{R}_{an}$) components can be distinguished in the expression of $\mathcal{R}_t$ and expressed as a function of δ. Under these conditions $\mathcal{R}_{is}$ and $\mathcal{R}_{an}$ can be written as

$$\mathcal{R}_{is} = \mathcal{R}_t(6 - 7\delta)/(6 + 6\delta) = \mathcal{R}_t C_f$$

where C_f is called the "Cabannes factor."

For natural light we have

$$\mathcal{R}_{an} = \mathcal{R}_t \frac{(3 - 4\delta)}{(3 + 3\delta)}$$
$$\mathcal{R}_{an} = \mathcal{R}_t(1 - C_f)$$

From both the measurement of the intensity scattered by a particle and its depolarization factor, its average polarizability and anisotropy can be deduced.

6.2.1.4. Interference Phenomenon of the Light Waves.

The radiation that is scattered by individual particles due to their interaction with incident light beam can interfere one with another. When such scattered waves reach a given point (occupied by a detector), the resulting electric field at this point is the sum of the fields produced by the sum of all the scattered radiation. Even if all the polarized waves can be considered as having a same wavelength and the same amplitude, they do not necessarily arrive at the detector at the same moment. If φ_i is the difference in phase between a wave i and the reference wave, the electric field produced by this wave can be written as

$$\mathbf{E}_i = \mathbf{E}_a \sin(2\pi\nu t - \varphi_i)$$

and for all the waves we have

$$\mathbf{E}_t = \sum \mathbf{E}_i = E_a \sum_{i=1}^{N_t} \sin(2\pi v t - \varphi_i)$$

Depending on the type of interference between the waves, the resulting field can be doubled (waves in phase) or canceled ($180°$—out of phase) or can take intermediate values (Figure 6.8).

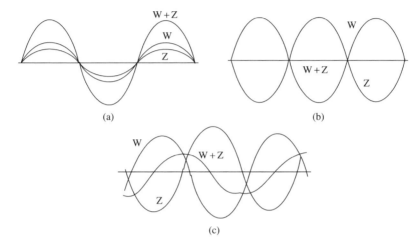

Figure 6.8. Interference between two waves ($W + Z$) in phase (a), $180°$(out of phase) (b), and with intermediate value (c).

Taking the phase differences into account, the scattering intensity from N randomly oriented individual particles can be written as

$$I_t = \frac{\tilde{c}\tilde{n}}{4\pi}|E_t|^2 = \frac{\tilde{c}\tilde{n}}{4\pi}v\int_0^{1/v}(\mathbf{E}_t \bullet \mathbf{E}_t)\,dt$$

The scalar product ($\mathbf{E}_t \bullet \mathbf{E}_t$), which is the product of two sums, can be expressed as

$$I_t = \frac{\tilde{c}\tilde{n}}{8\pi}E_a^2 \sum_{i=j}^{N_t}\sum_{j=1}^{N_t}\cos(\varphi_i - \varphi_j)$$

Should the waves exhibit the same phase φ, the double sum reduces to N_t^2; this means that the intensity of N_t in-phase waves is equal to that of only one times N_t^2.

6.2.1.5. Scattering by Large-Size Particles.

The above treatment is valid for molecules that are small ($<\lambda/20$) compared to the wavelength of the incident beam. Beyond a size that corresponds roughly to $\lambda/20$, a macromolecule cannot be considered as a single scatterer; it must instead be viewed like a collection of N_t scattering elements with the same polarizability and whose sizes are small compared to the wavelength of the incident light. Since the light scattered by one of these elements is capable of interfering with that scattered by another close element, the expression for the Rayleigh ratio of a large particle is similar to that established for an isolated particle, but includes a term taking into account the previously discussed interference phenomenon:

$$\mathcal{R} = \frac{16\pi^4}{\lambda_0^4}\alpha_n^2 \sin^2 v \sum_{i=1}^{N_t}\sum_{j=1}^{N_t}\cos(\varphi_i - \varphi_j)$$

For a macromolecule whose size is small compared to the wavelength of the incident light, φ_i angles take the same value; the double sum reduces to N_t^2; because the polarizability of such macromolecule is equal to $\alpha_n = N_t\alpha$, the previous equation reduces to that established for an isolated particle. This leads to the observation that the Rayleigh ratio is independent of the scattering angle (θ).

For a large-size particle, it is first of all necessary to consider the paths l_i and l_j followed by light when it is scattered by elements i and j of the macromolecule (Figure 6.9). These distances are arbitrarily chosen from the reference planes AA' and BB', and the path difference is $l_{ij} = l_j - l_i$.

Let $\mathbf{r}_{ij}$ be the vector connecting the element i to the element j, let $\mathbf{k}_i$ be the wave vector of the incident beam, let $\mathbf{k}_d$ be the wave vector associated with the scattered beam, and let θ be the scattering angle. The difference $\mathbf{k}_i - \mathbf{k}_d$ is the scattering vector q whose length is $q \div 2\sin(\theta/2)$.

The distance covered by the light scattered by i is equal to $l_i = a + b$; as for the path followed by the beam before being scattered by the element j, it is written

$$a + \mathbf{k}_i \cdot \mathbf{r}_{ij}$$

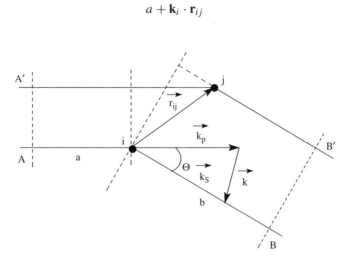

Figure 6.9. Difference in distance between the beams scattered by two elements i and j.

and between j and B' we have

$$b - \mathbf{k}_d \cdot \mathbf{r}_{ij}$$

Thus, l_{ij} reduces to

$$l_{ij} = l_j - l_i = (a + \mathbf{k}_i \cdot \mathbf{r}_{ij} + b - \mathbf{k}_d \cdot \mathbf{r}_{ij}) - (a + b) = \mathbf{k} \cdot \mathbf{r}_{ij}.$$

The existence of a path difference induces a phase difference between the two scattered beams, which can be related to l_{ij} by

$$\varphi_i - \varphi_j = 2\pi l_{ij}/\lambda$$

As the macromolecule and its constituting elements change their orientation at any moment, it is necessary to consider the average value of $\cos(\varphi_i - \varphi_j)$ over all the orientations taken by the vector $\mathbf{r}_{ij}$ with respect to the scattering vector ($\mathbf{k}$). Vectors r_{ij} and $\mathbf{k}$ respectively, express the possible conformations of the macromolecule and a specific orientation of the chain with respect to the scattered and incident beams. Introduction of the relation ($\varphi_i - \varphi_j = 2\pi l_{ij}/\lambda$) into the expression of the Rayleigh ratio, after calculation of the average (which will not be done here) gives

$$\mathcal{R} = \frac{16\pi^4}{\lambda_0^4} \alpha^2 \sin^2 v \left\langle \sum_{i=1}^{N_t} \sum_{j=1}^{N_t} \frac{\sin q r_{ij}}{q r_{ij}} \right\rangle$$

with

$$q = \frac{4\pi}{\lambda} \sin\left(\frac{\theta}{2}\right) \tag{6.14}$$

λ can be identified with the distance covered by the light in the medium of refractive index $\tilde{n}$ during the time $1/v$; λ is thus equal to $\lambda_0/\tilde{n}$. For a macromolecule of large size (higher than $\lambda/20$), the expression for the Rayleigh ratio shows that $\mathcal{R}$ depends on both of these factors—that is, on the scattering angle θ (through $\sin^2 v$) and on q.

When $\theta \to 0$, the $q r_{ij}$ term is close to 0 whatever the value of r_{ij}; consequently, the double summing reduces to N_t^2 and $\mathcal{R}_{\theta=0}$ becomes

$$\mathcal{R}_{\theta=0} = 8\pi^4 \frac{\alpha^2}{\lambda_0^4}$$

Under these conditions, effects due to the interference cancel each other.

If the angle of observation (θ) deviates from 0, an interference factor also called Zimm scattering function $P(\theta)$ can be defined as follows:

$$P(\theta) = \frac{\mathcal{R}_{(\text{observed})}}{\mathcal{R}_{(\text{without interference})}} = \frac{\mathcal{R}_\theta}{\mathcal{R}_{\theta=0}} = \frac{1}{N_t^2} \left\langle \sum_{i=1}^{N_t} \sum_{j=1}^{N_t} \frac{\sin q r_{ij}}{q r_{ij}} \right\rangle$$

$P(\theta)$, which tends toward 1 when $\theta \to 0$, is usually lower than unity. This expression shows that intramolecular interference causes a decrease in the scattered intensity. The envelope of the scattered intensity becomes unsymmetrical due to the path difference, which increases with θ: the "front" scattering is thus higher than the "rear" one. The previous expression also shows that the $P(\theta)$ function depends on the average shape of the macromolecules if the chains are flexible or on the particular geometry adopted by those with a persistent shape. For example, Debye established the expression for $P(\theta)$ of a Gaussian coil:

$$P(\theta) = (2/u^2)(u - 1 + e^{-u}) \qquad \text{with } u = q^2 \langle s^2 \rangle$$

where $\langle s^2 \rangle$ is the mean-squared radius of gyration of such a coil. In the same way, the following functions were derived for a sphere and a rod.

$$\text{Sphere}: \quad P(\theta) = \left[\frac{3}{u^3} [\sin u - u \cos u] \right]^2 \quad \text{with } u = q R_s$$

where R_s is the radius of this sphere.

$$\text{Rod (negligible thickness)}: \quad P(\theta) = \frac{1}{u_0} \int_0^{2u} \frac{\sin x}{x} dx - \frac{\sin^2(u/2)}{(u/2)^2} \quad \text{with}$$

$$u = qL,$$

where L is the rod length.

However, Guinier showed that $P(\theta)$ becomes independent of the shape of the macromolecules for $\theta \to 0$ as we have shown previously for $qr_{ij} \leq 1$. Indeed, for $qr_{ij} \leq 1$ (i.e., for q close to 0), the term $\sin qr_{ij}/qr_{ij}$ can be expressed as a MacLaurin series and can be written as

$$P(\theta) = \frac{1}{N_t^2} \left\langle \sum_{i=1}^{N_t} \sum_{j=1}^{N_t} 1 - \frac{q^2 r_{ij}^2}{3!} + \frac{q^4 r_{ij}^4}{5!} - \cdots \right\rangle$$

When limited to the first two terms, this expression gives

$$P(\theta) = 1 - \frac{q^2}{6N_t^2} \sum_{i=1}^{N_t} \sum_{j=1}^{N_t} \langle r_{ij}^2 \rangle$$

As shown in Chapter 5, $\langle r_{ij}^2 \rangle$ can be related to the radius of gyration ($\langle s^2 \rangle$) by the expression

$$\sum_{i=1}^{N_t} \sum_{j=1}^{N_t} \geq \langle r_{ij}^2 \rangle = 2 N_t^2 \langle s^2 \rangle$$

independently of the shape of the macromolecules.

After considering that all the scattering elements have the same mass, one obtains for $P(\theta)$

$$P(\theta) = 1 - q^2 \frac{\langle s^2 \rangle}{3} + \cdots$$

corresponding to

$$P(\theta) = 1 - \frac{16\pi^2}{3\lambda^2} \langle s^2 \rangle \sin^2 \left(\frac{\theta}{2} \right)$$

which gives for $P^{-1}(\theta)$

$$P^{-1}(\theta) = 1 + q^2 \frac{\langle s^2 \rangle}{3}$$

after observing that $\frac{1}{1-y} \cong (1 + y)$ when $y \leq 1$, where y is, in fact, given by

$$y = 6 \, q^2 \, \frac{\langle s^2 \rangle}{3}$$

These expressions show that the radius of gyration of a macromolecule can be deduced from the variation of $P(\theta)$ with q^2. Such a calculation is relevant only if the radius of gyration is at least higher than $\lambda/20$; indeed, under this condition, only the $q^2 \frac{\langle s^2 \rangle}{3}$ term can be non-negligible compared to 1, and $P(\theta)$ can significantly vary with θ. When undertaking such calculations, it is also necessary to bear in mind that the macromolecule should consist of identical scattering elements (which is not always the case—for example, as in block copolymers) and be homogeneous. Finally, when the dimensions of the macromolecule are very large, the $qr_{ij} \leq 1$ condition is not observed, and consequently $P^{-1}(\theta)$ does not vary linearly with $q^2\langle s^2 \rangle$. The curvature taken by the function $P^{-1}(\theta)$ then depends on the shape of the macromolecule under consideration.

The comparison between the experimental curve and those obtained theoretically for objects of specific geometrical shape (see Figure 6.10) gives an idea of the shape of the macromolecule. However, such a comparison should be used with care and only to get a tendency; indeed, $P^{-1}(\theta)$ can be affected by the dispersity of the chains.

6.2.1.6. Scattering Intensity by Macromolecular Solutions. In the preceding sections, the expression for scattering by an isolated particle and by extension by independent homogeneous particles was established; the possible interference by nonindependent particles with respect to the scattered intensity was also discussed. In a perfectly homogeneous system, the contribution of such particles to the total scattered intensity would cancel each other out due to the interference between the scattered waves. If matter scatters light, it is because of the fluctuation of the refractive index—reflecting a fluctuation of polarizability, density,

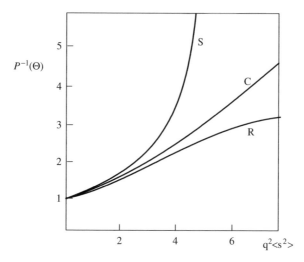

Figure 6.10. Variation of the $P^{-1}(\theta)$ function with $q^2 < s^2 >$.

concentration—on the molecular or supramolecular level. For instance, the blue of the sky or of large bodies of water results from the fluctuation in the spacing of gas or water molecules. The expression for Rayleigh ratio thus needs to be reformulated to take into account the existence of fluctuations of density, concentration, and so on, in real systems. The basic equations of light scattering by polymer solutions were established by Debye and Zimm, following a trail first explored by Smoluchowski and Einstein with their calculations of the fluctuations of refractive index in liquids.

According to the Smoluchowski–Einstein theory, the free energy (G) of a system increases spontaneously by a value (δG) in response to a fluctuation of any of its variables or characteristics. For the variable z of a system undergoing a fluctuation δz, δG is expressed as

$$\delta G = \frac{1}{2}\left(\frac{d^2 G}{dz^2}\right) z = z_0 \delta \bar{z}^2$$

Since

$$\delta \bar{G} = \frac{1}{2} kT$$

this becomes

$$\delta z^2 = \frac{kT}{(d^2 G/dz^2)_{z=z_0}}$$

Applied to a dilute gas containing N_t molecules in which an element of volume V of average dielectric constant ε would be subjected to a fluctuation $\Delta \varepsilon$, the

Smoluchowski–Einstein theory of fluctuations gives an expression for the Rayleigh ratio identical to that previously established [equation (6.12)] (see Appendix). This means that the intensity scattered by a collection of N_t particles of gas is merely the sum of the intensity scattered by individual molecules, with the scattering phenomenon resulting from the random presence/movement of a gas molecule.

APPENDIX: SCATTERING BY A GAS OF N_T MOLECULES THERMODYNAMIC APPROACH

In analogy with relation (6.11), the Rayleigh ratio for an element of volume V of average dielectric constant ε subjected to a fluctuation $\Delta\varepsilon$ can be written as

$$\mathcal{R} = \pi^2 V^2 (\Delta\varepsilon^2) \frac{\sin^2 \nu}{\lambda_0^4}$$

And for N_t elements of volume

$$\mathcal{R} = N_t \pi^2 V^2 (\Delta\varepsilon^2) \frac{\sin^2 \nu}{\lambda_0^4}$$

The determination of $\Delta\varepsilon^2$ requires the use of Smoluchowski–Einstein theory of fluctuations; the latter postulates that the free energy of a system increases spontaneously by a value $(\overline{\delta G})$ in response to a fluctuation of any of its characteristics. For the variable z of a system undergoing a fluctuation δz, its relation with δG is established as

$$\overline{\delta z^2} = \frac{kT}{(d^2 G/dz^2)_{z=z_0}}$$

If the variable is the density ρ and its fluctuation, the application of the above relation gives

$$\overline{\delta\rho^2} = \frac{kT}{d^2 G/d\rho^2}$$

After expressing the derivative of the free energy with respect to ρ, this becomes

$$\overline{\delta\rho^2} = kT\rho^2\beta/V$$

where β is the compressibility.

The fluctuation of the dielectric constant, which is related to the density by the relation $\delta\varepsilon = \frac{d\varepsilon}{d\rho}\,\delta\rho$, is given by

$$\overline{\delta\varepsilon^2} = \frac{kT\beta}{V}\left(\rho\frac{d\varepsilon}{d\rho}\right)^2$$

and thus the Rayleigh ratio becomes

$$\mathscr{R} = \frac{\pi^2}{\lambda_0^4} kT\beta \left(\rho \frac{d\varepsilon}{d\rho} \right)^2$$

Expressed as a function of the refractive index, it gives

$$\mathscr{R} = \frac{\pi^2}{\lambda_0^4} kT\beta \left(\tilde{n}\rho \frac{d\tilde{n}}{d\rho} \right)^2$$

For a gas consisting of N_t molecules, since the dielectric constant and the polarizability are related by

$$\varepsilon = 1 + 4\pi\alpha N_t$$

we obtain

$$\rho(d\varepsilon/d\rho) = 4\pi\alpha N_t$$

Thus the Rayleigh ratio reduces to

$$\mathscr{R} = N_t \frac{16\pi^2}{\lambda_0^4} \alpha^2$$

where $\rho = MN_t/\mathscr{N}_a$ and $\beta = 1/N_t kT$.

The same expression as (6.12a) is obtained, indicating that both the electromagnetic and thermodynamic approaches can be used to describe scattering by particles.

In a macromolecular solution both the density (ρ) and the concentration (c) are susceptible of fluctuations, inducing in turn those of the permittivity or dielectric constant and of the refractive index. The fluctuations of permittivity can thus be written as

$$\delta\varepsilon = \left(\frac{d\varepsilon}{d\rho} \right) \delta\rho + \left(\frac{d\varepsilon}{dc} \right) \delta c \qquad (6.15)$$

and its average quadratic values,

$$\overline{\delta\varepsilon^2} = \left(\frac{d\varepsilon}{d\rho} \right)^2 \delta\rho^2 + \left(\frac{d\varepsilon}{dc} \right)^2 \delta c^2 + 2 \left(\frac{d\varepsilon}{d\rho} \right) \left(\frac{d\varepsilon}{dc} \right) \delta\rho \, \delta c \qquad (6.15a)$$

As the fluctuations of density in the solution and in pure solvent are the same, only the second term corresponds to the excess of scattering due to the solute; as for the third term, it is equal to 0, with the fluctuations in density and concentration being independent of one another.

Since only the excess scattering is interesting for the experimenter, the Rayleigh ratio is expressed as the difference between those of the solute and of the pure solvent (6.15a):

$$\Delta \mathcal{R} = \mathcal{R}_{\text{solute}} - \mathcal{R}_{\text{solvent}} = \frac{\pi^2 V}{2\lambda_0^4} \left(\frac{d\varepsilon}{dc}\right)^2 \overline{\delta c^2}$$

The $(d\varepsilon/dc)^2$ term can be easily related to the refractive index knowing the relation $(\varepsilon = \tilde{n}^2)$ between the permittivity (ε) and the refractive index $(\tilde{n})$ and their variation with respect to the concentration:

$$\varepsilon = \varepsilon_1 + \left(\frac{d\varepsilon}{dc}\right)c \quad \text{and} \quad \tilde{n} = \tilde{n}_1 + \left(\frac{d\tilde{n}}{dc}\right)c$$

This gives

$$\left[\tilde{n}_1 + \left(\frac{d\tilde{n}}{dc}\right)c\right]^2 = \tilde{n}_1^2 + \left(\frac{d\varepsilon}{dc}\right)c$$

and thus

$$\left(\frac{d\varepsilon}{dc}\right)^2 = 4\tilde{n}_1 \left(\frac{d\tilde{n}}{dc}\right)^2$$

The mean-squared concentration fluctuation $\overline{\delta c^2}$ can be calculated using the expression relating the fluctuation of c to the free energy:

$$\overline{\delta c^2} = \frac{kT}{d^2 G/dc^2}$$

The variation of the free energy due to the change of the number of molecules of solvent (N_1) and solute (N_2) at constant pressure and temperature is defined as

$$dG = \mu_1 dN_1 + \mu_2 dN_2$$

where μ_1 and μ_2 are the chemical potentials of solvent and solute, respectively.

The concentration can be related to an element of volume V of the solution as follows:

$$c = \frac{N_2 M}{V}$$

By taking the partial molar volumes,

$$V = N_1 V_1^0 + N_2 V_2^0$$

and using the three relations given above, one obtains the following relations for the first- and second-order derivatives of the free energy with respect to the concentration c,

$$\left(\frac{dG}{dc}\right)_{T,P} = \frac{V}{M_2}\left[\mu_2 - \frac{V_2^0}{V_1^0}\mu_1\right]$$

$$\left(\frac{d^2G}{dc^2}\right)_{T,P} = \frac{V}{M_2}\left[\left(\frac{d\mu_2}{dc}\right)_{T,P} - \frac{V_2^0}{V_1^0}\left(\frac{d\mu_1}{dc}\right)_{T,P}\right]$$

With $d\mu_1$ and $d\mu_2$ being related by the Gibbs–Duhem equation

$$N_1 d\mu_1 + N_2 d\mu_2 = 0$$

the second-order derivative of the free energy with respect to the concentration c is thus expressed as

$$\left(\frac{d^2G}{dc^2}\right)_{T,P} = -\frac{V^2}{N_2 M_2 V_1^0}\frac{d\mu_1}{dc} = \frac{V}{V_1^0}\frac{d\mu_1}{dc}\frac{1}{c}$$

and the expression for the concentration fluctuation can be written as

$$\overline{\delta c}^2 = \frac{c}{V\left[-\frac{1}{kTV_1^0}\frac{d\mu_1}{dc}\right]}$$

By introducing the latter and that of $(d\varepsilon/dc)^2$ into the general formula of the Rayleigh ratio, one obtains

$$\Delta \mathcal{R} = \frac{2\pi^2}{\lambda_0^4}\frac{\tilde{n}_1^2\left(\frac{d\tilde{n}}{dc}\right)^2}{\left[-\frac{1}{kTV_1^0}\frac{d\mu_1}{dc}\right]}$$

The chemical potential can be related to osmotic pressure Π by

$$\Pi = (\mu_1 - \mu_1^0)/V_1^0$$

where μ_1^0 is the chemical potential of the pure solvent. Thus

$$\left(\frac{1}{V_1^0}\right)\frac{d\mu_1}{dc} = -\frac{d\Pi}{dc}$$

Light scattering can thus provide information about the osmotic pressure, which varies with the concentration according to

$$\Pi = RT \left(\frac{c}{M_2} + A_2 c^2 + A_3 c^3 + \cdots \right) \tag{6.16}$$

where M_2 is the molar mass of the solute and A_2 is the second virial coefficient expressing the deviation from ideality. For an ideal solution, this reduces to the van't Hoff equation (6.6).

By differenciating the osmotic pressure, one then obtains the expression for $d\mu_1/dc$:

$$\frac{d\mu_1}{dc} = -V_1^0 \, RT \left(\frac{1}{M_2} + 2A_2 c + 3A_3 c^2 + \cdots \right)$$

which, introduced in the above expression of the Rayleigh ratio, gives

$$\Delta \mathcal{R} = \frac{2\pi^2}{\lambda_0^4 \mathcal{N}_a} \tilde{n}_1^2 \left(\frac{d\tilde{n}}{dc} \right)^2 \frac{c}{\left(\frac{1}{M_2} + 2A_2 c + 3A_3 c^2 + \cdots \right)} \tag{6.17}$$

6.2.1.7. Determination of Molar Mass and Size of Macromolecules.
From the previous relation, a simpler expression of the Rayleigh ratio can be obtained with

$$K = \frac{2\pi^2}{\lambda_0^4 \mathcal{N}_a} \tilde{n}_1^2 \left(\frac{d\tilde{n}}{dc} \right)^2 \tag{6.18}$$

and thus

$$\Delta \mathcal{R} = K \frac{c}{\left(\frac{1}{M_2} + 2A_2 c + 3A_3 c^2 + \cdots \right)} \tag{6.19}$$

This expression shows that scattering increases with the concentration and the molar mass of the solute.

When the macromolecules of the solute are small ($<\lambda/20$) compared to the wavelength of the incident light, each macromolecule can be viewed as a unique scattering center. In absence of intermolecular interferences, no dissymmetry is observed at $45°$ and $135°$ and the scattering intensity can be measured only at one angle, generally $90°$.

The dissymmetry actually gives an idea of the size of the macromolecules (in the case $> \lambda/20$); this is why in a light-scattering experiment it is necessary to remove any dust because their size can be greater than that of the macromolecules and can alter the measurements—in particular, at wide angles.

At $90°$, the expression for the Rayleigh ratio reduces to

$$\frac{Kc}{\Delta \mathcal{R}_{90°}} = \frac{1}{M_2} + 2A_2 c + 3A_3 c^2 + \cdots \tag{6.20}$$

Extrapolation of $KC/\delta R_{90°}$ versus c to zero concentration allows the determination of $1/M_2$ and $2A_2$, with the two latter facts corresponding to the ordinate at the origin and the slope of this variation, respectively. For the reasons indicated earlier, the radii of gyration of macromolecules of size lower than $\lambda/20$ cannot be determined. With macromolecules of the same size as the light wavelength, intramolecular interference, which is a source of attenuation of the scattered intensity, have to be taken into account: the relation between the scattered intensity and the molar mass of the macromolecules thus requires correction, and $P(\theta)$ is introduced in the above expression to this end as shown previously. At infinite dilution ($c \to 0$), this relation reduces to

$$\frac{Kc}{\Delta R_{\theta(\text{observed})}} = \frac{1}{M_2 P(\theta)} \tag{6.21}$$

$$\Delta R_{(\text{without interferences})} = \frac{\Delta R_{\theta(\text{observed})}}{P(\theta)}$$

as $P^{-1}(\theta)$ can be expressed in the form

$$P^{-1}(\theta) = 1 + \frac{16\pi^2}{3\lambda_0^2}\langle s^2 \rangle \sin^2\left(\frac{\theta}{2}\right) = 1 + q^2 \frac{\langle s^2 \rangle}{3}$$

this leads to

$$\frac{Kc}{\Delta R_\theta} = \frac{1}{M_2}\left(1 + q^2 \frac{\langle s^2 \rangle}{3}\right) \tag{6.22}$$

This expression is valid only for $q\langle s^2 \rangle^{1/2} \leq 1$ in the Guinier approximation; indeed, only under this condition is $P(\theta)$ independent of the shape of the macromolecule as discussed earlier. For sufficiently small $\langle s^2 \rangle$, the condition $P(\theta) \to 1$ is met, but for larger $\langle s^2 \rangle$, q tends to 0 for $P(\theta)$ approaching unity.

As q represents $\frac{4\pi}{\lambda}\sin\left(\frac{\theta}{2}\right)$, data should be collected at small θ's or large λ's for q approaching 0.

At finite concentration, the preceding expression is written as

$$\frac{Kc}{\Delta R_\theta} = \frac{1}{M_2}[P^{-1}(\theta) + 2A_2 c] \tag{6.23}$$

which corresponds to

$$\frac{Kc}{\Delta R_\theta} = \frac{1}{M_2}\left(1 + \frac{q^2\langle s^2 \rangle}{3}\right) + 2A_2 c \tag{6.24}$$

Thus, when characterizing large molecules, it is necessary to extrapolate data to $q = 0$ so as to avoid the effects due to intermolecular interferences.

The molar mass can be determined from the $Kc/\Delta R_\theta$ plot through a double extrapolation, both to $c \to 0$ and to $q \to 0$.

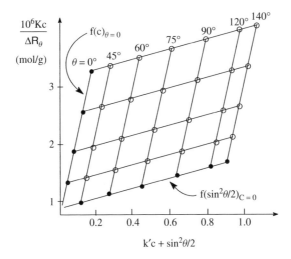

Figure 6.11. Typical Zimm plot; the slope of $Kc/\Delta\mathcal{R}_\theta$ versus c for $\theta \to 0$ gives the value of the second virial coefficient (A_2) and $f(\sin^2\theta/2)$, which allows the calculation of the radius of gyration (s).

The best way to carry out such a double extrapolation is by drawing the so-called "Zimm plot," where the variation of $Kc/\Delta\mathcal{R}_\theta$ is plotted against $(q^2 + k'c)$ (k' is an arbitrary constant whose sole purpose is to give a large spread to the grid-shaped plot) (Figure 6.11). Two series of straight lines can be drawn from experimental points, one corresponding to a same angle and the other to a same concentration; their extrapolation to $c = 0$ and to $\theta = 0$ gives $1/M_2$.

Two other pieces of information can be extracted from this plot: the slope of the straight line extrapolated to $c = 0$ yields the average radius of gyration, and the slope of the line extrapolated to $\theta = 0$ yields the second virial coefficient:

$$\frac{Kc}{\Delta\mathcal{R}_\theta}\bigg|_{\theta\to0} = \frac{1}{M_2} + 2A_2c$$

6.2.1.8. Average Values Obtained by Light Scattering. In an ideal dilute solution, intensities scattered by the various scattering centers are additive. The constant K has the same value for the same type of polymer. The Rayleigh ratio for a disperse system is given by $\mathcal{R}_\theta$:

$$\Delta\mathcal{R}_\theta = \sum_i \Delta\mathcal{R}_{\theta,i} = \sum_i Kc_iM_i\left(1 - \frac{q^2}{3}\langle s^2\rangle + \cdots\right)$$

which, after rearrangement, leads to

$$\frac{\Delta\mathcal{R}_\theta}{Kc} = \frac{\sum_i c_iM_i}{\sum_i c_i}\left(1 - \frac{q^2}{3}\frac{\sum_i c_iM_i\langle c^2\rangle}{\sum_i c_iM_i} + \cdots\right) \tag{6.25}$$

The first term of the product corresponds to the mass average molar mass and the radius of gyration is an average value.

The choice of the solvent is vital for the accuracy of the measurements. As the latter depends on $(d\tilde{n}/dc)^2$, it is essential to choose a solvent whose refractive index is quite different from that of the polymer. The increment of refractive index of a polymer in a solvent at a given wavelength can be determined by using a differential refractometer.

With the Guinier approximation ($q\langle s^2\rangle^{1/2} \leq 1$), the range of molar masses that are accessible to a light-scattering experiment is from a few thousands of g·mol^{-1} up to several millions, provided that the $d\tilde{n}/dc$ is high enough for low molar mass samples. As for the upper limit, $\langle s^2\rangle^{1/2}$ must correspond roughly to $\sim \lambda_0/2$; above that value a total destructive interference between the various parts of the same macromolecule is likely to occur.

With macromolecular systems consisting of more than one type of monomer units (statistical, block, or graft copolymers), the experimenter may be confronted with the problem of the compositional fluctuation of the macromolecules. In such a case the molar mass values measured are only apparent, if the increment of refractive index of the copolymer ($d\tilde{n}/dc_{copol}$) is deduced from the weight average of the two components A and B. Benoit suggested instead to carry out the measurements in three different solvents and to use the following relation:

$$M_{app} = \frac{(d\tilde{n}/dc)_A (d\tilde{n}/dc)_B}{(d\tilde{n}/dc)^2_{copol}} \overline{M}_w + \frac{(d\tilde{n}/dc)_A[(d\tilde{n}/dc)_A - (d\tilde{n}/dc)_B]}{(d\tilde{n}/dc)^2_{copol}} w_a \overline{M}_{w,A}$$

$$+ \frac{(d\tilde{n}/dc)_B[(d\tilde{n}/dc)_B - (d\tilde{n}/dc)_A]}{(d\tilde{n}/dc)^2_{copol}}(1 - w_a)\overline{M}_{w,B} \qquad (6.26)$$

where w_a is the mass fraction of component A.

6.2.1.9. Experimental Devices and Methodology.

A light-scattering apparatus consists of a light source, a thermostated cell placed on a goniometer which also accommodates a photomultiplier detector. The most recent apparatuses are equipped with monochromatic laser sources (He–Ne: $\lambda = 632.8$ nm or Ar: $\lambda = 488$ nm) and are able to measure the intensity of scattered waves in the range $10°-160°$. Before determining the Rayleigh ratio ($\mathcal{R}$) of a solution, it is necessary to calibrate the apparatus with a compound whose own Rayleigh ratio is known at a given wavelength; by doing this, one does not need to determine the intensity (I_0) of the incident beam. Generally, benzene is used as reference. If $\mathcal{R}_B$ is the Rayleigh ratio of this solvent and I_B is the scattered intensity at $90°$, one can write

$$\mathcal{R}_{B,90°} = \frac{I_{B,90°}}{(I_0/r^2)}$$

where I_0 is the intensity of the incident beam. Since $\mathcal{R}_B$ is a well-established number available in the literature, the Rayleigh ratio for an unspecified solution can be expressed as

$$\mathcal{R}_\theta = \frac{I_\theta}{(I_0/r^2)} = \frac{I_\theta}{I_B}\mathcal{R}_B \tag{6.27}$$

Without knowing the intensity of the incident beam (I_0), the experimenter can thus deduce $\mathcal{R}_\theta$ from the intensity (I_θ) scattered by his solution from the knowledge of $\mathcal{R}_B$, and after measuring the intensity scattered by benzene (I_B). As for $\Delta\mathcal{R}_\theta$, which corresponds to the contribution of macromolecular chains to the scattering (I_θ), its determination requires that the intensity scattered by the pure solvent (I_θ^0) be deduced and that the scattering volume be taken into account through the factor $\sin\theta$:

$$\Delta\mathcal{R}_\theta = (I_\theta - I_\theta^0)\sin\theta$$

6.2.1.10. Quasi-elastic Light Scattering.

In a static light-scattering experiment, the intensity of the light scattered is a space and time average reflecting the time average fluctuations of concentration which occur in a given volume; when examined over very short periods of time, the intensity of scattered light is actually subjected to sudden fluctuations, whose monitoring can provide invaluable information about the dynamics of the system. In particular, the analysis of the decay in intensity, which follows the scattering excess within a small element of volume, is very useful in determining the hydrodynamic properties of the polymer.

Photomultipliers equipped with very small windows to detect only the light scattered by a small element of volume (10^{-3} mm^3) are thus used for counting photons in very short intervals of time. The interval of time of each measurement is in the range of 1 μs. Under these conditions, there is a time dependence of the scattered light, which can be exploited to construct an intensity-time correlation function.

Remark. The fluctuations in the intensity of the scattered beam are inevitably accompanied by fluctuations in frequency. Interaction between the incident light and the scattering medium results in a scattered radiation whose optical spectrum has a finite width even if the incident beam is monochromatic. It is this phenomenon which is the origin of the term "quasi-elastic". If the scattered light and the incident beam have the same frequency, one can speak of "elastic" scattering; since the frequency of scattered radiation undergoes a small modification, the scattering is only "quasi-elastic."

An apparatus capable of counting photons in very short time intervals ($\Delta\tau$) was thus designed for this purpose; it compares signals recorded on such small lapses of time and measures their degree of similarity, this apparatus is called "correlator." The time between two successive countings is about 10 ns and its window of observation—comparison of two nonsuccessive signals—can be several decades. Correlators covering 12 orders of magnitude in time ($12 \, \text{ns}$ to 10^4s) now exist; significant progress was made in the comprehension of translational, internal, rotational, self-diffusing, cooperative, interdiffusive, and so on, movements (dynamics) in macromolecular systems.

If a signal recorded at moment t is compared to itself, one obtains a perfect correlation equal to 1; on the contrary, a value equal to 0 means the absence of any correlation. The comparison of two signals recorded at $(t + \Delta\tau)$, $(t + 2\Delta\tau)$, and so on, with the value measured at the origin (t) shows that the correlation often decreases in an exponential way against time down to a total decorrelation for $t = \infty$. To obtain this exponential function, it is necessary to multiply the intensities measured at t and $(t + \tau)$ and then to add them over a period of time corresponding to the observation window.

> **Remark.** The intensity-time correlation function does not go to zero. Actually, the correlation function decays to the square of the average value (as given below).

Thus, the correlator generates an autocorrelation function, which can be written as

$$G(t) = \lim_{\tau \to 0} \left[\frac{1}{T} \int_0^T I_\theta(t) I_\theta(t + \tau) d\tau \right] \tag{6.28}$$

where τ is the total time of the experiment.

This autocorrelation function can also be expressed in a normalized form:

$$g_2(t) = G(t)/G(0) \tag{6.29}$$

where $G(0)$ is the time average of the square of intensity ($\langle I_\theta^2(t) \rangle$).

The monitoring of the decrease of $g_2(t)$ with t provides information about the movement of macromolecules and, in particular, their diffusion coefficient.

In addition to the autocorrelation function of the scattered intensity, one can also define a temporal autocorrelation function of the electric field emitted by an oscillating dipole:

$$g_1(t) = \frac{\langle E(t)E(t + \tau) \rangle}{\langle I_\theta(t) \rangle} \tag{6.30}$$

where $E(t)$ is a Gaussian variable; the two autocorrelation functions $g_1(t)$ and $g_2(t)$ can be connected to each other by the Siegert formula

$$g_2(t) = 1 + \beta[g_1(t)]^2$$

where β is a constant of the instrument.

For a population of isometric objects moving through a Brownian movement and whose positions are completely uncorrelated, the variation of $g_1(t)$ can be expressed by the following exponential function:

$$g_1(t) = \exp(-\Gamma t) \tag{6.31}$$

where Γ is the relaxation frequency and corresponds to the product $q^2 D$, where q is the scattering vector length defined previously, and D is the diffusion coefficient of the objects. The diffusion coefficient depends on the concentration through the relation $D = D_0(1 + k_d c + \cdots)$, where D_0 is the diffusion coefficient at infinite dilution and k_d is a complicated function depending on both thermodynamic and hydrodynamic factors. The autocorrelation function of the scattered intensity becomes

$$g_2(t) = 1 + \beta \exp(-2\Gamma t) \qquad (6.32)$$

By successive adjustments, a correspondence between a simulated exponential curve and the experimental one $[g_1(t)]$ is used to determine D.

Since the sample consists of polymolecular chains, $g_1(t)$ can be described as a sum of exponential functions:

$$g_1(t) = \sum w_i \exp(-\Gamma_i t)$$

where w_i and Γ_i are the weight fraction of the species i and their characteristic relaxation frequency, respectively. Even if the principle is identical to the preceding case, complex algorithms are required to resolve the autocorrelation function $g_1(t)$ into a sum of exponential characteristics of objects of a given size and thus to find the best possible fitting.

With $\sum w_i \Gamma_i = \overline{\Gamma}$, the relation of the z-average diffusion coefficient D_z can be written as

$$\overline{\Gamma} = q_2 D_z \qquad (6.33)$$

The translational diffusion coefficient can also be expressed at infinite dilution as a function of the friction coefficient (ξ) through the Einstein relation:

$$D = kT/\xi \qquad (6.34)$$

where k is Boltzmann's constant. By applying Stokes' law for a spherical impermeable particle with a hydrodynamic radius R_h, one obtains

$$\xi = 6\pi \eta_0 R_h \qquad (6.35)$$

where η_0 is the viscosity of the solvent.

From a quasi-elastic light-scattering experiment, one can thus determine the hydrodynamic radius of macromolecules:

$$R_h = \frac{kT}{6\pi \eta_0 D_z} = \frac{RT}{\mathcal{N}_a 6\pi \eta_0 D_z} \qquad (6.36)$$

For sphere-shaped macromolecules, their hydrodynamic radius and volume are related by the expression $R_h = (3V/4\pi)^{1/3}$. Since their molar mass can be expressed as a function of their volume,

$$\overline{M} = \rho V \mathcal{N}_a \qquad \text{(where } \rho \text{ is the density)}$$

it is easy to show that D_z varies proportionally to $M^{-1/3}$ for sphere-shaped chains. For macromolecules with other shapes, another variation of D_z against $\overline{M}$ is observed and R_h is the hydrodynamic radius of the equivalent sphere. For coil-shaped polymers, D_z is proportional to $M^{-0.5}$ under θ conditions and is proportional to $M^{-0.6}$ in good solvents ($D_z \div KM^{-(\alpha+1)/3}$ with $\alpha = 1/2$ under θ conditions and $\alpha = 0.8$ in a good solvent). Under certain conditions, quasi-elastic scattering is thus useful for the determination of the shape of the macromolecules.

6.2.2. Small-Angle X-Ray and Neutron Scattering

In addition to elastic light scattering, neutrons and X-rays scatterings can also serve for the determination of long-distance spatial correlations. Visible light and X-rays are electromagnetic waves, and neutrons are matter waves. The three types of radiations are not redundant but are complementary.

First of all, these different types of radiation differ in the manner they interact with matter: the photons of visible light are scattered by atoms while X-rays interact with electrons and neutrons with atomic nuclei.

X-rays whose wavelength is about 0.1 nm provide interesting information through their scattering—in particular, when heavy atoms are present. The cloud of electrons of the latter is indeed more dense in the vicinity of atomic nuclei than for light elements, and therefore the scattering of X-rays will be stronger by them and hardly detectable by light elements such as hydrogen.

"Cold" (or "slow") neutrons are neutral particles with a wavelength of about 1 nm and a spin equal to 1/2, whose mass is nearly 2000 times heavier than that of the electron. These characteristics make them particularly penetrating particles that can traverse large thicknesses of matter. The intensity of neutron scattering depends on the interaction of their neutrons and spin with the concerned atomic nucleus; thus, isotopes of the same element will scatter neutron beams very differently: isotopes with a zero nuclear spin (^{12}C, ^{16}O) scatter neutrons in a "coherent" way, while other isotopes produce a "noncoherent" scattering. This noncoherent scattering occurs due to the fluctuation in the scattering amplitude from one nucleus to another; this fluctuation results from the presence of isotopes exhibiting a random spin orientation. Noncoherent scattering adds to the background noise; it is thus not useable in experiments; for instance, neutrons scattered by protons are noncoherent.

As for the coherent contribution, it results in a "forward" scattering whose angular variation can be used to determine the spatial correlations between atoms. The intensity of neutron scattering is generally expressed in terms of *cross section of effective scattering* ($d\sigma/d\Omega$) and of *scattering length* (b). $d\sigma/d\Omega$ is defined as the ratio of the scattered radiation to the incident neutron beam and is identical to the Rayleigh ratio in light scattering. It corresponds to the probability for a neutron to be scattered in a solid angle Ω per unit volume of the sample. The total scattering section can be split into two contributions, one coherent and the other noncoherent:

$$(d\sigma/d\Omega) = (d\sigma/d\Omega)_c + (d\sigma/d\Omega)_I \tag{6.37}$$

In an experiment of light or X-ray scattering, both scattered and incident radiations are always out of phase, which is also true for the neutron scattering by most of the isotopes but not for all; certain isotopes, like protons, scatter neutrons in phase.

In contrast, deuterium scatters neutrons out of phase and in a coherent manner. The concept of coherent scattering length was introduced to account for the phase of the scattered neutrons:

$$(d\sigma/d\Omega)_c = \left(\sum_i b_i \right)^2$$

b takes positive values for isotopes scattering out of phase and negative values for in-phase scattering centers. Thus, for hydrogen we have

$$b_H = -0.374 \times 10^{-12} \text{ cm}$$

and, for deuterium,

$$b_D = +0.667 \times 10^{-12} \text{ cm}$$

Under these conditions, the scattered intensity can be written as

$$I(\theta) \cong K_n M_0 c P(\theta)$$

where K_n is the contrast factor, which is expressed as

$$K_n = \left(\sum_i b_i - \sum b_j \right)^2 \frac{N_a}{M_u^2} \tag{6.38}$$

M_u is the molar mass of a repetitive unit. The first sum refers to repetitive units, and the second sum refers to solvent molecules. For neutron scattering, as in the case of light scattering, the total scattered intensity corresponds to the difference between scattering by the repetitive units and by the solvent molecules.

Recalling that in light scattering, the "contrast factor" K is written as

$$\frac{2\pi^2 \tilde{n}^2}{\lambda_0^4 N_a} (\frac{d\tilde{n}}{dc})^2$$

In X-ray scattering, the scattering cross section corresponds to

$$\frac{d\sigma}{d\Omega} = r_e^2 (\Delta Z)^2$$

where r_e is the radius of an electron (2.81×10^{-15} m) and ΔZ is the difference between the number of electrons of constitutive polymer atoms and those of solvent. ΔZ can be deduced from the electron density of the various components, $\rho_{e,u}$ (repetitive unit) and $\rho_{e,s}$ (solvent):

$$(\Delta Z)^2 = (\rho_{e,u} - \rho_{e,s})^2 V_u^2$$

where V_u is the volume of a repetitive unit. As indicated earlier, the intensity can be easily deduced,

$$I(\theta) \cong K_x M_u c P(\theta) \tag{6.39}$$

where K_x is the contrast factor, which is equal to

$$K_x = \gamma_e^2 (\rho_{e,u} - \rho_{e,s})^2 V_u^2 \frac{N_a}{M_u^2} \tag{6.40}$$

For the three scattering techniques a same overall expression is obtained for the scattered intensity

$$\frac{Kc}{I} = \frac{1}{M_w} \left(1 + \frac{\langle s^2 \rangle q^2}{3} + \cdots \right) \tag{6.41}$$

that can be used to determine $\overline{M}_w$, A_2, $\langle s^2 \rangle$. Due to their very short wavelength, neutrons and X-rays serve to measure small radii of gyration.

In order to get the same effect, the quantity corresponding to $P(\theta)$ must be the same for light scattering ($\lambda = 450$ nm) or X-ray ($\lambda = 0.1$ nm) scattering to avoid the effects due to interferences because they are detectable for sizes smaller than those measured in light scattering, that is, when $\langle s^2 \rangle^{1/2} \rangle \lambda_0/20$. This imposes that experiments be carried out at very small angles ($\theta \langle 2°\rangle$); thus in the two latter cases the detector has to be placed sufficiently away ($\rangle 1$ m) from the sample so that there is enough separation between the incident and the scattered beams.

On the other hand, these scattering techniques give information at a scale much smaller than that resulting from light scattering by photons; thus one can measure radii of gyration $\langle s^2 \rangle^{1/2}$ of $[1 - 9^2(<s^2)/3]$ much smaller macromolecules.

At these small angles, $P(\theta)$ can be approximated by the Guinier function $[1 - q^2 (\langle s^2 \rangle/3)]$, which allows the determination of the radius of gyration of macromolecules independently of their shape; the higher limit for measurements by SAXS, SANS, and light scattering is 5 nm, 20 nm, and 200 nm, respectively.

Synchrotron sources of radiation generate X-rays, whose wavelengths vary from 0.06 to 0.3 nm. With laboratory equipment, K_α radiation of Cu is used with a wavelength of 0.154 nm.

Neutrons are produced by nuclear reactors and have to be slowed down in order to obtain "cold" neutrons of about 1 nm wavelength.

6.3. MASS SPECTROMETRY APPLIED TO POLYMERS

Mass spectrometry is useful for the determination of molar masses of simple molecules. It is based on the vaporization and the ionization of the entities to be studied. Sent into an electric (or magnetic) field, each of these charged species undergoes a deflection proportional to its m/z ratio, where m represents the mass of the particle and z represents the number of charges carried.

Because of their thermal instability and their lack of volatility, polymers cannot be characterized by this technique.

The use of "MALDI" *(Matrix-Assisted Laser Desorption Ionization)* equipped with a *"time of flight"* (TOF) spectrometer offers an elegant and effective method for the precise and absolute determination of polymer molar masses. Initially designed for the characterization of biomolecules, MALDI-TOF spectrometry is now extensively applied to synthetic polymers; in this technique the polymer to be studied is dispersed in an organic matrix that is volatilized under the effect of a laser radiation whose wavelength is in the absorption range of the matrix. Owing to the specific molecular interactions between polymer and matrix—which plays an important role in the chain desorption—each polymer requires a suitable matrix. For example, 1,8,9-trihydroxyanthracene is well suited to poly(methyl methacrylate).

In order to avoid interchain entanglements, it is essential experimentally to keep the polymer concentration in the matrix very low. Upon volatilization of the matrix, the polymer species are desorbed and ionized (by a mechanism still ill-known), which makes them sensitive to the accelerating effect of an electric field. For polymer analysis, the positive mode—the one which causes the formation of a positive charge on the chain by interaction with a cationic species (proton or metal cation)—is generally used.

In order to obtain simple and easy to analyze spectrograms, chains should be monocationized. The separation of polymers ($pol - H^+$ or $pol - Met^+$) according to their molar mass is obtained through the time of flight necessary for each of the species to travel from the target to the detector. Although molar masses up to 1.5×10^6 g·mol^{-1} could be measured, most of the published studies pertain to polymers of molar mass ranging between 1000 and 2×10^4 g·mol^{-1}; this technique is an extremely sensitive one since quantities of matter in the range of femtomoles can be detected.

Problems encountered in MALDI-TOF mass spectrometry are due to:

- The low resolution beyond molar masses of about 2.5×10^4 g·mol^{-1}
- The difficulty to "cationize" certain nonpolar polymers
- The difficulty to desorb macromolecules of high molar masses

Remark. The upper limit of molar masses for well-resolved signals depends on the molar mass of the monomer unit. Beyond that limit, an envelope of the constituting signals is observed.

The example chosen to illustrate this technique of characterization is a sample of a macrocyclic polystyrene cyclized through an acetal function ($\overline{M}_n = 6900$ and $\overline{M}_w/\overline{M}_n = 1.5$) and represented here:

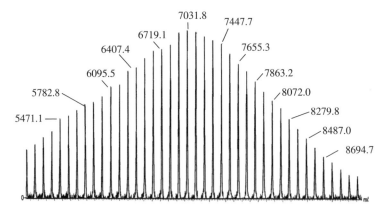

In the MALDI-TOF spectrogram shown in Figure 6.12, the exact degrees of polymerization (n) of the chains present in the samples can be seen without ambiguity through their mass and the various molecular groups carried by the chains identified; each signal corresponds to chains whose molar mass is given by the relation

$$M_{signal} = (104n + 2344 + 23)$$

where 23 represents the molar mass of Na^+ added into the matrix in order to cationize the chains.

The accuracy of this technique is of great interest for the study of the mechanisms of polymerizations.

Figure 6.12. MALDI-TOF mass spectrogram of a macrocyclic polystyrene whose formula is given above.

6.4. VISCOSITY OF DILUTE SOLUTIONS — MEASUREMENT OF MOLAR MASSES BY VISCOMETRY

Molar masses and molecular dimensions of polymers are accessible not only through scattering techniques but also from viscosity measurements. Indeed, the response of macromolecules to the application of hydrodynamic forces can give information about their volumes and their dimensions and thus indirectly about their molar masses. By definition, the viscosity of a liquid is proportional to the product of the flow time of a characteristic volume times its density:

$$\eta \div t \times \rho \tag{6.42}$$

The simplest case corresponds to a flow behavior described by the Newton law, in which the viscosity (η) is the ratio of the shear stress (σ) to the shear rate ($\dot{\gamma}$) and more specifically the slope of the straight line drawn from the variation of the shear stress versus the shear rate ($\eta = \sigma/\dot{\gamma}$). Liquids exhibiting such behavior are called Newtonian, and their viscosity is independent of the shear rate.

> **Remark.** Non-Newtonian liquids do not have this linear variation of σ versus $\dot{\gamma}$, in the entire range of shear rates.

Except for very high molar mass samples, viscosities of dilute polymer solutions are Newtonian, and the relations between the polymer molar mass and its viscosity are established in this context.

More than the proportionality constant, which relates the viscosity to the flow time, experimenters are interested in comparing the viscosity of a polymer solution with that of pure solvent (η_1). This has given rise to the notion of **relative viscosity** ($\eta_r = \eta/\eta_1$), **specific viscosity** [$\eta_{sp} = (\eta - \eta_1)/\eta_1 = \eta_\gamma - 1$], or **reduced viscosity** ($\eta_{red} = \eta_{sp}/c_2$).

The specific viscosity is an indicator of the increase in viscosity due to the addition of a polymer, whereas the reduced viscosity characterizes the propensity of a given polymer to increase the relative viscosity and is also called **intrinsic viscosity** [η] in the limit of infinite dilutions:

$$[\eta] = \left(\frac{\eta_{sp}}{c_2}\right)_{c_2 \to 0}$$

Thus, intrinsic viscosity has the dimension of a specific volume.

6.4.1. Variation of Viscosity with Concentration

The viscosity of a dispersion of sufficiently diluted rigid particles—so that their effect is thus simply additive—can be described by the Einstein viscosity relation:

$$\eta = \eta_1[1 + B_1\Phi_2 + B_2\Phi_2^2 \cdots] \tag{6.43}$$

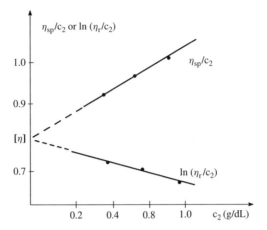

Figure 6.13. Example of extrapolation to zero concentration of the variation of η_{sp}/c_2 and $\ln(\eta_r/c_2)$ against concentration.

with $B_1 = 2.5$ when particles are nonsolvated rigid spheres and $B_2 = 14.1$; in this equation, B_1 and B_2 can take different values, depending on the shape and the size of the particles under consideration.

In the previous expression the volume fraction of the particles $\Phi_2 = V_2/V_1$ can be replaced by their mass concentration c_2, after observing that V_2 is equal to $N_2 V_H$ (i.e., the product of the number of particles times their hydrodynamic volume) and that the concentration (c_2) and the number (N_2) of these particles in the volume (V) are related by

$$N_2/V = c_2 \cdot \mathcal{N}_a/M$$

Thus, the reduced viscosity can be described by

$$\eta_{red} = 5/2(V_H \mathcal{N}_a/M) + B_2(V_H \mathcal{N}_a/M)^2 c_2 + \cdots$$

corresponding to

$$\eta_{red} \equiv \eta_{sp}/c = [\eta] + k_H[\eta]^2 c_2 + \cdots$$

where $[\eta]$ is equal to $5/2(V_H \mathcal{N}_a/M)$ and k_H, which is called the Huggins coefficient, is equivalent to $4B_2/25$. k_H can be obtained by plotting the linear variation of η_{sp}/c_2 against c_2. k_H takes values close to $1/3$ when the polymer is in a good solvent and can grow up to $0.5-1$ if a bad solvent is used. k_H is thus a criterion of the quality of solvent.

6.4.2. Relation Between Viscosity and Molar Mass of a Polymer

Long ago, Staudinger had the intuition that the molar mass of a polymer and its viscosity must be related and thus postulated that $[\eta]$ must be proportional to the

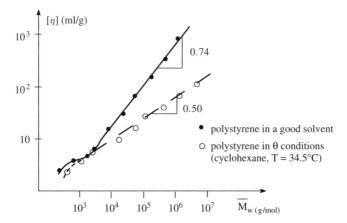

Figure 6.14. Variation of $[\eta]$ against molar mass in the case of polystyrene solutions in a good solvent and under θ conditions.

molar mass, which was subsequently proved true for rigid polymers (Figure 6.14). Actually, molar mass and viscosity could be related in an empirical manner through the Mark–Houwink–Sakurada (M–H–S) equation:

$$[\eta] = KM^\alpha \tag{6.44}$$

where K and α are constants varying with the polymer, solvent, and temperature under consideration. The value taken by the exponent α gives information about the conformation of the polymer in a given solvent and even its shape.

Thus α is equal to 0 for spheres, 0.5 for statistical coils in a nonperturbed state (θ conditions), ~0.8 for chains in solution in a good solvent, and 2 for rigid rods. For polymers with a wormlike shape, the coefficient α is intermediate between that of perturbed chains and rods—that is, between 0.8 and 2. Due to the non-Gaussian character of small chains—which are not statistical coils—the "M–H–S" equation applies strictly only to chains whose molar masses are higher than 2×10^4 g·mol^{-1}. The constants K and α for small chains in a given solvent are therefore different from those determined for Gaussian chains with the same repetitive units.

To know the type of average molar mass accessible by viscosity measurements, the following reasoning can be used; because η_{sp} is equal to $[\eta]c_2$ within the limit of low concentrations, η_{sp} can also be written as

$$\eta_{sp} = \sum_i \eta_{sp,i} = \sum_i [\eta_i]c_i$$

By dividing this equation by $c_2 \equiv \sum_i c_i$ and by considering that $w_i = c_i/c$, one obtains

$$\eta_{sp}/c_2 = [\eta] = \sum_i [\eta_i]w_i$$

The viscosity of a sample is thus the mass average of viscosities of the collection of chains present in the medium. Average molar masses obtained from viscosity experiments are called **viscosity average molar masses**.

$$\overline{M}_v = ([\eta]/K)^{1/\alpha} = \left(\sum_i w_i[\eta_i]/K\right)^{1/\alpha} = \left(\sum_i \omega_i M_i^{\alpha}\right)^{1/\alpha} \tag{6.45}$$

6.4.3. Determination of Molecular Dimensions from Intrinsic Viscosity Measurements

6.4.3.1. Case of Rigid Spheres. As previously shown, the intrinsic viscosity corresponds to a specific volume—that is, to the hydrodynamic volume of 1 g of the polymer analyzed within the limit of infinite dilutions. Insofar as relations can be established between the hydrodynamic volume of a polymer of a given conformation and its molar mass, use can be made of viscosity measurements to determine the molar mass of a sample. If the polymer analyzed is of spherical shape, its hydrodynamic volume corresponds to the volume of an equivalent sphere of radius R_{sph}:

$$[\eta] = \frac{5}{2} \frac{\mathcal{N}_a}{M} V_H = \frac{10\pi}{3} \frac{\mathcal{N}_a R_{sph}^3}{M} \tag{6.46}$$

Knowing the relation between the square of the radius of this sphere and its radius of gyration, we obtain

$$s = (3/5)(R_e^5 - R_i^5)/(R_e^3 - R_i^3)$$

where R_e and R_i are the outer and inner radii of a partially hollow sphere; in our case related to solid spheres, $R_i = 0$ and $R_e \equiv R_{sph}$; the conversion factor Q ($R_{sph} = Q_{sph,s}$) is thus equal to $(5/3)^{1/2}$ and then one obtains:

$$[\eta] = \frac{10\pi}{3} \frac{\mathcal{N}_a(5/3)^{3/2}s^3}{M} = \phi_{sph,s}\frac{s^3}{M} \tag{6.47}$$

with $\phi_{sph,s}$ equal to 13.57×10^{24} mol^{-1}.

Knowing that the relation between the molar mass of a spherical object and its radius of gyration is written as

$$s = (3/5)(4\pi\rho)\mathcal{N}_a^{1/3}M^{1/3} \tag{6.48}$$

it is easily shown that the intrinsic viscosity of a sphere is independent of its molar mass and depends only on its density.

6.4.3.2. Case of Statistical Coils. The same reasoning—that is, identifying a polymer (here a statistical coil) to an equivalent sphere of radius R—can be applied here:

$$[\eta] = \frac{5}{2}\frac{\mathcal{N}_a V_H}{M} = \frac{10\pi}{3}\frac{\mathcal{N}_a R_H^3}{M} = \frac{10\pi}{3}\frac{\mathcal{N}_a Q^3 s^3}{M} = \phi\frac{s^3}{M} \tag{6.49}$$

with $\phi = 10\frac{\pi}{3}\mathcal{N}_a Q^3$.

The problem is to establish the relation between the hydrodynamic radius and the radius of gyration of such equivalent sphere and thus to calculate Q, which is supposed to depend on the distribution of segments in the statistical coil.

Free Draining Model. In this model, known as the "Rouse model," the polymer is represented as beads interconnected by massless springs free of hydrodynamic interactions. The solvent passes freely through the statistical coil and exerts frictional forces on the segments, or the beads: each center or friction point moves independently as if the other points do not exist. The viscosity (η) is thus the product of the frictional coefficient ξ of a segment and a global factor F, which takes into account not only the friction undergone by each of the X segments but also the conformation effects:

$$F = (\rho\mathcal{N}_a/6)(\langle s^2\rangle_0/M)X$$

In dilute solutions, the density ($\rho \equiv m_{coil}/V_{coil}$) can be considered as identical to the concentration: $C_2 = m_2/V$, where m_2 is the mass of polymer and V its volume. Viscosity can be written as

$$\eta = \eta_1(\eta/\eta_1) \cong \eta_1(\eta - \eta_1)/\eta_1 = \eta_1\eta_{sp}$$

and

$$\eta/\rho \cong \eta_1\eta_{sp}/c = \eta_1[\eta]$$

This leads to the following equation:

$$[\eta] = \frac{\mathcal{N}_a\xi X}{6\eta_1}\frac{\langle s^2\rangle_0}{M} = \frac{\mathcal{N}_a\xi}{6\eta_1 M_{seg}}M = K_\eta M \tag{6.50}$$

With $\langle s^2\rangle_0/M$ being constant, it is included in K_η.

Thus, the free draining model predicts that the exponent α of molar mass is equal to 1 for chains without excluded volume as in the Staudinger empirical formula. Except for a few cases, this model is not very realistic because it neglects hydrodynamic interactions between elements of chains.

Unperturbed Statistical Coils. In the Kirkewood and Riseman model, the polymer is represented as a collection of beads interconnected by bonds of length L

and interacting with each other. This method involves the calculation of the perturbations due to the interactions between repeating units and to the long-range ones induced by the chains. Using the Oseen formula, these authors obtained

$$[\eta] = \pi^{2/3} \mathcal{N}_a (Q \cdot f(Q)) \frac{\langle s^2 \rangle_0^{3/2}}{M} = \phi \frac{\langle s^2 \rangle_0^{3/2}}{M} = \phi \left(\frac{\langle s^2 \rangle_0}{M} \right)^{3/2} M^{1/2} \qquad (6.51)$$

Through the function $Q \cdot f(Q)$, this model describes both the case of the free flow and that of impermeable chains. Depending on the degree of friction between the solvent and the chain, $Q \cdot f(Q)$ varies from values close to 0, in the case of free draining chains, to 1.26 for impermeable chains. In the latter case, ϕ is equal to 4.22×10^{24} mol^{-1}. Because the exponent α is not supposed to vary with the molar mass in this approach, it is simply assumed that the degree of permeability of the chains is related to their molar mass. The product $\Phi(\langle s^2 \rangle_0/M)^{3/2}$ then varies with M.

Statistical Coils in Perturbed Mode. This model is also called the Flory–Fox model. In the previous model, the effect of excluded volume was not taken into consideration. According to the Flory and Fox analysis, the latter model applies only to the case of unperturbed chains under θ conditions; the case of chains in a good solvent requires a separate treatment. The Flory–Fox model is based on the assumption that long-range interactions and the perturbations that they cause do not modify the flow of a solution:

$$[\eta]_\theta = \phi_\theta \frac{\langle s^2 \rangle_0^{3/2}}{M} = \phi_\theta \left(\frac{\langle s^2 \rangle_0}{M} \right)^{3/2} M^{1/2} = K_\theta M^{1/2} \qquad (6.52)$$

where ϕ_θ is a constant that is not supposed to vary with the molar mass and is thus said to be "universal." Indeed, the conversion factor, which relates the unperturbed radius of gyration to the hydrodynamic radius, is independent of the structure of the polymer, with the distribution of segments in an unperturbed polymer being independent of its structure. For the same chains under θ conditions, $\langle s_2 \rangle_0/M$ is also independent of the molar mass of the sample. Since $Q_\theta = R_{H,\theta}/s_0 = 0.87$ for such nonperturbed chains (in the case of spheres, $Q_{sph} = (5/3)^{1/2}$ as it was shown previously), ϕ_θ is equal to 4.22×10^{24} mol^{-1}, which corresponds to the value determined by Kirkewood and Riseman for impermeable chains. Figure 6.15 shows how ϕ varies in reality.

To take into consideration the continuous increase of the exponent α (in $[\eta] = KM^\alpha$) with the molar mass in a good solvent, Flory and Fox conditioned the degree of permeability of the chains to their mass, and they did it through the excluded volume effect. Considering that for perturbed chains s varies with M_2 as

$$\langle s^2 \rangle^{1/2} = KM^{3/5} \qquad (6.53)$$

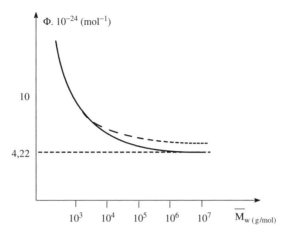

Figure 6.15. Variation of Φ, considered as an "universal" constant or Flory constant, versus the molar mass: case of polystyrene in good solvent (——) and under θ conditions (----).

which gives the following expression for the intrinsic viscosity,

$$[\eta] = \phi K^3 M^{9/5}\, M^{-1} = K_\eta M^{4/5} \tag{6.54}$$

In the latter model α can take values close to 0.8 for flexible chains in a good solvent. According to the de Gennes theory, the exponent in the relation between the molar mass and the radius of gyration is equal to 0.588 for chains with excluded volume and thus α is equal to 0.764. Because the phenomenon of excluded volume becomes more significant at higher molar masses, there is no universal value of ϕ for perturbed chains; thus ϕ differs from one system to another and decreases as the molar mass of the chains grows. The radius of gyration increases by a factor α in a good solvent according to the relation $\alpha_s = [\langle s^2 \rangle / \langle s^2 \rangle_0]^{1/2}$; the hydrodynamic radius undergoes the same phenomenon and one obtains $\alpha_H = [V_H/V_{H,\theta}]^{1/3}$. The preceding equation can thus be written as

$$[\eta] = \alpha_H^3 [\eta]_\theta = \alpha_H^3 \phi_\theta \frac{\langle s^2 \rangle_0^{3/2}}{M} = \alpha_H^3 \phi\theta = \frac{\langle s^2 \rangle_0^{3/2}}{M}(\alpha_H^3/\alpha_s^3) \tag{6.55}$$

α_H and α_s are different and can be obtained from scattering measurements for the radius of gyration and viscosity measurements for the hydrodynamic radius. For a sphere, one finds $\alpha_H^3 = \alpha_s^{2.43}$; for an ellipse, one finds $\alpha_H^3 = \alpha_s^{2.18}$; and so on.

6.4.3.3. Cases of Rods.

Examples of rigid rods among synthetic polymers are those which adopt a helical conformation either due to the size of their substituents [poly(triphenyl methacrylate), etc.] or to internal attraction forces [poly(L-γ-benzyl glutamate), etc.]. The volume of such cylindrical rods is the product of their length

L times the square of their radius R times π, that is,

$$V = \pi R^2 L$$

For such shapes, the radius of gyration is defined as

$$s^2 = \int_0^{L/2} \int_0^R \frac{2\pi y (x^2 + y^2)}{\pi R^2 L/2} dy \ dx = \frac{L^2}{12} + \frac{R^2}{12} \tag{6.56}$$

where x and y are longitudinal and radial axes, respectively. Because the volume corresponds to $V = M/N_a \rho$ (with $\rho = $ density) and since $L = V/\pi R^2$, the expression of the radius of gyration reduces to

$$s = (12^{1/2} \pi R^2 N_a \rho)^{-1} M \tag{6.57}$$

for R tending to 0. The radius of gyration of rods for which $L/R >> 0$ is thus proportional to M, which gives the following variation: $[\eta] \div M^2$.

6.4.3.4. Case of Branched Polymers.
Due to their branching points, branched polymers are characterized by hydrodynamic dimensions smaller than those of their linear counterparts. Their compactness can be assessed through the comparison of their radius of gyration with that of linear equivalents of the same molar mass under θ conditions,

$$g = \frac{\langle s^2 \rangle_{0, \text{branched}}}{\langle s^2 \rangle_{0, \text{linear}}} \tag{6.58}$$

The g parameter is thus always lower than 1. Because branched polymers are characterized by θ temperatures lower than those of their linear counterparts, the determination of $\langle s^2 \rangle_0$ by scattering techniques under θ conditions is not easy.

Comparing the radii of gyration determined in a good solvent is not necessarily the appropriate solution because branched and linear polymers swell differently. Indeed, the expansion coefficient (α) varies with the type of structure and $\alpha_{\text{branched}} < \alpha_{\text{linear}}$. In the case of star polymers, the closer to the core, the more stretched are the branch segments and at the same time a star polymer as a whole cannot swell as much as its linear equivalent.

Actually, viscosity measurements provide an easy means to compare intrinsic viscosities of polymers and to determine the compactness of branched ones. Under θ conditions, we have

$$g' = \frac{[\eta]_{\theta, \text{branched}}}{[\eta]_{\theta, \text{linear}}} \tag{6.59}$$

In good solvents, the expansion of a branched polymer is different from that of its linear equivalent as previously mentioned, and this has to be taken into account.

The sketch shown below illustrates the little impact caused by the introduction of two branches to a six-arm star on its hydrodynamic volume and hence on its

$[\eta]_{star}$; such a 33% increase (equivalent to two more branches) in the molar mass in a linear polymer would have caused a subsequent increase in $[\eta]_{lin}$.

Six-armed star polymer   Eight-armed star polymer

Various expressions predicting the variation of g' with the number of branches (f) have been proposed. The Fixman–Stockmayer model gives

$$g' = f^{2/3}[2 - f + 2^{1/2}(f - 1)]^{-3}$$

and for the Zimm and Kilb model we have

$$g' = (2/f)^{3/2}[0.39(f - 1) + 0.196]/0.586$$

These expressions yield either under- or overestimated values compared to the experimental ones. A more recent treatment that compares the intrinsic viscosity of f-branch stars to two-branch equivalent linear polymers provides better agreement:

$$G' = \frac{[\eta]_{st.}}{[\eta]_{lin.}} \tag{6.60}$$

Under θ conditions, the theoretical expression for G' can be written as

$$G' = g'(f/2)^{1/2} \tag{6.61}$$

where g' can be one of the functions described by the two previous models. Cassassa proposed a slightly modified version of the same approach that provides the best agreement with experiments. G' now becomes

$$G' = [\langle Y \rangle_{bran.}/\langle Y \rangle_{lin.}]^3 f \tag{6.62}$$

where $\langle Y \rangle$ represents $4(nfL^2/6)^{1/2} \Psi$. f corresponds to the number of branches of the star, n corresponds to the number of bonds of length L, and Ψ corresponds to a tabulated function.

For more complex and less well-defined branched structures, the following relation is the most commonly used:

$$g' \cong g^{0.5}$$

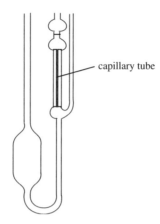

capillary tube

Figure 6.16. Ubbelohde viscometer.

Theoretical expressions predicting the variation of g for various branched architectures have been proposed. For a star polymer consisting of f branches of random length, we have

$$g = \frac{6f}{(f+1)\,(f+2)} \tag{6.63}$$

For a comb polymer including x points of branching of functionality f and p grafts, we obtain

$$g = \frac{6p^2 + (f-1)x(x^2-1)}{6p(p+1)(p+2)} \tag{6.64}$$

with $p = (f-1)x$.

6.4.4. Techniques of Measurement

It is imperative to work with solutions of precisely known concentration and free of any dust. The solutions should not be too dilute (lack of precision) or too concentrated (interchain interactions). The ideal concentration range for viscosity measurements in a good solvent corresponds to $1.1 \leq \eta_r \leq 1.4$. There are various types of capillary viscometers, with the Cannon–Ubbelohde viscometer being commonly used (Figure 6.16). Manual determination of the flow time ($>100\,\text{s}$) is sufficient for the viscosity measurements, but the greatest attention should be given to the temperature control.

6.5. APPLICATION OF SIZE EXCLUSION CHROMATOGRAPHY TO THE STUDY OF MOLAR MASSES AND THEIR DISTRIBUTION

Size exclusion chromatography (SEC) is the generic name given to the separation of macromolecules on the basis of their size, using liquid chromatography. This technique, also called "gel permeation chromatography," requires the use of a stationary phase that consists of cross-linked porous polymer beads solvated with a solvent or of an inorganic support (silica) filling a column.

Straightforward and easy to handle, SEC provides first-hand information about the distribution of molar masses of the sample analyzed. The result of an analysis by SEC is typically a graph representing the response of a concentration-sensitive or molar mass-sensitive detector (y-coordinate) placed at the column exit as a function of elution time (x-coordinate). After acquisition of the raw experimental data, there is a computer-assisted processing step that transforms the elution times into elution volumes, the latter into molar masses, and the detector response into polymer concentration and mass fraction. Figure 6.17 illustrates an SEC equipment.

While the mobile phase flows at a constant rate through a column filled with porous cross-linked polymer beads (stationary phase), the sample is injected into the mobile phase and enters the column at its top. Under the same gradient of pressure that forces the solvent to flow, the macromolecules constituting the sample also flow and travel through the column from top to bottom with residence times that vary with their size. Very large macromolecules do not penetrate the bead pores and thus travel through the interstitial volume (U_0) or exclusion volume: they are thus eluted first. As the stationary phase exhibits a broad distribution of pores of different sizes, macromolecules can thus be separated according to the pore size accessible to them. Smaller macromolecules thus reside, on average, longer times in pores than do bigger ones.

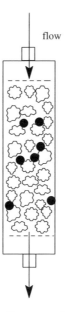

Figure 6.17. Diagrammatic representation of a SEC column.

The fractionation of macromolecules thus occurs between V_0 and V_t, which corresponds to the sum of the exclusion volume and the total pore volume. The smallest molecules are thus eluted at V_t. The elution volume of a given macromolecule of the i family is thus equal to

$$V_e = V_0 + j V_i$$

where j is the fraction of pores accessible to them.

From a theoretical point of view, any chromatography can be described through an equilibrium expressing the solute distribution between mobile and stationary phases.

Thus, any gradient of concentration of the solute that may exist implies their migration from one phase to another to restore the equal chemical potentials in the two phases. Such an equilibrium occurs in spite of the flow of the mobile phase because of the fast diffusion of the solute molecules in and out of the stationary phase. Under standard conditions, the equilibrium constant (K), which mirrors the ratio of concentrations of the solute in the two phases, can be written as

$$\Delta G^\circ = \Delta H^\circ - T \Delta S^\circ = -RT \ln K$$

In contrast to other modes of chromatography—particularly adsorption chromatography—the enthalpy term can be neglected in this case as the solute and the stationary phase are assumed not to exchange any interaction. The equilibrium constant (K) thus depends exclusively on the entropy term:

$$K \cong e^{\Delta S^\circ / R}$$

Since the number of possible conformations of a polymer chain inside a pore is lower than that in the interstitial volume (or in a dilute solution), an SEC experiment generates a decrease of entropy $(\Delta S < 0)$, with values of K lower than 1; under these conditions, solute molecules are expected to be eluted before the solvent front.

6.5.1. Techniques of Size Exclusion Chromatography

As shown in Figure 6.18, the equipment required to carry out SEC experiments is very simple. In addition to a solvent tank, this equipment comprises a pump to push the mobile phase through a filter, an injector, columns filled with the stationary phase, and finally detectors and recorders. It is through the injector that the polymer solution is introduced, which is then transported through the columns. The most commonly used detector is the concentration-sensitive differential refractometer, which detects differences in the refractive index between pure solvent and the eluted species. The response of such detector is proportional to the mass concentration of the polymer eluted independently of its molar mass. Certain SEC chromatographs are also equipped with UV, IR, or light-scattering detectors, which allow one to

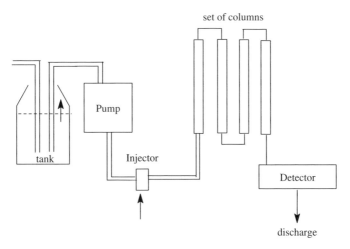

Figure 6.18. Diagrammatic representation of a SEC apparatus.

determine the molar concentration of functional groups carried by the chain or its absolute molar mass.

As mentioned above, the stationary phase used to fill the columns generally consists of beads of 5 to 10 μm size and 10 to 400 nm porosity. These beads are generally cross-linked polystyrene or modified silica for organic mobile phases. When the eluting solvent is water, poly(propylene oxide) beads or silica carrying hydrophilic groups on the surface are preferred. Typically, the experimenter injects about 25–100 μL of a polymer solution at a concentration lower than 1% and applies a flow rate of 1 mL/minute.

6.5.2. Treatment of SEC Data

The data provided by SEC—that is, the variation of the polymer concentration against the elution time/volume—can then be used to plot the curve of the sample molar mass distribution. To this end, it is necessary to establish the correspondence between the elution volume (or elution time) of polymer chains and their molar mass. This operation requires the prior establishment of a calibration curve from elution volumes of standard samples as a function of their molar mass. Standard samples of polystyrene for polymers soluble in organic solvents or poly(ethylene oxide) for water-soluble polymers are used for such calibration curves (see Figure 6.19).

Practically, the experimenter divides the curve of the detector response as a function of the elution volume into small slices of height h_i; he then utilizes the previously established calibration curve to determine the M_i corresponding to a particular V_e (see Figure 6.20). Since h_i is proportional to the concentration c_i of the eluted polymer and hence to the product $N_i M_i$ (where N_i is the number of

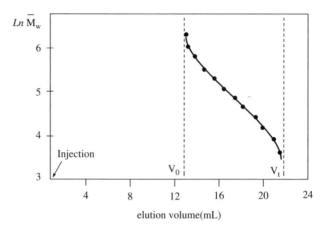

Figure 6.19. Molar mass calibration curve.

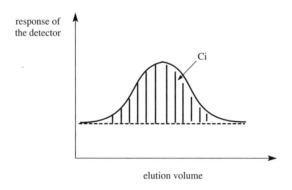

Figure 6.20. Distribution curve given by a concentration detector.

macromolecules of mass M_i), we obtain

$$\overline{M}_n = \frac{\sum_i N_i M_i}{\sum_i N_i} = \frac{\sum_i h_i}{\sum_i h_i / M_i} \tag{6.65}$$

$$\overline{M}_w = \frac{\sum_i N_i M_i^2}{\sum_i N_i M_i} = \frac{\sum_i h_i M_i}{\sum_i h_i} \tag{6.66}$$

Modern SEC apparatuses are equipped with software that calculates the average molar masses and the various moments of the distribution from a calibration curve provided by the experimenter. However, this curve applies only for a given polymer–solvent system.

Aware of this limitation, Benoit proposed a method known as "universal calibration," which is independent of the molecular structure and chemical nature of the polymer being analyzed. From the consideration that the hydrodynamic volume (V_H) of a macromolecule is proportional to the product $[\eta]M$ (more precisely,

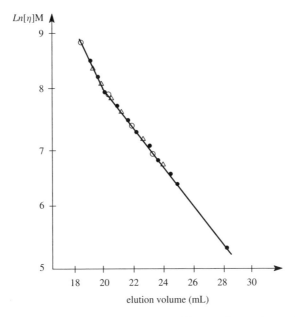

Figure 6.21. Universal calibration curve obtained from various polymer structures.

$[\eta] = c_2 \frac{V_H}{M}$) and that V_H determines the retention time in the columns, Benoit proposed to use the product $[\eta]M$ as a calibration parameter; he showed that in the $\ln[\eta]M$ versus elution volume plot, all the points fall in a unique and same curve whatever their structure (Figure 6.21). Thus, two polymers with identical elution volumes satisfy this expression:

$$[\eta]_1 M_1 = [\eta]_2 M_2$$

Using the Mark–Houwink–Sakurada expression and since K_1 and α_1 are known for the standard polymer 1 and K_2 and α_2 for the polymer analyzed, one can easily deduce M_2 with the following expression:

$$\ln M_2 = \frac{1}{1+\alpha_2}\ln \frac{K_1}{K_2} + \frac{1+\alpha_1}{1+\alpha_2}\ln M_1 \qquad (6.67)$$

However, the most recent SEC apparatuses are equipped with molar mass-sensitive detectors (of viscometer or light-scattering type), which directly yield the molar mass of the polymer analyzed; in the latter case the universal calibration is not necessary.

6.6. STUDY OF REGULAR CONFORMATIONAL STRUCTURES — PRINCIPLE OF FIBER PATTERN

The characterization of crystalline structures in polymers by X-ray diffractometry is considerably more complex than that of single crystals. Indeed, semicrystalline polymer systems do not develop three-dimensional crystalline organization of macroscopic size. On the other hand, macromolecular single crystals are so small that the conventional methods generally used with simple molecular or ionic crystals cannot be applied to them. In addition, structural irregularities at the molecular level as well as imperfections due to the inability of chains to disentangle induce complex phenomena, which makes such analyses difficult. In order to prevent the multidirectional orientation of the crystal zones in semicrystalline polymers, the sample is stretched under appropriate thermal conditions. This brings about an alignment in the direction of stretching of both the crystallites in their large dimension and the chain segments of amorphous zones; a semicrystalline fiber is thus obtained with all the related structural and physical characteristics.

Figure 5.40 illustrates the transformation undergone by an unoriented spherulite into a fibrous structure under the effect of an unidirectional stretching. Figure 6.22 schematically sketches the orientation of crystalline zones and that of the chains in such a fiber along the fiber axis; this figure also shows that crystallites preserve two degrees of freedom: one along the fiber axis and the other one perpendicular to the said axis.

If a monochromatic X-ray beam is directed perpendicularly to the axis of such a fiber, the angle between the incident beam and the crystal planes depends on the orientation of the crystallite with respect to the fiber axis. The only families of crystal planes giving rise to diffraction are those satisfying the Bragg relation,

$$2d\sin\theta = n\lambda \tag{6.68}$$

in which d is the interplanar spacing for a family of crystal plane, θ is the angle of the incident X-ray beam with the set of crystal planes, and λ is the wavelength of the incident radiation.

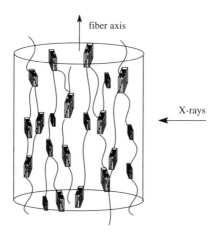

Figure 6.22. Schematic representation of the orientation of crystallites in a fiber resulting from the stretching of a semicrystalline polymer (see Figure 5.40).

The situation is thus analogous to that of a rotating crystal whose rotating axis would be identical to a crystallographic one; in the case of fibers, their axis serves as director for the crystallographic axis and the number of crystallites traversed by the beam is so high that all the orientations of crystal planes can be found—in particular, those fulfilling the Bragg conditions. The corresponding fiber pattern is thus similar to a diffraction pattern of a rotating crystal (Laue method) and can be exploited by the same guiding principles.

From the position of diffraction spots, the Miller indices (h, k, l) of the various families of crystal planes can be established and the corresponding crystal space group can be deduced. Using the parameters of the unit cell, its volume can be calculated upon combination of the volume with the crystal density, the number of constituting units per unit cell can be determined.

By measuring the intensity of the diffraction spots, one can identify the nature of the diffracting elements and thus the position of each atom of the constitutive unit. The location of other constituting units can be deduced from symmetry operations.

Obviously, X-ray diffraction allows one to determine not only the regular conformational structure of individual chains, but also that of their packing.

However, uncertainties due to the imperfect orientation of the crystallites and their defects limit the number of spots and can make their identification very difficult. In such a case, the various atoms are first assumed to occupy certain positions after considering the steric hindrance of the molecular groups and the interatomic distances; then, diffracted intensities are calculated and the values obtained are compared with experimental results. If there is no agreement, the assumed structure is false; if the agreement is satisfactory, the assumption made is close to reality. The structure first assumed is then modified until obtaining a perfect agreement (trial-and-error method).

6.7. DETERMINATION OF THE DEGREE OF CRYSTALLINITY (X)

All the specific properties of the crystalline state differ to variable extent from those of the amorphous state of the same molecular structure—that is, chemical, optical, thermal, and mechanical properties, and so on. If $\overline{P}_a$, $\overline{P}_c$, and $\overline{P}$ are the specific properties of the amorphous, crystalline, and semicrystalline states—the latter with a degree of crystallinity (X)—one can write

$$\overline{P} = X\overline{P}_c + (1 - X)\overline{P}_a$$

from which

$$X = \frac{\overline{P} - \overline{P}_a}{\overline{P}_c - \overline{P}_a} \tag{6.69}$$

can be drawn.

A priori, all the properties of matter can be used for the determination of X; in reality, only those that can be measured and are sufficiently different to afford a precise measurement are utilized.

6.7.1. X-Ray Diffraction

It is one of the rare absolute methods that can be used for the determination of the degree of crystallinity. *A priori*, it does not require any calibration since the molar transmission of X-ray beams is about the same for both amorphous and crystalline states; the beam is scattered by amorphous matter and diffracted by crystallized matter; thus, the simple comparison between diffracted and scattered intensities allows the evaluation of the degree of crystallinity (X).

Since a semicrystalline polymer sample can be assimilated to a microcrystalline powder in which chain segments of irregular conformation would play the role of amorphous cement (binding material), the Debye–Scherrer method can be used for such determination. A typical diffraction pattern is presented in Figure 6.23. The degree of crystallinity is equal to the ratio of the diffracted intensity I_c to the total intensity transmitted in the form of a coherent radiation I, integrated over the reciprocal lattice:

$$X = \frac{\displaystyle\int_0^\infty q'^2 I_c ds}{\displaystyle\int_0^\infty q'^2 I ds}$$

where

$$q' = \left(\frac{2}{\lambda}\right) \sin\theta \qquad (6.70)$$

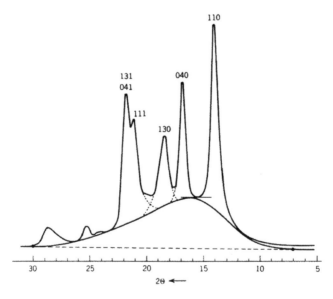

Figure 6.23. Debye–Scherrer pattern of a semicrystalline isotactic polypropylene sample. Each peak of the diffracted intensity corresponds to a family of crystal planes.

and θ is the angle of transmission of either the diffracted or the scattered beam and λ the wavelength of incident radiation.

The applicability of the above relation strictly implies the following:

1. The crystallites are randomly oriented.
2. The crystalline zones are entirely three-dimensional.
3. The scattered or diffracted intensities can be separated from the background noise.
4. The disorder function (D) resulting from thermal vibrations and the existence of defects in crystallites is known.

It was shown that this disorder function can be written in the form

$$D = \exp(-kq'^2) \tag{6.71}$$

where k is a constant that takes into consideration the nature of the imperfections of the crystalline phase.

Actually, this method of determination of the degree of crystallinity is extremely complex and inaccurate—in particular, due to the difficulty of evaluating the intensity of the background noise. Moreover, because it is long and expensive, it is rather rarely used.

6.7.2. Volumetric Method

This is a relatively precise method when applied to the comparison of polymers with the same molecular structure. In addition, it requires only simple equipment after calibration. It is based on the difference in density between the amorphous zones of the sample and the crystalline ones.

For a semicrystalline sample, one can write

$$V = V_c + V_a$$

where V_a, V_c, and V are the specific volumes of the amorphous and crystalline phases and that of the whole polymer, respectively; the preceding relation can be put in the form

$$\overline{V}_m = \overline{V}_c m_c + \overline{V}_a m_a$$

$$\overline{V} = \overline{V}_c \frac{m_c}{m} + \overline{V}_a \frac{m_a}{m}$$

where m_a, m_c, and m represent the corresponding masses.

Expressing $m_c/m = X_m$ as the degree of crystallinity, the previous relation becomes

$$\overline{V} = \overline{V}_c X_m + \overline{V}_a (1 - X_m)$$

or

$$X_m = \frac{\overline{V}_a - \overline{V}}{\overline{V}_a - \overline{V}_c} \tag{6.72}$$

With several crystalline structures

$$\overline{V} = X_{m1}\overline{V}_{c1} + X_{m2}\overline{V}_{c2} + (1 - X_{m1} - X_{m2})\overline{V}_a \tag{6.73}$$

Remark. Mass additivity of amorphous and crystalline phases instead of additivity of volumes leads to *densities* ($\overline{\rho}_c$, $\overline{\rho}_a$, $\overline{\rho}$) and *volumic degree of crystallinity* (X_v).

There is a simple relation between X_m and X_v:

$$X_m = \frac{\overline{V}}{\overline{V}_c}X_v = \frac{\overline{\rho}_c}{\overline{\rho}_a}X_v$$

The volumetric method requires the knowledge of the densities of a totally amorphous sample and that of a completely crystalline one.

To obtain $\overline{\rho}_a$, the values of the density (or of the specific volume) of a given polymer measured in the molten state need to be extrapolated to the ambient temperature (Figure 6.24).

$\overline{\rho}_c$ can be determined using an absolute method of measurement of X or can be calculated from the knowledge of the conformational structure and of the crystal cell parameters.

Due to the strong difference in density between amorphous and crystalline phases (Table 6.1), the determination of X by the volumetric method is relatively precise.

The determination of the density can be carried out with a pycnometer or using columns with gradient of density. The latter technique is currently used to measure the density of polyolefins.

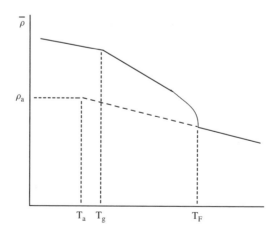

Figure 6.24. Variation of density of a semicrystalline polymer with the temperature.

Table 6.1. Density of some synthetic polymers in the amorphous state and in the crystalline state (values at 20°C in g·cm^{-1})

Polymer	$\bar{\rho}_a$	$\bar{\rho}_c$	$\Delta\bar{\rho}$
Polyethylene	0.854	1.007	0.153
Isotactic polypropylene	0.852	0.937	0.085
Syndiotactic polypropylene	0.856	0.900	0.044
Polyoxymethylene	1.215	1.491	0.276
Isotactic polystyrene	1.06	1.111	0.051
Syndiotactic poly(vinyl chloride)	1.402	1.530	0.128
Polytetrafluoroethylene	2.00	2.34	0.34
Poly(ethylene terephtalate)	1.335	1.515	0.180

6.7.3. Calorimetric Methods

Below the glass transition temperature, the heat capacities of amorphous and crystalline phases of polymers (C_{pa} and C_{pc}, respectively) are roughly identical:

$$C_p \approx C_{pa} \approx C_{pc}$$

Above T_g, a linear variation of C_p with the degree of crystallinity is sometimes observed, which provides a means to determine X.

$$C_p = X C_{pc} + (1 - X) C_{pa}$$

The determination of C_{pa} can be effected by extrapolating to room temperature the values measured beyond the melting point; that of C_{pc} can be done on single crystals, on extended chains, or for polymers having simple molecular structure by extrapolating the values obtained on model molecules. For polyethylene, the following values were obtained:

$$C_{pa} = 2.391 \ J \cdot g^{-1} \cdot degree^{-1}$$
$$C_{pc} = 1.881 \ J \cdot g^{-1} \cdot degree^{-1}$$

In this case, C_{pc} was determined from values measured on a series of n-alkane models.

In spite of the strong difference between the values of C_{pa} and C_{pc} and the precision that can be expected for the determination of X, methods based on heat capacities for temperatures ranging between T_g and T_m are scarcely utilized because of operational difficulties.

In the vicinity of the melting point, a term taking into account the enthalpy of melting (ΔH_m) has to be added to the preceding relation giving C_p:

$$C_p = X C_{pc} + (1 - X) C_{pa} + \frac{d\chi}{dT} \Delta H_m \tag{6.74}$$

Table 6.2. Enthalpy of melting of a series of n-alkanes [$C_nH_{(2n+2)}$]

n:	19	25	36	100
ΔH_F (J·g^{-1}):	222	238	255	276

In the melting zone, this term indeed becomes largely predominant, resulting in the relation

$$X = \frac{\Delta H'_m}{\Delta H_m} \tag{6.75}$$

where $\Delta H'_m$ is the enthalpy of melting of the sample studied and ΔH_m is the enthalpy of melting of a perfect single crystal of the same polymer. The main difficulty of this method lies in the determination or the estimation of ΔH_m. Indeed, various parameters come into play in the determination of ΔH_m—in particular, the thickness of the crystal layers—due to the conditions of crystallization. This could be shown on n-alkane models of polyethylene (see Table 6.2). The value of ΔH_m that is generally taken is the one extrapolated to $n = $ infinity. For PE, $\Delta H_m = 293.3$ J/g^{-1}.

Differential scanning calorimetry is the technique most often used to measure ΔH_m, because it is sensitive, precise, and fast.

6.7.4. Spectroscopic Methods

6.7.4.1. Broadband NMR. According to the fundamental relation of NMR, namely

$$h\nu = h\gamma \frac{H_0}{2\pi} = \frac{\mu}{I} H_0$$

where ν is the frequency of the electromagnetic radiation, γ is the gyromagnetic ratio that is characteristic of the nucleus, H_0 is the orienting magnetic field, μ is the magnetic moment of the nucleus, and I is its spin, a resonance phenomenon occurs at a well-defined frequency (ν) and an isolated nucleus yields a narrow signal if the relaxation process is efficient.

The situation is different for a collection of nuclei; indeed, each one of them has one magnetic moment μ, which generates a secondary magnetic field for the nuclei of its vicinity. Any nucleus is thus placed in a magnetic field (H) that corresponds to the combination of the local field created by the neighboring nuclei with the orienting magnetic field (H_0). For example, in the simple case of two nuclei, the magnetic interaction between them is given by the relation (for $I = 1/2$)

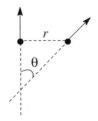

$$H_{local} = \pm 2\mu r^{-3}(3\cos^2\theta - 1)$$

In the case of an amorphous polymer, the random orientation of the dipoles results in a compensation of the magnetic effects, which gives rise to a narrow resonance signal. In the crystal state, this compensation of magnetic effects for pairs of nuclei is only partial and can lead to a broadening of the resonance signal as

$$h(v + \Delta v) = \frac{\mu(H_0 + H_{local})}{I}$$

Semicrystalline polymers can thus be assimilated to a two-phase system:

- With an amorphous phase giving a narrow signal
- With a crystalline phase yielding a broad signal

If enough time is given to nuclei for relaxation, the absorbed intensity is proportional to the number of nuclei. Upon deconvolution of the signals and a comparison of the resulting surfaces, an absolute value of the degree of crystallinity (Figure 6.25) can be determined.

6.7.4.2. Infrared Spectrometry.

Theoretical studies of the effect of the degree of crystallinity on vibration spectra account only partially for the appearance or the disappearance of certain absorption bands when comparing amorphous polymers with semicrystalline ones. However, these differences are admittedly related both to the appearance (or the disappearance) of intra- or intermolecular interactions and also to stricter selection rules for the crystalline state. In the latter case, absorption bands corresponding to interactions are better defined and thus narrower; they are fewer due to stricter rules of selection. For example, in the case of polyamide-6,6, a linear variation of the intensity of two characteristic absorption bands is observed as a function of the density and hence degree of crystallinity (Figure 6.26). One band is due to the crystalline state, and the other one is due to the amorphous state.

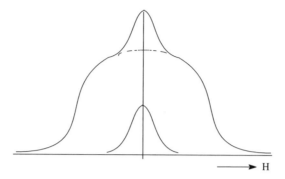

Figure 6.25. Low-resolution NMR signal of the ^1H nuclei of a polymer with high degree of crystallinity. Dashed lines represent two signals resulting from the deconvolution.

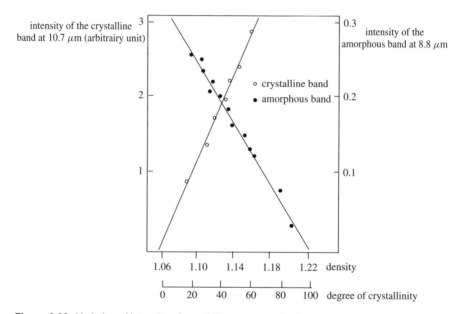

Figure 6.26. Variation of intensity of two different absorption bands for a series of polyhexam-ethyleneadipamide (PA-6,6) as a function of their density.

6.7.5. Chemical Methods

These methods are based on the higher reactivity of amorphous zones compared to that of crystalline zones. Indeed, the lesser cohesion of amorphous zones allows a better penetration of the polymer sample by the reagent.

However, each system is a particular case; for instance, acid hydrolysis of cellulose only concerns the amorphous zones and leads to a degree of crystallinity close to that measured by other methods. One can also mention:

- The selective oxidation of polyethylene by nitric acid
- The selective hydrolysis of polyamides and that of linear polyesters, and so on. The corresponding techniques are delicate to carry out.

6.8. STUDY OF SPHERULITES

6.8.1. Light Scattering

Small-angle elastic light scattering using a laser source is a convenient means to analyze the structures in the micrometer range (0.5 to 100 μm). In semicrystalline solid polymers, the crystalline parts exhibit a fluctuation in their orientation that is responsible for Rayleigh scattering. When the scattering by spherulitic structures is observed with crossed polarizer and analyzer, a pattern (H_v pattern) with a shape of a four-leaf clover is obtained. When the scattering is observed with parallel polarizer and analyzer, it arises from the fluctuations in orientation and from those in density (V_v pattern). Figure 6.27 illustrates the two types of patterns observed. These scattering patterns can be used to measure the size of spherulites. Thus, from the H_v pattern and using the equation V_V pattern H_V pattern

$$(1/\lambda)4\pi R \sin(\theta_{max}/2) = 4.1 \qquad (6.76)$$

where θ_{max} is the maximum scattering angle and λ is the wavelength of the incident radiation, one can determine the average radius of spherulites (R).

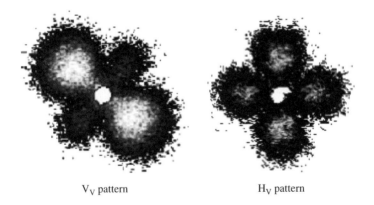

V_V pattern H_V pattern

Figure 6.27. V_v and H_v scattering patterns of a semicrystalline polyethylene sample.

6.8.2. Optical Microscopy in Polarized Light

This technique is currently used to monitor and to study the crystallization of simple molecules. Applied to semicrystalline polymers, it provides further detail regarding the chain orientation in spherulites. A typical example of micrography is shown in Figure 6.28. The pattern shows an aggregation of spherulites observed between

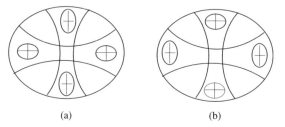

(a)   (b)

Figure 6.28. Optical sign of spherulites. (a) Positive spherulite and (b) negative spherulite.

crossed Nicols. They are delimited by a rectilinear border; a black Maltese cross oriented in a preferential direction crosses each of them. The branches of the cross correspond to the orientation of the analyzer and the polarizer. The optical sign of the observed textures is determined by the difference of radial and tangential refractive indexes of the spherulite. This difference is due to the preferential chain orientation—that is, the orientation of lamellae in the spherulite.

Figure 6.28 schematically shows two situations that can be practically found: the spherulite is said to be "positive" when $n_{radial} > n_{tangential}$ and "negative" in the opposite case; this optical sign is determined using an optical retardation plate. In any case, these images reflect a tangential orientation of the chain main axis.

6.9. DETERMINATION OF CHAIN ORIENTATION

The average chain orientation in transparent or translucent anisotropic materials can be determined through the measurement of their birefringence. Among the three refractive indices, which correspond to the three directions of space, at least two are different in an anisotropic material. The difference between two of these refractive indexes is the birefringence.

Infrared dichroism is the technique used to determine the polymer anisotropy; it does not require transparent samples and is very useful for the characterization of solid state materials. A particular vibration is said to be active as regards infrared absorption only if it is accompanied by a change in the dipole moment of the molecule solicited. Moreover, the dipole must vibrate in the same direction as that of the incident radiation for the absorption to occur.

In anisotropic materials, if vibrators are directed toward a certain direction, the absorption will depend on the respective orientation of the dipoles and the polarization of the incident beam. The dichroism ratio (R_{zy}) is defined as the ratio of the absorbances along two directions (z,y), from which the average chain orientation can be deduced. In the case of randomly orientated vibrators, there is no preferential orientation for absorption at the macroscopic level.

One can also get access to the orientation function using techniques such as wide-angle X-ray scattering, acoustic spectroscopy, NMR spectroscopy, and so on, but they are difficult to utilize.

LITERATURE

R. Xu, *Particle Characterization: Light Scattering Methods*, Kluwer Academic Publishers, Dordrecht, 2000.

S. F. Sun, *Physical Chemistry of Macromolecules—Basic Principles and Issues*, Wiley-Interscience, New York, 1994.

7

STEP-GROWTH POLYMERIZATIONS

Synthetic polymers can be derived either from the chemical modification of existing ones or through **polymerization** of simple molecules (called **monomers**). Such conversions of monomers into polymers can be obtained either through the basic condensation/addition reactions of organic chemistry or *via* chain reactions whose requirements and kinetic treatment are quite different. It leads to two categories of polymerization: step-growth polymerization and chain polymerization, which will be presented in this chapter and Chapter 8, respectively.

7.1. GENERAL CHARACTERISTICS OF STEP-GROWTH POLYMERIZATION

Step-growth polymerization involves the condensation or addition reactions of molecular entities carrying X and Y antagonist functional groups. In other words, $\mathcal{A}-\mathcal{B}$ molecules would be formed by condensation of monovalent $\mathcal{A}-X$ and $Y-\mathcal{B}$ species, and polycondensates would result from the reaction between monomers of valence equal to or greater than 2 ($v \geq 2$). Two situations are encountered in the latter case:

- The antagonist functional groups are carried by the same monomer molecules denoted $X-\mathcal{A}-Y$, whose polymerization can be schematically represented as

$$X-\mathcal{A}-Y + X-\mathcal{A}-Y \rightleftharpoons X-\mathcal{A}-\mathcal{A}-Y + XY^{\uparrow}$$

Organic and Physical Chemistry of Polymers, by Yves Gnanou and Michel Fontanille
Copyright © 2008 John Wiley & Sons, Inc.

- The two types of antagonist functions are carried by two different families of monomers {X–$\mathcal{A}$–X and Y–$\mathcal{B}$–Y}; their mutual reactions can be schematically represented as

$$X-\mathcal{A}-X + Y-\mathcal{B}-Y \rightleftharpoons X-\mathcal{A}-\mathcal{B}-Y + XY\uparrow$$

then

$$X-\mathcal{A}-\mathcal{B}-Y + X-\mathcal{A}-X \rightleftharpoons X-\mathcal{A}-\mathcal{B}-\mathcal{A}-X + XY\uparrow, \quad \text{etc.}$$

with –$\mathcal{A}$–$\mathcal{B}$–then being regarded as the resulting monomeric unit.

In the above examples, all reactants are in equilibrium with each other and with the leaving X–Y molecules. Polymerizations involving the formation of low molar mass X–Y species are called **polycondensations**.

In many cases the shift of the reaction equilibrium towards the product implies that the by-product generated be gradually eliminated as soon as it is formed. For example, in the amide formation

$$R-CO-OH + H_2N-R' \rightleftharpoons R-COO^-, {}^+NH_3-R'$$

$$\rightleftharpoons R-CO-NH-R' + H_2O\uparrow$$

the equilibrium is shifted toward the right-hand side by the elimination of water at high temperature; condensation can then proceed until total consumption of the reagents.

Thus, the **polymerizability** of a system is first of all determined by the reactivity of the X and Y antagonist sites, one with respect to the other, as well as by the propensity of the reaction under consideration to be completely driven toward the formation of the product. Equilibrium reactions that can be used to generate polycondensates are those lending themselves to the elimination of either volatile or insoluble by-products. Such a requirement considerably restricts the possibilities of choice.

There are also examples of nonequilibrium reactions. As will be shown later, it is essential that the reaction between X and Y functional groups proceed to completion for polymers to attain a high degree of polymerization. There are certain reactions, such as those leading to formation of urethanes,

$$R-OH + O=C=N-R' \longrightarrow R-O-CO-NH-R'$$

that spontaneously and even at ambient temperature comply with this criterion. Such polymerizations with no leaving by-products are called **polyadditions**.

Remarks

(a) Both polycondensations and polyadditions start with the reaction of monomers with each other and proceed subsequently through the random condensation/addition of oligomers with themselves and with residual monomers. Polymer growth in both polycondensations and polyadditions occur through multiple steps, each of them consuming two reactive functional groups:

$$X-\mathcal{A}_m-Y + X-\mathcal{A}_n-Y \longrightarrow X-\mathcal{A}_{m+n}-Y + XY\uparrow$$

with m and $n = 1, 2, 3$, etc.

The term that best captures such a polymer assembly process is **step-growth** or **stepwise polymerization**, which is characterized by a very specific relationship between the monomer conversion and the degree of polymerization.

(b) The category of polyadditions also includes substitution reactions like those occurring in cross-linking of diepoxides by primary diamines.

Whatever their size, all the polycondensates formed carry X and Y reactive groups at their end that can undergo intramolecular cyclization under dilute conditions:

In step-growth polymerizations the possibility thus exists for macrocyclic polycondensates to arise next to linear ones. Their proportion is generally negligible due a low probability of intramolecular collision between X and Y, especially when n takes high values.

When the two antagonist functional groups are carried by the same initial monomer molecule $X-\mathcal{A}-Y$ and are separated only by a limited number of bonds, cyclization may even occur preferentially. The influence exerted by the distance between $-X$ and $-Y$ is illustrated with the following series of α-hydroxy, ω-carboxylic acids; one can note in this series that the first cycle to appear is the one containing five bonds; the cycles of smaller size are too strained to arise spontaneously.

Thus, for the step-growth polymerization of $X-\mathcal{A}-Y$ to give linear polymerization, it is crucial that $-X$ and $-Y$ antagonist functional groups be located at a large distance from one another.

$$HO-(CH_2)_n-COOH$$

For $n = 1$ HO—C(=O)—OH $\longrightarrow$ O=C—O—C=O (Glycolide) + $2\,H_2O$

$n = 2$ HO—CH$_2$—C(=O)—OH $\longrightarrow$ CH$_2$=CH—C(=O)—OH (Acrylic acid) + H_2O

$n = 3$ HO—C(=O)—OH $\longrightarrow$ γ-Butyrolactone + H_2O

$n = 4$ HO—C(=O)—OH $\longrightarrow$ δ-Valerolactone + H_2O

$n = 5$ HO—C(=O)—OH $\longrightarrow$ ε-Caprolactone + polymer + H_2O

$n > 5$ HO$-(CH_2)_n-$C(=O)—OH $\longrightarrow$ Polymer

Remark. Polymers arising from cyclization of monomer molecules can also be obtained by ring-opening chain-polymerization of the corresponding heterocycles; thus

$$n \;\;\text{(γ-butyrolactone)} \;\longrightarrow\; \text{poly(ester)}$$

(see Sections 8.4 and 8.5)

The **average valence** of a reaction mixture, $\bar{v} = \Sigma_i N_i v_i / \Sigma_i N_i$, determines eventually the dimensionality and thus the structural and physical characteristics of the resulting polycondensate formed. Indeed, if $\bar{v} = 2$, the resulting polymer is monodimensional (either linear or branched) whereas if $\bar{v} > 2$, it is three-dimensional.

It is particularly important at this stage to make a distinction between **valence** and **functionality** of a given reactive functional group. In step-growth polymerizations, a same functional group can indeed exhibit a valence that can vary with the reaction under consideration. For instance, primary diamines (bifunctional molecules) behave as divalent species in polyamidations and as tetravalent species in the cross-linking of diepoxides. In the same way, cyclic acid anhydrides react as a monovalent species in imidation and as a divalent reactant in esterification. The corresponding reaction mechanisms are given in Section 7.4.

7.2. POLYMERIZATION OF BIVALENT MONOMERS

7.2.1. Carothers Relation for Self-Condensation of X–A–Y-Type Monomers

To establish the relation—also called the Carothers relation—between the number average degree of polymerization $(\overline{X}_n)$ and the *extent of reaction* (p), it is necessary first to define these two quantities. The extent of reaction is defined as the fraction of reacted functional groups at that instant (t). If all molecules are bifunctional and bivalent and if N_0 and N_t are the number of (monomer **or** polymer) molecules present in the reaction medium at time 0 and t, respectively, the extent of reaction can be written as

$$p = 2(N_0 - N_t)/2N_0$$

which gives

$$N_t = N_0(1 - p)$$

Since the number average degree of polymerization $(\overline{X}_n)$ is equal to the ratio of the total initial number of monomeric units (which itself is equal to the number of initial monomer molecules) to the total number of molecules present in the system at time t (N_t) irrespective of their degree of polymerization, then

$$\overline{X}_n = \frac{N_0}{N_t}$$

and

$$\overline{X}_n = \frac{1}{1 - p}$$

The Carothers relation shows that a high degree of polymerization can be obtained only for an extent of reaction close to unity. Thus, to obtain

$\overline{X}_n = 20$, p should attain a value of $p = 0.950$
$\overline{X}_n = 50$, p should attain a value of $p = 0.980$
$\overline{X}_n = 200$, p should attain a value of $p = 0.995$

Thus, properties specific of the macromolecular state can only be observed for high degrees of polymerization. In other words, only polyadditions reaching high conversion or equilibrium reactions that can be driven toward the complete formation of the product can be used in step-growth polymerizations. Volatile (H_2O, HCl, etc.) or insoluble by-products must be eliminated in the latter case.

Side reactions and the presence of reactive monovalent molecules in the reaction medium (see page 220) must be avoided if long chains and hence a high degree of polymerization $(\overline{X}_n)$ are desired.

The **distribution of molar masses** of the polycondensates formed throughout polymerization can be calculated using statistical methods. Since all unreacted functional groups are assumed to be equally reactive and behave independently of one another, the probability that a reaction occurs to link any $-X$ to any $-Y$ is equal to the extent of reaction (p)

$$p = \frac{N_0 - N}{N_0}$$

and the probability that no link exists is equal to $(1 - p)$.

Following a similar reasoning, the probability of existence or appearance of i-mers in the reaction medium must be equal to their molar fraction. These i-mers contain $(i - 1)$ bonds that result for the reaction of $(i - 1)$ X groups (with as many Y), and only one X group (and one Y group) is left unreacted. Hence the probability of existence of the i-mers is given by the product of the probability ($p^{(i-1)}$) of formation of $(i - 1)$ bonds through consumption of i monomer molecules and the probability $(1 - p)$ for one group left unreacted.

$$f_i = p^{(i-1)}(1 - p)$$

N_t being the total number of reactive molecules present at time t, the number of i-mers is equal to

$$N_i = N_t f_i = N_t p^{(i-1)}(1 - p)$$

and since

$$N_t = N_0(1 - p)$$

then

$$N_i = N_0 p^{(i-1)}(1 - p)^2$$

From this relation, the proportion of residual monomer molecules left unreacted for a given extent of reaction p can be deduced by taking $i = 1$ (corresponding to monomer molecules). Thus, for $p = 0.90$, 1% of unreacted monomer is present in the reaction medium.

If the molar mass of the monomer units is denoted by M_0, one can calculate the mass fraction w_i of the i-mers:

$$w_i = \frac{i M_0 N_i}{M_0 N_0} = i \frac{N_i}{N_0} = i p^{(i-1)}(1 - p)^2$$

Figures 7.1 and 7.2 represent the variation of the number (N_i) and mass (w_i) distributions of macromolecules present in the system as a function of their degree of polymerization, respectively.

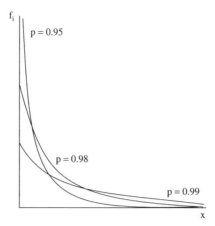

Figure 7.1. Variation of the molar fraction of *i*-mers as a function of their degree of polymerization for various extents of reaction.

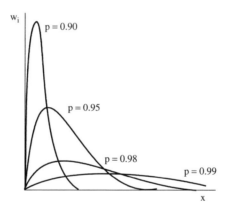

Figure 7.2. Variation of the mass fraction of the *i*-mers as a function of their degree of polymerization for various extents of reaction.

When p increases, one can observe that the molar mass distribution, which is best expressed by the polymolecularity (dispersity) index $D_M = \overline{M}_W/\overline{M}_n$, tends to widen.

$$\overline{M}_n = \frac{\sum_{i=1}^{i=\infty} i M_0 N_i}{\sum_{i=1}^{i=\infty} N_i} = \frac{M_0}{1-p}$$

knowing that

$$\sum_{i=1}^{i=\infty} i p^{(i-1)} = \frac{1}{(1-p)^2}$$

and

$$\overline{M}_w = \frac{\sum_{i=1}^{i=\infty} i^2 M_0^2 N_i}{\sum_{i=1}^{i=\infty} M_0 N_i} = \frac{M_0(1+p)}{1-p}$$

since

$$\sum_{i=1}^{i=\infty} i^2 p^{(i-1)} = \frac{1+p}{(1-p)^3}$$

thus

$$D_M = \overline{M}_w / \overline{M}_n = (1+p)$$

and tends toward 2 when p tends toward 1.

The addition of monovalent molecules, such as X–$\mathcal{A}$ or Y–$\mathcal{B}$, induces a deviation from the Carothers relation because they form nonreactive chain ends of $\sim\sim\mathcal{A}$ or $\sim\sim\mathcal{B}$ type. Addition of monovalent molecules is currently used to limit the degree of polymerization to a previously selected value. If N_1 represents the number of $\mathcal{B}$–Y monovalent molecules used in excess and $r = N_1/N_2$ the stoichiometric imbalance with respect to N_2, then the number average degree of polymerization can be written as

$$\overline{X}_n = (1+r)/(1-p+r)$$

7.2.2. Polymerization of Bivalent Monomers of (X–$\mathcal{A}$–X + Y–$\mathcal{B}$–Y) Type

Let us consider a reaction mixture containing N_X molecules of X–$\mathcal{A}$–X and N_Y molecules of Y–$\mathcal{B}$–Y in a ratio (r) corresponding to the stoichiometric imbalance ($r = N_X/N_Y$). The extent of reaction of the X and Y functional groups is defined as

$$p_X = 2\left(\frac{N_X - N_{X,t}}{N_X}\right) \quad \text{and} \quad p_Y = 2\left(\frac{N_Y - N_{Y,t}}{N_Y}\right)$$

After each step of polymerization, there is a disappearance of one molecule from the reaction medium and as a result growth of one chain. Each link created between X and Y contributes to the consumption of one X functional group and one Y functional group. Therefore,

$$N_X p_X = N_Y p_Y$$

so that the relation between the total numbers of molecules initially present (N_0) and present at time t(N_t) can be written as

$$N_0 = N_X + N_Y = N_X(1 + p_X/p_Y)$$

$$N_t = N_X(1 - 2p_X + p_X/p_Y)$$

Again it will be assumed for the calculation of the size distribution, which is developed below, that all functions of the same type are equally reactive, that they react independently of one another whether they are carried by simple molecules or by large ones, and that no cyclization reactions occur in finite species. Flory used combinatorial arguments to perform this calculation. In this type of step-growth polymerization, three varieties of polycondensates may be formed in the reaction mixture:

- Polycondensates composed of an even number of repeating units
- Polycondensates containing an odd number of repeating units and ended by X functional groups
- Polycondensates composed of an odd number of repeating units and ended by Y functional groups

For instance, in the family of polycondensates comprising i X–$\mathcal{A}$–X monomeric units, three kinds of molecules can be found in the medium as shown below:

$$\text{X–}\mathcal{A}\text{–X–(Y–}\mathcal{B}\text{–Y–X–}\mathcal{A}\text{–X)i} - 1\text{–Y–}\mathcal{B}\text{–Y}$$

$$\text{X–}\mathcal{A}\text{–X–Y–(}\mathcal{B}\text{–Y–X–}\mathcal{A}\text{–X–Y–}\mathcal{B}\text{–Y)i} - 2\text{–X–}\mathcal{A}\text{–X}$$

$$\text{Y–}\mathcal{B}\text{–Y–(X–}\mathcal{A}\text{–X–Y–}\mathcal{B}\text{–Y–)i} - 1\text{X–}\mathcal{A}\text{–X–Y–}\mathcal{B}\text{–Y}$$

Remark. To better explain the subsequent calculation and demonstration, monomeric units constituting the polycondensates are represented with the functional groups X and Y they carried before the reaction unlike previously used. This notation will be limited to this paragraph only.

The calculation of the distribution function begins with the determination of the probability (π_a) that a X–$\mathcal{A}$–X monomer molecule selected at random is part of a particular polycondensate. In the example that follows, a X–$\mathcal{A}$–X monomeric unit is selected at random in a polycondensate containing a total of 3 of these X–$\mathcal{A}$–X units.

$$\text{X–}\mathcal{A}\text{–X–Y–}\mathcal{B}\text{–Y–}\textbf{X–}\mathcal{A}\textbf{–X–}\text{Y–}\mathcal{B}\text{–Y–X–}\mathcal{A}\text{–X–Y–}\mathcal{B}\text{–Y}$$

$$\underbrace{\quad}_{1-p_X}\ \underbrace{\quad}_{p_Y}\ \underbrace{\quad}_{p_X}\ \underbrace{\quad}_{p_X}\ \underbrace{\quad}_{p_Y}\ \underbrace{\quad}_{p_X}\ \underbrace{\quad}_{1-p_Y}$$

Since the monomeric unit selected happens to be a X–$\mathcal{A}$–X, the probability that its two functional groups react and establish a linkage is p_X^2, and the probability that the two Y groups introduced react is p_Y^2.

The probability that any given X–$\mathcal{A}$–X monomeric unit is a component of such an even hexamer is therefore

$$\pi_{3,\text{even}} = 3 \cdot 2 p_X^3 p_Y^2 (1 - p_X)(1 - p_Y)$$

where the factor 6 represents all the possibilities of constructing such a hexamer starting from a X–A–X monomer molecule.

By an identical reasoning, one can determine the probabilities ($\pi_{3,\text{odd,X}}$ and $\pi_{3,\text{odd,Y}}$) that X–A–X leads to or is a component of a polycondensate consisting of 3 X–A–X units and a total of 5 or 7 units:

$$\pi_{3,\text{odd,X}} = 3p_X^2 p_Y^2 (1 - p_X)^2$$

$$\pi_{3,\text{odd,Y}} = 3p_X^4 p_Y^2 (1 - p_Y)^2$$

By generalizing this reasoning to polycondensates containing i X–A–X units, we obtain for the probability (π_i) that any given X–A–X is contained in that particular family

$$\pi_{i,\text{even}} = 2ip_X^i p_Y^{(i-1)} (1 - p_X)(1 - p_Y)$$

$$\pi_{i,\text{odd,X}} = ip_X^{(i-1)} p_Y^{(i-1)} (1 - p_X)^2$$

$$\pi_{3,\text{odd,Y}} = ip_X^{(i+1)} p_Y^{(i-1)} (1 - p_Y)^2$$

The determination of the molar fraction of the three kinds of species containing i X–A–X units requires the prior calculation of the total number of units that belong to that family of polycondensates. The number (N_i,even) of polycondensates that contain i X–A–X units and contain an even number of units can be written as

$$iN_{i,\text{even}} = N_X \pi_{i,\text{even}}$$

$$N_{i,\text{even}} = 2N_X p_X^i p_Y^{(i-1)} (1 - p_X)(1 - p_Y)$$

The molar fraction of these particular species and that of the family of polycondensates with i X–A–X units can then be easily deduced:

$$f_{i,\text{even}} = \frac{N_{i,\text{even}}}{N_t} = \frac{2p_X^i p_Y^{(i-1)} (1 - p_X)(1 - p_Y)}{(1 - 2p_X + p_X/p_Y)}$$

For $f_{i,\text{odd,X}}$ and $f_{i,\text{odd,Y}}$ we obtain

$$f_{i,\text{odd,X}} = \frac{p_X^{(i-1)} p_Y^{(i-1)} (1 - p_X)^2}{1 - 2p_X + p_X/p_Y}$$

$$f_{i,\text{odd,Y}} = \frac{p_X^{(i+1)} p_Y^{(i-1)} (1 - p_Y)^2}{1 - 2p_X + p_X/p_Y}$$

f_i, the molar fraction of the family containing i X–A–X units, can be written as

$$f_i = f_{i,\text{even}} + f_{i,\text{odd X}} + f_{i,\text{odd Y}} = \frac{(p_X p_Y)^{(i-1)} p_Y (1 - p_X p_Y)^2}{p_X (1 - p_Y) + p_Y (1 - p_X)}$$

The sum of all f_i $\left(\sum_{i=1}^{\infty} f_i\right)$ to which f_y has to be added is equal to 1 by defini-tion. f_y which represents the molar fraction of residual monomer Y–$\mathscr{B}$–Y, can be written as

$$f_Y = \frac{N_Y(1 - p_Y)^2}{N_t} = \frac{p_X/p_Y(1 - p_Y)^2}{(1 - 2p_X + p_X/p_Y)}$$

The expressions for the mass fractions of the three kinds of polycondensates that contain i units originating from X–$\mathscr{A}$–X can be obtained in a similar way:

$$w_{i,\text{even}} = \frac{f_{i,\text{even}} N_t (i M_{0,X} + i M_{0,Y})}{N_X M_{0,X} + N_Y M_{0,Y}} = \frac{f_{i,\text{even}} N_t (i M_{0,X} + i M_{0,Y})}{N_X [M_{0,X} + M_{0,Y}(N_Y/N_X)]}$$

where $M_{0,X}$ and $M_{0,Y}$ correspond to the molar mass of monomer units originating from X–$\mathscr{A}$–X and Y–$\mathscr{B}$–Y, respectively.

$$w_{i,\text{odd},X} = \frac{f_{i,\text{odd}\,X} N_t [i M_{0,X} + (i - 1) M_{0,Y}]}{N_X [M_{0,X} + M_{0,Y}(N_Y/N_X)]}$$

$$w_{i,\text{odd},Y} = \frac{f_{i,\text{odd}\,Y} N_t [i M_{0,X} + (i + 1) M_{0,Y}]}{N_X [M_{0,X} + M_{0,Y}(N_Y/N_X)]}$$

The mass fraction of all species containing i monomer units originating from X–$\mathscr{A}$–X reads as

$$w_i = 2(1 - p_X)(1 - p_Y)(p_X p_Y)^i \frac{i M_{0,X} + i M_{0,Y}}{p_Y M_{0,X} + p_X M_{0,Y}}$$

$$+ (1 - p_X)^2 p_X^{(i-1)} p_Y^i \frac{i M_{0,X} + (i - 1) M_{0,Y}}{p_Y M_{0,X} + p_X M_{0,Y}}$$

$$+ (1 - p_Y)^2 p_X^{(i+1)} p_Y^i \frac{i M_{0,X} + (i + 1) M_{0,Y}}{p_Y M_{0,X} + p_X M_{0,Y}}$$

The mass fraction of residual Y–$\mathscr{B}$–Y molecules is given by

$$w_y = \frac{p_X(i - p_Y)^2 M_{0,Y}}{p_Y M_{0,X} + p_X M_{0,Y}}$$

From these quantities, expressions for the number average $(\overline{M}_n)$ and mass average $(\overline{M}_W)$ molar masses can be easily established as

$$\overline{M}_n = \frac{\text{Total mass of reactants}}{\text{Total number of molecules at time } t}$$

$$\overline{M}_n = \frac{N_X M_{0,X} + N_Y M_{0,Y}}{N_X(1 - 2p_X + p_X/p_Y)} = \frac{p_X M_{0,Y} + p_Y M_{0,X}}{p_Y - 2p_X + p_Y + p_X}$$

Knowing that

$$\overline{M}_w = \frac{\sum_{i=1}^{\infty} N_i M_i^2}{\sum_{i=1}^{\infty} N_i M_i}$$

and that

$$N_i = N_{i,\text{even}} + N_{i\text{odd},X} + N_{i,\text{odd},Y}$$

$\overline{M}_W$ can be written as

$$\overline{M}_w = \frac{1 + p_X p_Y}{1 - p_X p_Y}\left(\frac{p_Y M_{0,X}^2 + p_X M_{0,Y}^2}{p_Y M_{0,X} + p_X M_{0,Y}}\right) + \frac{4 p_X p_Y M_{0,X} M_{0,Y}}{(1 - p_X p_Y)(p_Y M_{0,X} + p_X M_{0,Y})}$$

which also takes into account the contribution of residual molecules Y–$\mathcal{B}$–Y.

Figure 7.3 shows the importance of using one of the two monomers in excess. The polycondensates formed under these conditions are end-capped with the two same functional groups originating from the monomer used in excess; it is a practical tool for limiting the chain length to a precise molar mass.

Under stoichiometric conditions for which $N_X = N_Y$ and $p_X = p_Y = p$, very simple expressions are obtained:

$$\overline{M}_n = \left(\frac{M_{0,X} + M_{0,Y}}{2}\right)\left(\frac{1}{1 - p}\right) \quad \text{and} \quad \overline{M}_w = \left(\frac{M_{0,X} + M_{0,Y}}{2}\right)\left(\frac{1 + p}{1 - p}\right)$$

In this case, the expression for the dispersity index (D_M) is identical to that established for a polymerization of the X–$\mathcal{A}$–Y type, namely $D_M = (1 + p)$.

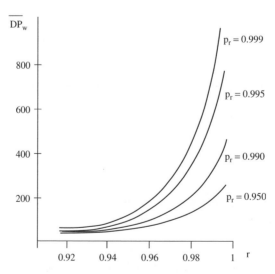

Figure 7.3. Variation of the mass average degree of polymerization as a function of the stoichiometric imbalance ratio (r) for a step-growth polymerization of {X–$\mathcal{A}$–X + Y–$\mathcal{B}$–Y} type and various extents of reaction.

7.3. POLYMERIZATION OF MONOMERS WITH AVERAGE VALENCE HIGHER THAN 2 — "GEL POINT"

The kinetic treatment developed in Section 7.4 for divalent monomers applies to systems with $\bar{v} > 2$. In the latter case, macromolecular chains grow simultaneously in many directions and each one carries a growing number of reactive functional groups. A three-dimensional network is eventually formed after the conversion reaches a critical value called the **gel point**. In such a tridimensional sample, only a small fraction of polymer, which decreases with conversion, is soluble and the insoluble part is considered to have infinite molar mass.

> **Remark.** For monomer molecules whose molar mass does not exceed 100 or 200 g·mol^{-1}, it is difficult to attain high extents of reaction for $\bar{v} > 3$. Indeed, the network density increases rapidly and hinders the movement of the reactive groups, thus preventing progress of the reaction beyond the gel point.

The gel point can be approximately predicted through an approach similar to that used to establish the Carothers relation.

Taking into consideration the multivalence of monomers and the fact that each reaction between two reactive functional groups decreases the number of molecules (N) present in the reaction medium by one unit, the extent of reaction can be written as

$$p = \frac{2(N_0 - N)}{N_0 \bar{v}}$$

where $2(N_0 - N)$ is the number of reacted functional groups and $N_0 \bar{v}$ is the number of initial reactive functional groups. Solving for N gives

$$N = N_0 \left(1 - \frac{p\bar{v}}{2}\right)$$

The number average degree of polymerization is equal to

$$\overline{X}_n = \frac{N_0}{N}$$

and, thus, using the expression for N above, we obtain

$$\overline{X}_n = \frac{N_0}{N_0 \left(1 - \dfrac{p\bar{v}}{2}\right)} = \frac{2}{2 - p\bar{v}}$$

Thus, for $\bar{v} = 2.1$ and $p = 0.90$, $\overline{X}_n = 18$ (instead of $\overline{X}_n = 10$ for the polymerization of divalent monomers) and tends to infinity for $p = 2/\bar{v} = 0.952$.

The discrepancy between the experimental observation and these predictions led several authors to formulate a better prediction of the gel point, by using a reasoning similar to that previously described for $(X-\mathcal{A}-X+Y-\mathcal{B}-Y)$ bivalent systems. Let us consider the particular case of a system constituted of bi- and trivalent monomers,

$$X-\mathcal{A}-X, \qquad X-\mathcal{A}-X \qquad Y-\mathcal{B}-Y$$
$$\qquad\qquad\qquad\;\; | $$
$$\qquad\qquad\qquad\; X$$

where ρ is the molar fraction of groups $\mathcal{A}$ belonging to the trivalent molecule.

In this calculation, all functional groups are assumed to exhibit an equal reactivity and cyclization reactions will be overlooked. The probability that a functional group X belonging to a branching point leads to another branching point *via* a chain formed only by two $X-\mathcal{A}-X$ units can be expressed as

$$\sim\sim\sim\sim X-\mathcal{A}-X-Y-\mathcal{B}-Y-X-\mathcal{A}-X-Y-\mathcal{B}-Y-X-\mathcal{A}-X-Y-\mathcal{B}-Y-X-A-X\sim\sim$$

$$\alpha_2 = p_X p_Y \rho [p_X p_Y (1-\rho)]^2$$

In a same fashion, the probability for a functional group X carried by a branching point to be connected to another branching point *via* a chain formed by i $X-\mathcal{A}-X$ units can thus be written as

$$\alpha_i = p_X p_Y \rho [p_X p_Y (1-\rho)]^i$$

Summing over all the possibilities corresponding to $0 \leq i \leq \infty$, we obtain the probability for a branching point to be connected to another one by means of a chain:

$$\bar{\alpha} = p_X p_Y \rho \sum_{i=0}^{\infty} [p_X p_Y (1-\rho)]^i$$

Since $\sum_{i=0}^{\infty} [p_X p_Y (1-\rho)]^i$ is equal to $1/[1 - p_X p_Y (1-\rho)]$, $\bar{\alpha}$ may be written as

$$\bar{\alpha} = \frac{p_X p_Y \rho}{1 - [p_X p_Y (1-\rho)]}$$

Consider the general case of polymerizations involving molecules of valence v. The number of directions of growth that result from the reaction of a growing polycondensate with a v-valent molecule can be written as

$$\bar{\alpha}(v-1)$$

To reach the gel point and eventually obtain a macromolecule of infinite molar mass, it is necessary that at least one chain emerges from the v-valent molecule or branching point just added:

$$\bar{\alpha}(v - 1) \geq 1$$

$$\bar{\alpha} \geq \frac{1}{v - 1}, \quad \text{or} \quad \alpha_{\text{critical}} = \frac{1}{v - 1}$$

Taking into consideration the expression for $\bar{\alpha}$, we obtain

$$\left\{ \frac{p_X p_Y \rho}{1 - [p_X p_Y (1 - \rho)]} \right\}_{\text{critical}} = \frac{1}{v - 1}$$

In the case of $\rho = 1$, which corresponds to a same number of X and Y functional groups in the reaction medium, and $p_X = p_Y$, the expression for the extent of reaction at the gel point can be written as follows:

$$p_{\text{critical}} = \left(\frac{1}{v - 1} \right)^{1/2}$$

In reality, this expression is sometimes not obeyed because of cyclization reactions that consume functional groups without contributing to the growth of the polycondensates.

APPENDIX: CALCULATION OF AVERAGE MOLAR MASSES USING A MARKOVIAN APPROACH

(a) Polymerization of {X–$\mathcal{A}$–X + Y–$\mathcal{B}$–Y} Type

As mentioned before, Flory laid out the basic relations for the size distribution of finite macromolecules as a function of the extent of reaction; but for cases of practical importance, these distribution functions become quite complex. Macosko and Miller described a simple method for calculating average physical quantities, such as average molar masses, gel point, soluble fraction, and so on. This method is based on an elementary law of conditional probability and on the recursive nature of a step-growth polymerization.

For the calculation of average characteristics, it retains the same three simplifying assumptions as in Flory's approach, namely,

- all functional groups are equally reactive.
- all groups react independently of one another.
- absence of cyclization reactions.

If A is an event and its complement is $\overline{A}$, and Y is a random variable, $E(Y)$ its mathematical expectation (or average value), and $E(Y/A)$ its conditional expectation given that the A event has occurred, then the law of total probability for the expectations is written as

$$E(Y) = E(Y/A) \cdot P(A) + E(Y/\overline{A}) \cdot P(\overline{A})$$

where $P(A)$ and $P(\overline{A})$ are the probabilities that an event A and its complement ($\overline{A}$) occur, respectively. Let us illustrate this method with the case of a step-growth polymerization of N_X X$-\mathcal{A}$$-$X with N_Y Y$-\mathcal{B}$$-$Y which continues until a rate of conversion p_X as a function of X and p_Y as function of Y is attained. This case is illustrated as follows:

$$
\text{X—A—X} + \text{Y—B—Y} \xrightarrow{\qquad} \sim\sim\sim\text{B—Y—X—A—}\overset{\xleftarrow{\text{int.}}}{\text{X}}\text{—Y—B—Y—X—A—X—Y—B}\sim\sim
$$

$$
\underset{\underset{1}{\xrightarrow{}}}{\text{out}}\ \underset{\underset{2}{\xrightarrow{}}}{\text{int.}} \qquad \underset{\xrightarrow{}}{\text{out}}\ \underset{\xrightarrow{}}{\text{int.}} \qquad \underset{\xrightarrow{}}{\text{out}}
$$

After a fraction p_X and p_Y of the X and Y groups have reacted, let us pick up randomly a function X called X$'$ and determine the mass $W_{X'}^{out}$ of the chain attached to X$'$ in the outer direction $\underline{1}$, looking out from its parent molecule. This mass is equal to 0 if X$'$ has not reacted and to W_Y^{int} in the opposite case—that is, the mass of the chain linked to Y in direction $\underline{2}$. Thus,

$$W_{X'}^{ext} = 0 \qquad \text{if X}' \text{ did not react}$$

$$W_{X'}^{ext} = W_{Y'}^{int} \qquad \text{if X}' \text{ reacted with Y}'$$

In terms of mathematical expectation, this may expressed as

$$E(W_{X'}^{ext}) = p_X E(W_{Y'}^{int}) + (1 - p_X) \cdot 0 \tag{7.1}$$

Proceeding along the structure shown, its recursive nature brings us back to the above equation:

$$E(W_Y^{int}) = M_{0,Y} + E(W_Y^{ext}) \tag{7.2}$$

$$E(W_Y^{ext}) = p_Y E(W_X^{int}) + (1 - p_Y) \cdot 0 \tag{7.3}$$

$$E(W_X^{int}) = M_{0,X} + E(W_X^{ext}) \tag{7.4}$$

Let W_{XAX} be the total mass of the polycondensate to which the randomly selected X$'$ function belongs. Its expectation may be written as

$$E(W_{XAX}) = E(W_X^{int}) + E(W_X^{ext}) \tag{7.5}$$

Analogously, for W_{YBY} a randomly chosen function Y:

$$E(W_{\text{YBY}}) = E(W_{\text{Y}}^{\text{int}}) + E(W_{\text{Y}}^{\text{ext}}) \tag{7.6}$$

In fact, equation (7.5) means that the mass of the polycondensate where X' was randomly chosen is the sum of the masses of two parts, one standing on the left- and the other on the right-hand side of this X' group. Similarly for equation (7.6).

Solving these various equations, we obtain

$$E(W_{\text{XAX}}) = M_{0,\text{X}} + 2p_{\text{X}}\frac{M_{0,\text{Y}} + p_{\text{Y}}M_{0,\text{X}}}{1 - p_{\text{X}}p_{\text{Y}}}$$

$$E(W_{\text{YBY}}) = M_{0,\text{Y}} + 2p_{\text{Y}}\frac{M_{0,\text{X}} + p_{\text{X}}M_{0,\text{Y}}}{1 - p_{\text{X}}p_{\text{Y}}}$$

To obtain the mass average molar mass of the sample, one just needs to pick up one monomeric unit at random and compute the molar mass of the polycondensate to which it belongs:

$$\overline{M}_w = W_{\text{XAX}}E(W_{\text{XAX}}) + W_{\text{YBY}}E(W_{\text{YBY}}) \tag{7.7}$$

where W_{XAX} and W_{YBY} denote the mass fractions of X–$\mathcal{A}$–X and Y–$\mathcal{B}$–Y units:

$$W_{\text{XAX}} = \frac{M_{0,\text{X}}N_{\text{X}}}{M_{0,\text{X}}N_{\text{X}} + M_{0,\text{Y}}N_{\text{Y}}}$$

$$W_{\text{YBY}} = \frac{M_{0,\text{Y}}N_{\text{Y}}}{M_{0,\text{Y}}N_{\text{Y}} + M_{0,\text{X}}N_{\text{X}}}$$

Introducing the expressions for $E(W_{\text{XAX}})$ and $E(W_{\text{YBY}})$ into equation (7.7), we obtain the same expression for $\overline{M}_w$ as that established by the Flory combinatorial method. For $\overline{M}_n$, we just need to write the ratio of the total mass of X–$\mathcal{A}$–X and Y–$\mathcal{B}$–Y units to the total number of molecules at time t.

(b) Polymerizations Implying a Monomer with a Valence Higher than 2

Unlike Flory's combinatorial approach, a Markovian analysis such as that proposed by Macosko and Miller leads to easy derivations of expressions for $\overline{M}_n$ and $\overline{M}_w$ for nonlinear polymers. Before generalizing to a reaction implying multivalent molecules having valence v, the case of a step-growth polymerization involving X_4 tetravalent molecules and N_{Y} Y–$\mathcal{B}$–Y molecules will be considered.

The below diagram features one part of the reaction medium after a fraction p_{X} of X functions has reacted with a p_{Y} fraction of Y functions. Since Y can only react with X and *vice versa*, p_{X} and p_{Y} are related by the equation

$$4X_4 p_{\text{X}} = 2N_{\text{Y}}p_{\text{Y}}$$

$$
\begin{array}{c}
\text{X} \\
| \\
\text{X}\!-\!\mathcal{A}\!-\!\text{X} \\
| \\
\text{X} \downarrow \text{out} \quad 1 \\
| \\
\text{Y} \downarrow \text{int.} \quad 2 \\
| \\
\mathcal{B} \\
| \\
\text{Y} \downarrow \text{out} \\
| \\
\text{out } \text{X} \downarrow \text{int.} \\
| \\
\scriptstyle\sim\sim -\text{Y}-\text{X}-\mathcal{A}-\text{X}-\text{Y}-\scriptstyle\sim\sim\sim\sim \\
\text{out.} \downarrow \text{X} \quad \text{out} \rightarrow \\
| \\
\text{Y} \\
|
\end{array}
$$

$$
\begin{array}{c}
\text{X} \\
| \\
\text{X}\!-\!\mathcal{A}\!-\!\text{X} \quad + \quad 2\ \text{Y}\!-\!\mathcal{B}\!-\!\text{Y} \quad \longrightarrow \\
| \\
\text{X}
\end{array}
$$

or in the case of molecules X with a valence v:

$$vX_v p_X = 2N_Y p_Y$$

As in the preceding treatment an X function called X' is randomly chosen in a polycondensate. Again, the problem is to determine the mass $W_{X'}^{\text{out}}$ which is attached to the polycondensate in direction $\underset{\rightarrow}{1}$. If X' has not reacted, $W_{X'}^{\text{out}}$ would be equal to 0 and if X' has reacted:

$$W_{X'}^{\text{out}} = W_{Y'}^{\text{int}}$$

In terms of mathematical expectation, this gives

$$E(W_{X'}^{\text{out}}) = E(W_{X'}^{\text{out}}/X'^{\text{has reacted}})P(X'^{\text{has reacted}})$$

$$+ E(W_{X'}^{\text{out}}/X'^{\text{did not react}}) \cdot P(X'^{\text{did not react}})$$

Then
$$E(W_{X'}^{\text{out}}) = E(W_Y^{\text{int}})p_X + (1 - p_X) \cdot 0 \qquad (7.8)$$

In a similar way as before, the expected masses can be derived following the arrows in scheme 2 until attaining an X function, the recursive nature of the polycondensate bringing us back to equation (7.8):

$$E(W_Y^{\text{int}}) = M_{0,Y} + E(W_Y^{\text{out}})$$

$$E(W_Y^{\text{out}}) = p_Y E(W_X^{\text{int}})$$

$$E(W_X^{\text{int}}) = M_{0,X4} + (4 - 1)E(W_X^{\text{out}})$$

In the general case of molecules X with valence v, the latter equation becomes

$$E(W_X^{\text{int}}) = M_{0,X_v} + (v - 1)E(W_X^{\text{out}})$$

To determine the expectation of the molar mass of the polycondensate W_{X_v} to which the randomly chosen X' function belongs, the same reasoning as for the $\{X-\mathcal{A}-Y + Y-\mathcal{B}-Y\}$ system will be followed, that is,

$$E(W_{X_v}) = E(W_X^{int}) + E(W_X^{out})$$

In the same way, for the Y functions we obtain

$$E(W_Y) = E(W_Y^{int}) + E(W_Y^{out})$$

which corresponds to

$$E(W_{X_v}) = M_{0,X_v} + vp_X \left[\frac{M_{0,Y} + p_Y M_{0,X_v}}{(1 - p_X p_Y)(v - 1)} \right]$$

$$E(W_{YBY}) = M_{0,Y} + 2p_Y \left[\frac{M_{0,X_v} + p_X(v - 1)M_{0,Y}}{(1 - p_X p_Y)(v - 1)} \right]$$

The mass average molar mass can then be simply written as

$$\overline{M}_w = W_{X_v} E(W_{X_v}) + W_{YBY} E(W_Y)$$

where W_{Xv} and W_{YBY} are, respectively, the mass fractions of the X_v-valent units and the Y-$\mathcal{B}$-Y units.

$$W_{X_v} = \frac{M_{0,X_v} X_v}{M_{0,X_v} X_v + M_{0,Y} N_Y} \quad \text{and} \quad W_{YBY} = 1 - W_{X_v}$$

This gives

$$\overline{M}_w = \frac{(2p_Y/p_X v)(1 + p_X p_Y)M_{0,X_v}^2 + [1 + (v - 1)p_X p_Y]M_{0,Y}^2 + 4p_Y M_{0,X_v} M_{0,Y}}{[2(p_Y M_{0,X_v}/p_X v) + M_{0,Y}][(1 - p_Y/p_X)(v - 1)p_X^2]}$$

and

$$\overline{M}_n = \frac{M_{0,X_v} X_v + M_{0,Y} N_Y}{X_v + N_Y - vp_X X_v}$$

The simplicity of the Markovian approach for step-growth polymerization can be profitably used to establish the expressions for $\overline{M}_w$ and $\overline{M}_n$ in cases taking into account possible cyclizations, functional groups with different reactivity, and so on.

7.4. KINETICS OF STEP-GROWTH POLYMERIZATIONS

In the expression derived above, it was assumed that the reactivity of the antagonist functional groups does not depend on the degree of polymerization of the chain carrying them. In reality, the gradual increase in viscosity of the reaction medium and its possible vitrification as polymerization proceeds may hinder the segmental motion of chains and, consequently, decreases the frequency of reactive collisions between antagonist sites. When vitrification sets in (T_g of the polycondensate $\leq$ cure temperature), reactions are diffusion-controlled and their rate decreases dramatically. In contrast, the gelation of plurivalent systems does not necessary slow down their kinetics of reaction beyond the gel point, provided that it occurs before vitrification. The kinetic treatment described here thus pertains to polymerizations kinetically controlled and overlooks phenomena such as vitrification

As in the case of simple organic reactions, three types of kinetics can be distinguished whether the reaction is

- spontaneous,
- catalyzed by an external catalyst (most of the cases),
- self-catalyzed by one of the reactive groups.

These three kinetic behaviors are described below for the case of *bivalent monomers* used in stoichiometric conditions.

In the case of *uncatalyzed reactions*, the rate of consumption of X and Y can be expressed as

$$-\frac{d[X]}{dt} = k[X][Y] = k[X]^2 = k[Y]^2 = -\frac{d[Y]}{dt}$$

After integration, the relation

$$\frac{1}{[X]} - \frac{1}{[X]_0} = kt$$

is obtained. Substituting $[X]_0(1-p)$ for $[X]$ since $p = \{[X]_0 - [X]\}/[X]_0$

$$\frac{1}{[X]_0(1-p)} - \frac{1}{[X]_0} = kt$$

leads to $1/(1-p) - 1 = kt[X]_0$, which corresponds to $X_n = k_t[X]_0 + 1$. Under such conditions, the number average degree of polymerization grows linearly with time.

In the case of *catalyzed reactions*, the rate of consumption of X functions can be written as

$$-\frac{d[X]}{dt} = k'[cat][X][Y]$$

since $k'[\text{cat}]$ takes a constant value (A)

$$-\frac{d[X]}{dt} = A[X]^2$$

and the kinetic treatment is the same as in the preceding case.

The case of *self-catalyzed reactions* is rather frequent. For example, the condensation of monomers carrying carboxylic acids with antagonist monomers is self-catalyzed (by acids). In the absence of a strong acid, we obtain

$$-\frac{d[X]}{dt} = k''[X][X][Y] = k''[X]^3$$

After integration, this leads to

$$\frac{1}{[X]^2} - \frac{1}{[X]_0^2} = 2k''t$$

Substituting $[X]_0(1-p)$ for $[X]$, one obtains

$$\frac{1}{(1-p)^2} = 2k''t[X]_0^2 + 1 = \overline{X}_n^2$$

For a self-catalyzed system, the number average degree of polymerization grows proportionally with $\sqrt{t}$.

Remark. It is important to stress that the rate constant of a reaction (k) depends on whether the catalyst is present or not.

7.5. MAIN REACTIONS USED IN STEP-GROWTH POLYMERIZATION

The aim of this paragraph is to stress the constraints generated by the application of some common reactions of organic chemistry to polymer synthesis. The effect of side reactions on the structure of the resulting polymers will be also considered.

In any case, it is important to note the necessity of driving reaction equilibria toward the products and maintaining a perfect stoichiometry between reactive groups to obtain polymers with high molar mass.

7.5.1. Nucleophilic Substitution Reactions on Carbonyls

The synthesis of polyesters and polyamides are based on such nucleophilic substitution reactions. The mechanism of such reactions is schematically represented as with

$$X = -OH, -OR', -NH_2, -O-CO-R', -Cl$$

$$Y = R'O^-, R'OH, R'NH_2, R'CO-O^-$$

R and R' can be either the alkyl or aryl groups. By combining various X and Y in any order, one can obtain a vast choice of polycondensates, some of them of considerable economic importance.

The metastable intermediate resulting from the addition of Y onto the carbonyl group leads to three possibilities:

- Elimination of the nucleophilic group Y, which justifies the existence of a reaction equilibrium,
- Elimination of X, which corresponds to the desired reaction,
- Capture of a proton

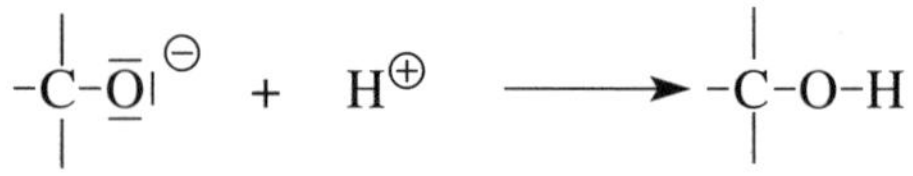

The path taken by the reaction depends on the nature of X and Y, which also in turn determines the value of the various rate constants of the process.

This reaction can be catalyzed by (Lewis or protonic) acids through more or less complex mechanisms. In the case of a catalysis by protonic acids, various mechanisms can be written depending upon the nature of X and Y. For example, the esterification of alcohols is known to be self-catalyzed by antagonist carboxylic acids as shown below:

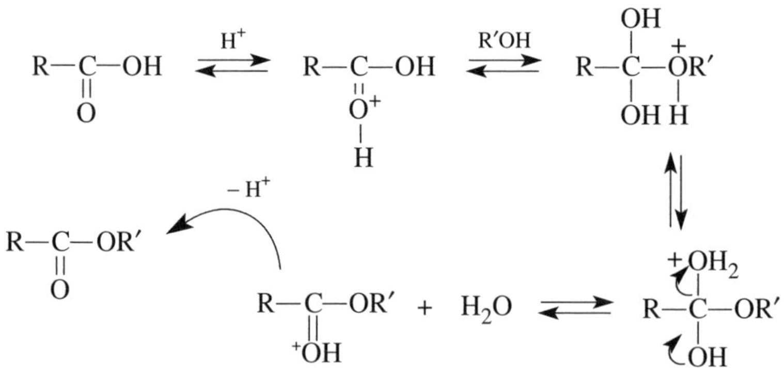

External catalysts are used in this type of condensation, generally Lewis acids.

(a) The different methods of preparation of polyesters from bivalent reagents are described below:

- By direct esterification

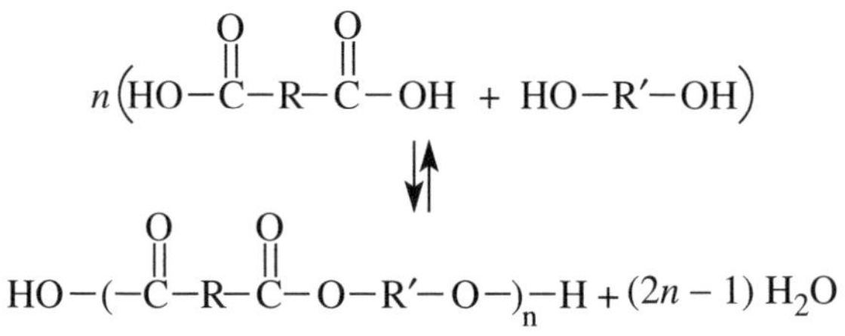

$$n\left(HO-\overset{O}{\overset{\|}{C}}-R-\overset{O}{\overset{\|}{C}}-OH \ + \ HO-R'-OH\right)$$

$$\updownarrow$$

$$HO-(-\overset{O}{\overset{\|}{C}}-R-\overset{O}{\overset{\|}{C}}-O-R'-O-)_n-H+(2n-1)\,H_2O$$

A diol that can be easily eliminated by distillation is often preferred because it can serve to increase the initial rate of reaction when used in excess and be eliminated by volatilization at the end of polymerization to retain the stoichiometry. The water generated upon condensation of alcohols with acids is gradually removed from the reaction medium;

- By transesterification

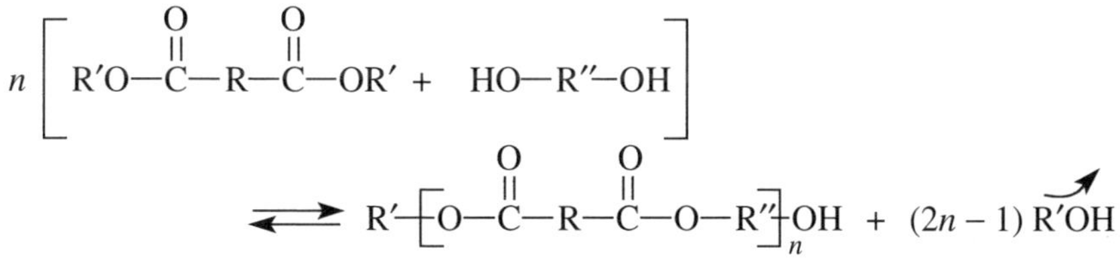

$$n\left[R'O-\overset{O}{\overset{\|}{C}}-R-\overset{O}{\overset{\|}{C}}-OR' \ + \ HO-R''-OH\right]$$

$$\rightleftarrows \ R'\left[O-\overset{O}{\overset{\|}{C}}-R-\overset{O}{\overset{\|}{C}}-O-R''\right]_n OH \ + \ (2n-1)\,R'OH$$

It is essential in the latter method that R'OH be volatile;

- By esterification of anhydrides with diols

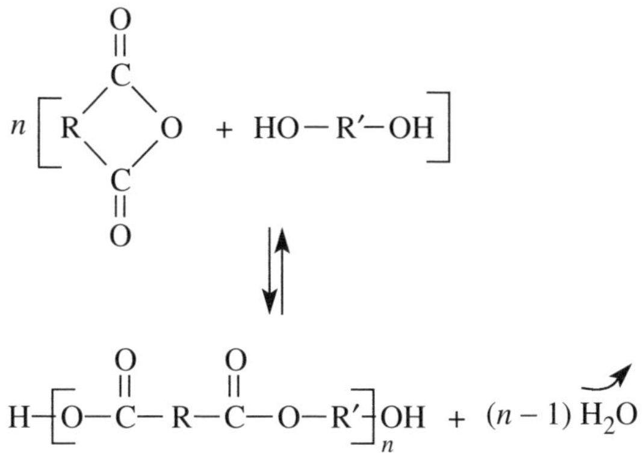

$$n\left[R\underset{\overset{C}{\overset{\|}{O}}}{\overset{\overset{O}{\overset{\|}{C}}}{\diamondsuit}}O \ + \ HO-R'-OH\right]$$

$$\updownarrow$$

$$H\left[O-\overset{O}{\overset{\|}{C}}-R-\overset{O}{\overset{\|}{C}}-O-R'\right]_n OH \ + \ (n-1)\,H_2O$$

- By esterification of acid dichlorides with diols (Schotten–Baumann reaction when acid chloride is aromatic)

$$n \left[\underset{\substack{\parallel \\ O}}{Cl-C}-R-\underset{\substack{\parallel \\ O}}{C}-Cl \ + \ HO-R'-OH \right] \rightleftharpoons$$

$$Cl\left[\underset{\substack{\parallel \\ O}}{C}-R-\underset{\substack{\parallel \\ O}}{C}-O-R'-O\right]_n H \ + \ (2n-1) \ HCl \nearrow$$

This reaction is spontaneously driven toward the formation of the polyester, and it is carried out in the presence of a base whose role is to neutralize the HCl formed throughout polymerization. It is a very fast reaction that can be used, in particular, in interfacial polycondensations.

(b) Reactions of the same type can be utilized to synthesize **polyamides**:

- By direct amidation of primary amines by carboxylic acids. In an initial phase, this leads to formation of an ammonium carboxylate

$$n \left[HOOC-R-COOH \ + \ H_2N-R'-NH_2 \right] \rightleftharpoons n \left[\begin{array}{c} {}^-OOC-R-COO^- \\ \overset{+}{H_3N}-R'-\overset{+}{NH_3} \end{array} \right]$$

whose purification by recrystallization in water brings about a perfect stoichiometry. The salt formed is then dehydrated by heating

$$n \left[\begin{array}{c} {}^-OOC-R-COO^- \\ \overset{+}{H_3N}-R'-\overset{+}{NH_3} \end{array} \right] \rightleftharpoons (2n-1) \ H_2O \ + \ n \ {}^-O\left[\underset{\substack{\parallel \\ O}}{C}-R-\underset{\substack{\parallel \\ O}}{C}-\underset{\substack{| \\ H}}{N}-R'-\underset{\substack{| \\ H}}{N}\right]_n \overset{+}{H_2}$$

In the case of amino-acids, an ammonium carboxylate is formed by intramolecular reaction;

- By reaction of acid chlorides with primary diamines (Schotten–Baumann reaction); it is a reaction analogous in all points with that affording polyesters

$$n \left[\underset{\substack{\parallel \\ O}}{Cl-C}-R-\underset{\substack{\parallel \\ O}}{C}-Cl \ + H_2N-R'-NH_2 \right] \rightleftharpoons (2n-1) \ HCl \nearrow$$

$$+ \ Cl\left[\underset{\substack{\parallel \\ O}}{C}-R-\underset{\substack{\parallel \\ O}}{C}-\underset{\substack{| \\ H}}{N}-R'-\underset{\substack{| \\ H}}{N}\right]_n H$$

This reaction of polyamidation is also utilized in interfacial polycondensations.

Other methods of polyamidation such as (a) the amidation of esters by primary amines and (b) amidation of nitriles by primary amines are much less used.

7.5.2. Addition Reactions on Carbonyls

These reactions have found applications in all step-growth polymerizations using aldehydes, more particularly formaldehyde. They are mainly utilized for the preparation of phenolic resins and amino resins.

7.5.2.1. Reaction of Phenols with Formol (Phenolic Resins). Depending upon the type of catalysis (acidic or basic), the reaction mechanisms and the resulting products will be different; in both cases the reaction is carried out in aqueous solution.

Acid Catalysis. This method is used to produce resins called "novolacs" by reaction of an excess of phenol (trivalent molecule) with formol (bivalent molecule). In aqueous solution, formol exists mainly as an oligomer of α,ω-dihydroxypoly (oxymethylene), whose major component is $HO-CH_2-OH$. In acidic medium, methylene glycol generates a strongly electrophilic carbocationic species

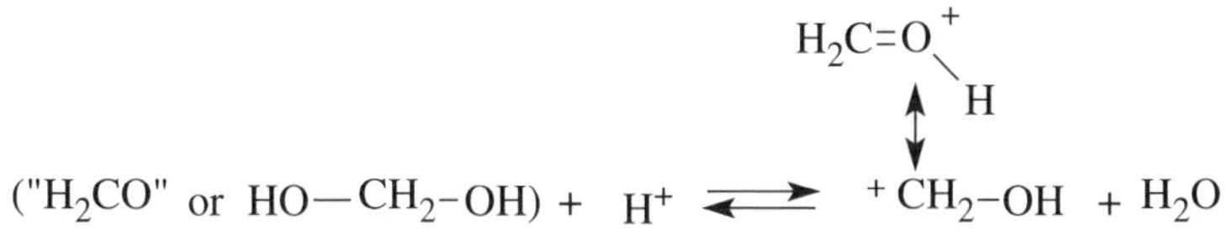

$$H_2C=O^+\diagdown H$$

$$("H_2CO" \text{ or } HO-CH_2-OH) + H^+ \rightleftharpoons {}^+CH_2-OH + H_2O$$

that brings about an electrophilic substitutions on aromatic rings such as phenols. For example:

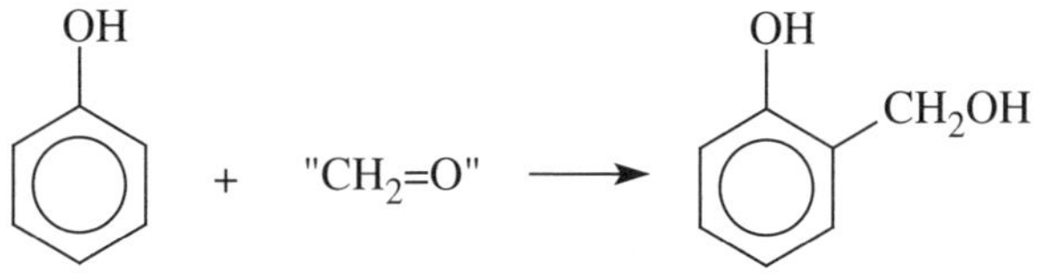

o-Methylolphenol

Such electrophilic substitutions seldom proceed beyond monomethylolphenols to afford plurimethylphenols, because the reactions shown below are much faster than the previous reaction.

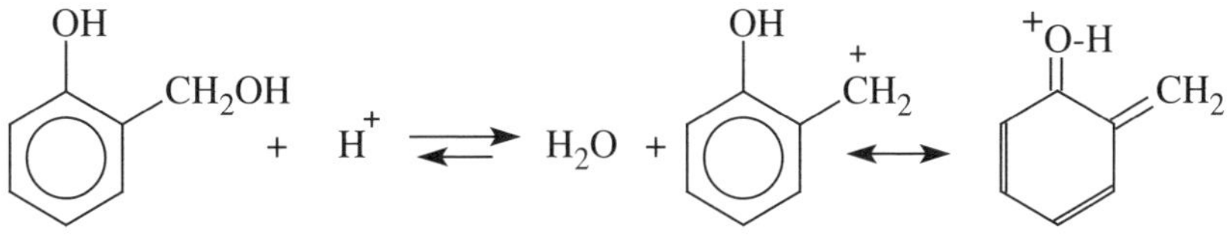

then

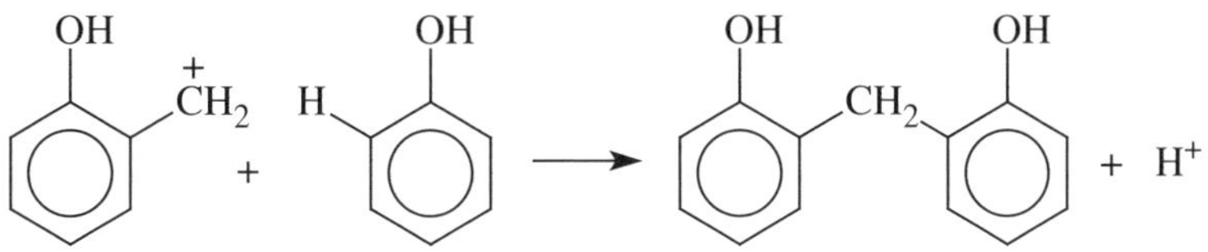

This series of reactions continues and eventually affords oligomers mainly terminated with phenol groups.

To cross-link the oligomers formed, it is necessary to adjust the stoichiometry by introducing additional formaldehyde, in the form of hexamethylenetetramine whose hydrolysis generates CH_2O.

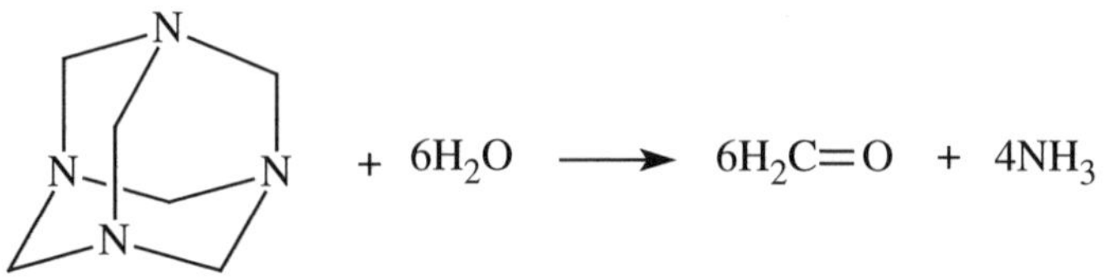

$$+ \; 6H_2O \longrightarrow 6H_2C{=}O \; + \; 4NH_3$$

Cross-linking occurs through reaction of formaldehyde with the oligomers at their free ortho- and para- positions.

Base Catalysis. In the presence of an excess of formaldehyde {commonly[formaldehyde]/[phenol] $= 1.2$}, such a catalysis affords oligomers called **resols**.

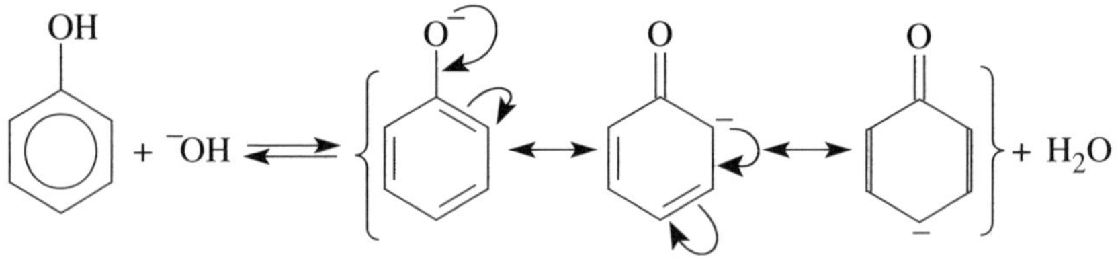

The main reaction consists in a nucleophilic addition on formaldehyde, a very reactive species that is in equilibrium with dihydroxypoly(oxymethylene)s and is therefore present in minute concentration. The most commonly accepted mechanism is as follows:

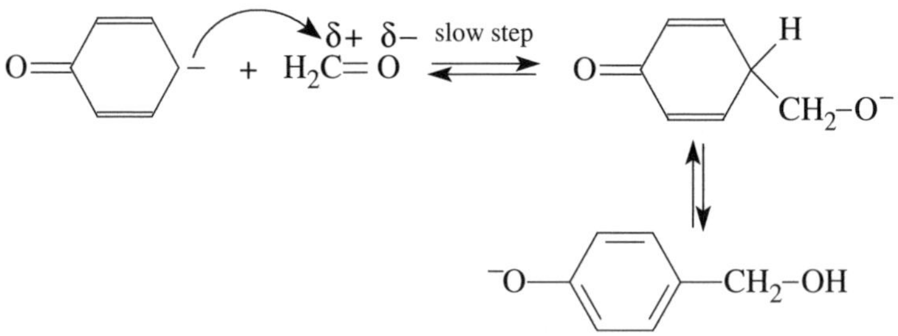

The products formed after this step are generally mono-, di-, and trimethylolphenols. As soon as they appear in the reaction medium, they tend to self-condense via the mechanism below, especially at elevated temperature:

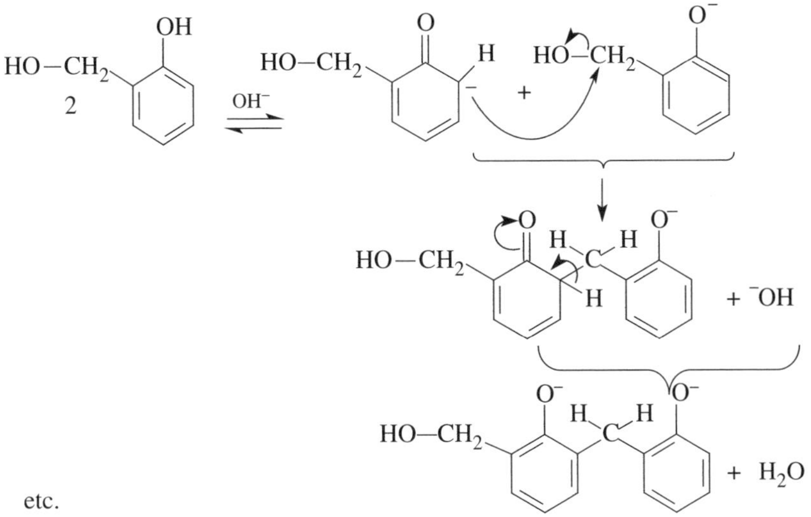

etc.

Another mechanism that leads to the formation of methylene bridges has also been proposed:

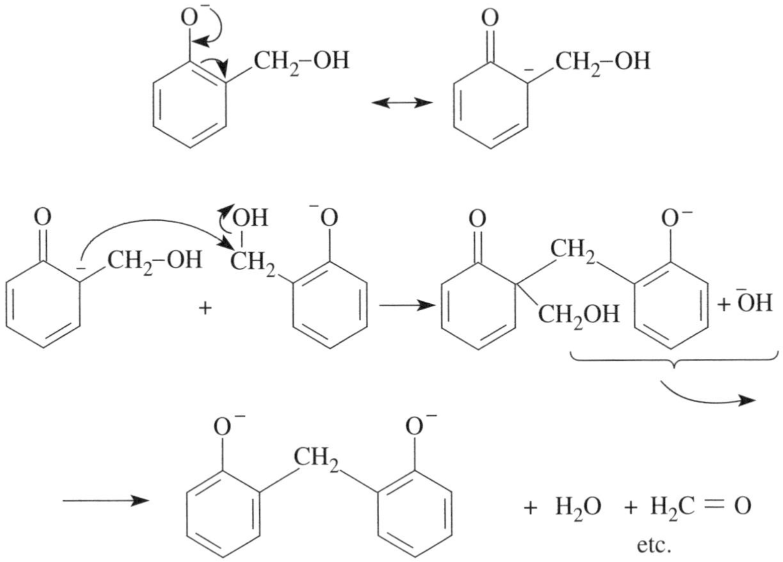

Formaldehyde thus regenerated can react further with phenolic moieties; since the formaldehyde used in excess generates a mixture of mono-, di-, and trifunctional species, the condensation of the latter leads to a three-dimensional network.

7.5.2.2. Reaction of Amines with Formaldehyde (Amino Resins). Such resins mainly result from the reactions of amino groups carried by

Urea $\quad O=C\begin{smallmatrix}NH_2\\[4pt]NH_2\end{smallmatrix}\quad$ and $\quad$ Melamine

Melamine structure (triazine ring with three NH_2 groups)

with formaldehyde in slightly basic medium (pH $\sim$ 8–9).

Remark. A basic pH is essential to avoid an uncontrolled spontaneous polycondensation.

The overall mechanism of such a polycondensation is independent of the type of amine used. The first step is similar to that previously described for the formation of resols:

$$O=C\begin{smallmatrix}\overset{-}{N}H\\[4pt]NH_2\end{smallmatrix} + H_2C=O \longrightarrow O=C\begin{smallmatrix}NH-CH_2-O^-\\[4pt]NH_2\end{smallmatrix} \xrightarrow[\text{(ii) H}^+]{\text{(i) CH}_2O} O=C\begin{smallmatrix}NHCH_2OH\\[4pt]NHCH_2OH\end{smallmatrix}$$

$$\text{and } O=C\begin{smallmatrix}NHCH_2OH\\[4pt]N\overset{\displaystyle CH_2OH}{\underset{CH_2OH}{\big\langle}}\end{smallmatrix}$$

No tetrafunctional species is formed.

The step corresponding to the condensation is carried out under slightly acidic conditions (pH $\sim$ 5) to better control the evolution of the reaction medium. The detailed sequence of reactions is ill-known, but the mechanism proposed below is the most generally accepted one:

$$H_2N-\overset{\displaystyle O}{\overset{\|}{C}}-NH-CH_2-OH + H^+ \rightleftharpoons H_2N-\overset{\displaystyle O}{\overset{\|}{C}}-NH-CH_2-\overset{+}{O}H_2$$

$$H_2N-\overset{\displaystyle O}{\overset{\|}{C}}-N=CH_2 + H_3O^+$$

$$\text{then } 3H_2N-\overset{\displaystyle O}{\overset{\|}{C}}-N=CH_2 \longrightarrow$$

(triazine ring product with three $C=O$ and NH_2 substituents)

This sequence of reactions might continue as long as there are free amino groups, but, in an acidic medium and at high temperature, competing condensations by etherification could also occur as shown below:

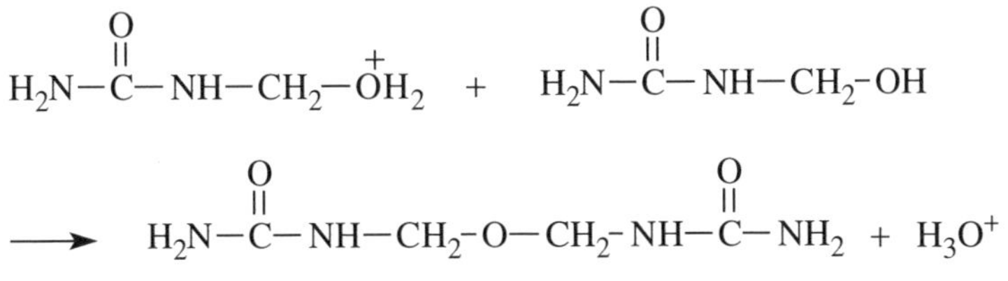

$$H_2N-\overset{\overset{\displaystyle O}{\|}}{C}-NH-CH_2-\overset{+}{O}H_2 \quad + \quad H_2N-\overset{\overset{\displaystyle O}{\|}}{C}-NH-CH_2-OH$$

$$\longrightarrow \quad H_2N-\overset{\overset{\displaystyle O}{\|}}{C}-NH-CH_2-O-CH_2-NH-\overset{\overset{\displaystyle O}{\|}}{C}-NH_2 \ + \ H_3O^+$$

The formed species may either remain in their original form or eliminate $H_2C=O$ to give methylene bridges

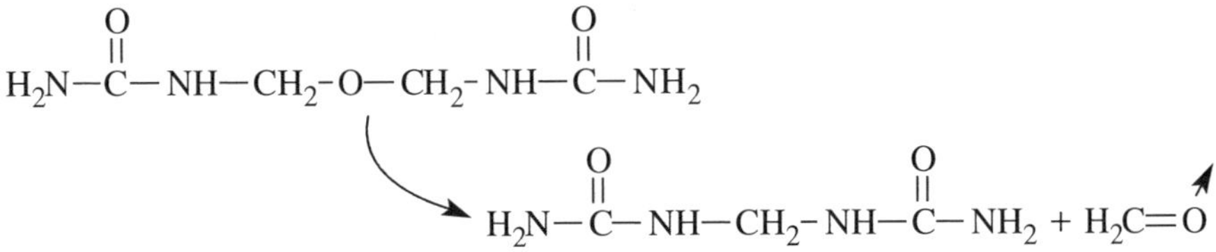

$$H_2N-\overset{\overset{\displaystyle O}{\|}}{C}-NH-CH_2-O-CH_2-NH-\overset{\overset{\displaystyle O}{\|}}{C}-NH_2$$

$$\longrightarrow \quad H_2N-\overset{\overset{\displaystyle O}{\|}}{C}-NH-CH_2-NH-\overset{\overset{\displaystyle O}{\|}}{C}-NH_2 \ + \ H_2C=O \nearrow$$

and so on. These two reaction mechanisms occur together, which accounts for the real composition of amino resins.

In the case of melamine-based resins (a trifunctional hexavalent molecule), the hexamethylol derivative shown below is formed in a first step:

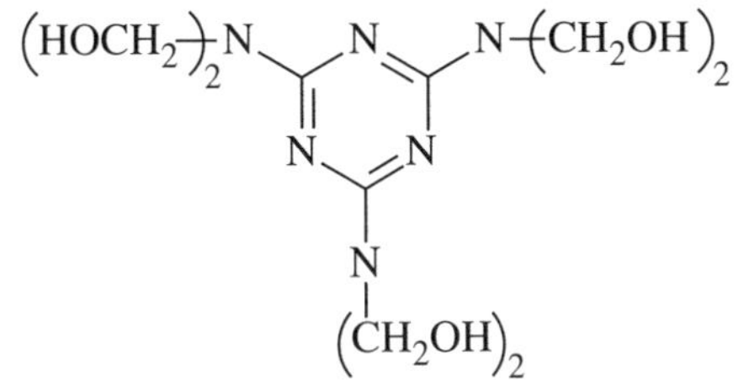

$$\left(HOCH_2\right)_2 N - \underset{\underset{\displaystyle N}{\overset{\displaystyle N}{\bigtriangleup}}}{\overset{N}{\bigtriangleup}} - N\left(CH_2OH\right)_2$$

with N bearing $\left(CH_2OH\right)_2$

7.5.3. Addition Reactions on C=N Double Bonds

Reaction of hydroxyl with isocyanate is primarily an addition which leads to polyurethanes and no small molecule is eliminated in this case. By no means should this reaction be confused with a chain reaction because the product formed is unable to react with the initial antagonist reagent. The reaction scheme is as follows:

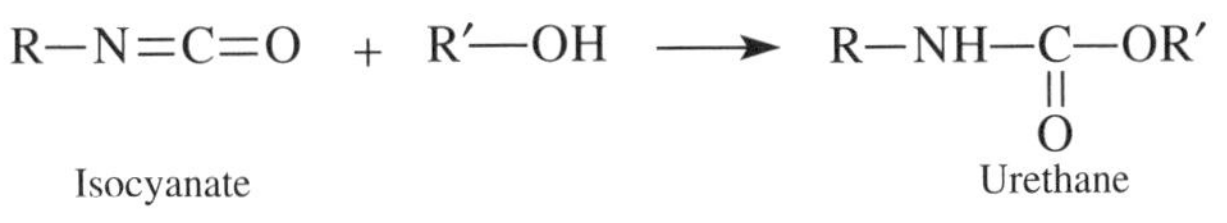

$$R-N=C=O \ + \ R'-OH \ \longrightarrow \ R-NH-\overset{\overset{\displaystyle \ }{\underset{\underset{\displaystyle O}{\|}}{C}}}{}-OR'$$

Isocyanate   Urethane

In the absence of any by-product eliminated, the reaction is naturally driven toward the product and high molar mass polymers can be obtained by satisfying the Carothers relation.

Polyurethanes are also called polycarbamates (esters of the carbamic acid $R-NH-COOH$) and can be viewed as amido-esters of carbonic acid $HO-CO-OH$.

The detailed mechanism of the reaction of alcohols with isocyanates is not known. In particular, the steps in the attack of the strongly polarized carbon atom of the isocyanate group by the oxygen atom of the hydroxyl are not well understood.

$$\sim\sim\sim R-\overset{\delta-}{\underset{}{N}}=\overset{\delta+}{C}=\overset{\delta-}{O} \quad + \quad \overset{\delta+}{H}-\overset{\delta-}{O}-R'\sim\sim\sim$$

$$\sim\sim\sim R-\overset{\ominus}{N}-C\overset{O}{\underset{\underset{H\,\oplus\,R'\sim\sim\sim}{O}}{}} \quad \longleftrightarrow \quad \sim\sim\sim R-N=C\overset{\overset{O^{\ominus}}{}}{\underset{\underset{H\,\oplus\,R'\sim\sim\sim}{O}}{}}$$

$$\sim\sim\sim R-NH-\overset{O}{\overset{\|}{C}}-O-R'\sim\sim\sim$$

The presence of hydrogen bonds complicates the kinetics of the reaction and makes its interpretation difficult. Even if this reaction can be carried out to completion at room temperature, it is generally catalyzed; to this end, use can be made of light, tertiary amines, lithium alkoxides, and organometallic compounds such as tin salts.

- Catalysis by tertiary amines:

$$\sim\sim R-\overset{}{N}=C=O + |NR''_3 \rightleftharpoons \sim\sim R-\overset{\ominus}{\underset{\oplus NR''_3}{N}}-C=O \longleftrightarrow \sim\sim R-\overset{\delta-}{\underset{\oplus NR''_3}{N}}=\overset{\delta+}{C}-O^{\ominus}$$

$$\text{then} \quad \sim\sim\sim\sim R-\overset{\delta-}{\underset{\oplus NR''_3}{N}}=\overset{\delta+}{C}-O^{\ominus} + \sim\sim\sim\sim R'-\overset{\delta-}{O}-\overset{\delta+}{H} \rightleftharpoons$$

$$\sim\sim\sim\sim R-NH-\overset{OR'\sim\sim\sim}{\underset{\oplus NR''_3}{C}}O^{\ominus} \rightleftharpoons \sim\sim\sim\sim R-NH-\overset{O}{\overset{\|}{C}}-OR'\sim\sim\sim + R_3''N|$$

Another mechanism implying the activation of the hydroxy group by the tertiary amine has also been proposed:

$$R-O\overset{H.....NR''_3}{\diagup}$$

- Catalysis by metal salts:

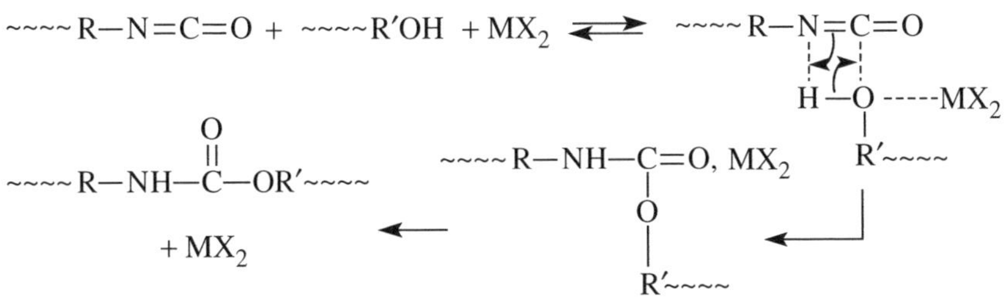

7.5.4. Nucleophilic Substitution Reactions

Many examples of step-growth polymerizations are based on this type of reactions, but only the most important among them—the preparation and polymerization of the diepoxide precursors—will be discussed here.

The preparation of the diepoxide precursors is carried out in basic medium using diol and epichlorhydrin:

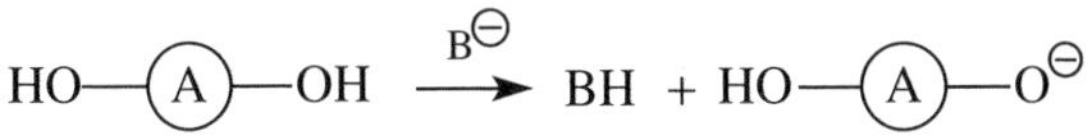

then

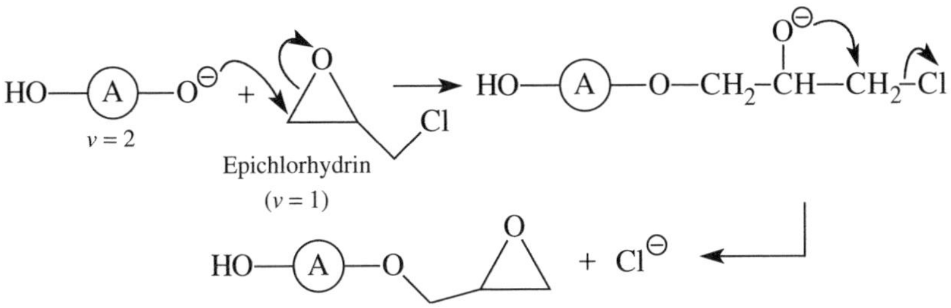

Through two successive nucleophilic substitutions an **oxirane** (epoxide) function can be introduced at the expense of the initial diol. When the reaction is carried out with a diol as precursor, a diepoxide is formed:

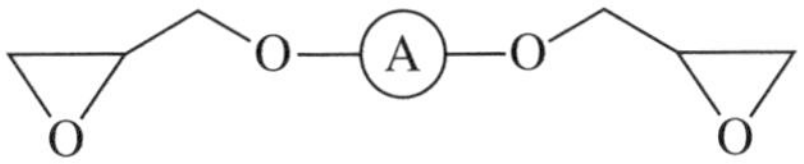

In the presence of a base, the latter can react with the remaining diol to afford a precursor of higher degree of polymerization.

Depending upon the composition of the reaction medium, mixtures of oligomers of variable $\overline{X}_n$ can be obtained, the majority of them exhibiting an n value between

0 and 2:

Bisphenol A

is the most used diol.

To form networks, these oligomeric precursor can be:

- Polymerized by a chain process using a Lewis acid (cationic polymerization) or a tertiary amine (anionic polymerization) as activator (see Chapter 8); in this case, the bifunctional diepoxide behaves as a tetravalent monomer;
- Polymerized by a stepwise process using an antagonist molecule of valence higher than 2. Only this second case will be described below.

In the **reaction with multiamines**, the valence of the latter is determined by the number of reactive hydrogens; thus $v = 4$ for a primary diamine. The mechanism generally accepted is as follows:

Self-catalysis by the amines themselves is observed at the onset of the polymerization.

Both the reactive carbon atoms of the epoxide group and the nucleophilicity of the nitrogen atom undergo such an activation until the hydroxyls gradually formed upon polymerization also participate in the activation process.

The tertiary amines generated by the epoxide/amine reaction play a role of activator in this nucleophilic substitution, for primary and secondary amines.

The **reaction of carboxylic acids with anhydrides** is also commonly used for the cross-linking of diepoxide prepolymers. The reaction has to be carried out at higher temperature than previously and, in the presence of tertiary amines, occurs through ring-opening of the oxiranes by the carboxylate generated from anhydrides:

$$R\text{-}COO^{\ominus} + \quad \overset{R'}{\triangle}_O \longrightarrow R\text{-}COO\text{-}CH_2\text{-}\underset{\underset{O^{\ominus}}{|}}{CH}\text{-}R'$$

At high temperature other reactions occur, which make the structure of the resulting networks extremely complex. Indeed, secondary hydroxyls formed upon ring-opening of epoxides can in turn react with the oxiranes of the precursor:

$$ROH \; + \quad \overset{R'}{\triangle}_O \; \rightleftharpoons \quad RO\text{—}\underset{HO}{\diagup}\text{—}R'$$

Such reactions increase the density of cross-linking of the network formed.

7.5.5. Substitution Reactions on Silicon Atoms

Only a minor portion of industrially produced polysiloxanes is obtained by chain polymerization of cyclosiloxanes (octamethylcyclotetrasiloxane). Most are synthesized by water-induced hydrolysis of dialkyldichlorosilanes followed by self-condensation of the disilanol formed. The starting monomer is dimethyldichlorosilane, which is prepared by copper-catalyzed reaction of methyl chloride on metal silicon. The hydrolysis of the chlorinated derivative

$$\underset{\underset{CH_3}{|}}{\overset{\overset{CH_3}{|}}{Cl\text{—}Si\text{—}Cl}} \; + \; 2H_2O \; \rightleftharpoons \; 2HCl \; + \; \underset{\underset{CH_3}{|}}{\overset{\overset{CH_3}{|}}{HO\text{—}Si\text{—}OH}}$$

corresponds to a nucleophilic substitution.

In the presence of bases, the condensation occurs by nucleophilic substitution, and the result of the self-condensation of silanol groups is poly(dimethylsiloxane):

$$n\,\underset{\underset{CH_3}{|}}{\overset{\overset{CH_3}{|}}{HO\text{—}Si\text{—}OH}} \; \rightleftharpoons \; H\!\!\left[\!O\text{—}\underset{\underset{CH_3}{|}}{\overset{\overset{CH_3}{|}}{Si}}\!\right]_n\!\!OH \; + \; (n-1)H_2O$$

Depending on whether the silanol function is carried by a mono-, di-, or trivalent monomer, one may have termination, polymerization, or cross-linking. The reactivity of silanols is closely related to the nature of the alkyl groups, the number of hydroxyls carried by the silicon atom, and the size of the polysiloxane carrying them. Thus, $(CH_3)_2Si(OH)_2$ is the most reactive among dialkylsilanediols. The mechanism of the acid-catalyzed condensation of silanols (by HCl originating from the first step) can be represented by

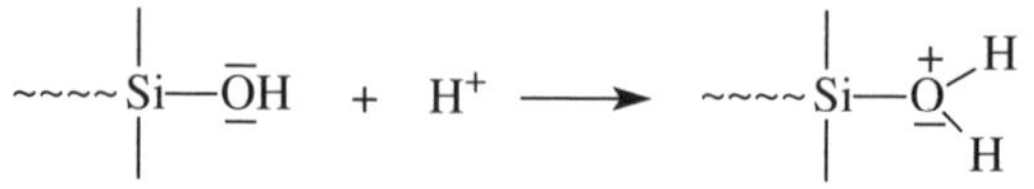

Then

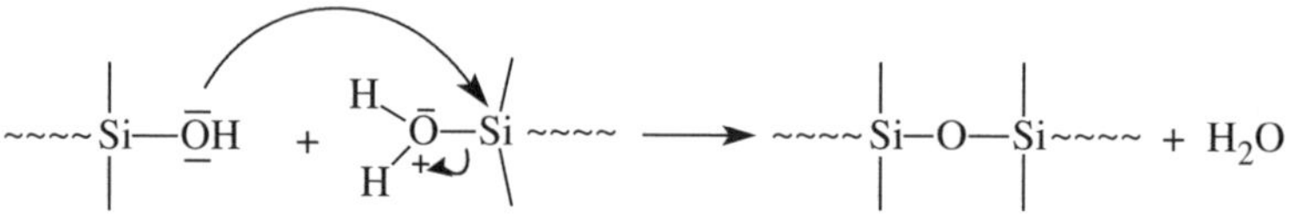

As the oligodimethylsiloxanediols gain in size, the reactivity of their terminal silanols decreases due to their tendency to establish intramolecular hydrogen bonding.

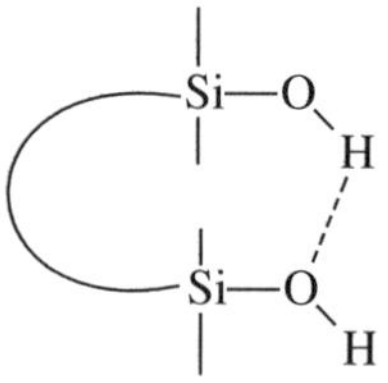

Such interactions only exist after the condensing oligomer has reached a certain size, with the cyclization being impossible when they are still too small.

In the presence of bases (Et_3N), condensation proceeds by nucleophilic substitution:

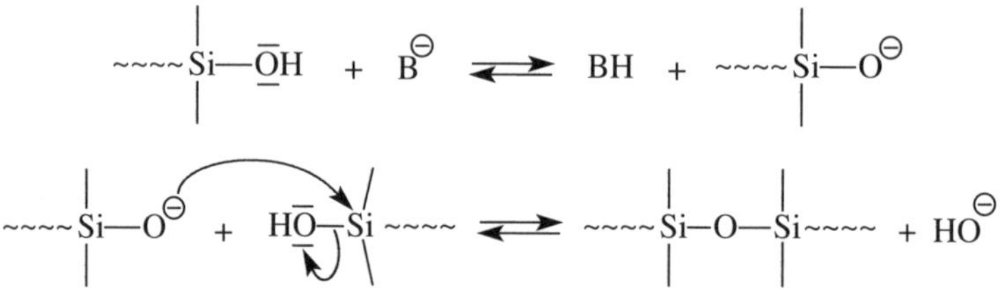

7.5.6. Chain-Growth Polycondensation

Conventional step-growth polymerizations occurs in the initial phase through condensation/addition of monomers with each other and then proceeds *via* reactions of all size oligomers with themselves and with monomers. In such a process the precise control of the polycondensate molar mass is elusive—in particular, in the initial and intermediate stages where only oligomers are formed. The polycondensate molar mass indeed builds up only in the final stage and its dispersity index increases up to 2.

In an attempt to better control both molar masses and the dispersity in polycondensates, a new concept of polycondensation has been recently proposed that proceeds in a chain polymerization manner (Chapter 8). In a context where monomers would have little option but to react first with an "initiating" site and then with the polymer end-group and would be prevented from reacting each other, all the requirements would be met to bring about so-called chain-growth polycondensations. Under such conditions, the polycondensate would increase linearly with conversion and be controlled by the [monomer]/[initiator] ratio and its mass dispersity index would be close to unity.

Yokosawa and co-workers have proposed two approaches to such chain-growth polymerization of X−A−Y-type monomers:

(a) Specific activation of propagating end-groups and concomitant deactivation of those carried by the monomer through substituent effects;
(b) Phase-transfer polymerization with the monomer being stored in a separate solid phase.

The polycondensation of phenyl-4-(alkylamino)-benzoate carried out in the presence of phenyl-4-nitrobenzoate acting as initiator and a base is a perfect illustration of approach (a) theorized by Yokozawa.

$$\bar{M}_n \leq 22{,}000 \quad \bar{M}_w/\bar{M}_n \leq 1.1$$

Reaction mechanism:

The base serves to abstract a proton from the monomer and generate an aminyl anion, which in turn deactivates its phenyl moiety. This anion reacts preferentially with the phenyl ester group of phenyl-4-nitrobenzoate and the amide group formed has a weaker electron-donating character than the aminyl anion of the activated monomer. The reaction of monomers with each other was thus efficiently prevented so that well-defined aromatic polyamides could be obtained up to 22,000 g/mol molar mass and with a dispersity index of 1.1.

The case of solid monomers that are progressively transferred to an organic phase with the help of a phase transfer catalyst and thus placed in a situation to react with the polymer end group is an illustration of approach (b).

This concept of chain growth polycondensation is new in synthetic polymer chemistry but not in Nature. In the biosynthesis of many natural polymers, Nature takes indeed full advantage of this concept: for instance, DNA is obtained via a polycondensation of deoxyribonucleoside of 5′-triphosphate with the 3′-hydroxy terminal group of polynucleoside with the help of DNA polymerase.

LITERATURE

G. Odian, *Principles of Polymerization*, 4th edition, Wiley-VCH, New York, 2004.

M. E. Rodgers and T. E. Long (Eds.), *Synthetic Methods in Step-Growth Polymers*, Wiley, New York, 2003.

8

CHAIN POLYMERIZATIONS

8.1. GENERAL CHARACTERS

Chain polymerizations proceed differently from these occurring by **step growth**. In the latter case, polymers grow by reaction (condensation or coupling) with either a monomer molecule, an oligomer, another chain, or any species carrying an antagonist functional group. Each condensation/addition step results in the disappearance of one reactive species (whatever its size) from the medium, so that the molar mass of such a "condensation polymer" is due to increase in an inverse proportion to $(1 - p)$, where p is the *extent of reaction*. The reaction between these antagonist functional groups that can be carried indifferently by monomer molecules or growing polymer chains brings about the formation of the constitutive units of polycondensates through covalent bonding. Two reactive functional groups are consumed after each condensation/addition step.

Unlike the case of polycondensations and polyadditions, in **chain-growth polymerizations**, very long macromolecules can be formed just after induction of the reaction, and active centers are generally carried by the growing chains. The general scheme describing chain growth is the same as for other chain processes: after production of a primary active center (P^*) by an initiator (I) or a supply of energy to the system, this species activates a monomer molecule (M) through transfer of its active center on the monomer unit thus formed:

$$A \longrightarrow P^*$$

$$P^* + M \longrightarrow PM^*$$

Organic and Physical Chemistry of Polymers, by Yves Gnanou and Michel Fontanille
Copyright © 2008 John Wiley & Sons, Inc.

This first step called *initiation* and often consisting of two phases is followed by a *propagation* (or *growth*) step, during which macromolecules grow by chain addition of monomer molecules to the newly formed PM* species. Upon reaction with a "fresh" monomer molecule, the active center carried by the growing chain is transferred to the last generated monomeric unit, and so on:

$$PM^* + M \longrightarrow PMM^* \text{ (written } PM_2^*)$$

$$PM_n^* + M \longrightarrow PM_{(n+1)}^*$$

In most systems, propagation is very fast and corresponds to an exothermic phenomenon whose overall activation energy is generally positive; in some cases, the reaction may run out of control and even become explosive. Termination reactions, when existing, may self-inhibit polymerizations getting out of control by deactivating growing chains. These terminations occur irrespective of the degree of polymerization of the growing chains:

$$PM_n^* \longrightarrow PM_n$$

In addition to terminations, certain systems can undergo other chain-breaking reactions, such as chain transfer represented as follows:

$$PM_n^* + T \longrightarrow PM_n + T^*$$

$$T^* + M \longrightarrow TM^*$$

$$TM^* + nM \longrightarrow TM_{n+1}^*$$

T is called *transfer agent*, but transfer can occur to monomer, polymer, initiator, or any molecule present in the reaction medium.

This transfer phenomenon blocks active chains in their growth and generates new active centers (T*) that are able to initiate the formation of novel macromolecules. Chain transfer prevents the obtainment of polymeric chains of high molar masses but can be used to control molecular dimensions when targeting oligomers or samples of low molar masses. In certain conventional chain polymerizations, the three steps of initiation, propagation, and termination as well as transfer can occur simultaneously, which means that each initiated chain propagates and undergoes termination or perhaps transfer, independently of events occurring in its surrounding. In other words, the time required for the formation of a chain can be lesser than one second in certain systems, whereas the corresponding half-polymerization time can be equal to several hours.

Chain-growth polymerizations are distinguished from one another, depending upon the types of active centers that initiate and propagate the polymerization process. Thus, four families of chain polymerizations are generally considered:

- **Free radical polymerizations**, whose propagating active centers involve free radicals,

- **Anionic polymerizations**, which require nucleophilic reactive species,
- **Cationic polymerizations** ("symmetrical" of the preceding ones), whose propagating species are electrophiles,
- **Coordination polymerizations**, whose active centers are complexes formed by coordination between monomer molecules and transition metal atoms.

These four important methods of polymerization exhibit their own peculiarities. Certain monomers can be polymerized (until today) by only one of them; this is the case, for example, of vinyl acetate or acrylic acid, which can be polymerized only by free-radical means. On the contrary, styrene can be polymerized by any of the aforementioned methods of polymerization.

8.2. POLYMERIZABILITY

Polymerizability is the faculty of an organic compound (monomer molecule) to undergo polymerization. Two conditions must be fulfilled to this end:

- Compliance with thermodynamic constraints
- Existence of an adequate reaction

The polymerizability of a monomer can be evaluated by means of the rate constant of polymerization which varies with the method of polymerization chosen.

8.2.1. Compliance with Thermodynamic Constraints

Like any other reaction of organic chemistry, chain polymerizations are equilibrium reactions that can be schematically represented as follows:

$$PM_n^* + M \xrightleftharpoons{K} PM_{n+1}^*$$

The equilibrium between growing polymer chains and the monomer is determined by the thermodynamic conditions. By definition, at equilibrium

$$\Delta G = 0$$

Therefore, one has

$$\Delta G = \Delta G^0 + RT \ln K = \Delta H^0 - T\Delta S^0 + RT \ln K = 0$$

where ΔG^0, ΔH^0, and ΔS^0 represent the standard variations of free energy, enthalpy, and entropy, respectively, corresponding to the transition undergone by monomer molecules in their standard state (pure liquid, gas, or unimolar solution) becoming the monomer units of polymeric chains, in their novel standard state (amorphous solid state or solution in unimolar concentration).

From the above equilibrium, the equilibrium constant can be written as

$$K = [PM_{n+1}^*]/\{[PM_n^*][M]\}$$

If the concentrations of species PM_n^* and PM_{n+1}^* are assumed practically identical, which is reasonable (at a first approximation) at equilibrium for values of n higher than a few monomeric units, one can write

$$K = 1/[M]$$

which gives

$$RT \ln[M] = \Delta H^0 - T \Delta S^0$$

$$R \ln[M] = (\Delta H^0/T) - \Delta S^0$$

and corresponds to

$$T_c = \Delta H^0/\{\Delta S^0 + R \ln[M]_{equ}\}$$

or

$$\ln[M]_{equ} = (\Delta H^0/RT_c) - (\Delta S^0/R)$$

In these last two equations, the c index after T denotes *ceiling conditions* corresponding to the monomer concentration at equilibrium $[M_{equ}]$. Indeed, in most of polymerizations, the variation of entropy is negative since the transition from the monomer to the polymer state corresponds to a decrease in the degrees of freedom of the system; thus, the entropy term is unfavorable to the polymerization process. For the latter to occur, it should be compensated by a negative value of the polymerization enthalpy, which implies that chain polymerization reactions are exothermic processes. When the temperature is raised, the entropy term increases as well until becoming equal, in absolute value, to the enthalpy term. The polymerization can then no longer proceed.

The maximum temperature beyond which the monomer concentration cannot be lower than a reference value, taken in general equal to the concentration of the pure monomer, is called *ceiling temperature*. For example, in the case of styrene, it corresponds to $8.6 \, mol \cdot L^{-1}$. It should be emphasized that certain authors take a monomer concentration of $1 \, mol \cdot L^{-1}$ as reference value, which entails a value of T_c higher than the one resulting from the preceding convention. The definition of the ceiling temperature is thus fully arbitrary since there exists for any temperature considered a certain monomer concentration in equilibrium with the growing chains.

In the case of liquid vinyl monomers and related ones, the value of the enthalpy of polymerization is generally in the range -30 to $-155 \, kJ \cdot mol^{-1}$; it is definitely lower (in absolute value) for heterocycles.

The two terms (enthalpy and entropy) affect the value of the ceiling temperature, but for different reasons; in the case of vinyl and related monomers, ΔH^0—which reflects the energy difference between the π bonds in the monomer molecule and the σ bonds in the polymer chain—closely depends on the number and the nature of the substituents carried by the double bond; these substituents determine the rigidity of the polymer chain and, in turn, the value of the entropy term. However, the relative variations of the entropy term with the nature of the polymer are less significant than those characterizing the enthalpy term, and hence the latter is more prominent.

The values of ΔH^0 and ΔS^0 found in handbooks (*Polymer Handbook*, *Comprehensive Polymer Science*, etc.), were actually taken from primary publications. However, these values often correspond to states of matter which differ from one monomer to another and, in addition, were determined by different means. It is thus inappropriate to present these values in a same table since they cannot be valuably compared. The readers willing to determine either the ceiling temperature or equilibrium concentration under given conditions for a particular monomer are requested to refer to primary publications whose references can be found in *Polymer Handbook*. As an example, the well-known case of α-methyl styrene is discussed below from data drawn from the article in *Journal of Polymer Science*, **25**, 488, 1957:

$$\Delta H^0 = -29.1 \text{ kJ·mol}^{-1}, \qquad \Delta S^0 = -103.7 \text{ J·mol}^{-1}\cdot\text{K}^{-1}$$

$$[M]_{\text{bulk}} = 7.57 \text{ mol.L}^{-1}$$

Based on these values, a ceiling temperature of 334 K (i.e., 61°C) was calculated for pure α-methyl styrene in the total absence of polymer.

The above example illustrates the necessity to carry out certain polymerizations at relatively low temperatures for conversions to reach completion. Should a particular monomer be characterized by a rather low ceiling temperature, the thermal decomposition of the corresponding polymer would occur at low to moderate temperature.

8.2.2. Reaction Processes Compatible with Chain Polymerizations

A chain polymerization implies that the active species formed upon addition or insertion of the last monomer molecule is of the same nature as the original one. Such chain growth also entails the formation of at least two covalent bonds between other monomer units. In view of the previously mentioned thermodynamic constraints, a negative variation of the free enthalpy of polymerization is another imperative to fulfill. These two conditions considerably restrict the variety of the organic compounds that can be polymerized, and only two main categories of monomers meet these criteria:

- Monomers carrying unsaturated groups whose high negative value of ΔH is due to the transformation of π bonds into σ bonds under the effect of an addition reaction:

- Cyclic strained monomers, which can be opened under action of an active center, by nucleophilic substitution, addition–elimination on carbonyls, and so on; here, the negative enthalpy of polymerization results from the release of the cycle strain:

Oxiranes → Polyethers
Lactams → Polyamides
Cyclosiloxanes → Polysiloxanes
Cycloalkenes → Polyalkenamers, etc.

Depending upon the electronic structure of the molecular group responsible for the polymerization, monomer molecules can be susceptible to an attack by free radicals, nucleophilic species, electrophilic species, or coordination complexes. In all cases, the polymerizability (measured by the rate constant of propagation, k_p) is determined not only by the reactivity of the monomer (M) but also by that of the active center PM^*_{n+1} resulting from its insertion,

$$PM^*_n + M \xrightarrow{k_p} PM^*_{n+1}$$

The effects induced by the substituents of the polymerizable function on the two reactivities play often in opposite directions. Generally, the reactivity of the monomer outweighs that of the corresponding active center; in other words, the higher the monomer reactivity and the lower that of the active center, the higher the corresponding rate constant of propagation. The reasons will be discussed when considering each type of polymerization.

It is indisputable that, at the present time, vinyl and related monomers are by far the most used (in particular from the economic point of view); this is why examples will be generally taken from this family of compounds.

8.3. STEREOCHEMISTRY OF CHAIN POLYMERIZATIONS

A vinyl monomer possesses a plane of symmetry and is thus achiral. Upon polymerization, sp^2-hybridized carbon atoms are transformed into sp^3 ones, and this process generates an asymmetry that is particularly noticeable when the inserted monomer molecule is located at the growing chain end:

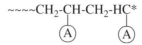

However, this asymmetry is only observed for active centers $\sim\sim\sim$HAC* in their final configuration—that is, when the carbon atom carrying the active center is sp^3-hybridized. It is therefore not the case of active centers such as carbon centered free radicals or free ions, which are sp^2-hybridized. For such systems, the final configuration of tertiary carbon atoms is fixed only after insertion of a monomer molecule—that is, next to the asymmetrical carbon atom of the penultimate unit.

Among the parameters that determine the final configuration of the last unit inserted at chain end (or the penultimate one if the final configuration is not attained), the stereochemistry of previously inserted monomeric unit is obviously a significant one. Two repeating units are necessary to define any such stereochemistry that requires conditional probabilities. Depending upon the number of preceding monomer units exerting an influence on the configuration of the last unit added, the mechanism of monomer addition indeed follows either zeroth-, first-, or second-order Markovian statistics. If a simple probability, P_m, is sufficient to describe the various additions and the structures formed—either meso (m) or racemic (r)—the process is said to follow zeroth-order Markov (or Bernouillian) statistics. If the last linkage in the chain—either m or r—controls the addition and the stereochemistry of the monomer to be added, the mechanism is called first-order Markov process. Limiting our discussion to the case of zeroth-order Markovian (or, more usually, "Bernouillian") statistics, one can thus define P_m as the probability of formation of an m dyad—that is, the insertion of two successive units of the same configuration ([R] or [S])—and define P_r as the probability of formation of an r dyad, with

$$P_r = (1 - P_m)$$

and

$$(P_r + P_m) = 1$$

From these equations, one can write the probability of existence of longer sequences, such as that of triads, and so on.

For isotactic triads (mm $= i$), one has $P_i = P_m{}^2$
For syndiotactic triads (rr $= s$), one has $P_s = (1 - P_m)^2$
For heterotactic triads (mr $=$ rm $= h$), one has $P_h = 2\,P_m\,(1 - P_m)$
For rmrr pentads, one has $P = P_m\,(1 - P_m)^3$, etc.

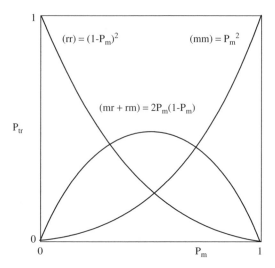

Figure 8.1. Probability (P_{tr}) of formation of *i*, *h*, and *s* triads, according to *Bernouillian statistics*.

In Figure 8.1 are plotted the variations of the probabilities of existence of the various types of triads against P_m. To check whether the addition of such monomer follows Bernouillian statistics, one generally resorts to NMR and compare the relative intensity of meso and racemic dyads and iso-, syndio-, and heterotactic triads with calculated values.

When the relative configuration of the last added unit is controlled by the configuration of the last inserted dyad (and not that of the last monomeric unit), the statistics is more complex, reflecting a peculiar mechanism of polymerization.

With ionic or coordination polymerizations, different families of active species (free ions and ion pairs, for example) may be simultaneously involved, each one of them propagating with its own statistics. Analysis of the probabilities of existence of the various sequences is even more difficult to interpret in these cases.

The identification of the type of configurational statistics thus gives extremely valuable information about the intimate mechanism of the propagation.

8.4. "LIVING" AND/OR "CONTROLLED" POLYMERIZATIONS

In certain conventional polymerizations, the three steps of initiation, propagation, and termination occur simultaneously, in a ceaseless movement that ends with the total consumption of the initiator and/or the monomer. In other words, new chains appear at all times, grow, and eventually stop growing as a result of one of the chain-breaking reactions (termination or transfer). The lifetime of a propagating center can be very short compared to the total polymerization time. Such a situation causes inevitably a great disparity in the degrees of polymerization of the various chains constituting a sample.

A completely different situation arises when the propagating active centers are not subject to transfer or termination and the initiation step is short compared to that of propagation. Although contemplated a long time ago, it was only in the 1950s that Szwarc succeeded in his search for termination/transfer-free polymerizations with his work on anionic polymerization. He called such systems "living," assimilating the initiation of polymerization to the "birth" of chains, the propagation to their "growth," and the termination/deactivation of growing species to their "demise"; Szwarc carried even further the analogy with biological systems, identifying temporarily inactive species to "dormant" ones.

If the efficiency of the initiating system is total and the time necessary to create the chains is short compared to that of the propagation, all the chains "are born" and "grow" simultaneously until all the monomer is consumed. Under such conditions the polymer samples formed exhibits a little dispersity of their molar masses.

Because the number of chains is determined by the number of molecules of initiator, the degree of polymerization of chains can be easily expressed as a function of the monomer conversion (p), the initial monomer concentration $[M]_0$, and the (monovalent) initiator concentration $[I]$:

$$\overline{X}_n = \frac{[M]_0 p}{[I]}$$

If the monomer conversion is total, this expression reduces to

$$\overline{X}_n = \frac{[M]_0}{[I]}$$

which represents

$$\overline{M}_n = \frac{\text{mass of produced polymer}}{\text{number of moles of initiator}}$$

Remark. When an initiator gives rise to a bivalent propagation (bivalent initiator or its precursor), the predictable molar masses are equal to the double of those calculated using the above relations.

As for conversion, its expression can be easily deduced assuming that all active species $[M_n^*]$ exist and propagate under an unique form and all the conditions mentioned above are respected. The rate of monomer (M) consumption can then be written as

$$-\frac{d[M]}{dt} = k_p[M] \sum_{n=1}^{\infty} [M_n^*]$$

where k_p is the rate constant corresponding to addition of the monomer molecule onto the growing species. In this equation, the total concentration in active species is written for simplicity $[M^*]$ without any index and is identified with $[I]$:

$$\sum_{n=1}^{\infty}[M_n^*] = [M^*] = [I]$$

which gives, after integration,

$$\ln \frac{[M]_0}{[M]} = k_p[M^*]t$$

Conversion can be easily deduced:

$$p = 1 - \exp\left\{-k_p[M^*]t\right\}$$

since

$$p = \frac{[M]_0 - [M]}{[M]_0}$$

thus

$$\ln \frac{[M]_0}{[M]} = -\ln(1 - p)$$

The persistence of active centers even after consumption of all the monomer allows one to trigger further chain growth by incremental addition of monomer and/or to synthesize complex macromolecular architectures that would be inaccessible by conventional polymerizations.

For systems that are not (strictly speaking) "living" and may be subject to chain-breaking reactions with possible interruption of chain growth, most of the advantages of truly "living" polymerizations may, however, be preserved, provided that transfer and termination are minimized. Indeed, if the latter reactions occur only to a limited extent and the initiation step is short compared to that of propagation, polymer chains of controlled size and relatively well-defined complex architectures can nonetheless be obtained. Such polymerizations are called "controlled."

> **Remark.** "Living" polymerizations are not necessarily "controlled." Polymerizing systems subject to a slow or incomplete initiation, as well as those with a propagation step faster than the homogeneous mixing of the reagents or faster than the rate of exchange between different active species, enter in this category of uncontrolled and yet living polymerizations. A high dispersity in the size of the resulting polymer chains is observed.

Obviously, polymerization systems that exhibit at the same time a "living" character and afford chains and architectures of controlled size and structure are in great demand. Among the specific characteristics, one can mention a low dispersity of chains. Such a narrowing of molar mass distributions with the degree of polymerization can be calculated.

Let $[M_1^*]$, $[M_2^*]$, $[M_3^*]$, ..., $[M_n^*]$ be the concentrations of active species corresponding to degrees of polymerization indicated in index, the rate of disappearance of the species M_1^* can be written

$$-\frac{d[M_1^*]}{dt} = k_p[M][M_1^*]$$

with the same for species $[M_2^*]$ and $[M_n^*]$:

$$+\frac{d[M_2^*]}{dt} = k_p[M]\{[M_1^*] - [M_2^*]\}$$

$$+\frac{d[M_n^*]}{dt} = k_p[M]\{[M_{n-1}^*] - [M_n^*]\}$$

On the other hand, the average number $(d\bar{v})$ of monomeric units consumed by an active chain during an interval of time dt can be expressed as follows:

$$d\bar{v} = k_p[M]dt = -\frac{d[M]}{[M^*]}$$

Identifying $k_p[M]dt$ with $d\bar{v}$ in the expression of the rate of disappearance of $[M_1^*]$ species, one obtains

$$\frac{d[M_1^*]}{[M_1^*]} = -d\bar{v}$$

which gives, upon integration,

$$\int_I^{M_1^*} \frac{d[M_1^*]}{[M_1^*]} = -\int_0^t d\bar{v} \Rightarrow \ln\frac{[M_1^*]}{[I]} = -\bar{v}$$

and thus

$$[M_1^*] = [I]e^{-\bar{v}}$$

Introducing $[I]e^{-\bar{v}}$ in the expression of the rate of disappearance of the species $[M_2^*]$ gives

$$d[M_2^*] = \{[I]e^{-\bar{v}} - [M_2^*]\}d\bar{v}$$

which is a differential equation of the following type:

$$dy = (ae^{-x} - y)\,dx \quad \text{or} \quad \frac{dy}{dx} + y = ae^{bx}$$

whose solution is

$$y = axe^{bx}$$

The variation of $[M_2^*]$ as a function of $\bar{v}$ can then be written

$$[M_2^*] = [I]\bar{v}e^{-\bar{v}}$$

Repeating the same reasoning for the variation of $[M_3^*]$, one obtains

$$d[M_3^*] = \left\{[I]\bar{v}e^{-\bar{v}} - [M_3^*]\right\}d\bar{v}$$

an expression which is of the type

$$\frac{dy}{dx} + y = P(x)ae^{bx}$$

whose solution is

$$y = e^{ax}\int P(x)\,dx$$

Thus, the variation of $[M_3^*]$ can be deduced as follows:

$$[M_3^*] = [I]\left(\frac{\bar{v}^2}{2}\right)e^{-\bar{v}}$$

and in the general case of $[M_i^*]$

$$[M_i^*] = [I]\frac{\bar{v}^{(i-1)}e^{-\bar{v}}}{(i-1)!}$$

The molar fraction of the species of degree of polymerization i is thus written

$$f_i = \frac{[M_i^*]}{[I]} = \frac{\bar{v}^{(i-1)}e^{-\bar{v}}}{(i-1)!}$$

As for the mass fraction of the species having a degree of polymerization i, it can be easily deduced [if one identifies the mass of fragment I of the initiator (M_a) with that of a repetitive unit (M_0)]:

$$W_i = \frac{M_0(i+1)f_i}{M_0\sum_{i=1}^{\infty}(i+1)f_i} = \frac{(i+1)f_i}{\bar{v}+1}$$

which can be also written

$$W_i = \frac{\bar{v}e^{-\bar{v}}}{(\bar{v}+1)} \cdot \frac{\bar{v}^{(i-1)}}{(i-1)!}(i+1)$$

The expressions of $\overline{M}_n$ and $\overline{M}_w$ can thus be easily deduced:

$$\overline{M}_n = M_0\overline{v} + M_a = M_0(\overline{v} + 1)$$

$$\overline{M}_w = M_0\overline{X}_w = M_0 \sum_{i=1}^{\infty} i w_i = M_0 \frac{\overline{v}e^{-\overline{v}}}{\overline{v}+1} \sum_{i=1}^{\infty} \frac{i^2\overline{v}^{(i-2)}}{(i-1)!}$$

The above sum can also be written $\left(\overline{v} + 3 + \frac{1}{\overline{v}}\right)e^{\overline{v}}$, giving for $\overline{M}_w$

$$\overline{M}_w = M_0 \frac{(\overline{v}^2 + 3\overline{v} + 1)}{(1 + \overline{v})}$$

and thus

$$D_M = \frac{\overline{M}_w}{\overline{M}_n} = 1 + \frac{\overline{v}}{(\overline{v}+1)^2}$$

The samples obtained under such conditions exhibit a Poisson-type distribution of their molar masses. This type of distributions is obtained when one distributes in a random way m objects in n boxes, with $m >>> n$.

8.5. FREE RADICAL POLYMERIZATION

8.5.1. Reminders on Free Radical Reactions

Free radicals can be regarded as resulting from the homolytic rupture of covalent bonds. They are generated by using either physical (thermal, radiative, etc.) exci-tation or chemical (oxydo-reduction, free radical addition, etc.) means. If they are not stabilized by particular substituents, their lifetime (about one second in normal polymerization conditions) is extremely short due to a very high reactivity. Their hybridization state is generally trigonal (sp^2) except for those carrying substituents of large size developing steric hindrance.

Free radicals can be involved in the following six reactions, all occurring in the polymerization processes:

Combination	$R^{\bullet} + {}^{\bullet}R' \longrightarrow R-R'$
Disproportionation	$2\ R-CH_2-CH_2{}^{\bullet} \longrightarrow R-CH_2-CH_3 + R-CH=CH_2$
Abstraction/transfer	$R^{\bullet} + R'X \longrightarrow RX + {}^{\bullet}R'$
Addition	$R^{\bullet} + H_2C=CR_1R_2 \longrightarrow R-CH_2-{}^{\bullet}CR_1R_2$
Fragmentation	$RA^{\bullet} \longrightarrow R^{\bullet} + A$
Rearrangement	$R''R'R^{\bullet} \longrightarrow {}^{\bullet}R''R'R$

Free radicals can be stabilized by resonance and electron-withdrawing effects. When their stabilization is sufficient—in particular, due to the existence of many canonical forms—they can become persistent and be isolated, like the following free radicals:

Diphenylpicrylhydrazyl (DPPH)

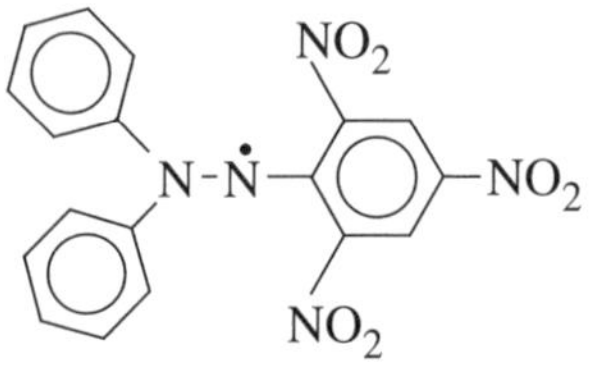

Tetramethylpiperidyloxyl (TEMPO)

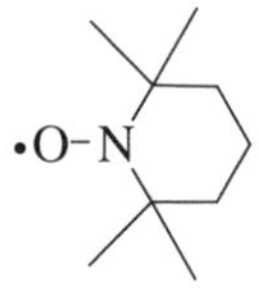

For TEMPO, the most suitable representation features a 3-electron N-O bond which explains why this free radical cannot dimerize by its nitrogen or oxygen atom. The free radicals of this family (known as "nitroxyl" radicals) are usually employed to reversibly trap growing transient radicals and thus ensure a control of the propagation step (see Section 8.5.8).

Free radicals have thus a marked tendency to participate in chain reactions, more particularly in addition and abstraction reactions.

8.5.2. General Kinetic Scheme of Free Radical Polymerization

This kinetic scheme describes the initiation step by a molecule (initiator I) releasing free radicals by homolytic rupture of a covalent bond (dissociation reaction).

Initiation:

Activation

$$I \xrightarrow{k_d} 2R^\bullet$$

then

$$R^\bullet + M \xrightarrow{k_i} R-M^\bullet$$

with

$$R^\bullet + R^\bullet \xrightarrow{k_c} R-R \qquad \text{as side reaction.}$$

Because of their proximity when they appear in the reaction medium and the high value of the rate constant of combination (k_c), a non-negligible fraction of $R^\bullet$

radicals generated by the initiator are lost in termination reactions and thus do not initiate polymeric chains: the proportion really active is called *efficiency factor* or *efficiency* (*f*) of this initiator.

To establish the kinetic equations, it is considered—what was experimentally established—that all reactions occurring in free radical polymerizations are first-order with respect to each reactive species.

For the initiation step:

$$R_d = -d[I]/dt = k_d[I] = \tfrac{1}{2}d[R^\bullet]/dt$$
$$R_i = +d[RM^\bullet]/dt = k_i[R^\bullet][M]$$

Since the first "leg" of the initiation step (activation of initiator) is generally slow due to a high value of the corresponding energy of activation (about $120\,kJ\cdot mol^{-1}$), it determines the global kinetics of initiation; thus

$$R_i = +d[RM^\bullet]/dt = 2f\,R_d = 2f k_d[I]$$

The coefficient 2 takes into account the fact that two $R^\bullet$ radicals are simultaneously formed by decomposition of one molecule of initiator (I).

Propagation:

$$RM^\bullet + M \longrightarrow RMM^\bullet$$

$$RM_n^\bullet + M \xrightarrow{k_p} RM_{n+1}^\bullet$$

In a first approximation—confirmed experimentally—one assume that the rate constant of propagation (k_p) is nearly independent of the degree of polymerization.

The rate of propagation (R_p) is roughly equal to the total rate of polymerization (R_{pol}) since all monomer molecules except one per chain (that implied in initiation) are consumed during this step:

$$R_p = k_p \left[\sum_n RM_n^\bullet \right][M]$$

Because of the low selectivity of free radical reactions, the chain growth can be stopped at any moment by **termination** reactions; two polymeric radicals are neutralized either by combination or by disproportionation in the process:

$$RM_n^\bullet + RM_m^\bullet \underset{k_{disp}}{\overset{k_c}{\rightleftarrows}} \begin{array}{c} RM_{n+m} \\ RM_n + RM_m \end{array}$$

The two reactions can occur simultaneously and thus the rate constant of termination (k_t) corresponds to a weight average of the individual rate constants (k_c, k_{dis}).

The overall rate of termination (R_t) is given by

$$R_t = k_t \left[\sum_n RM_n^\bullet \right]^2$$

Remarks

(a) In the above equation, k_t is the rate constant of the two types of bimolecular termination reactions.

(b) A factor of 2 is often found in the literature in the expression of the rate of termination to take into account the fact that 2 polymeric radicals are consumed by a same termination reaction. This reasoning is unjustified and is equivalent to count twice the reactive species participating in the termination process, which is appropriately described through the square of the concentration in free radicals. It is recommended to be careful when using k_t values found in the literature.

Because of the respective values of rate constants of termination ($k_t \sim 10^7$ to 10^8 L·mol^{-1}·s^{-1}) and propagation ($k_p \sim 10^2$ to 10^4 L·mol^{-1}·s^{-1} at 60°C) reactions, it is recommended to work with particularly low instantaneous concentrations in free radicals ([RM$_n^\bullet$] $\sim 10^{-8}$ M), in order to favor propagation over termination reactions. It is difficult to measure such low value of [RM$_n^\bullet$], except by using a spectrometric technique as sensitive as electron spin resonance (ESR). Assuming that the number of active chains remains constant—which is true only during short intervals of time—, one can calculate the rate of polymerization even if [RM$_n^\bullet$] is experimentally inaccessible and thus unknown. This assumption implies that the rate of appearance of RM$_n^\bullet$ is equal to their rate of disappearance, which corresponds to *steady-state* conditions; one can accordingly write $R_i = R_t$, which corresponds to

$$2 f k_d [I] = k_t \left[\sum_n RM_n^\bullet \right]^2$$

This equation can be solved for [$\sum_n RM_n^\bullet$]:

$$\left[\sum_n RM_n^\bullet \right] = \{ 2f[I]k_d/k_t \}^{1/2}$$

Introducing this expression into the equation of the rate of polymerization gives

$$R_p = k_p \left\{ 2f[I]\frac{k_d}{k_t} \right\}^{1/2} [M] \sim R_{pol} = -d[M]/dt$$

where R_{pol} is the overall rate constant of polymerization. The above equation can be rewritten as

$$-\frac{d[M]}{[M]} = k_p \left\{ 2 f [I] k_d / k_t^{1/2} \right\} dt$$

which corresponds to

$$\ln \frac{[M]_0}{[M]} = k_p \left\{ 2 f [I] \frac{k_d}{k_t} \right\}^{1/2} t$$

assuming that [I] is constant and does not vary over the period t. The general equation of polymerization can also be expressed under the form

$$R_{pol} = \text{constant } [I]^{1/2}[M]$$

or

$$R_{pol} = \text{constant } [M] R_i^{1/2}$$

The last equation shows that the overall rate of polymerization is primarily determined by the rate of initiation. This is not surprising and can be accounted for by examining the energies of activation of the various steps:

$$R_p^2 = 2(k_p^2 / k_t) k_d f [I][M]^2$$

from which one obtains

$$2E_{ao} = 2E_{ap} + E_{ad} - E_{at}$$

or

$$E_{ao} = E_{ap} + E_{ad}/2 - E_{at}/2$$

where E_{ao}, E_{ap}, E_{ad}, and E_{at} represent the energies of activation of the overall polymerization, the propagation, the dissociation (first part of initiation), and the termination, respectively. The values of E_{ap} and E_{at} could be experimentally determined for a certain number of systems. For example, in the case of the free radical polymerization of styrene one has:

$$(E_{ap} - E_{at}/2) \sim 27 \text{ kJ·mol}^{-1}$$

The energy of activation of dissociation for most of the initiators functioning by homolytic rupture of a covalent bond ranges between 120 and 170 kJ·mol^{-1}; since the experimental determination of E_{ao} gives a value in the range of $\sim$100–125 kJ·mol^{-1}, one can deduce that it is E_{ad} which contributes the most to the value of E_{ao}. The generation of primary free radicals is thus the step that determines the global kinetics of the whole process.

> **Remark.** Energies of activation of reactions that generate free radicals by redox systems ($\sim$50 kJ·mol^{-1}) are much lower than those corresponding to homolytic ruptures; polymerization kinetics are likely to be affected by such a difference.

8.5.3. Initiation of Free Radical Polymerizations

8.5.3.1. Generation of Initial ("Primary") Free Radicals. Most of the free radical initiators (generators) used are unstable molecules that can homolytically dissociate ($I \rightarrow 2\,R^{\bullet}$) under thermal effect, due to the presence of a weak covalent bond.

The homolytic dissociation of a covalent bond is all the weaker since:

- The electronegativity of the covalently bonded elements is high.

$$E_{d_{O-O}} < E_{d_{N-N}} < E_{d_{C-C}}$$

- The stabilization (by electron-donor and/or resonance effects) of the radicals resulting from the dissociation is high (see Table 8.1).

Table 8.1. Dissociation energy of C–H bonds and stabilization energy of the corresponding hydrocarbon free radicals

Molecule R–H	Dissociation energy E_d (C–H) (kJ·mol^{-1})	Stabilization Energy of Radical $R^{\bullet}$ (kJ·mol^{-1})
H–CH$_3$	426	0
H–CH$_2$–CH$_3$	393	33
H–C(CH$_3$)$_3$	376	50
H–CH$_2$–CH=CH$_2$	324	102
H—CH$_2$—⬡	322	104

It must be stressed that radicals generated by dissociation are all the more reactive because their formation is difficult; thus, in Table 8.2, methyl radical is the most reactive among the represented radicals.

Organic peroxides and **hydroperoxides** are very commonly used at the laboratory scale as well as at the industrial level. Their instability can be characterized by their half-life time ($t_{1/2}$)—that is, the time necessary to their half-decomposition at a given temperature—or by the temperature at which they exhibit a given half-life time (see Table 8.2); from these half-life times it is possible to easily find the corresponding value of k_d using the kinetic equation of decomposition of the initiator

$$[I] = [I]_0 \exp(-k_d t)$$

corresponding to

$$\ln[I]_0/[I] = k_d t$$

Table 8.2. Half-life times of organic peroxides

	Temperature (°C) for a Half-Life of			Half-Life Times (hours) for various temperatures											
	10 h	1 h	1 min	40°	50°	60°	70°	80°	90°	100°	110°	120°	130°	140°	150°
2,5-Di(*tert*-butylperoxy)-2,5-dimethylhexyne	128	149	191	—	—	—	—	—	—	—	90	27	8.1	1.7	0.9
Tert-butyl peroxide	126	147	186	—	—	—	—	—	—	—	75	22	6.0	1.4	0.6
Cumyl hydroperoxide	122	147	200	—	—	—	—	—	—	115	42	13	4.8	2.0	0.8
Tert-butyl hydroperoxide	121	140	179	—	—	—	—	—	—	165	42	12	3.2	1.0	0.3
2,5-Di(*tert*-butylperoxy)-2,5-dimethylhexane	118	137	172	—	—	—	—	—	—	135	33	8.5	2.3	0.7	0.2
Dicumyl peroxide	117	134	170	—	—	—	—	—	—	100	25	6.6	1.7	0.5	0.1
Tert-butyl perbenzoate	109	125	163	—	—	—	—	—	135	30	7.8	2.2	0.6	0.2	—
2-2-Bis(*tert*-butylperoxy)butane	103	125	168	—	—	—	—	—	55	16	4.8	1.6	0.6	—	—
Tert-butyl diperphthalate	103	122	160	—	—	—	—	—	78	19	4.8	1.4	0.5	—	—
Tert-butyl peracetate	101	120	160	—	—	—	—	165	43	11	3.3	1.1	0.3	—	—
2,5-Dibenzoylperoxy-2,5-dimethylhexane	100	119	158	—	—	—	—	135	37	10	2.8	0.9	0.3	—	—
Tert-butyl permaleate	81	105	155	—	—	—	35	12	4.2	1.3	0.6	0.3	—	—	—
Tert-butyl perisobutyrate	78	95	130	—	—	120	28	6.7	1.8	0.5	0.1	—	—	—	—
Bis(4-chlorobenzoyl) peroxide	76	94	133	—	—	67	18	5.5	1.5	0.5	0.2	—	—	—	—
Tert-butyl per-2-ethylhexanoate	76	92	124	—	—	140	28	5.7	1.4	0.4	0.1	—	—	—	—
Benzoyl peroxide	72	92	133	—	—	45	13	3.7	1.2	0.4	0.1	—	—	—	—
Succinyl peroxide	67	90	142	—	—	19	7.0	2.5	1.0	0.4	—	—	—	—	—
Acetyl peroxide	68	86	122	—	—	30	8.0	2.2	0.6	0.2	—	—	—	—	—
Propionyl peroxide	64	81	118	—	80	17	4.5	1.2	0.4	0.1	—	—	—	—	—
Lauroyl peroxide	62	80	116	—	70	15	3.7	1.0	0.3	—	—	—	—	—	—
Decanoyl peroxide	62	79	115	—	67	13	3.5	1.0	0.3	—	—	—	—	—	—
Octanoyl peroxide	62	78	114	—	63	13	3.3	0.9	0.3	—	—	—	—	—	—
Bis-(3,5,5-trimethylhexanoyl) peroxide	60	77	112	—	47	9.9	2.6	0.7	0.2	—	—	—	—	—	—
Tert-butyl perpivalate	55	73	110	—	19	5.0	1.5	0.4	—	—	—	—	—	—	—
Bis-(2,4-dichlorobenzoyl) peroxide	55	72	106	—	20	4.7	1.4	0.3	—	—	—	—	—	—	—
Tert-butyl per-neodecanoate	48	66	99	40	8.0	1.7	0.4	—	—	—	—	—	—	—	—
Isopropyl peroxydicarbonate	47	62	95	30	6.0	1.2	0.3	—	—	—	—	—	—	—	—
Cyclohexyl peroxydicarbonate	45	60	93	26	4.2	0.9	0.2	—	—	—	—	—	—	—	—
Acetylcyclohexane-sulfonyl peroxide	32	46	67	5	0.4	0.05	—	—	—	—	—	—	—	—	—

and for $[I]_0/[I] = 2$ we have

$$k_d = 0.693/t_{1/2}$$

In general, free radicals initiators are used under conditions of half-life times of about 10 hours. The decomposition of peroxides can be single-step or multistep; for example, dicumyl peroxide (DICUP) decomposes as follows:

whereas the decomposition of benzoyl peroxide requires two successive steps with a last fragmentation step:

The same is true for *tert*-butyl peroxide:

$$(CH_3)_3C-O-O-C(CH_3)_3 \xrightarrow{k_d} 2(CH_3)_3C-O^\bullet \longrightarrow 2(H_3C^\bullet + CH_3-CO-CH_3)$$

In the last two cases, the two reactions can occur successively only if the initially generated radicals did not succeed in adding a monomer molecule to initiate the polymerization.

When the reaction medium requires the initiator to be water soluble (polymerizations in emulsion, in aqueous solution, etc.), mineral peroxides such as potassium persulfate are often utilized:

$$K^+, {}^-O_3S-O-O-SO_3{}^-, K^+ \xrightarrow{k_d} 2SO_4^{\bullet-}, K^+$$

Azo compounds also are very much used; and depending on whether they carry hydrophilic groups or not, they can be water- or organosoluble; it is the case of azobis(isobutyronitrile) (AIBN)

which is organosoluble, whereas its dicarboxylic homolog

is water-soluble.

It is sometimes necessary to generate free radicals at low temperature, which implies that reactions with low activation energy such as **oxydo-reductions** are used. Depending upon the needs of the reaction medium, one can use either hydrophilic, hydrophobic, or mixed systems of initiation; for example,

$$S_2O_8{}^{2-} + S_2O_3{}^{2-} \longrightarrow SO_4{}^{2-} + SO_4{}^{-\bullet} + S_2O_3{}^{-\bullet}$$

$$Fe^{2+} + H_2O_2 \longrightarrow Fe^{3+} + HO^- + HO^\bullet$$

$$R\text{--}OH + Ce^{4+} \longrightarrow RO^\bullet + Ce^{3+} + H^+$$

$$RO\text{--}OH + Fe^{2+} \longrightarrow Fe^{3+} + HO^- + RO^\bullet$$

or even

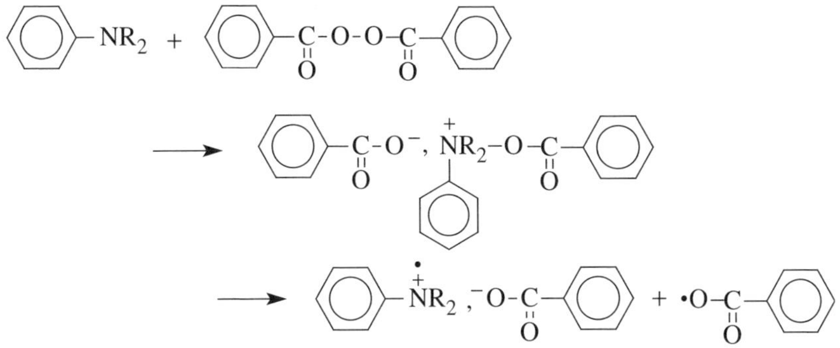

Photochemical initiation resulting from the activation of monomer molecules by photons alone is difficult to achieve; generally, a molecule is added in the reaction medium which will be used as intermediate between the photon and the monomer molecule to activate. Free radicals can be generated by intramolecular scission; an example is given below for benzoin ethers:

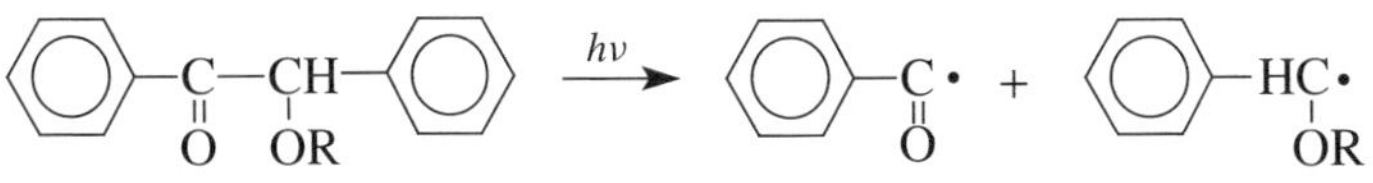

Another possibility is the intra- or intermolecular abstraction of $H^\bullet$ as with benzophenone in association with an amine:

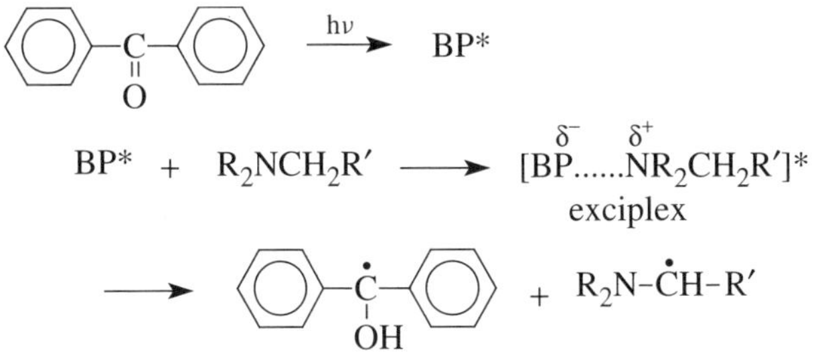

A system made of an onium salt associated with a H• donor can also be used; the reaction pathway suggested is as follows:

$$Ar_3S^+, A^- \xrightarrow{h\nu} [Ar_3S^+, A^-]^* \longrightarrow Ar^\bullet + Ar_2S^{+\bullet}, A^-$$

Triarylsulfonium salt

then, in the presence of a donor (DH), we have

$$Ar_2S^{+\bullet}, A^- + DH \longrightarrow Ar_2HS^+, A^- + D^\bullet$$

followed by

$$Ar_2HS^+, A^- \longrightarrow Ar_2S + H^+, A^-$$

Initiation can also be obtained by direct physical activation of monomer molecules as shown below.

Thermal initiation is widely used, but, in general, the outcome is not "pure" because it may be disturbed by the presence of impurities in the reaction medium (atmospheric oxygen, in particular) which participate in the uncontrolled generation of "primary" free radicals. Thermal activation is utilized in industry to polymerize styrene; due to thermal agitation, collision between monomer molecules brings about a complex mechanism that ends up with the formation of two monoradicals as described hereafter; each one of these free radicals can initiate a polymeric chain:

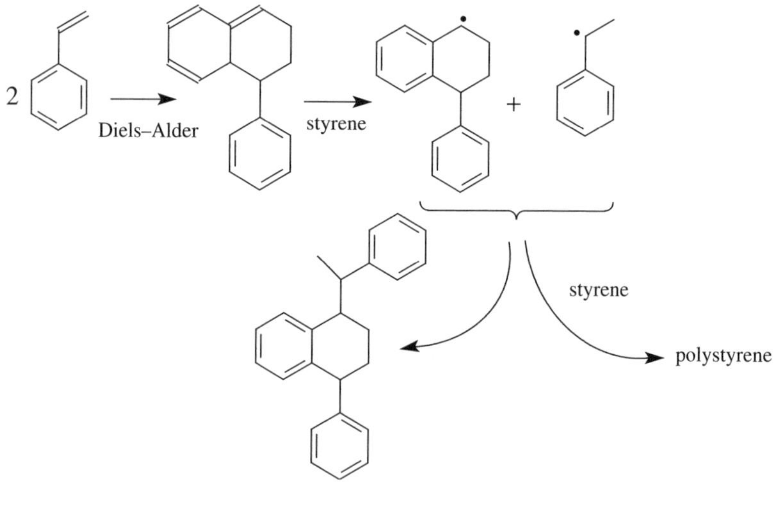

Ionizing **radiations** (β, γ) can also initiate free radical polymerizations through two mechanisms involving any molecule (AB) present in the reaction medium (monomer, solvent...):

- Excitation similar to that mentioned for photochemical initiation

- Ionization, which is responsible for many side reactions:

$$AB \xrightarrow{\gamma} {}^{+}AB^{\bullet} + e^{-}$$
$${}^{+}A + B^{\bullet}$$
$$AB + e^{-} \longrightarrow {}^{\bullet}AB^{-} \longrightarrow {}^{\bullet}A + B^{-}$$

Both free radicals and (positive or negative) radical ions can be formed upon radiation and initiate polymerizations.

Depending upon the nature of the monomer and the experimental conditions of the reaction medium, polymerization can be either ionic, free radical, or mixed; however, polymerizations triggered by irradiation are generally carried out at room temperature, which favors free radical processes. In all cases, like for photochemical polymerization, the rate of generation of free radicals is proportional to the intensity of the radiation that is absorbed by the system. In the absence of solvent, one can write

$$R_i = f I [M]$$

where I is the intensity of the radiation and f is its efficiency.

In steady-state conditions one has

$$R_i = f I [M] = R_t = k_t \left[\sum_n RM_n^{\bullet} \right]^2$$

$$\left[\sum_n RM_n^{\bullet} \right] = \left\{ f I [M] / k_t \right\}^{1/2}$$

$$R_{pol} \# R_p = k_p \left[\sum_n RM_n^{\bullet} \right] [M] = k_p \left\{ f I / k_t \right\}^{1/2} [M]^{3/2}$$

This is why the monomer concentration appears in the above kinetic equation with a power 3/2.

8.5.3.2. Monomer Addition by a "Primary" Radical (Initiation).

The activation of polymerization by free radical means is currently utilized with vinyl and related monomers:

$$R^{\bullet} + \underset{/}{\overset{\backslash}{C}} = C \overset{\nearrow}{\underset{\backslash}{}} \xrightarrow{k_i} R - \overset{|}{\underset{|}{C}} - \overset{|}{\underset{|}{C}} ^{\bullet}$$

The concentration in "primary" free radicals ($R^{\bullet}$) is not only determined by [I] and k_d but also by their efficiency (f). For a given radical $R^{\bullet}$, f reflects its relative aptitude to add on the monomer double bond rather than to self-terminate. Thus, f depends on both the ratio of the rate constant of initiation to the rate constant of termination of "primary" free radicals ($k_i/k_{t_R\bullet}$) and the monomer concentration. For example, for low monomer concentration, free radicals resulting from the decomposition of benzoyl peroxide can mutually deactivate and give

Phenylbenzoate

and/or another possible reaction

Biphenyl

Generally, the rate constants of recombination of free radicals are very high (in the range of $10^8 \, \text{L·mol}^{-1}\text{·s}^{-1}$) and the main reason for the "survival" of $R^\bullet$ before it could add onto a monomer molecule is its low instantaneous concentration. The rate constant of initiation (k_i) is determined not only by the reactivity of the free radical ($R^\bullet$) but also by that of the monomer (M). These two parameters will be successively analyzed.

Steric hindrance and electronic (inductive and resonance) effects are involved in the intrinsic reactivity of $R^\bullet$. The two same effects also determine the reactivity of the monomer (M). To evaluate the proper reactivity of M irrespective of that of $R^\bullet$, it can be measured by what is called *methyl affinity*. By convention, this affinity (a) is taken equal to the ratio of the rate constant of addition (k_i) onto the monomer double bond to the rate constant of a reference reaction, which is the transfer reaction (k_{tr}) of $^\bullet CH_3$ to isooctane (abstraction reaction):

$$^\bullet CH_3 + C_8H_{18} \xrightarrow{k_{tr}} CH_4 + {}^\bullet C_8H_{17}$$

$$a = k_i/k_{tr}$$

Addition of $^\bullet CH_3$ to the monomer double bond is all the easier as the new radical formed $RM^\bullet$ is more stabilized, as shown by the values given in Table 8.3. The energy level of the final state is thus one of the parameters that determine the intrinsic reactivity of a given monomer.

Steric effects can also play an important role on the affinity of the double bonds for methyl free radicals, and the values shown in Table 8.4 show that the disubstitution of a double bond in β-position reduces considerably its reactivity. The increase in reactivity that is observed when the monomer is disubstituted in α-position is due to the stabilising effect of alkyl groups on the radical formed.

Depending upon the respective reactivity of $R^\bullet$ and M and the polymerization conditions, the efficiency factor varies from 0.30 to 0.95 with most commonly

Table 8.3. Affinity for methyl radical of some ethylenic monomers. Effect of the stabilizing capacity

Monomer $H_2C-CH-A$	Stabilizing group of $RM^\bullet$ (A)	Methyl affinity $a = k_i/k_{tr}$
Ethylene	—	17
Propylene	$-CH_3$	22
Isobutene	$-(CH_3)_2$	36
Styrene	$-C_6H_5$	792
Butadiene	$-CH=CH_2$	1008

Table 8.4. Methyl affinity of some styrene monomers — steric effects

Monomer	Formula	Methyl affinity $a = k_i/k_{tr}$
Styrene		792
α-Methylstyrene		926
cis-β-Methylstyrene		40
trans-β-Methylstyrene		92.5
α,β-Dimethylstyrene		66
α_1,β_1,β_2-Trimethylstyrene		20

used initiators. The efficiency is higher in the absence of solvent because effective collisions between free radicals and monomer molecules are more likely. On the other hand, in the presence of a solvent, free radicals resulting from the decomposition of the initiator can be subject to a "cage effect" by solvent molecules. They react together before target monomer molecule could be reached, which contributes to decreasing the efficiency.

Remark. Vinyl and related monomers are those whose polymerization is the result of an addition reaction onto C=C double bond: vinyl, acrylic, diene, allyl, etc., monomers.

The rate constants of addition of R$^{\bullet}$ onto M vary from 10 to 10^5 L·mol^{-1}· s^{-1}. To reach a high value of the efficiency factor an optimization of the reaction conditions is required for the competition between the two processes to be in favor of initiation.

8.5.4. Free Radical Propagation

Polymer chains grow during this step; it is thus an essential step for both the structure of the resulting polymer and the properties of the material formed.

Like the initiation step, the reaction mechanism is of the "free-radical-addition" type:

As for the **reactivity** (polymerizability), the situation is different from that of initiation. Indeed, the new free radical active center formed after monomer insertion in the polymeric chain is roughly identical to the last formed one. Thus, its formation does not entail an increase of stability. The negative variation of the free enthalpy is only due to the exothermic transformation of the monomer molecule into a monomeric unit. The stabilizing power of the substituent A carried by the double bond is exerted not only on the active center formed after addition but also on the monomer molecule. Logically, a progressive decrease of k_p is observed with the increase of the stabilizing power of substituents A (see Table 8.5).

Table 8.5. Polymerizability (rate constant of propagation) of some ethylenic monomers at 25°C

Monomer	k_p (L·mol^{-1}·s^{-1})	Monomer	k_p (L·mol^{-1}·s^{-1})
Vinylidene chloride	9	Methyl methacrylate	1,010
Styrene	35	Vinyl chloride	3,200
Chloroprene	228	Acrylamide	18,000
Acrylic acid	650	Acrylonitrile	28,000

Remark. Even though the reactivity of the propagating species is known to vary with the degree of polymerization for the shortest oligomers, it can be considered that the monomer polymerizability (k_p) remains essentially constant throughout the propagation step.

As for the **regioselectivity** of the process, some irregularities in the placement of the monomeric units can be observed; their proportion depends closely on the stabilizing effect of the substituent A on the growing free radical (see Table 8.3): the higher the stabilization of growing free radicals, the lower will be the proportion of irregular (head-to-head or tail-to-tail) placements.

In the case of conjugated dienes the free radical is subject to resonance and can thus react through the last carbon atom of the chain or the antepenultimate one; monomeric units of respectively 1,4- or 1,2- (or 3,4-, depending on the regioselectivity in the case of monosubstituted dienes) type are formed. The free radical polymerization of butadiene leads to approximately 80% of 1,4-type units.

The **stereoselectivity** of free radical polymerizations is generally poor because of the indetermination of the free radical configuration (sp^2 hybridization) at the time of its addition to the entering monomer (see Table 8.6).

For certain monomers, however, steric and electrostatic effects can play a role and induce a slight difference in the activation energy of the addition reaction, thus favoring the formation of racemic dyads (r) compared to that of meso dyads (m) (see Table 8.7). However, for polymerizations carried out under usual conditions of temperature, this small difference in the activation energy between $E_{a(m)}$ and $E_{a(r)}$ leads essentially to atactic polymers with a slight prominence of r dyads (Table 8.6). With certain encumbered or polarized monomers, a more marked

Table 8.6. Tacticity of several vinyl polymers obtained by free-radical polymerization

Polymer	Polymerization Temperature (in °C)	$P_m = (1 - P_r)$
Poly(vinyl chloride)	40	0.40
Poly(vinyl fluoride)	37	0.42
Polyacrylonitrile	35	0.52
Polymethacrylonitrile	60	0.41
Poly(methyl acrylate)	60	0.49
Poly(methyl methacrylate)	60	0.20
Polystyrene	80	0.46
Poly(vinyl acetate)	80	0.46

Table 8.7. Influence of the temperature of free radical polymerizations on the tacticity of poly(methyl methacrylate)

T (°C)	$P_m = (1 - P_r)$	T (°C)	$P_m = (1 - P_r)$
−35	0.12	60	0.20
−15	0.16	110	0.22
10	0.17	160	0.25
30	0.18	250	0.27

tendency to syndiotacticity is observed for polymerizations carried out at low temperature (Table 8.7).

8.5.5. Termination Step

This step implies the collision of two growing free radicals which, considering their low instantaneous concentration, is less probable than their collision with monomer molecules. Under normal conditions of polymerization—that is, for $[\sim M^{\bullet}] \sim 10^{-8}\,mol \cdot L^{-1}$—propagation is thus favored compared to termination. The latter occurs nevertheless and determines the kinetic behavior of the systems and—in absence of transfer reactions—the average molar mass of the sample.

From a mechanistic point of view, the termination reactions between "primary" and/or macromolecular free radicals can occur by two different processes. In the absence of steric constraints, the collision of two macromolecular free radicals results in their combination by the scheme shown below:

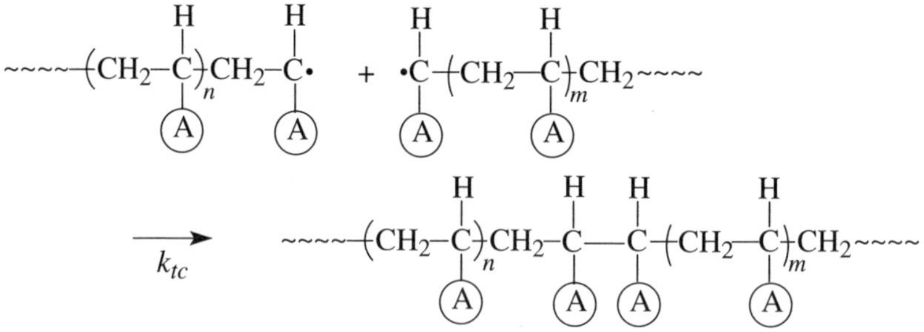

Monomers whose termination occurs exclusively by this process do not possess a bulky substituent **A** whose position and size would destabilize the resulting covalent bond. It is the case for most vinyl monomers, and Table 8.8 shows the very high values of the corresponding rate constants; such high values are associated with a low value of the energy of activation of the process ($E_a \sim 10\,kJ \cdot mol^{-1}$).

Table 8.8. Rate constants of termination by combination (k_{tc}) for some vinyl and related monomers, at 60°C

Monomer	$10^{-7}\, k_{tc}$ $(L \cdot mol^{-1} \cdot s^{-1})$
Ethylene	54
Styrene	5.8
Methyl acrylate	1.8
Acrylonitrile	8.0
Acrylic acid	0.9
Vinyl acetate	5.8
Vinyl chloride	150

Table 8.9. Rate constants of termination of active polystyrene chains versus their degree of polymerization, X_n (at 60°C)

X_n	10^{-6} k_t (L·mol^{-1}·s^{-1})	X_n	10^{-6} k_t (L·mol^{-1}·s^{-1})
5	200	100	10
20	50	500	2
50	20	1000	1

Depending upon the composition of the reaction medium, the methods used for the determination and the authors, values of k_t can be very different. This is not a surprise since these values are strongly influenced by the size of the reactive species concerned (see Table 8.9), more than for other free radical reactions implying macromolecular chains. The rate constant (k_t) is also influenced by the viscosity of the reaction medium, this last aspect having a major incidence on the aptitude of vinyl and related monomers to undergo polymerization in bulk (see Section 8.8.2). For example, in the polymerization in bulk of styrene at 100°C, k_t decreases by four orders of magnitude beyond 30% of conversion for free radical active chains having the same degree of polymerization.

When steric hindrance prevents their combination, the collision of two free radicals can result in their disappearance by a disproportionation reaction. This reaction is described to occur in two half-steps. The first one corresponds to a β-elimination under the effect of the collision, and the second one corresponds to the combination with the ejected H$^•$:

In the radical polymerization of methyl methacrylate, termination by combination (also called coupling) is demanding because it requires that each carbon atom forming the resulting bond be tetrasubstituted. Thus, the relative importance of the two processes varies with temperature since their activation energies are appreciably different:

$$k_t = 1.9 \times 10^7 \text{L·mol}^{-1}\text{·s}^{-1} \quad \text{at } 25°C \text{ (32\% coupling)}$$

$$k_t = 2.5 \times 10^7 \text{L·mol}^{-1}\text{·s}^{-1} \quad \text{at } 60°C \text{ (15\% coupling)}$$

With nonhindered monomers, the energy of activation of the coupling reaction is comparable to that of disproportionation and is in the range of about $20\,kJ\cdot mol^{-1}$; thus the two modes of termination can coexist. In the latter case, only the apparent overall rate constants are considered.

8.5.6. Transfer Step

The phenomenon of transfer of growing active centers onto other species present in the reaction medium is extremely frequent in free radical polymerization:

$$\sim\sim\sim M_n^{\bullet} + T \xrightarrow{k_{tr}} \sim\sim\sim M_n + T^{\bullet}$$

It can be regarded as a true transfer only if the formed species ($T^{\bullet}$) is able to react with a monomer molecule and initiate a new chain:

$$T^{\bullet} + M \longrightarrow TM^{\bullet} \xrightarrow{nM} T-M_n-M^{\bullet}$$

The phenomenon of transfer can be either provoked or undergone by the system; in the former case it is used to control average molar masses. The influence of transfer on the value taken by the average molar masses of the resulting polymers is treated in Section 8.5.8. However, one can intuitively deduce that the higher the ratio of the rate of transfer to the rate of propagation, the shorter the chains resulting from polymerization. This leads to defining a *constant of transfer* for each species participating in the transfer process, each one indicated by C_{tr}, which is equal to the ratio k_{tr}/k_p. This value is a measure of the importance of transfer in the polymerization process (see Tables 8.10 and 8.11).

For transfer to be thermodynamically possible, it is necessary that the species ($T^{\bullet}$) generated be better stabilized than the initial one ($\sim\sim\sim M_n^{\bullet}$). Addition of $T^{\bullet}$ to M is thus generally slower than addition of $\sim\sim\sim M_n^{\bullet}$ to M. In spite of that, the

Table 8.10. Values of transfer constants to monomer (C_{trm}) for some vinyl and related monomers at 60°C

Monomer	$10^4\,C_{trm}$
Styrene	0.6
Vinyl acetate	1.8
Methyl acrylate	0.4
Methyl methacrylate	0.1
Acrylamide	0.6
Acrylonitrile	0.3
Methacrylonitrile	5.8
Vinyl chloride	11
Allyl acetate	1600 (80°C)
Allyl chloride	700 (80°C)

Table 8.11. Some values of transfer constants (C_{trY}) relative to polymerization of styrene at 60°C

Transfer agent	$10^4 C_{trY}$
Heptane	0.4
Benzene	0.03
Toluene	0.15
Ethyl acetate	6
Acetone	0,32
Tetrachloromethane	110
Tetrabromomethane	2200
Bromotrichloromethane	650,000
1-Chlorobutane	0.04
Triethylamine	7
Benzoyl peroxide	0.08
Cumyl peroxide	0.01
2,2'-Azobis(isobutyronitrile)	0.1
n-Butanethiol	210,000

incidence of chain transfer in the overall kinetics of polymerization is generally neglected because it represents the insertion of only one monomer molecule compared to the large number of monomer molecules consumed during the chain growth.

A transfer agent (T) can be considered from two different aspects: that of its chemical function or that of its role in the process. Any chemical group capable of affording a new free radical that would be more stable than the growing radical can cause chain transfer: groups with labile hydrogen atoms, alkyl halides, peroxides, disulfides, and so on. Chain transfer involves all types of molecules present in the reaction medium whatever their role in the process: monomer, polymer, solvent, initiator, additive, or impurity. Examples hereafter describe these two aspects.

Remark. If several molecules play the role of transfer agent in the same reaction medium, their effects add up in proportion to their concentration and to their reactivity.

(a) Transfer to Initiator. Such a transfer consumes molecules of initiator and thus decreases its efficiency. The reaction described below shows that hydroperoxides can behave differently from peroxides.

$$\sim\sim\sim\sim M_n^{\bullet} + \text{R-O-O-H} \longrightarrow \sim\sim\sim\sim M_nH + \text{R-O-O}^{\bullet}$$

$$\sim\sim\sim\sim M_n^{\bullet} + \text{R-O-O-R} \longrightarrow \sim\sim\sim\sim M_n\text{-O-R} + \text{R-O}^{\bullet}$$

In the case of transfer to hydroperoxides by $H^{\bullet}$ abstraction, the reinitiation by $ROO^{\bullet}$ generates a peroxide which, in turn, can either undergo transfer or thermal scission to generate two new free radicals.

The transfer to initiator is sometimes used in macromolecular synthesis. For example, hydrogen peroxide (H_2O_2) (which reacts like a peroxide) is used to initiate the polymerization of butadiene and to functionalize chain ends with hydroxyl groups. Dihydroxytelechelic polybutadiene is obtained in this way to be used as precursor of polyurethane:

$$H_2O_2 \xrightarrow{k_d} 2\ HO^\bullet$$

$$HO^\bullet + n\ H_2C{=}CH{-}CH{=}CH_2 \xrightarrow{k_a} HO{-}(H_2C{-}CH{=}CH{-}CH_2)^\bullet_n$$

$$HO{-}(H_2C{-}CH{=}CH{-}CH_2)^\bullet_n + H_2O_2$$

$$\xrightarrow{k_{tr}} HO{-}(H_2C{-}CH{=}CH{-}CH_2)_n{-}OH + HO^\bullet \qquad \text{etc.}$$

(b) Transfer to Monomer (see Table 8.10). When a monomer is prone to transfer, its polymerizability is impaired; this is, for example, the case for propene, which is subject to transfer through its allyl hydrogen atom:

$$\text{\~\~\~}CH_2{-}\ ^\bullet CH(CH_3) + H_2C{=}CH{-}CH_3 \longrightarrow \text{\~\~\~}CH_2{-}CH_2{-}CH_3 + H_2C{=}CH{-}^\bullet CH_2$$

Allyl radical

The transfer reaction shown hereafter does occur but its extent is such that it does not affect the polymerizability of the corresponding monomer:

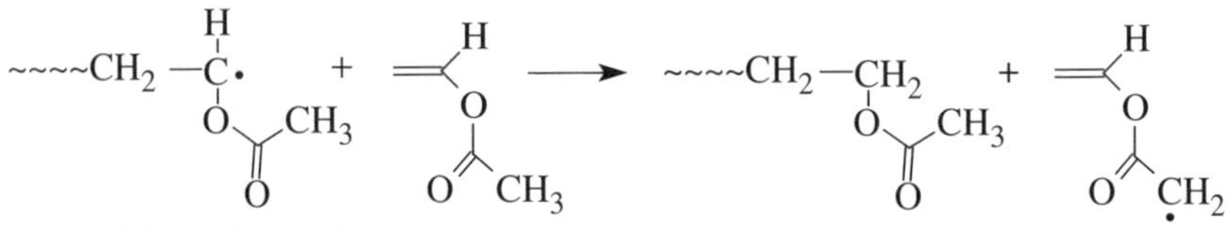

Poly(vinyl acetate)

(c) Transfer to Solvent. The propensity of certain solvents to induce transfer reactions prevents one from using them as reaction medium, more particularly when high molar masses are targeted:

$$\text{\~\~\~\~}M^\bullet_n + R{-}OH \longrightarrow \text{\~\~\~\~}M_nH + RO^\bullet$$

$$\text{\~\~\~\~}M^\bullet_n + C_6H_6 \longrightarrow \text{\~\~\~\~}M_nH + {}^\bullet C_6H_5$$

$$\text{\~\~\~\~}M^\bullet_n + C{-}Cl_4 \longrightarrow \text{\~\~\~\~}M_nCl + Cl_3C^\bullet$$

(d) Transfer to Polymer. The best-known example of transfer to polymer is that corresponding to the high-pressure free radical polymerization of ethylene:

$$\text{\textasciitilde\textasciitilde\textasciitilde\textasciitilde CH}_2-\text{CH}_2 \text{\textasciitilde\textasciitilde\textasciitilde\textasciitilde CH}_2 - \overset{\bullet}{\text{CH}}_2 \longrightarrow \text{\textasciitilde\textasciitilde\textasciitilde\textasciitilde CH}_2 - \overset{\bullet}{\text{CH}}\text{\textasciitilde\textasciitilde\textasciitilde\textasciitilde CH}_2-\text{CH}_3$$

$$\text{\textasciitilde\textasciitilde\textasciitilde\textasciitilde CH}_2-\underset{\underset{\underset{\text{CH}_2-\text{CH}_2-\text{CH}_2\text{\textasciitilde\textasciitilde\textasciitilde\textasciitilde CH}_2-\overset{\bullet}{\text{CH}}_2}{|}}{\overset{|}{\text{CH}_2}}}{\text{CH}}\text{\textasciitilde\textasciitilde\textasciitilde\textasciitilde CH}_2-\text{CH}_3 \qquad \nearrow_n\text{CH}_2=\text{CH}_2$$

etc.

Transfer to polyethylene occurs because of the very high reactivity of the initial free radicals ("primary" radical) and the greater stability of the free radicals formed after transfer ("secondary" radical). The transfer to polymer does not generate new chains but instead generates branches on the existing polymeric chains; intramolecular transfer promotes preferentially the formation of short branches (due to the high probability for an early termination between growing free radical and carried by neighboring monomeric units), and intermolecular reactions promotes that of long branches.

(e) Transfer to Transfer Agents. In free radical polymerization, the molar mass of a sample can be conveniently controlled by addition to the reaction medium of molecules carrying reactive sites prone to transfer. For instance, the degree of polymerization of synthetic polydienes is modulated through the use of thiols (mercaptans):

$$\text{\textasciitilde\textasciitilde CH}_2\text{-CH=CH-CH}_2^{\bullet} + \text{C}_{12}\text{H}_{25}\text{SH} \xrightarrow{k_{tr}} \text{\textasciitilde\textasciitilde\textasciitilde CH}_2\text{-CH=CH-CH}_3 + \text{C}_{12}\text{H}_{25}\text{S}^{\bullet}$$

This family of compounds, as well as sulfides, disulfides, and alkyl halides are commonly used for transfer purpose. They also serve to generate functional sites at the chain ends.

When the rate constant of transfer is high ($k_{tr} \gg k_p$) and so is the concentration in transfer agent, the chain length is considerably reduced and end-functionalized oligomers are eventually formed by a process termed *telomerization*.

Telomers ($\text{Y}-\text{M}_n-\text{X}$) result from a series of reactions of chain propagation of monomer molecules and transfer to molecules called *telogens*, as schematized below:

$$\text{R}^{\bullet} + n\text{M} \longrightarrow \text{RM}_n^{\bullet}$$
$$\text{RM}_n^{\bullet} + \text{XY} \longrightarrow \text{RM}_n\text{X} + \text{Y}^{\bullet}$$
$$\text{Y}^{\bullet} + n\text{M} \longrightarrow \text{Y}-\text{M}_n^{\bullet}$$
$$\text{Y}-\text{M}_n^{\bullet} + \text{XY} \longrightarrow \text{Y}-\text{M}_n\text{X} + \text{Y}^{\bullet} \qquad \text{etc.}$$

8.5.7. Inhibition and Retardation

In both cases, the polymerization of a given monomer is temporarily prevented from occurring.

Inhibition thwarts any inopportune polymerization (fortuitous, thermal, or photochemical initiation) during the monomer storage.

Retardation corresponds to an initiation of the polymerization that is delayed because of the presence of a retarder. In an experimental setup containing all the components of the reaction mixture including an initiator generating "primary" free radicals, the polymerization begins only after all the *retarder* has been consumed.

The principle is the same for both inhibition and retardation phenomena, and two general methods are used. The first one takes advantage of the propensity of free radicals to undergo near-diffusion-controlled bimolecular coupling reactions; it consists in introducing stable free radicals into the reaction medium. These species are highly stabilized and are thus unable to initiate polymerization, but they are able to recombine with active radicals and discontinue chain growth. Tetramethylpiperidyloxyl (TEMPO) and diphenylpicrylhydrazyl (DPPH) radicals (see Section 8.5.1) are generally used for this purpose. The corresponding reaction pathways are as follows:

At high temperatures these reagents are not completely efficient because the coupling reaction is reversible.

The second method consists in transforming a reactive free radical into an unreactive free radical which is then too stable to initiate the polymerization. Compounds of various structures can be used as inhibitors; only the most significant ones will be mentioned here.

Phenols and polyphenols—and even aromatic amines—reacting with growing free radicals undergo hydrogen abstraction

$$\sim\sim\sim\sim\sim\sim M_n^{\bullet} + ArOH \longrightarrow \sim\sim\sim\sim\sim\sim M_nH + ArO^{\bullet}$$

and give rise to aryloxyl radicals whose reactivity is too low to efficiently re-initiate polymerization.

Quinones react by a completely different mechanism:

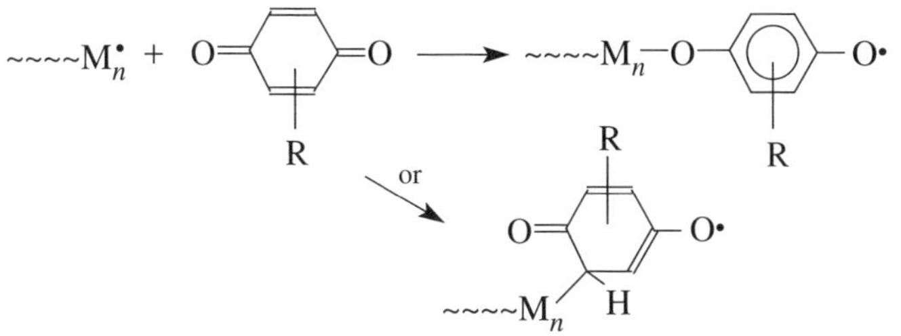

Hydroquinone is very frequently used as inhibitor, although the mechanism of inhibition is still matter of controversy; indeed, this compound can react like a regular phenol but also give benzoquinone in the presence of molecular oxygen and react as such.

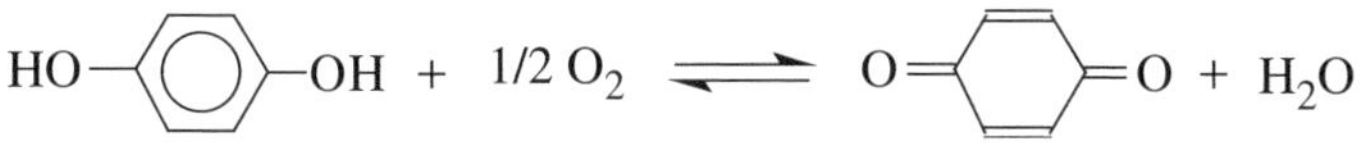

Some nitro-aromatic derivatives are also commonly used as inhibitors, their reaction with active radicals being shown hereafter:

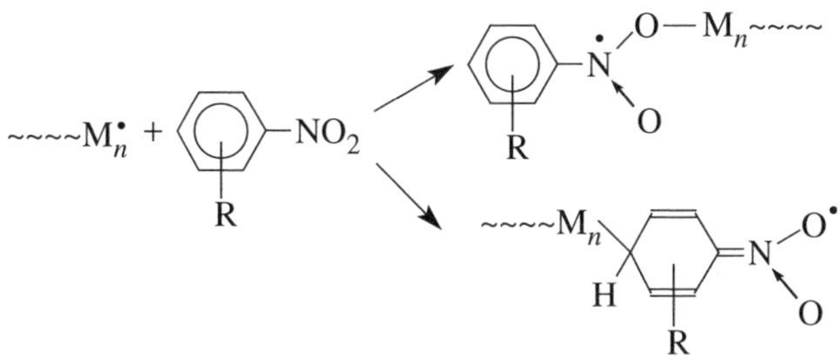

Molecular oxygen by itself has also an inhibiting effect, as shown in particular for (meth)acrylic growing radicals. A peroxyl radical is formed

$$\sim\sim\sim M_n^\bullet + O_2 \longrightarrow \sim\sim\sim M_n O-O^\bullet$$

that is unable to reinitiate in this case the polymerization process. This characteristic feature is exploited industrially for producing adhesives whose viscosity builds up soon after molecular oxygen could not be renewed at the interface. With other more reactive monomers, reinitiation occurs, introducing peroxy groups in the polymer chain which can thus be viewed as a copolymerization between monomer and molecular oxygen:

$$\sim\sim\sim CH_2-CHA\text{-}O\text{-}O^\bullet + CH_2=CHA \longrightarrow \sim\sim\sim CH_2\text{-}CHA\text{-}O\text{-}O\text{-}CH_2 - CHA^\bullet$$

It may happen that in the reaction medium, the presence of compounds possessing labile hydrogen atom causes transfer (if the radical R$^{\bullet}$ formed is sufficiently reactive)

$$\sim\!\sim\!\sim\!\sim CH_2-CHA\text{-}O\text{-}O^{\bullet} + RH \longrightarrow \sim\!\sim\!\sim\!\sim CH_2\text{-}CHA\text{-}O\text{-}OH + R^{\bullet}$$

or termination.

8.5.8. Chain Length in Free Radical Polymerization

Because of the extreme fugacity of free radicals, a parameter such as the instantaneous concentration in growing species $[M_n^{\bullet}]$ is likely to undergo important variations between the beginning and the end of the polymerization. $[M_n^{\bullet}]$ can be considered as being constant only during short intervals of time, so that the kinetic parameters established below (kinetic chain length, $\overline{X}_n$) do have a meaning only at a given moment. Using the kinetic expressions appearing below, the relation between these parameters and the molar masses of a sample and their distribution can be established. It is also possible to utilize these equations to adjust the composition of the reaction medium and modify the polymerization conditions and thus to obtain the required molecular dimensions.

The kinetic chain length ($\overline{v}$) is defined as being the number of molecules of monomer consumed by a just created active center (RM$^{\bullet}$). Thus $\overline{v}$ is equal to the ratio of the rate of consumption of monomer (R_p) to the rate of initiation (R_i):

$$\overline{v} = R_p/R_i$$

From equations giving R_p and R_i (see Section 8.3), one can write

$$\overline{v} = \frac{k_p}{(2fk_dk_t)^{1/2}} \frac{[M]}{[I]^{1/2}}$$

In the absence of any transfer reaction and for a termination occurring by disproportionation only (which is exceptional), the kinetic chain length is equal to the number average degree of polymerization:

$$\overline{X}_n = \frac{k_p}{(2fk_dk_t)^{1/2}} \frac{[M]}{[I]^{1/2}}$$

If termination occurs by coupling of the free radicals and still in absence of transfer, the number average degree of polymerization corresponds to twice the kinetic length:

$$\overline{X}_n = \frac{k_p\sqrt{2}}{(fk_dk_t)^{1/2}} \frac{[M]}{[I]^{1/2}}$$

If transfer reactions are to be taken into account, it is necessary to count chains created by transfer as well as those initiated by primary radicals:

$$\overline{X}_n = \frac{R_p}{R_i + R_{\text{trm}} + R_{\text{tri}} + R_{\text{trs}} + R_{\text{trt}}}$$

$(\overline{X}_n = R_p/\text{sum of the rates of creation of new chains})$

Remark. Transfer to polymer is not to be taken into account since it does not generate new chains but only branches.

Assuming that the efficiency of free radicals generated by transfer is equal to unity and expressing k_{trm}, k_{trs}, k_{tri} and k_{trt} as the rate constants of transfer to monomer, solvent, initiator, and a transfer agent, respectively, one can write

$$\frac{1}{\overline{X}_n} = \frac{R_i}{R_p} + \frac{R_{\text{trm}}}{R_p} + \frac{R_{\text{trs}}}{R_p} + \frac{R_{\text{tri}}}{R_p} + \frac{R_{\text{trt}}}{R_p}$$

In the case of a termination by pure recombination, we have

$$\frac{1}{\overline{X}_n} = \frac{(fk_d k_t[\text{I}])^{1/2}}{\sqrt{2}k_p[\text{M}]} + \frac{k_{\text{trm}}}{\sqrt{2}k_p} + \frac{k_{\text{trs}}[\text{S}]}{\sqrt{2}k_p[\text{M}]} + \frac{k_{\text{tri}}[\text{I}]}{\sqrt{2}k_p[\text{M}]} + \frac{k_{\text{trt}}[\text{T}]}{\sqrt{2}k_p[\text{M}]}$$

or

$$\frac{1}{\overline{X}_n} = \frac{(fk_d k_t[\text{I}])^{1/2}}{\sqrt{2}k_p[\text{M}]} + \frac{k_{\text{trm}}}{\sqrt{2}k_p} + \sum_y \frac{k_{\text{try}}}{\sqrt{2}k_p} \frac{[\text{Y}]}{[\text{M}]}$$

where Y represents any transfer agent other than monomer.

To establish the existence of possible transfer reactions, a direct method can be used which consists in characterizing molecular groups resulting from transfer and which are necessarily located at chain ends. Because of their low proportion, it is necessary to utilize extremely sensitive techniques and, for example, ^{3}H- or ^{14}C-labeled molecules.

Although less accurate and direct, a kinetic method can also be used. In the preceding equation (Mayo equation), numerator and denominator of the first term can be multiplied by the factor $\{k_p k_t[\text{M}]\}$, and the whole equation can be rewritten under the form

$$\frac{1}{\overline{X}_n} = \frac{k_t}{\sqrt{2}k_p^2} \frac{R_P}{[\text{M}]^2} + \frac{k_{\text{trm}}}{k_p} + \sum_y \frac{k_{\text{try}}}{k_P} \frac{[\text{Y}]}{[\text{M}]}$$

or

$$\frac{1}{\overline{X}_n} = \frac{k_t}{\sqrt{2}k_P^2} \frac{R_P}{[\text{M}]^2} + C_{\text{trm}} + \sum_y C_{\text{try}} \frac{[\text{Y}]}{[\text{M}]}$$

If there is no transfer, that is,

$$1/\overline{X}_n = c^{\text{st}} R_p$$

with

$$c^{\text{st}} = (k_t/\sqrt{2}k_p^2)\{1/[\text{M}]^2\}$$

the plot of $1/\overline{X}_n$ versus (R_p) results in a straight line passing through the origin.
If there is transfer to monomer, that is,

$$1/\overline{X}_n = c^{\text{st}} R_p + C_{\text{trm}}$$

the value taken by $1/\overline{X}$ versus (R_p) plot at the origin gives C_{trm}.
In the case of transfer to initiator,

$$1/\overline{X}_n = c^{\text{st}} R_p + C_i[\text{I}]/[\text{M}]$$

considering the general equation of free radical polymerization

$$[\text{I}]^{1/2} = R_p/\{k_p[\text{M}](2fk_d/k_t)^{1/2}\}$$

and

$$1/\overline{X}_n = c^{\text{st}} R_p + C_i R_p^2/\{k_p^2[\text{M}]^3(2fk_d/k_t)$$

$1/\overline{X}_n$ versus (R_p) plot of this quadratic equation must show an upward curvature.

To determine the **distribution of molar masses**, the following reasoning based on statistical probability can be used:

For a growing free radical, the probability (P) that it adds a monomer molecule is equal to the ratio of the rate of propagation to the sum of the rates of all the reactions (propagation, transfer, termination) that can undergo this very same free radical:

$$P = \frac{R_p}{R_p + R_t + R_{\text{tr}}}$$

The probability of discontinuing the chain growth is equal to $(1 - P)$ and thus the probability for obtaining (or existing) an i-mer is equal to

$$P_i = P^{(i-1)}(1 - P)$$

If the polymeric sample considered consists of N chains, the number of i-mers present is equal to

$$N_i = NP^{(i-1)}(1 - P)$$

The total number of chains is equal to the product of the total number of monomeric units N_0 by the probability of discontinuing the growth, that is, $(1 - P)$:

$$N = N_0(1 - P)$$

which gives

$$N_i = N_0 P^{(i-1)}(1 - P)^2$$

If M_0 is the molar mass of the monomeric unit, the mass fraction of the i-mers is given by

$$W_i = N_i i M_0 / N_0 M_0$$

which can be written as

$$W_i = i P^{(i-1)}(1 - P)^2$$

By summing over i the number and the mass of all i-mers, it is possible to find the values of the number- and mass-average molar masses of the resulting polymers:

$$\overline{M}_n = \sum_i N_i M_i / \sum_i N_i = N_0 M_0 / \sum_i N_i = N_0 M_0 / \left\{ N_0 \sum_i P^{(i-1)}(1 - P)^2 \right\}$$

which is a relation of the same type as that established for the $\overline{M}_n$ values of polycondensates (see Section 7.2.2) and for which it was shown that

$$\sum_i P^{(i-1)}(1 - P)^2 = (1 - P)$$

for high values of i.

For $\overline{M}_n$ one obtains

$$\overline{M}_n = M_0 / (1 - P)$$

In the same way, one can write

$$\overline{M}_w = \sum_i N_i M_i^2 / \sum_i N_i M_i = M_0 \sum_i i^2 P^{(i-1)}(1 - P)^2 / \left\{ \sum_i i P^{(i-1)}(1 - P)^2 \right\}$$

Since

$$\sum_i i P^{(i-1)}(1 - P)^2 = 1$$

and

$$\sum_i i^2 P^{(i-1)}(1 - P)^2 = (1 + P)/(1 - P)$$

one obtains for $\overline{M}_w$:

$$\overline{M}_w = M_0(1 + P)/(1 - P)$$

The dispersity index can be written

$$D_M = \overline{M}_w/\overline{M}_n = (1 + P)$$

which means that D_M tends to 2 for systems of high molar masses.

> **Remark.** The calculated values of average molar masses do not take into account the possible dimerization of chains by termination (recombination); when such a dimerization occurs, the average molar masses must be multiplied by 2.

In the case of "controlled" free radical polymerizations (see Section 7.5.8), the initiation period is short compared to that of propagation and all the chains created grow simultaneously at the same rate. The polymer formed under such conditions exhibits a narrow dispersity of molar masses corresponding to a Poisson distribution (see Section 8.4).

8.5.9. Determination of Rate Constants

To predict the kinetic behavior and the molar mass of a sample, it is essential to know the various rate constants of reactions occurring during a free radical polymerization.

8.5.9.1. Rate Constants of Decomposition of the Initiator (k_d).
They can be measured separately, out of the polymerization medium by a conventional method, but it is preferable to carry out these measurements in the reaction conditions. The residual initiator can be titrated by a spectrometric technique or by measuring the amount of gas released by decomposition against time; for example, CO_2 is produced by decomposition of benzoyl peroxide (in the absence of monomer) or molecular nitrogen is produced by that of azo-bis(isobutyronitrile).

8.5.9.2. Efficiency (f).
The precise determination of f is difficult and often inaccurate. The method most often used consists in comparing the number of primary free radicals formed by decomposition of the initiator with the number of primary free radicals having generated a chain through the use of ^{3}H- or ^{14}C-labeled initiators. Such labeled polymeric chains can be easily separated from the side products generated by the initiator decomposition by simple selective precipitation. An alternative method relies on the determination of $\overline{M}_n$ to count the chains. This second method requires both the knowledge of the mechanisms of termination (disproportionation and/or recombination) and the absence of transfer reactions. The values measured are very dependent on the reaction conditions.

8.5.9.3. Rate Constants of Propagation and of Termination (k_p and k_t).

The expression of the rate of polymerization can be written as

$$R_p = \frac{k_p}{k_t^{1/2}} R_i^{1/2}[M]$$

or

$$\frac{k_p^2}{k_t} = \frac{R_p^2}{R_i[M]^2}$$

Monomer conversion can be measured by different means: dilatometry, spectrometry, sampling, and gravimetry for variable reaction times; R_p is equal to the slope of the straight line drawn from the plot of the monomer conversion *versus* the reaction time. Knowing R_i, it is relatively easy to obtain k_p^2/k_t. This value is an important feature of chain polymerizations; indeed, for a same rate of initiation, this value is directly proportional to the rate of polymerization of the monomer considered; it is also proportional to the average length of the polymer chains formed. Some typical values are given in Table 8.12.

To determine k_p, one generally resorts to a photochemical initiation by successive flash irradiations using a laser beam. The intervals of time between two consecutive flashes are very short compared to the average duration of chain growth. A chromatographic (or better MALDI mass spectrometry) analysis of the polymer sample resulting from several successive flashes provides the number of monomer molecules added by each initial active center, in between the interval of time separating two successive flashes. Knowing the monomer concentration, the rate constant of propagation (k_p) can be easily deduced. Using k_p and the number average molar mass of a polymer obtained under steady-state conditions, it is possible to establish the average lifetime of active centers (δ); since under steady-state conditions

$$\delta = \frac{[M_n^\bullet]}{R_t} = \frac{[M_n^\bullet]}{k_t[M_n^\bullet]^2} = \frac{1}{k_t[M_n^\bullet]}$$

Table 8.12. k_p^2/k_t values for some vinyl monomers

Monomer	Temperature (°C)	k_p^2/k_t (L·mol^{-1}·s^{-1})
Styrene	60	0.0007
Methyl methacrylate	60	0.028
Vinyl acetate	60	0.185
Vinyl acetate	30	0.030

and

$$R_p = k_p[\text{M}_n^\bullet][\text{M}]$$

where $[\text{M}_n^\bullet] = R_p/k_p[\text{M}]$, one can write

$$k_p/k_t = \delta R_p/[\text{M}]$$

From the knowledge of k_p, R_p, and $[\text{M}]$, k_t can be deduced.

Several values of k_p are given in Table 8.5. Those relative to k_t appear in Tables 8.8 and 8.9.

8.5.10. Controlled Free Radical Polymerization

In a regular free radical polymerization, the initiation, chain growth, and chain termination steps occur concomitantly and repeatedly until both the monomer and/or the initiator are totally consumed. The choice of experimental conditions—and particularly that of temperature—is determined by the necessity to favor propagation over termination in order to achieve the complete consumption of monomer within reasonable reaction times. The expression of the rates of termination and propagation given below clearly indicate the reason for a slow initiation step in a regular free radical polymerization:

$$R_t = k_t[\text{M}_n^\bullet]^2, \qquad R_p = k_p[\text{M}][\text{M}_n^\bullet]$$

Should the period of initiation and thus of chain formation be reduced to short lapses of time, a high concentration in free radicals would be created, causing irreversible terminations to such an extent that polymerization would be irremediably stopped.

By contrast, the concept of controlled free radical polymerization requires that two *a priori* antagonist objectives be reconciled, namely:

- A short period of chain initiation compared to that of propagation
- At the same time, minimized termination reactions.

In other words, for a controlled free radical polymerization to occur, one has to find the means to generate the chains in a short lapse of time and yet maintain an instantaneous concentration in free radicals ($[M_n^\bullet]$) in the range of 10^{-8} mol·L^{-1}, for propagation to be favored over termination.

The use of stable nitroxyl radicals provides a satisfactory answer to these apparently contradictory constraints. Due to their peculiar electronic structure, these stable radicals are unable to add onto C=C double bonds but can be used to trap transient free radicals and thus identify free radical intermediates. They indeed react by recombination with carbon-centered radicals at diffusion-controlled rates.

In the presence of both the monomer and a fast-decomposing radical initiator, such stable radicals can serve to trap reactive free radical species with a rate constant comparable to that of irreversible recombination.

If the "primary" free radicals generated by the initiator or those resulting from the first addition steps are effectively trapped by these intentionally added nitroxides, stable alkoxyamines are produced and polymerization can come to a standstill.

Actually, the alkoxyamines formed by recombination between a nitroxide and a transient alkyl radical are able to dissociate upon thermal activation and generate again the initial radicals as schematized in Figure 8.2. Depending upon the structure of the two free radicals involved, this dissociation, which occurs by homolytic rupture of the C–ON bond, will be observed in a varying range of temperatures.

The challenge here is to find out the domain of temperature suitable for the formation of a minute fraction ($<10^{-8}$ mol·L^{-1}) of active species by dissociation of the alkoxyamines; in addition, one has to establish the conditions for the species remaining under the dormant alkoxyamine form (typically $\sim10^{-3}$ mol·L^{-1}) to participate in the polymerization and consume monomer. This last criterion can be satisfied only if the dissociation time of the dormant species is short compared to the average time of activation of a growing chain within one such cycle of activation/deactivation; it is also essential that the rate of capture of free radicals be higher than that of irreversible termination.

$$\tfrac{1}{2}\,A \longrightarrow R^{\bullet} \quad / \quad {}^{\bullet}O-NR_1R_2 \quad / \quad M$$

$$\sim 10^{-2}\ \text{mol·L}^{-1} \qquad \sim 10^{-2}\ \text{mol·L}^{-1} \qquad \sim 8\ \text{mol·L}^{-1}$$

$$\text{(Initiator)} \qquad\qquad \text{(Stable nitroxide)} \qquad \text{(Monomer)}$$

If all these conditions are fulfilled, samples exhibiting both controlled molar mass and narrow molar mass distribution ($D_M \sim 1$) can be obtained. However, taking into consideration the respective rate constant of free radical capture and that of irreversible termination, one can wonder how the condition of a rate of reversible termination notably higher than that of their irreversible deactivation could be reasonably satisfied. The only way to favor the former reaction against termination is to increase the nitroxide concentration in the reaction medium to a level higher

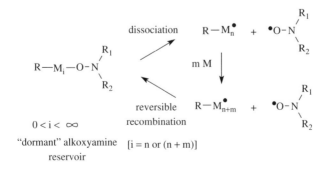

Figure 8.2. Scheme of chain growth in a stable nitroxide-mediated radical polymerization.

than that of polymeric radicals. Under such conditions only, will the probability of a polymeric radical to react with a nitroxide be higher and will the chain grow by repeated propagation–deactivation–activation steps, before their irreversible termination.

If one does not pay attention to this aspect and care to introduce into the reaction medium a concentration in nitroxide higher than that of the transient free radicals generated by the initiator, the system by itself will produce the excess of nitroxide necessary to control the polymerization by consenting the irreversible termination of a certain percentage of its chains. By doing so, it limits the deactivation of further chains and exhibits a "controlled" character during the rest of the polymerization. Elucidated and theorized by Fisher (name of the researcher who proposed this mechanism), this phenomenon is termed "persistent radical effect" or "Fischer effect" (see Figure 8.4).

Several families of monomers could be polymerized under living/controlled conditions using nitroxide-mediated free radical polymerization, including styrene, alkyl acrylates, acrylamides among others.

Besides nitroxides, the Kharasch reaction was also exploited to control free radical polymerizations. In a Kharasch reaction also termed "atom transfer radical addition," a radical generated from an organic halide in the presence of a transition metal catalyst adds to an unsaturated compound to form an adduct carrying a carbon–halogen bond; the transition metal catalyst successively plays the role of a reducing agent for the carbon–halogen bond of the precursor molecule and plays that of an oxidizing agent for the free radical adduct formed: this catalyst thus undergoes a reversible one-electron oxidation upon abstraction of the halogen, followed by a one-electron reduction through transfer of a halogen back to the resulting radical.

Polymer chemists have realized only lately the benefit they could draw from this reaction in the field of controlled free radical polymerizations. Using catalytic systems such as CuX–bipyridyl (Bipy) and a halogenated compound serving as initiator, various families of monomers such as styrenes, alkyl (meth)acrylates, and acrylamides could be polymerized in a controlled manner. The mechanism proceeds by transfer of a halogen carried by the initiator (RX) to the metal (the latter being oxidized in CuX_2) and then to the growing free radical species ($M^{\bullet}$) (see Figure 8.3), which temporarily returns to the ω-halogenated dormant form. The

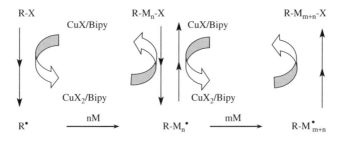

Figure 8.3. Mechanistic scheme of chain growth in an atom transfer radical polymerization.

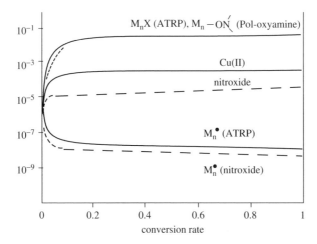

Figure 8.4. Variation of the concentration of the various species constituting the reaction medium in atom transfer radical polymerization (ATRP: ——) and in nitroxide-mediated polymerization (– –).

chain grows between each cycle of reduction/oxidation undergone by its end(s). The role played by the ligand is essential and therefore its choice is decisive: it must form a strong complex with the metal and increases its oxidability by its electron-donating character. Such a mechanism of propagation was termed "atom transfer radical polymerization" (ATRP) (see Figures 8.3 and 8.4).

Recently, another technique of control of free radical polymerization was disclosed; it involves the use of thiocarbonylthio compounds reacting by reversible addition–fragmentation transfer (RAFT) with the growing polymeric radicals.

The effectiveness of these RAFT reagents as radical traps depends on the nature of their Z and R groups, with Z determining their reactivity toward entering radicals and R determining their aptitude for fragmentation. The key step in a successful RAFT process is the transfer of growing radicals to the RAFT reagents and the subsequent formation of an intermediate radical. Upon undergoing fragmentation, the latter releases a radical ($R^\bullet$) that can initiate polymerization. The entering polymer chain is transformed into a dormant species that can be reactivated through its terminal [S=C(Z)S] moiety

Polymerization proceeds through repetition of the same cycle of RAFT deactivation/activation by an entering radical/propagation.

Because of the protection provided by these techniques of control to the growing free radicals through a "dormant" form, the latter retain a "living" character that can be advantageously used in the domain of macromolecular engineering. As established for "living" polymerizations, the expression of the degree of polymerization of chains prepared by controlled radical polymerization can be written as

$$\overline{X}_n = [M]_0 P/[I]$$

where I is the radical initiator used in nitroxide-mediated polymerization, the organohalide used in ATRP, and the thiocarbonylthio compound used in RAFT.

Architectures such as block copolymers or star polymers, which could be obtained until now only by "living" ionic polymerizations, are now accessible by controlled free radical polymerization (see Chapter 9).

8.5.11. Free Radical Copolymerization

Only copolymerizations involving two comonomers homogeneously mixed and leading to statistical copolymers will be considered here. The propagation step in a free radical copolymerization is the only step that differs from that of a homopolymerization.

8.5.11.1. Copolymerization Equation–Monomer Reactivity Ratios. During the copolymerization of two comonomers (A and B), the chain can grow by the occurence of the four following reactions that differ from one another by the nature of the free radical and the inserted monomer:

$$\sim\sim\sim\sim\sim\sim A^\bullet + A \xrightarrow{k_{AA}} \sim\sim\sim\sim\sim AA^\bullet \tag{8.1}$$

$$\sim\sim\sim\sim\sim\sim A^\bullet + B \xrightarrow{k_{AB}} \sim\sim\sim\sim\sim AB^\bullet \tag{8.2}$$

$$\sim\sim\sim\sim\sim\sim B^\bullet + B \xrightarrow{k_{BB}} \sim\sim\sim\sim\sim BB^\bullet \tag{8.3}$$

$$\sim\sim\sim\sim\sim\sim B^\bullet + A \xrightarrow{k_{BA}} \sim\sim\sim\sim\sim BA^\bullet \tag{8.4}$$

Free radicals of $\sim\sim A^\bullet$ type appear in the reaction medium by primary initiation

$$R^\bullet + A \longrightarrow RA^\bullet$$

and by reaction (8.4); they disappear by termination and reaction (8.2). The same can be said about $\sim\sim B^\bullet$ radicals which are in a symmetrical situation to the preceding ones. Under steady-state conditions, the rates of appearance and disappearance of these radicals are equal and one can write

$$k_{AB}[\sim\sim A^{\bullet}][B] = k_{BA}[\sim\sim B^{\bullet}][A]$$

Because monomer A is consumed by reactions (8.1) and (8.4) and monomer B by reactions (8.2) and (8.3), this situation leads to the following equations:

$$-d[A]/dt = k_{AA}[\sim\sim A^{\bullet}][A] + k_{BA}[\sim\sim B^{\bullet}][A]$$

$$-d[B]/dt = k_{BB}[\sim\sim B^{\bullet}][B] + k_{AB}[\sim\sim A^{\bullet}][B]$$

Defining the reactivity ratios as

$$r_1 = k_{AA}/k_{AB} \quad \text{and} \quad r_2 = k_{BB}/k_{BA}$$

and taking into account the steady-state conditions,

$$[\sim\sim A^{\bullet}]/[\sim\sim B^{\bullet}] = (k_{BA}[A])/(k_{AB}[B])$$

one obtains

$$\frac{d[A]}{d[B]} = \frac{[A]}{[B]} \frac{r_1[A] + [B]}{r_2[B] + [A]}$$

This equation, known as *Mayo–Lewis equation*, gives the instantaneous composition of the copolymer formed, as a function of the instantaneous composition of the comonomers mixture and the reactivity ratios (r_1 and r_2). In copolymerizations carried out in a continuous flow process, this equation can be used to predict the composition of the reaction mixture (*feed*) that is necessary to obtain a constant copolymer composition.

The knowledge of these reactivity ratios is essential not only because the copolymer composition can be predicted, but also because the relative reactivity of monomers A and B with respect to $\sim\sim A^{\bullet}$ and $\sim\sim B^{\bullet}$ free radicals can be measured. For example, $r_1 < 1$ means that monomer B is more easily added by $\sim\sim A^{\bullet}$ than monomer A, whereas $r_2 > 1$ means that B is added by $\sim\sim B^{\bullet}$ more easily than A.

In addition to the determination of the relative reactivities of both the comonomers and the growing free radicals, the knowledge of r_1 and r_2 can also serve to predict the frequency of AB or BA alternations in a statistical copolymer for a given composition of the feed. In the case of a Bernouillian statistics of distribution of monomeric units—that is, when the type of dyad formed is only determined by the nature of the growing chain end (which is the case in radical copolymerization)—it is relatively easy to predict the structure of the sequences appearing in a copolymer as a function of the composition of the feed and the reactivity ratios.

If P_A, P_{AA}, P_{AAA}, P_{AAB}, and so on, are the probabilities of finding a monomeric unit A, a dyad AA, a triad AAA or AAB, and so on, respectively, one can write

$$P_A = d[A]/\{d[A] + d[B]\}$$

$$\frac{1}{P_A} = 1 + \frac{d[B]}{d[A]} = \left\{1 + \frac{[B]}{[A]} \cdot \frac{r_2[B] + [A]}{r_1[A] + [B]}\right\}$$

According to the Bernouillian statistics, one has

$$P_{AA} = P_A^2$$

$$P_{AAA} = P_A^3$$

$$P_{AAB} = P_A^2 P_B = P_A^2(1 - P_A)$$

or

$$P_{AAA} = \left\{1 + \frac{[B]}{[A]} \cdot \frac{r_2[B] + [A]}{r_1[A] + [B]}\right\}^{-3}$$

$$P_{AAB} = \left\{1 + \frac{[B]}{[A]} \cdot \frac{r_2[B] + [A]}{r_1[A] + [B]}\right\}^{-2} \left\{1 + \frac{[A]}{[B]} \cdot \frac{r_1[A] + [B]}{r_2[B] + [A]}\right\}^{-1} \quad \text{and so on.}$$

Five different and typical situations can be considered, depending upon the values taken by r_1 and r_2, with the [A] and [B] instantaneous concentrations being taken equal by convention:

- For $r_1 \sim r_2 \sim 1$, the rate constants of propagation and cross addition are approximately equal and the insertion of the monomeric units in the copolymer occurs randomly; the frequency of alternations depends only on the relative concentration of the two comonomers.
- For r_1 and $r_2 < 1$, $k_{AB} > k_{AA}$ and $k_{BA} > k_{BB}$, $\sim\sim A^\bullet$ growing chains tend to incorporate preferently monomer B and, conversely, a marked tendency to alternations results;
- For r_1 and $r_2 > 1$, the situation is opposite with respect to the preceding case and $k_{AA} > k_{AB}$ and $k_{BB} > k_{BA}$; the system tends to form sequences of AAA... and BBB... types whose length depends on the values of r_1 and r_2; no system corresponding to this case is known in radical copolymerization;
- For $r_1 > 1$ and $r_2 < 1$, $k_{AA} > k_{AB}$ and $k_{BB} < k_{AB}$; the polymerization of A is favored and the first copolymer formed contains little proportion of B;
- For $r_1 \sim r_2 \sim 0$, the tendency to alternation is almost perfect.

If $r_1 r_2 = 1$, which corresponds to $k_{AA} k_{AB} = k_{BA} k_{BB}$, the two types of active centers have a same capacity to add A or B; such systems are known as "ideal."

When the composition of the copolymer formed is equal to that of the feed, the copolymerization is termed "azeotropic." This means that

$$d[A]/d[B] = [A]/[B]$$

and

$$\frac{r_1[A] + [B]}{r_2[B] + [A]} = 1$$

which corresponds to

$$\frac{[A]}{[B]} = \frac{1 - r_2}{1 - r_1}$$

Table 8.13 gathers some values of reactivity ratios illustrating the various situations described above. The styrene/butadiene couple ($r_1 r_2 = 1.09$) corresponds roughly to an "ideal" system.

When a copolymerization is carried out in batch and in a closed reactor and without additional incorporation of comonomers, the composition of the copolymer formed at complete conversion will be obviously equal to that of the initial mixture of comonomers whatever the values of reactivity ratios. This implies that the composition of the mixture of comonomers changes progressively with the extent of the copolymerization reaction and thus the instantaneous composition of the resulting copolymer also. A copolymer formed at the beginning of the reaction can be very rich in A monomer as a consequence of the very high reactivity of the latter monomer; the copolymer formed at the end of the reaction will therefore be rich in B monomer, with the low reactivity of this monomer being compensated by its high concentration when A is almost totally consumed.

Figure 8.5 illustrates the influence of the values of both the reactivity ratios and the instantaneous composition of the comonomers mixture on the instantaneous composition of the copolymer formed. The case corresponding to r_1 and $r_2 \ll 1$ is not represented because no such example is known in free radical copolymerization.

Table 8.13. Reactivity ratios (r_1 and r_2) of some couples of ethylenic monomers

A	r_1	B	r_2
Methyl acrylate	0.84	Vinylidene chloride	0.99
Styrene	0.58	Buta-1,3-diene	1.40
Methacrylonitrile	0.15	α-Methylstyrene	0.21
Acrylonitrile	5.5	Vinyl acetate	0.06
Styrene	42	Vinyl acetate	0.01
Styrene	0.05	Maleic anhydride	0.005

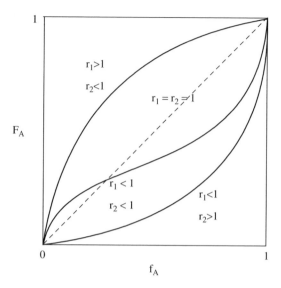

Figure 8.5. Variation of the copolymer composition (F_A) versus the proportion of comonomer A (f_A) for various values of r_1 and r_2.

The total composition of a copolymer prepared in a closed batch reactor, for a yield of reaction lower than unity, can be calculated by integration of the composition equation written in a slightly different form from the preceding one and through the use of the definitions of F_A and f_A given in Figure 8.5:

$$\frac{F_A}{1 - F_A} = \frac{[A]}{[B]} \cdot \frac{r_1[A] + [B]}{r_2[B] + [A]} = \frac{f_A}{f_B} \cdot \frac{r_1 f_A + f_B}{r_2 f_B + f_A}$$

For a system comprising M moles of monomers (A + B), the amount dM of monomer consumed corresponds to $F_A dM$ of the monomeric units A incorporated in the polymeric chain and thus consumed. Thus one has $(M - dM)(f_A - df_A)$ moles of A in the reaction mixture; and by virtue of the reaction balance, one can write that the amount of A monomer disappeared is equal to that of A polymerized:

$$F_A dM = M f_A - (M - dM)(f_A - df_A)$$

If the term $dM \cdot df_A$ is neglected, one obtains

$$dM/M = df_A/(F_A - f_A)$$

or

$$\int_{f_{A0}}^{f_A} \frac{dM}{M} = \ln \frac{M}{M_0} = \int_{f_{A0}}^{f_A} \frac{df_A}{(F_A - f_A)} = \ln(1 - p)$$

for $F_A > f_A$ (which corresponds to $r_1 > r_2$) and if p is the overall extent of reaction $[p = (M_0 - M)/M_0]$.

Using the composition equation, one can express F_A as a function of f_A, r_1, and r_2 and thus integrate the preceding equation (known as the *Skeist equation*) by a graphical method or a numerical method.

8.5.11.2. Determination of r_1 and r_2 Reactivity Ratios.

All the methods of determination of reactivity ratios are based on the determination of the composition of the copolymer obtained from various mixtures of comonomers. Several techniques can be used for the data analysis.

The first method was proposed by Lewis and Mayo. It consists of linearizing the relationship between the reactivity ratios; starting from the copolymerization equation previously established, we obtain

$$r_2 = \frac{[A]}{[B]} \left\{ \frac{d[A]}{d[B]} \left(1 + \frac{r_1[A]}{[B]} \right) - 1 \right\}$$

and defining y and x as

$$y = F_A/F_B \quad \text{and} \quad x = [A]/[B]$$

one obtains

$$y = x \frac{r_1 x + 1}{r_2 + x} \quad \text{or} \quad r_2 + x = r_1 \frac{x^2}{y} + \frac{x}{y}$$

This equation can be rewritten as

$$x - \frac{x}{y} = r_1 \frac{x^2}{y} - r_2$$

and defining G and F as

$$G = x - (x/y) \quad \text{and} \quad F = x^2/y$$

one obtains the following equations:

$$r_2 = Fr_1 - G \qquad \text{(Mayo–Lewis)}$$

$$G = Fr_1 - r_2 \qquad \text{(Finemann–Ross)}$$

Because the instantaneous composition of the copolymer is directly inaccessible, the conversion must be limited to 5% for the composition of the mixture of comonomers to remain sensibly equal to the initial one. To utilize the Finemann–Ross equation, one has to plot G against F for various experimental values, where the slope of the resulting line corresponds to r_1 while the intercept with the ordinate corresponds to r_2.

Although frequently used, the Finemann–Ross method is not accurate for low values of [A]/[B] (or [B]/[A]) and is not suitable for a wide range of concentrations. The method described by Kelen and Tüdos is often preferred; it consists of dividing all the terms of the Finemann–Ross equation by $(\alpha + F)$, with the value of the constant α being taken equal to $(F_{min}/F_{max})^{1/2}$:

$$\frac{G}{(\alpha + F)} = r_1 \frac{F}{(\alpha + F)} - \frac{r_2}{(\alpha + F)}$$

which corresponds to

$$\frac{G}{(\alpha + F)} = \left(r_1 + \frac{r_2}{\alpha}\right) \frac{F}{(\alpha + F)} - \frac{r_2}{\alpha}$$

Defining ξ and η as

$$\xi = \frac{F}{(\alpha + F)} \quad \text{and} \quad \eta = \frac{G}{(\alpha + F)}$$

and plotting η as a function of ξ, one obtains a straight line (Figure 8.6) whose intercept with the ordinate is equal to $-r_2/\alpha$ and whose value on Y-axis (ordinate) is equal to r_1 for $\xi = 1$.

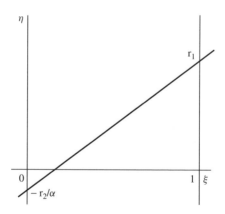

Figure 8.6. Kelen–Tüdos diagram for the determination of the reactivity ratios.

8.5.11.3. Q–e Scheme. Systems whose r_1 and r_2 values are much lower than unity exhibit a strong tendency to the alternation of the comonomeric units, and it can be roughly considered that the product $r_1 \cdot r_2$ is inversely proportional to the frequency of alternations. After examining in more detail the situation, Alfrey and Price proposed an **empirical** expression for the rate constant of "cross-addition" (k_{AB}); they assumed that this constant is all the higher as the reactivity of the copolymerizing species is high and their polarity is different. Regarding this last aspect, they assumed that the tendency to alternation is all the more pronounced as an electrostatic interaction exists between $\sim\sim A^\bullet$ and B. They suggested

$$k_{AB} = P_A Q_B \exp(-e_A e_B)$$

a relation in which

PA is proportional to the reactivity (or inversely proportional to the stabilization by resonance) of the active center $\sim\sim\sim\sim A^\bullet$.

Q_B is proportional to that of monomer B.

e_A and e_B measures the polarities of both radicals ($\sim\sim\sim\sim A^\bullet$ and $\sim\sim\sim\sim B^\bullet$) and monomers (A and B), respectively.

It is normal that the polarity of the radical and that of the monomer are defined by the same value since this polarity is determined by the electronic effect of the substituent which is the same for A and $\sim\sim A^\bullet$.

Using the above relation, one can write

$$k_{AA} = P_A Q_A \exp(-e_A^2)$$

from which one can deduce

$$r_1 = \frac{k_{AA}}{k_{AB}} = \frac{Q_A}{Q_B} \exp[-e_A(e_A - e_B)]$$

and

$$r_2 = \frac{k_{BB}}{k_{BA}} = \frac{Q_B}{Q_A} \exp[-e_B(e_B - e_A)]$$

Thus to each monomer corresponds a Q value and an e value from which r_1 and r_2 values can be calculated for a couple of comonomers. One must stress the fact that the *Alfrey–Price relation* does not take into account steric effects and that the $Q-e$ scheme does not apply to 1,2-disubstituted ethylenic monomers presenting a relatively low ceiling temperature.

Alfrey and Price took **styrene** as **reference monomer** and they assigned values of

$$e = 0.80 \quad \text{and} \quad Q = 1.0$$

to this molecule by convention.

Also by convention $\Delta e > 0$ when both the monomer double bond and the corresponding free radical undergo an increase in positive polarization and conversely. As for Q, only the stabilization by resonance is taken into account. In a copolymerization reaction, the reactivity of vinyl and related monomers increases in the following order; the more stabilized newly formed free radical, the higher the reactivity.

$$-O-CO-CH_3 < -Cl < -CO-O-CH_3 < -CN < -CH=CH \quad \text{and} \quad -C_6H_5$$

Obviously, the stabilization by resonance has an opposite effect on the reactivity of free radicals; this means that polystyryl radicals $\sim\sim\sim\sim CH_2-(C_6H_5)HC^\bullet$ have a low reactivity if compared to other free radicals.

The values of Q and e gathered in Table 8.14 illustrate the two effects of substituents. The $Q-e$ system is an empirical system whose validity is limited; it can, however, be used to predict the overall behavior in copolymerization reactions.

Table 8.14. Q and e values for a series of ethylenic monomers

Monomer	Q	e	Monomer	Q	e
Isoprene	3.33	−1.22	Styrene	1.0	−0.80
Butadiene	2.39	−1.05	Vinyl chloride	0.044	0.20
Isobutene	0.033	−0.96	Methyl methacrylate	0.74	0.40
Propene	0.002	−0.78	Vinylidene cyanide	20.13	2.58

8.5.12. Processes Utilized for Radical Polymerization

Free radical polymerization is a versatile method of polymerization that is extensively utilized in industry for the preparation of a variety of polymeric materials; it can be applied to a large variety of vinyl and related monomers under various conditions and processes because of its compatibility with many functional groups and its tolerance of water and protic media. Five main techniques with their advantages and their drawbacks are commonly utilized. A monomer that can be polymerized by radical means can be subjected to one or several of these techniques.

8.5.12.1. Polymerization in Bulk. Polymerization in bulk—or polymerization of neat monomer melts—is *a priori* the most economical one because it requires neither solvent nor emulsifier and affords high-purity polymers. However, the *gel effect*, with its consequences (high viscosity, low diffusion rates, small heat conductivity) that will be described hereafter makes difficult its control and very few are the monomers that can be polymerized in bulk. The initiator must be soluble in the monomer melt, and organic peroxides are most commonly utilized.

The term of *polymerization in bulk* is utilized whenever the polymer formed is monomer-soluble; the latter plays the role of solvent for a reaction medium that witnesses a very strong viscosity buildup as the yield increases; the kinetics can then take an explosive mode, accelerated by the *gel effect* (or *Trommsdorff effect*). This phenomenon is actually a self-acceleration of the rate of polymerization (possibly until becoming explosive) followed by a strong deceleration of the process (see Figure 8.7).

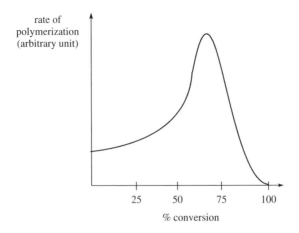

rate of
polymerization
(arbitrary unit)

25   50   75   100

% conversion

Figure 8.7. Representation of the gel effect for bulk polymerization.

This self-acceleration can hardly be anticipated because it occurs simultaneously with a decrease of the monomer concentration; it has two causes that self-maintain the process:

1. The high viscosity of the reaction medium causes a pronounced decrease of the values of the rate constant of termination (k_t) that are diffusion-controlled; this contributes to an increase in the concentration of polymeric free radicals and to a rise of the temperature in the reaction medium due to the exothermicity of the polymerization;

2. Because of the high value of the energy of activation of the dissociation step (about $130 \, kJ \cdot mol^{-1}$), this increase in temperature favors an even faster decomposition of the initiator, adding more free radicals into the medium.

The increase in viscosity, which is the first cause for the Trommsdorff effect, makes difficult the stirring of the reaction medium and thus the removal of the heat produced upon polymerization. With a continuous increase of the concentration in active centers, the system is no longer under steady-state conditions and the rate of polymerization as well as the temperature get out of control. Only when the viscosity buildup is such that monomer molecules can hardly move in the reaction medium does the rate of polymerization slow down until reaching negligible values. To avoid this situation, polymerizations in bulk are generally discontinued at relatively low conversions and are either continued in suspension or in thin films for a better thermal transfer. A particular case is that of the high-pressure ($\sim$200 kPa) polymerization of ethylene: the latter is under supercritical conditions, and its density ($\sim$0.5) is such that its behavior is close to that of liquids. Initiation is carried out by simultaneous introduction into the reactor of molecular oxygen and a peroxide. Polymerization is carried out at a temperature higher than the melting point of polyethylene, and the polymer is "swollen" by the monomer. For the recovery of the PE formed, which represents 30–40% of the reactor content, pressure has to be released.

Photochemical polymerizations are only performed in bulk, generally through thin-film irradiation.

8.5.12.2. Solution Polymerization.

Solution polymerizations are frequently used in research laboratories because high yields can be reached without facing the difficulties associated with the "gel effect." The solvents that can be used to this end must solubilize the polymer formed and be inert toward free radicals—in particular, not induce transfer reactions. The initiator, like all other reagents, must be soluble in the reaction medium, which is thus homogeneous. At the end of the polymerization, the polymer is recovered by evaporation of the solvent or by precipitation in a nonsolvent.

Free radical polymerization in solution is seldom used in industrial processes due to the cost of the solvent, the cost of its removal, and the toxicological and environmental concerns it causes; it is, however, sometimes utilized to obtain high-purity products or when the solvent is water (for example, for the polymerization of acrylamide).

8.5.12.3. Dispersion Polymerization.

In these polymerizations, the initial reaction medium (monomer, solvent, initiator, additives) is homogeneous, but the polymer formed is insoluble and precipitates progressively as the polymerization proceeds. The polymer particles formed can continue to grow in size in the precipitated phase (by adsorption of monomer and/or initiator on the particles), but their precipitation provokes in general a limitation of the sample molar masses. Due to this precipitation of the polymer, a low viscosity of the medium can be maintained up to high yields and the gel effect can be avoided.

When formed, the precipitate can be progressively recovered from continuous polymerization reactors in order to maintain approximately a constant viscosity of the reaction medium.

Dispersion polymerizations can be carried out either in bulk or in the presence of a solvent. Dispersion bulk polymerizations are industrially utilized to polymerize vinyl chloride, acrylonitrile, and vinylidene chloride. Polymerizations requiring the presence of a solvent are those whose monomer is solid in the polymerization conditions; a solvent that favors phase separation can also be useful; for example, styrene and methyl methacrylate are polymerized in dispersion in the presence of an alcohol inducing phase separation. When an adequate suspending agent (hydroxypropylcellulose, etc.) is added in the reaction medium, dispersion polymerization affords particles of uniform size (about 3–30 μm).

8.5.12.4. Suspension Polymerization.

One way to remove the heat that builds up in bulk polymerizations due to the gel effect is to disperse water-insoluble monomers as small droplets (10–500 μm) in an aqueous phase containing "suspending" agents. Pictured as a water-cooled bulk process, a suspension polymerization occurs in the microreactors that are the monomer droplets, and the initiator that is used is monomer-soluble.

Reverse suspension polymerizations can also be carried out using an organic solvent as continuous medium and an insoluble monomer.

When the proportion of surfactant used is relatively high, the size of the particles produced can be small and the term *"micro-suspension"* is utilized.

The control of temperature is obtained by stirring the dispersed phase and cooling the walls of the reactor. The size of the particles or beads formed is determined by the effectiveness of the stirring and by the presence of water-soluble "suspending agents" such as poly(ethylene oxide), poly(vinyl alcohol), hydroxypropylcellulose, and so on, and surfactants (represented by ○−in Figures 8.8 and 8.9); the polymer serves to increase the viscosity of the medium and the surfactant (in low proportion) to improve the stabilization of the particles.

Beyond a certain yield, the particles can agglomerate and induce a coalescence of the medium that can be prevented by addition of mineral powders (talc, calcium carbonate, barium sulfate, etc.) settling at the interphase. Figure 8.8 represents monomer-swollen pearls of polymer in a suspension polymerization.

At the end of such polymerization, the polymer particles require a work-up step to wash out most of the water-soluble polymer used as additive and the mineral powders; they are recovered by filtration.

8.5.12.5. Emulsion Polymerization.

In spite of some analogies with the suspension polymerizations, emulsion polymerizations proceed in a completely different way and result in the formation of particles with size ranging between 0.05 and 5 μm. Most of emulsion polymerizations are carried out in water as continuous medium, but, as mentioned for suspension polymerizations, it is also possible to carry out "reverse" emulsion polymerizations in organic solvents in which the monomer is insoluble.

Aqueous emulsion polymerizations of hydrophobic monomers emulsified by surface active compounds are initiated by water-soluble initiators ($S_2O_8K_2$, H_2O_2,

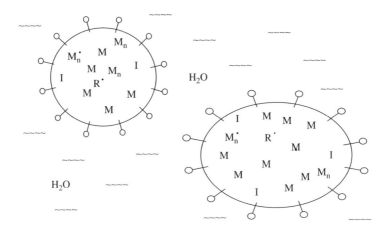

Figure 8.8. Diagrammatic representation of suspension polymerization.

mineral redox system, etc.). Other than the solubility of the initiator in the continuous water phase, another major difference with suspension polymerization lies in the presence of a emulsifier or surfactant in a proportion (from 1 to 2 weight%) well above its critical micellar concentration (CMC).

A simplified representation of the medium before initiation of the polymerization is shown in Figure 8.9. The continuous dispersing aqueous medium (①) contains monomer droplets (M) of large size (1–10 μm) (②) which are stabilized by the surfactant (o—); the surfactant molecules associate in water and form monomer-swollen micelles (15–100 surfactant molecules/micelle) (⑥), but a small fraction of them are "dissolved" in aqueous phase (③); due to their low solubility in water, only a small proportion of monomer molecules (④) are also solubilized in the aqueous phase—in contrast with the initiator (⑤), which is entirely soluble. Because of their large number (typically 10^{20} monomer-swollen micelles of about 10 nm per liter against 10^{12} monomer droplets of about 5 μm), the micelles have a much higher specific surface (external envelope) (typically 2×10^{21} mm²/L) than that of the droplets (10^{20} mm²/L). Upon decomposition of the initiator ⑤, hydrophobic radicals are generated which add to the monomer present in the aqueous phase; as more monomer molecules are added, the growing radical becomes increasingly hydrophobic and surface-active. It then tends to adsorb to the nearest hydrophobic interface available and has thus much higher probability to penetrate a micelle than be captured by a droplet. In fact, the latter play the role of monomer "tank," and their size decreases progressively with monomer depletion until they disappear.

Remark. The term "emulsion" is normally used for liquids dispersed in liquids; upon emulsion polymerization, the product can be called "latex."

Several theories have been proposed to explain the mechanism occurring in an emulsion polymerization; the one proposed by Smith and Ewart, which is also the most accepted, is based on the following simplifying assumptions:

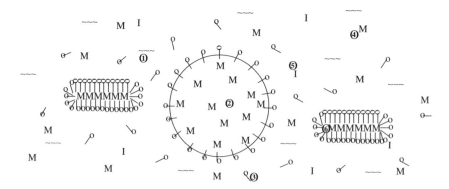

Figure 8.9. Diagrammatic representation of an initial system of emulsion polymerization.

Initiation occurs in aqueous phase as previously mentioned, and the resulting oligomers $RM_n^\bullet$ (n small) penetrate into the micelles and consume the monomer available inside until arrival of a second radical. In such a confined volume, the two radicals present recombine or disproportionate and chain growth is discontinued.

Polymerization resumes as soon as a third radical enters this hydrophobic compartment, which has in the meantime become a polymer particle swollen with the monomer diffusing from other droplets.

The expression of the rate of polymerization can be written as

$$R_p = k_p[RM_n^\bullet][M]_{\text{part}}$$

with $[M]_{\text{part}}$ the monomer concentration inside the particles expressed in moles per liter of swollen particles. Because of the permanent diffusion of monomer from the droplets to the particles, due to depletion in the particle, $[M]_{\text{part}}$ is assumed to remain constant from the beginning up to 70–80% conversion. The rate of polymerization is thus independent of the total monomer concentration $[M]$ expressed in moles per liter. As for the concentration in free radicals, it can be easily deduced from the knowledge of the number of particles (N_p) present in the medium through the following reasoning, identifying the capture of free radicals by the polymer particles to the filling of containers by the drops of rain! At any moment, each one of these containers featuring these particles has received either an even or an odd number of drops. Insofar as the numbers of containers and drops are large, the number of containers that have received an even number of drops must be imperceptibly equal to that with an odd number of drops. Using this analogy, one can deduce that half of the particles have captured an even number of radicals and the other half an odd number. At a given moment, only those that have received an odd number of radicals witness polymerization so that $[RM_n^\bullet] = N_p/2$, where N_p is the number of particles per unit of volume of emulsion.

The expression of the rate of polymerization (R_p) and that of the kinetic chain length ($\bar{v}$) can thus be written:

$$R_p = k_p[M]_{\text{part}}\frac{N_p}{2}$$

$$\bar{v} = \frac{k_p N_p[M]_{\text{part}}}{2d[RM^\bullet]/dt}$$

where $d[RM^\bullet]/dt$ is the rate of chain initiation ($\equiv 2fk_d[I]$ for a homolytic decomposition).

As for the number of particles (N_p), it depends on the concentration of both surfactant and initiator, but its calculation is complex. With the simplifying assumptions of the Smith–Ewart theory, this calculation is made easier using the following reasoning.

Polymer particles are being created throughout the so-called "nucleation" period during which micelles are progressively transformed into latex particles. Once formed, the particle is replenished with monomer replacing the just consumed one. The increase in volume ($\mu = dV/dt$) that results drives the surfactant of the empty micelles toward the external envelope of the particles; micelles thus disappear from the medium by becoming particles or by supplying surfactant molecules to the already formed particles. As soon as all the micelles have been used up by one of the mechanisms mentioned above (first period in Figure 8.10)—corresponding to approximately 15% monomer conversion—the number of particles (N_p) can be considered constant until the end of the polymerization. The rate of polymerization during this second period in Figure 8.10 can be expressed by the relation

$$R_p = k_p[M]_{\text{part}} \frac{N_p}{2}$$

Thus it will be constant up to 70–80% conversion. Assuming that free radicals are generated at constant rate ($d[RM^\bullet]/dt = \rho = \text{const}$)) and that all of them serve to create particles, at the time t_1 corresponding to the total disappearance of micelles, N_p can be written as

$$N_p = \rho t_1$$

At t_1, a particle created at t_0 will exhibit the volume

$$V(t_1, t_0) = \mu(t_1 - t_0)$$

with its volume at t_0 (when it was a micelle) being negligible.

The surface of its external envelope can be easily deduced from its volume:

$$a(t_1, t_0) = (36\pi)^{1/3}[\mu(t_1 - t_0)]^{2/3}$$

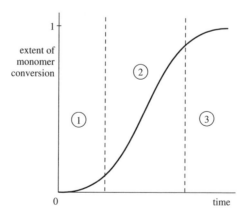

Figure 8.10. Kinetics of the monomer conversion for an emulsion polymerization in a closed batch reactor.

Because the number of particles generated for the period of time dt is ρdt, the total external surface at time t_1 can be written

$$A_{t_1} = \int_0^{t_1} a(t_1, t_0)\rho \, dt = (36\pi)^{1/3} 0.6\rho\mu^{2/3} t_1^{5/3}$$

Because this total surface can be directly related to the concentration [S] of the surfactant and to its molar surface (a_s),

$$A_{t_1} = a_s[S]$$

one obtains the following for the expression of N_p:

$$N_p = 0.53 \left(\frac{\rho}{\mu}\right)^{0.4} \{a_s[S]\}^{0.6}$$

The Smith–Ewart model describes satisfactorily the polymerization of styrene, isoprene, and methyl methacrylate; for these systems, it can be used to predict the size of the latex particles and the corresponding molar masses. In contrast, it is unsuited for the case of monomers partially water-soluble or polymers insoluble in their monomer—that is, polymerization of vinyl chloride and vinyl acetate. It accounts neither for the fact that styrene can be polymerized in absence of surfactant nor for the fact that free radicals $(RM^\bullet)$ can equally penetrate into a micelle or in an already formed particle during the initial phase.

Fitch has thus proposed another model which considers that initiation and the early stages of the propagation occur in the aqueous phase, with the chains precipitating only when a critical size is reached—that is, for degree of polymerization of a few units to a few tens depending upon the hydrosolubility of the oligomer formed.

8.6. ANIONIC POLYMERIZATION

This type of polymerization is a very old one, used at the beginning of the twentieth century in Germany to produce a well-known synthetic rubber named "Buna". However, it is only in the middle of the 1950s that anionic polymerization took all its importance when Szwarc shed a new light on this field and discovered that it can be carried out in the absence of any transfer and termination. Szwarc called such polymerizations "living" (see Section 8.4), and his discovery triggered an intense research activity that culminated in the synthesis of unprecedented complex macromolecular architectures (block copolymers, stars, etc.).

8.6.1. General Characteristics

The anionic polymerization is a chain reaction that can be schematized by

$$\sim\sim\sim\sim M_n^-, Met^+ + M \longrightarrow \sim\sim\sim\sim M_{n+1}^-, Met^+$$

where $\sim\sim\sim\sim M_n^-$ represents a negatively charged or polarized species carried by the growing chain, and Met^+ is a positive counterion (or a polarized species), generally a metallic cation. Whatever the precise mechanism involved in this type of polymerization, it proceeds via repeated nucleophilic reactions. In the case of **vinyl and related monomers**, for the propagation to occur by nucleophilic addition, an activation of the monomer double bond is generally required (see, however, "Remark," page 312). Electron-withdrawing substituents ($-CO-OR$, $-CN$, etc.) or those inducing a strongly positive polarization of the β-carbon atom of the double bond, when neared by a nucleophilic active species,

fulfill this condition.

Anionic polymerization also applies to **heterocyclic monomers**. In this case, it can occur either by nucleophilic substitution or by nucleophilic addition onto a carbonyl group followed by an elimination (mechanism $B_{AC}2$), and so on. A negative enthalpy of polymerization is a necessary condition for the monomer to be polymerized, and thus heterocyclic monomers must be strained enough to undergo ring-opening and polymerization. Another constraint of prime importance that affects the polymerizability of monomers—in particular, that of ethylenic ones—is the extreme reactivity of species that propagate the process. In a first approach—and without mistaking between the notions of nucleophilicity and basicity—a carbanionic species can be considered as the conjugate base of a protonic acid whose pK_a can be evaluated. Thus, the species formed in the anionic polymerization of styrene

$$\sim\sim\sim CH_2-(C_6H_5)HC^-$$

is the conjugate base of the species

$$\sim\sim\sim\sim CH_2-(C_6H_5)HCH$$

whose pK_a is around 41, a high value that mirrors an extremely low acidity. The corresponding conjugate base is thus particularly strong; the comparison of a pK_a of 41 with that of water which is the conjugate acid of metal hydroxides gives an idea of the very high reactivity of the conjugate base. This imposes that monomers are free of electrophilic species that could potentially react with nucleophilic growing chains. Depending upon the nature of the substituent A, this nucleophilicity varies to a large extent.

Anionic polymerization is utilized only when the "living" character of the chain growth can be ensured. In addition, initiators are selected for their ability to give a complete initiation ($f \sim 1$) and a short period of initiation compared to that of propagation, allowing a controlled polymerization to occur. This situation is exploited in macromolecular engineering to synthesize polymeric chains with well-defined structure and narrow molar mass distribution.

8.6.2. Structure of the Propagating Species

The "living character" of the growing species formed in carbanionic polymerization offers an opportunity to study comprehensively their structure. The concentration of the reactive centers being always extremely low in the polymerization medium, it is easier to carry out such structural studies on simple organometallic models of the "living" ends. Some of these are used to initiate the polymerization and the knowledge of the parameters that determine their reactivity is interesting by itself.

There is a close relationship between the structure of organometallic species ($\sim\sim M_n^-$, Met$^+$) and their reactivity. In the case of species responsible for the polymerization of ethylenic monomers, their nucleophilicity and thus their reactivity are strongly determined by the electron density on the carbanionic site

$$\sim\sim\sim CH_2 - HC^{\delta-}, Met^{\delta+}$$
$$|$$
$$\mathbf{A}$$

This electron density depends on the polarization of the C–Met bond and the possible delocalization of the negative charge on the substituent **A**. So, the parameters that control the structure of the active centers responsible for the anionic polymerization of ethylenic monomers are:

- The nature of the substituent(s) carried by the double bond,
- The nature of the counterion associated with the carbanionic species,
- The nature of the solvent in which the reaction is carried out and the presence of possible additives.

8.6.2.1. Effect of the Substituent A. If the substituent promotes a delocalization of the negative charge [as is the case for styrene, vinylpyridines, (meth)acrylates, etc.], it entails a decrease of the intrinsic reactivity of the carbanionic species. Thus, in the case of acrylates, the active center is an enolate of rather low reactivity:

$$\sim\sim\sim\sim CH_2 - CH \underset{CH_3-O}{\overset{}{\diagdown}} = O \quad -, Met^+$$

The intrinsic reactivity of carbanionic active centers is increased by the presence of electron-donating substituents and is conversely decreased by that of electron-withdrawing ones. However, in the case of acrylates, the monomer double bond is more activated by the electron-withdrawing character of its substituent than the reactivity of the corresponding enolate is lowered by the same substituent; this explains the very high anionic polymerizability of these monomers. Thus, the intrinsic reactivity of the monomer determines the global reactivity of the system—that is, its polymerizability. For example, methacrylic monomers (methyl methacrylate is shown hereafter) are characterized by a lower polymerizability than that of acrylates, in spite of the electron-donating effect of their methyl group presumed to increase the electron density on the active center and thus its reactivity; as a matter of fact, this $-CH_3$ group in α-position prevents (by its donor effect) a full polarization of the double bond and thus decreases the monomer reactivity.

Methyl methacrylate

Styrene and butadiene are the two reference monomers in anionic polymerization. Their high polymerizability is primarily due to the virtue of their double bonds to undergo a positive polarization and an electron shift toward their substituent when neared by a negatively charged active center.

> **Remark.** Ethylene is a monomer with no possibility of activation of its double bond. However, it can be polymerized by nucleophilic addition but its anionic polymerizability is very low, the absence of any stabilizing substituent next to the carbanionic site making the latter particularly reactive.

8.6.2.2. Effect of the Nature of the counterion.

Examples of polymerizations that can be carried out with nonmetallic counter-ions (quaternary ammonium, phosphonium ions, etc.) are scarce, the vast majority of them requiring the use of alkali or alkaline-earth cations.

Lithium and magnesium cations exhibit a small ionic radius which explains the partial covalent character of their bond with carbon atoms in nonpolar solvents, provided that the carbanion is not too delocalized.

With cations of higher ionic radius, the interionic distance favors the separation of charges, and thus the corresponding species can be considered totally ionized.

In polar solvating media as well as in the presence of solvating additives, the ionic radius of the counterion affects its capacity to be solvated.

Large cations like cesium can by no means be solvated even by solvents known for their strong solvating power.

Lithium is by far the most used counterion known; this is primarily due to the practical and synthetic ease that is associated with the utilization of butyllithium as initiator, but also to the virtue of this cation to generate different configurational structures in the polymers formed. Indeed, lithium cations can generate either partially covalent or totally ionic species with different regio- and stereospecificity, depending upon the solvent in which it is dispersed.

8.6.2.3. Effect of the Nature of the Solvent and that of Potential additives.
Because of the very high reactivity of anionic reactive species, the solvents used in anionic polymerization should not exhibit any acidic character; thus basic or neutral solvents are generally chosen.

The functions of a solvent are manifold and, depending upon its structure, it can fulfill one, two or three of these functions.

The first function is that of a **diluent**; the simultaneous generation of carbanionic initiating/propagating sites and the monomer consumption by the latter can liberate a considerable heat in the reaction medium that can be better removed if a solvent is present. Solvents used as diluents are always aliphatic or aromatic hydrocarbons; they do not modify or only to a little extent the structure of active centers.

Organolithium compounds are aggregated species whose degrees of aggregation vary with the nature of the carbanion and sometimes with the range of concentration. For instance, polystyryllithium ion pairs are aggregated as dimers like shown below:

In the latter case, only non-aggregated species—in equilibrium with aggregated ones—are reactive and contribute to the propagation:

$$\left(\text{~~~~PS}^-, \text{Li}^+\right)_2 \underset{K_{ag}}{\overset{}{\rightleftarrows}} 2 \text{ ~~~~PS}^-, \text{Li}^+$$

Non active   Active

The second potential function of a solvent is that of a **solvating agent**. Solvents used for that purpose are ethers or tertiary amines whose basic character—according to Lewis definition—entails a coordination to the Lewis acids that are the metal cations associated with the nucleophilic species. This role of solvating agent can also be played by additives (crown-ether, cryptands, tertiary diamines, etc.) used in small amount in a hydrocarbon serving as diluent. Depending upon their geometry or their concentration, such additives can either solvate externally the ion pairs [see hereafter the case of polybutadienyllithium in the presence of tetramethylethylene-diamine (TMEDA)],

TMEDA

or cause a stretching of the carbon–metal bond (see hereafter the case of polystyryl-lithium solvated by a crown-ether):

$$\text{~~~~PS~~~~CH}_2\text{--HC}^-,$$

Depending upon the size of the cation and the geometry of the solvating agent, such solvation may be more or less effective. Stretching ion pairs increases considerably their reactivity due to the easier insertion of monomer between the anion and the cation.

When the dielectric constant (permittivity) of the solvent is sufficiently high, it can play the role of **dissociating agent**. Such a solvent can then cause the charges to separate more markedly and induce a partial dissociation of ion pairs into free ions,

$$\text{~~~~~M}_n^-, \text{S}_x,\text{Met}^+ \xrightleftharpoons{K_{\text{diss}}} \text{~~~~~M}_n^- + \text{S}_x,\text{Met}^+$$

where S_x corresponds to X molecules of solvent coordinating to the metal cation.

The relation between the dissociation equilibrium constant (K_{diss}) and the permittivity of the reaction medium (ε) can be written as

$$- \ln K_{\text{diss}} = - \ln K_{\text{diss}}^0 + \frac{e^2}{(r_1 + r_2)\varepsilon k T}$$

where K_{diss}^0 is the constant of dissociation of ion pairs in a medium of infinite permittivity, r_1 and r_2 are the ionic radius of cation and anion, respectively, and e is the electron charge. This relation shows that by increasing the apparent ionic radius of the cation and that of the interionic distance, the solvating effect favors the dissociation; most of high permittivity solvents exhibit also a strong solvating power. The reactivity of free ions resulting from the dissociation of ion pairs is extremely high and, even at relatively low concentration, they have a major impact on the global kinetics of polymerization. In contrast to the case of free radical polymerization, the same monomer can generate various propagating species depending upon the nature of the initiator and that of the surrounding medium. The various reactive species are ranked hereafter in the increasing order of their reactivity,

$$\text{R--Met} < (\text{R}^{\delta-}\text{--Met}^{\delta+})_x < \text{R}^{\delta-}\text{--Met}^{\delta+} < \text{R}^-,\text{Met}^+\text{S}_y$$

$$< \text{R}^-,\text{Met}^+ < \text{R}^-,\text{S}_y\text{Met}^+ < \text{R}^- + \text{S}_y\text{Met}^+$$

Generally inactive species   *Highly reactive species*

where R represents the polymer chain and S represents a solvent molecule.

8.6.3. Initiation Step

The initiator has to be selected with care so as to ensure a short initiation step (compared to that of propagation) and the absence of side reactions. The preference must go to initiators that are more nucleophilic than the active species resulting from their addition onto a monomer molecule.

Two types of reaction can be utilized to generate primary active centers.

The first one resorts to an electron transfer from a metal atom (generally an alkali metal) to a molecule whose electron affinity is sufficiently high. The role of the electrophilic entity can be played by the monomer molecule and, in this case, the transfer of ns electrons from the alkali metal results in the formation of a radical-anion based on the monomer molecule:

$$Met + \overset{\displaystyle\diagup\!\!\diagdown}{\underset{(A)}{}} \longrightarrow \left[\overset{\displaystyle\diagup\!\!\diagdown}{\underset{(A)}{}} \right]^{\overset{\bullet}{-}} Met^+$$

Radical-anion

The species obtained can be represented by the electron distribution shown hereafter,

$$\bullet CH_2 - \underset{(A)}{HC^-}, Met^+$$

and by recombination of the two free radical sites, a bicarbanionic species is formed:

$$2 \ \bullet CH_2 - \underset{(A)}{HC^-}, Met^+ \longrightarrow Met^+, \ ^- \underset{(A)}{HC} - CH_2 - CH_2 - \underset{(A)}{HC^-}, Met^+$$

This direct initiation is rarely utilized because the formation of such a radical-anion through the reaction between a solid (metal) and a liquid (monomer) is generally slow. To overcome this limitation, an organic intermediate that cannot polymerize itself but can accommodate electrons by transfer is generally utilized. More often, these intermediates are polycyclic aromatic hydrocarbons; for example, naphthalene is commonly used for this purpose; the reaction between naphthalene (in solution) and sodium (solid) is schematized hereafter:

$$Na + naphthalene \rightleftharpoons \ ^\bullet(naphthalene)^-, Na^+$$

The reaction must be carried out in a sufficiently solvating solvent (tetrahydrofuran, dimethoxyethane, etc.) for the electron transfer to occur, and, after elimination of the metal in excess, a homogeneous and quasi-instantaneous initiation step can be obtained upon addition of monomer:

The bicarbanionic species formed is persistent under conditions of "living" polymerization.

Remark. Since two molecules of initiator lead to the formation of a single chain, the relationship giving the degree of polymerization as a function of the conversion must be modified. In the case of a monofunctional initiation, the relation is $\overline{X}_n = [M_{pol}]/[I]$, whereas for a difunctional initiation we obtain $\overline{X}_n = 2[M_{pol}]/[I]$ [M_{pol}], representing the concentration of monomer polymerized.

More usually—and in particular in industry—initiation is obtained by the means of strongly nucleophilic **Lewis bases**. They are usually monofunctional and monovalent organometallic species; compounds like benzylsodium or phenylisopropylpotassium (cumylpotassium) may be utilized in research laboratories, but in industry it is exclusively the isomers of butyllithium (*n*-, *sec*-, *tert*-) that are employed. They are strongly aggregated in hydrocarbon media (*n*-BuLi is hexameric, *tert*-BuLi is tetrameric, etc.) and they react only under their "unimeric" (nonaggregate) form, the latter being in equilibrium with aggregates:

$$\left(n\text{-BuLi} \right)_6 \quad \underset{\text{...}}{\overset{K_{ag}}{\rightleftharpoons}} \quad 6 \ n\text{-BuLi}$$

$$n\text{-BuLi} \ + \quad \overset{k_i}{\longrightarrow} \quad n\text{-Bu} \diagdown_{A}^{-}, \text{Li}^+$$

In hydrocarbon solution, the initiation is rarely instantaneous and mixed aggregates can be formed which adds to the complexity of the kinetic expression of the initiation step and that of the first propagation steps. In polar solvents or in the presence of certain additives, such aggregation disappears and very reactive species (free ions, solvated ion pairs, etc.) are generated that may provoke side reactions.

The above Lewis bases can also be utilized to initiate the polymerization of heterocyclic monomers. However, the high reactivity of the latter authorizes the use of weaker bases (for example, KOH) than those required for the polymerization of vinyl and related monomers. In this way it is possible to limit side reactions and thus to preserve the "living" character of the polymerization:

$$\text{HO}^-, \text{K}^+ \ + \quad \triangle_{\text{O}} \quad \longrightarrow \quad \text{HO} \diagdown\diagup_{\text{O}^-, \text{K}^+}$$

8.6.4. Propagation Step

With vinyl and related monomers, the mechanism of propagation is the same as that of the initiation by Lewis bases:

$$\text{\textasciitilde\textasciitilde\textasciitilde\textasciitilde}M_n\text{\textasciitilde\textasciitilde\textasciitilde\textasciitilde}CH_2-HC^-, \text{Met}^+ \quad + \quad \overset{}{\diagdown}_A$$

$$\longrightarrow \quad \text{\textasciitilde\textasciitilde\textasciitilde\textasciitilde}M_{n+1}\text{\textasciitilde\textasciitilde\textasciitilde\textasciitilde}CH_2-HC^-, \text{Met}^+$$

In hydrocarbon solvents, active centers have a strong tendency to be aggregated; as previously seen, it is the case for the polymerization of styrene initiated by an organolithium compound, in bulk or in a hydrocarbon solvent:

$$(\text{\textasciitilde\textasciitilde\textasciitilde\textasciitilde\textasciitilde\textasciitilde}PS^-,\text{Li}^+)_2 \quad \overset{K_{ag}}{\rightleftharpoons} \quad 2 \ \text{\textasciitilde\textasciitilde\textasciitilde\textasciitilde\textasciitilde\textasciitilde}PS^-,\text{Li}^+$$

Because only nonaggregated species are active, the **kinetic equation** for the **propagation step** can be easily established:

$$K_{ag} = [(\text{\textasciitilde\textasciitilde\textasciitilde}PS^-,Li^+)_2]/[\text{\textasciitilde\textasciitilde\textasciitilde}PS^-,Li^+)]^2$$

$$[\text{\textasciitilde\textasciitilde\textasciitilde\textasciitilde}PS^-,Li^+] = \{[(\text{\textasciitilde\textasciitilde\textasciitilde\textasciitilde}PS^-,Li^+)_2]/K_{ag}\}^{1/2}$$

$$R_p = -d[S]/dt = k_{p,app}[\text{\textasciitilde\textasciitilde\textasciitilde\textasciitilde}PS^-,Li^+][S] = k_p\{[(\text{\textasciitilde\textasciitilde\textasciitilde\textasciitilde}PS^-,Li^+)_2]/K_{ag}\}^{1/2}[S]$$

where $k_{p,app}$ is the apparent rate constant of propagation and k_p is the rate "constant" of propagation, both of which vary with the concentration in potentially active centers—that is, with [Li].

Because the equilibrium constant of aggregation (K_{ag}) is very high, it cannot be measured generally so that the above equation can be simplified as

$$R_p = k_p[Li]^{1/n}[S]$$

where n is the degree of aggregation.

Remark. In the anionic polymerization of dienes, the degree of aggregation varies with the concentration in organometallic species, which complicates the kinetic treatment.

The addition of solvating agents in the reaction system causes a total disaggregation, without appreciably modifying the permittivity of the medium. All organometallic species are then active and the kinetics becomes first order in active centers:

$$R_p = k_p[Li][S]$$

The situation is similar if the polymerization is carried out in a solvent exhibiting some solvating power but no dissociating effect. It is the case for dioxane, which is particularly suitable for the study of the influence of the ionic radius of the cation on the reactivity of the corresponding ion pairs. Because of its structure, dioxane has indeed a chelating power and thus generates externally solvated ion pairs with bigger alkali cations, modifying only to a little extent the interionic distance. A similar behavior can be observed in a nonpolar solvent containing chelating additives. Table 8.15 shows how an increase in the ionic radius of the cation strongly increases the reactivity of ion pairs.

Indeed, an increase in the interionic distance decreases the electrostatic interaction between the two electric charges and favors insertion of the monomer. If the solvent exhibits both a solvating power and a dissociating capacity, the two effects play a role. For instance, in tetrahydrofuran (THF), whose permittivity is equal to 7.8 at 20°C, the solvation of ion pairs generates "loose" ion pairs, with each cation being surrounded by several molecules of THF. Ion pairs are thus in equilibrium

Table 8.15. Rate constants of propagation of styrene for various alkali counterions (solvent: dioxane, $T = 25°$)

Counterion (Met^+)	$k_{p\pm}$ ($L \cdot mol^{-1} \cdot s^{-1}$)
Li^+	0.94
Na^+	3.4
K^+	19.8
Rb^+	21.5
Cs^+	24.5

with free ions formed by dissociation, and the resulting overall reactivity combines the contribution of the various species, each with its own reactivity:

$$\sim\sim\sim M_n^-,Met^+ + x THF \underset{}{\overset{K_{solv}}{\rightleftarrows}} \sim\sim\sim M_n^-, xTHF,Met^+$$

$$\updownarrow K_{diss}$$

$$\sim\sim\sim M_n^- + xTHF,Met^+$$

If i families of reactive species are simultaneously present in the reaction medium, each one contributes through its own propagation kinetics:

$$R_{pi} = k_{pi}[C_i^*][M]$$

the additivity of their contribution being expressed by

$$R_p = \sum_i R_{pi} = [M] \sum_i k_{pi}[C_i^*] = k_{app}[C^*][M]$$

with $[C^*] = \sum_i [C_i^*]$.

If only free ions (C_-^*) and (solvated or not) ion pairs ($C_\pm^*$) are taken into consideration,

$$[C^*] = [C_\pm^*] + [C_-^*]$$

the dissociation equilibrium

$$\sim\sim\sim\sim\sim M_n^-,Met^+ \underset{}{\overset{K_{diss}}{\rightleftarrows}} \sim\sim\sim\sim M_n^- + Met^+$$

can be written as

$$K_{diss} = [C_-^*]^2/[C_\pm^*] = [C_-^*]^2/\{[C^*] - [C_-^*]\}$$

Equilibrium constants of dissociation can be measured by conductometry from solutions containing different concentrations in organometallic species; they vary with the charge density of the anion, the interionic distance, and the permittivity of reaction medium and are generally very low. $[C_-^*]$ can thus be considered negligible as compared to $[C^*]$ and $[C_\pm^*]$ can be assimilated to $[C^*]$, which means that

$$[C_-^*] = \{K_{\mathrm{diss}}[C^*]\}^{1/2}$$

This leads to

$$R_p = -d[M]/dt = k_{\mathrm{app}}[C^*][M] = \{k_{p\pm}[C^*] + k_{p^-} K_{\mathrm{diss}}^{1/2}[C^*]^{1/2}\}[M]$$

which can be written as

$$R_p = \underbrace{\{k_{p\pm} + k_{p^-} K_{\mathrm{diss}}^{1/2}[C^*]^{-1/2}\}}_{k_{\mathrm{app}}}[C^*][M]$$

The above relation shows that k_{app} varies with $[C^*]$; as for $k_{p\pm}$ and k_{p^-} they can be determined from the k_{app} versus $[C^*]$ plot (Figure 8.11) if K_{diss} is known. As mentioned above, K_{diss} can be measured under given experimental conditions by conductimetry on active solutions. For instance, the rate constant of propagation of ion pairs, the rate constant of free ions, and the constant of dissociation for polystyrylsodium (PS^-,Na^+) in tetrahydrofuran solution at 25°C are

$$k_{p\pm} = 80\,\mathrm{L\cdot mol^{-1}\cdot s^{-1}}$$
$$k_{p^-} = 65{,}000\,\mathrm{L\cdot mol^{-1}\cdot s^{-1}}$$
$$K_{\mathrm{diss}} = 1.5 \times 10^{-7}\,\mathrm{mol\cdot L^{-1}}$$

Thus, the proportion of free ions can be deduced from these values: for $[C^*] \sim 10^{-4}$ mol·L, free ions represent only about 4%; in spite of that and because of their very high reactivity, free ions contribute to an extent of ~97% to propagation.

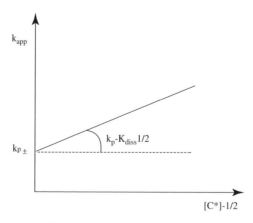

Figure 8.11. Determination of the rate constants of propagation on ion pairs ($k_{p\pm}$) and on free ions (k_{p^-}).

Remarks

(a) The value of $k_{p\pm}$ measured in THF is different from that measured in dioxane (Table 8.15). Indeed, in this last solvent, ion pairs are externally solvated, which modifies only very little the interionic distance, whereas in THF, ion pairs not only can be externally solvated but can partially also be stretched ("loose" ion pairs) under the effect of the solvent and these are more reactive.

(b) Addition of homoionic species in the reaction medium by using soluble and highly dissociated salts causes a retrogradation of the dissociation equilibrium of reactive ion pairs; it entails a deceleration of the propagation step. By combining the results of the kinetic study and the value of the added homoionic salt dissociation constant, it is possible to calculate the propagation rate constant of free ions (k_{p-}).

The **stereochemistry** of the propagation step closely depends on the polarity of the solvent and on the nature of the counterion. For polymerizations involving free ions, the sp^2 hybridization of carbanionic species prevents any marked stereoregulation of the propagation step and the resulting polymers are atactic. In nonpolar solvents, the carbanionic species exhibit generally an sp^3 hybridization; with Li^+ and Mg^{++} as counterions, (meth)acrylic and similar monomers (2-vinylpyridine, etc.) polymerize under stereoregulating conditions through the combined effect of monomer coordination and steric hindrance. High contents in either isotactic or syndiotactic triads (mm or rr < 0.90) can be obtained.

As for the propagation step of **heterocyclic monomers** (oxiranes, thiiranes, lactones, lactams, etc.), the general phenomena are very similar to those observed with vinyl and related monomers. Although propagating species differ by the nature of the nucleophilic entities (oxoanions, thioanions, nitranions, etc.) involved, the corresponding ion pairs can also be prone to aggregation, solvation, and dissociation, depending upon the nature of the solvent or that of additives introduced into the reaction medium. Since the propagating species in the polymerization of heterocycles are much less reactive than pure carbanions, it is often necessary to activate them in order to bring about sufficiently high rates of polymerization. In general, the kinetics is complex because of the aggregation of active centers independently of the reaction media, and the degrees of aggregation vary with their concentration. Moreover, the energies of activation of the propagation reaction are appreciably different for ion pairs and free ions; depending upon the temperature, the contribution of the various active species to the propagation can vary in a large extent with respect to the kinetics.

As for the nature of the propagating active centers, the polymerization of oxiranes (epoxides) and of ε-caprolactone occurs through alkoxides, that of thiiranes (episulfides) occurs through thiolates, and that of strained lactones occurs through carboxylates. The corresponding reaction mechanisms are well-established (nucleophilic substitution, nucleophilic addition on carbonyl, etc.). For example, as established in the polymerization of ethylene oxide initiated by a potassium derivative, the propagation occurs by the following mechanism (nucleophilic substitution):

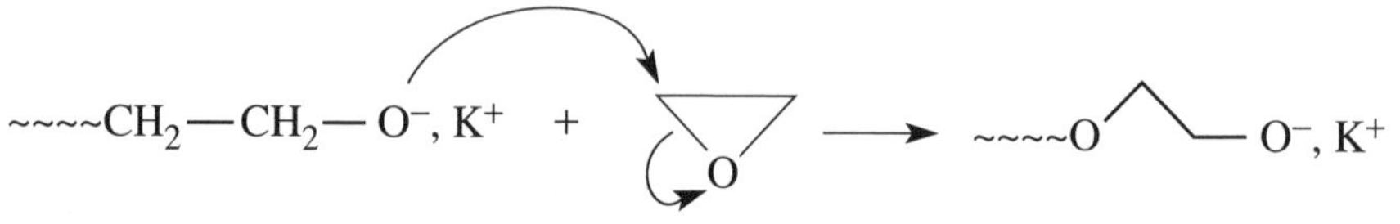

The polymerization proceeds differently in the case of lactams, with the nucleophilic species being carried alternatively by the monomer and the growing chains. For the monomer to be inserted at the chain end, it needs to be activated through proton abstraction and charge transfer before it can add onto the end-standing carbonyl group; such a mechanism is called *activated monomer* polymerization, and it is an unusual process in chain polymerization:

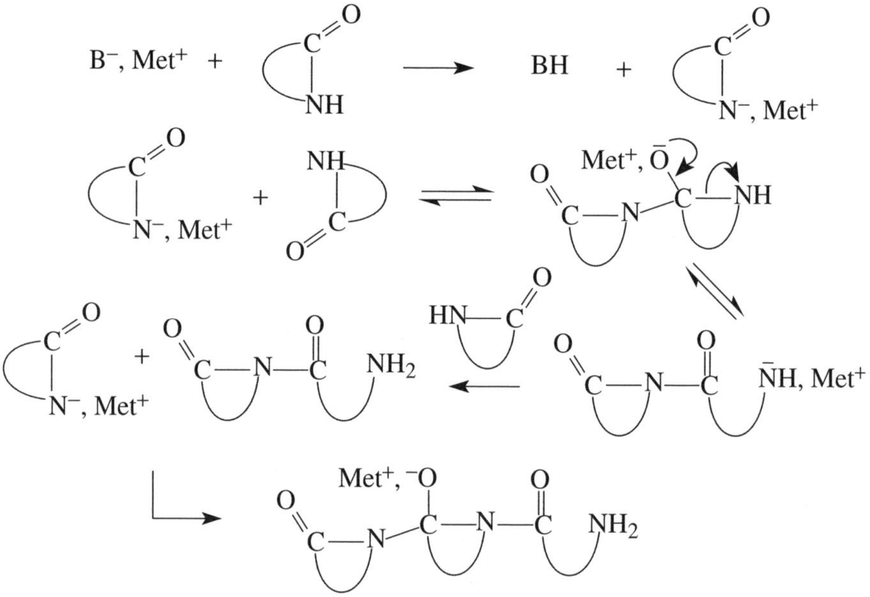

Anionic polymerization of caprolactam is industrially used to produce polyamide-6 (PA-6).

Anionic polymerization of N-carboxyanhydrides (Leuch's anhydrides—NCA) affords polypeptides; it proceeds by nucleophilic addition onto the carbonyl function, followed by an elimination releasing CO_2:

Relatively weak bases (primary or secondary amines, alkoxides, etc.) are used to initiate the polymerization of such monomers.

8.6.5. Anionic Copolymerization

In this section, only "statistical" copolymerizations will be considered.

Differences in the reactivity of the growing species are more pronounced in anionic polymerization than in free radical polymerization. In particular, alkoxides, thiolates, carboxylates, and so on, generally do not initiate the polymerization of (and do not copolymerize with) vinyl monomers. Even in the latter family, the differences between the reactivity of active centers are such that only a very few of them give "statistical" copolymers. Among those, the styrene/butadiene system is the best known, being industrially produced (in solution) under the trade name of SBR.

The existence of active centers under the form of various structures for each comonomer, with each of these structures having its own reactivity, complicates the kinetic treatment of such copolymerizations in comparison to the case of free radical polymerization. The **reactivity ratios** formalism can be utilized, but only apparent values of rate constants, valid only under specific experimental conditions, can be obtained. It is thus unrealistic to discuss the meaning of these values.

As a matter of fact, anionic copolymerization is essentially utilized for the preparation of block copolymers (see Section 9.2); in general, one operates by sequential addition of the comonomers in the order of increasing electroaffinity.

8.6.6. Termination Reactions

Being mainly known and utilized for its "living" character, it may appear at first glance misleading to mention the existence of termination reactions in the anionic polymerization of vinyl and related monomers. They indeed occur, and the conditions have to be found when necessary to minimize them so as to obtain the control of the polymerization. Otherwise, these termination reactions can also be exploited for the purpose of functionalization of the chain ends.

8.6.6.1. Spontaneous Termination Reaction. They mainly depend on the molecular structure of the active centers considered. With polystyryl sodium in tetrahydrofuran solution, a hydride β-elimination is observed in a first step:

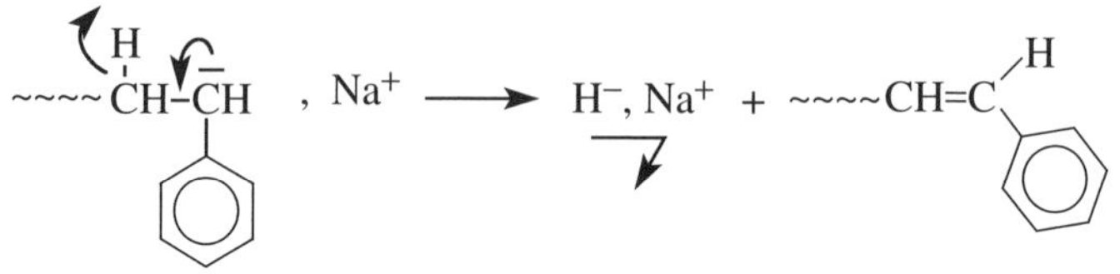

This results in the formation of a labile H. Because of its acidity, this hydrogen atom can react with $\sim\sim\sim PS^-, Na^+$ still present:

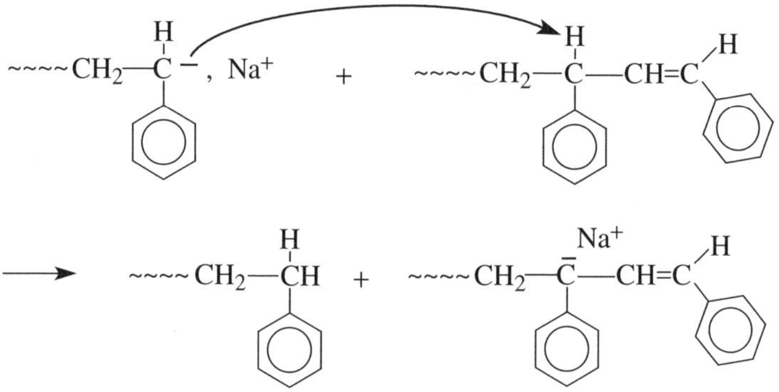

The newly formed carbanion is particularly well stabilized and is unable to reinitiate and propagate the polymerization.

The mechanism of spontaneous termination occurring in the polymerization of (meth)acrylic monomers is completely different; the termination affecting $\sim\sim\sim\sim MMA^-, Li^+$ active chains is mainly an addition reaction onto the carbonyl groups of the antepenultimate units by the growing enolates, followed by an elimination reaction:

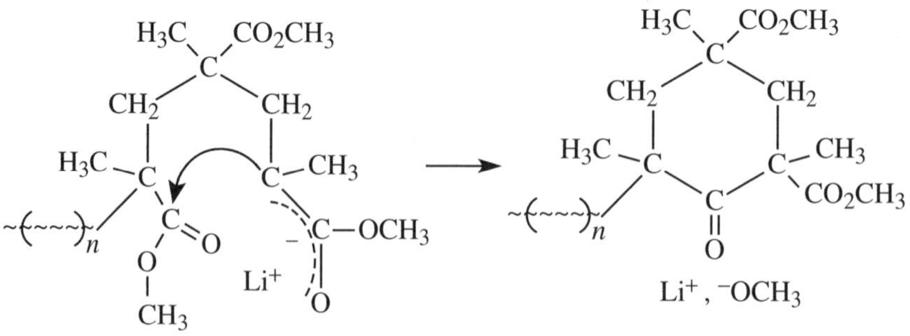

Because the resulting alkoxide is unable to add onto the double bond of MMA, the net result is a termination of the chain growth process.

A way to limit these reactions consists of replacing Li^+ by a bulkier cation (quaternary ammonium, phosphonium, etc.) or by adding in the reaction medium, a solvating agent that increases the apparent radius of the lithium ion; the attack of carbonyl groups by ion pairs can thus be thwarted, the probability of termination reduced, and the control of polymerization improved in this way.

8.6.6.2. Reaction with Termination Reagents.

Because of their very high reactivity, carbanionic species react with many compounds—in particular, those exhibiting an acidic character:

$$\sim\sim\sim M_n^-, Met^+ + A-H \longrightarrow \sim\sim\sim M_n H + A^-, Met^+$$

Several atmospheric components can be utilized as terminating reagents:

with CO_2 $\sim\sim\sim M_n^-, Met^+ + CO_2 \longrightarrow \sim\sim\sim M_n COO^-, Met^+$

with O_2 $\sim\sim\sim M_n^-, Met^+ + O_2 \longrightarrow \sim\sim\sim M_n^\bullet + O_2^{\bullet\,-}, Met^+$

then: $2\sim\sim\sim M_n^\bullet \longrightarrow \sim\sim\sim M_n\text{-}M_n\sim\sim\sim$

or $\sim\sim\sim M_n^\bullet + O_2 \longrightarrow \sim\sim\sim M_n\text{-O-O}^\bullet$

$\sim\sim\sim M_n^\bullet + O_2^{\bullet\,-}, Met^+ \longrightarrow \sim\sim\sim M_n\text{-O-O}^-, Met^+$

The existence of such reactions imposes a reaction medium free of any protic and electrophilic impurity so that the polymerization can be carried out under "living" conditions; the techniques of purification providing such purity are now well-developed.

8.6.7. Group Transfer Polymerization

It is now widely admitted that group transfer polymerization, which was unveiled in 1983 by a team from DuPont de Nemours, belongs to the category of anionic polymerization. It applies to (meth)acrylic monomers whose Li-based anionic polymerization suffers from the termination by attack onto carbonyl groups as previously shown.

Group transfer polymerization of these monomers exhibits a "living" character at ambient temperature and under normal experimental conditions.

The initiator is a silylated acetal of dimethylketene (1-methoxy-2-methyl-1-trimethylsiloxypropene, indicated by TMS) which is active only in the presence of a "catalyst." The reaction pathway is represented hereafter:

The degree of polymerization obtained is determined by the molar ratio of [monomer] to the [initiator (TMS)], with the "catalyst" concentration determining the rate of polymerization. The "catalyst" can be a nucleophilic entity, with the best effects being obtained with fluoride (F^-) or bifluoride (HF_2^-) anions derived from salts soluble in the reaction medium, such as tris(dimethylamino)sulfonium bifluoride and tetrabutylammonium fluoride. These "catalysts" are particularly well suited to the polymerization of methacrylic monomers.

Lewis acids such as zinc halides or a dialkylaluminium chloride (AlR_2Cl) are preferentially used to catalyze the polymerization of acrylics.

When strongly nucleophilic entities are utilized as catalysts, the active centers have been identified as enolates (as with alkali counterions). Due to the nature of counterions—in particular, their size—their reactivity is strongly reduced compared to that of enolates associated with lithium; moreover, they are only present in low concentration, the major part of the active species being in a "dormant" silylacetal form in fast exchange with the reactive enolates. The mechanism occurring in such polymerizations can be represented as below:

8.6.7.1. *Initiation*

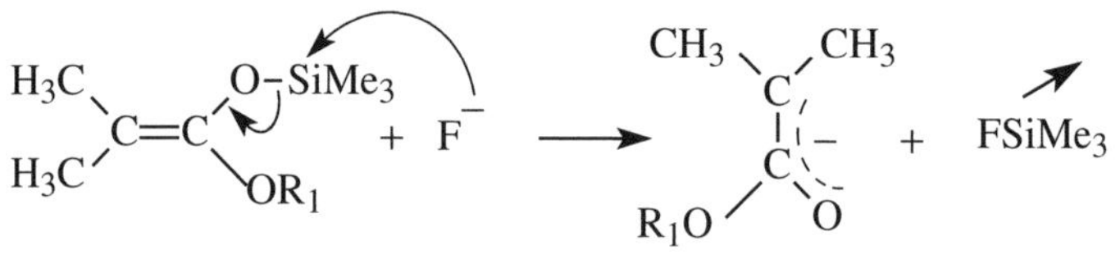

then,

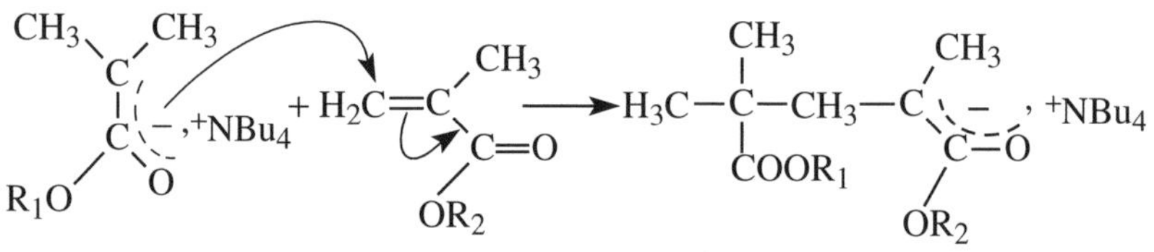

Remark. The mechanism of initiation reaction reveals the consumption of the "catalyst." It does not function as a true catalyst but rather like a "co-initiator" whose presence is essential for the activation of the initiator.

8.6.7.2. Propagation. It must be stressed here that the propagation reaction which is represented hereafter occurs through monomer addition by the carbon form of the enolate whereas the exchange of trimethylsilyl groups between dormant and reactive chains resorts to the oxygen form (due to "oxophilicity" of silicon atom).

The mechanism shown below is described as "dissociative" because the reactive species are fully ionized. With weak nucleophilic catalysts, an "associative" mechanism was proposed implying a pentacoordination of the silicon atom (discussed) and mainly covalent active species.

$$\underset{\underset{\text{COOR}^1}{|}}{\overset{\underset{\text{CH}_3}{|}}{\text{H}_3\text{C}-\text{C}}}-\text{CH}_2-\underset{\overset{|}{\underset{\overset{\diagdown}{\text{C}=\text{O}}}{\diagup\text{OR}^2}}}{\overset{\overset{\text{CH}_3}{\diagup}}{\text{C}}}\ \overset{-}{,}\ ^+\text{NBu}_4\quad +\ n\ \ \text{H}_2\text{C}=\underset{\underset{\text{OR}^2}{\diagup}}{\overset{\overset{\text{CH}_3}{\diagup}}{\text{C}}}\diagdown_{\text{C}=\text{O}}$$

$$\underset{\underset{\text{COOR}^1}{|}}{\overset{\underset{\text{CH}_3}{|}}{\text{H}_3\text{C}-\text{C}}}\!\!\left(\!\text{CH}_2-\underset{\underset{\text{COOR}^2}{|}}{\overset{\underset{\text{CH}_3}{|}}{\text{C}}}\!\right)_{\!\!n}\!\!\text{CH}_2-\underset{\overset{|}{\underset{\overset{\diagdown}{\text{C}=\text{O}}}{\diagup\text{OR}^2}}}{\overset{\overset{\text{CH}_3}{\diagup}}{\text{C}}}\ \overset{-}{,}\ ^+\text{NBu}_4$$

$$\underset{\text{or TMS}}{\underset{\sim\sim\text{H}_2\text{C}}{\overset{\text{H}_3\text{C}}{\diagup}}\text{C}=\text{C}\underset{\diagdown\text{OR}^2}{\overset{\diagup\text{O}-\text{SiMe}_3}{}}}$$

propagation

$$\underset{\underset{\text{COOR}^1}{|}}{\overset{\underset{\text{CH}_3}{|}}{\text{H}_3\text{C}-\text{C}}}\!\!\left(\!\text{CH}_2-\underset{\underset{\text{COOR}^2}{|}}{\overset{\underset{\text{CH}_3}{|}}{\text{C}}}\!\right)_{\!\!n}\!\!\text{CH}_2-\underset{\underset{\text{OR}^2}{\diagup}}{\overset{\overset{\text{CH}_3}{\diagup}}{\text{C}}}\diagdown_{\text{C}-\text{OSiMe}_3}$$

Dormant species

In the case of catalysis by Lewis acids found to be suitable for acrylates, the mechanism is completely different and propagation occurs by monomer activation.

8.6.8. Application of Anionic Polymerization to Macromolecular Synthesis

The "living" character of anionic polymerizations combined with a short initiation step—compared to that of propagation—affords polymers with low dispersity index ($D_M = M_w/M_n < 1.1$). An additional advantage of these systems is the accurate control of the degree of polymerization of the polymers obtained, which merely reflects the high efficiency (f) of the initiators employed.

Such a good definition of the molecular dimensions associated with the persistence of the active centers was extensively applied for the purpose of so-called macromolecular engineering, to design and construct precision macromolecular architectures. An account of the various possibilities is described in Chapter 9, including those based on other "living" polymerizations.

8.6.9. Techniques of Anionic Polymerization

Due to the generation of the totality of active centers at the onset polymerization and the very high polymerizability of vinyl and related monomers, there is no option but to carry out anionic polymerizations in solution. Generally, the solvents used are hydrocarbons (aromatic or aliphatic) acting as diluents. At a smaller scale in laboratories, ethers (tetrahydrofuran, dioxane, dimethoxyethane, etc.) are sometimes used, for their solvating and dissociating effects in addition to their role as a diluent.

The extreme sensitivity of (carb)anionic active centers toward electrophilic impurities, along with their utilization in low concentration to obtain high molar masses, requires a thorough purification of all reagents. Initially, this purification step appeared to be a limitation to the industrial development of anionic polymerization; now, it is not anymore the case as shown by the increasing number of industrial applications of anionic polymerization.

Studies carried out recently on the control of the reactivity of propagating species indicate that solvent-free processes may well be developed in the near future.

8.7. CATIONIC POLYMERIZATION

Cationic polymerization has witnessed an intense development in the middle of the twentieth century after it could be successfully applied to polymerize certain ethylenic hydrocarbons such as isobutene, carbonyl monomers such as formaldehyde, or cyclic ethers such as oxiranes, tetrahydrofuran, and cyclosiloxanes.

Because of the very high reactivity of the cationic propagating species—in particular, with ethylenic monomers—the polymerization systems that are commonly used in industry often entail side reactions and frequent structural irregularities in the polymers formed.

The discovery of compositional and experimental conditions affording "living" cationic polymerizations has attracted much interest in particular because some of unsaturated and heterocyclic monomers concerned can only be polymerized by cationic means.

8.7.1. General Characters

A cationic polymerization can be defined in a way exactly symmetrical to anionic polymerization as schematized hereafter:

$$\sim\!\sim\!\sim\!\sim\!M_n^+ ,A^- + M \longrightarrow \sim\!\sim\!\sim\!\sim\!M_{n+1}^+ ,A^-$$

In this equation, $\sim\!\sim\!\sim\!M_n^+$ represents a positively charged (or polarized) species carried by growing chains, and A^- represents a negative counterion (or a negatively polarized species) ensuring the neutralization of the positive charge. With **ethylenic monomers**, the propagation reaction is an electrophilic addition onto the polymerizable double bond; the first step is the coordination of the double bond onto the carbocationic site:

$$\sim\sim\sim M_n-\overset{|}{\underset{|}{C}}-\overset{/}{\underset{\backslash}{C}}{}^+, A^- \quad + \quad \overset{\backslash}{\underset{/}{C}}=\overset{/}{\underset{\backslash}{C}} \longrightarrow \sim\sim\sim M_n-\overset{|}{\underset{|}{C}}-\overset{/}{C}{}^+, A^-$$

$$\sim\sim\sim M_n-\overset{|}{\underset{|}{C}}-\overset{|}{\underset{|}{C}}-\overset{|}{\underset{|}{C}}-\overset{/}{C}{}^+, A^-$$

This reaction is all the more facile as the nucleophilic character of the monomer is pronounced: electron-donating substituents increase the cationic polymerizability, and in turn the intrinsic reactivity of the carbocationic site formed is reduced by the effect of such substituents. Thus, as in anionic polymerization, an increase of reactivity of the monomer has more influence on its polymerizability than a decrease in reactivity of the corresponding active center. Because of the strong Lewis acid character of the active species, for a monomer to be polymerized, strongly nucleophilic sites must not be present.

The mechanism is similar in the case of **carbonylated monomers** or **n-donor heteronuclear double bonds** with an attack by the cationic active center onto the oxygen atom of the carbonyl group:

$$\sim\sim\sim M_n-O-\overset{/}{\underset{\backslash}{C}}{}^+, A^- \quad + \quad \overset{/}{\underset{\backslash}{O}}=C \longrightarrow \sim\sim\sim M_n-O-\overset{|}{\underset{|}{C}}-O-\overset{/}{C}{}^+, A^-$$

Cationic polymerization is also utilized with **heterocyclic monomers**. In this case, disregarding the thermodynamic constraints, polymerization proceeds by nucleophilic attack of the hetero-element of a monomer molecule on the electron-deficient α-carbon atom of the onium ion:

$$\sim\sim\sim M_n\!\!-\!\!\overset{+}{X}\!\!\bigcirc , A^- \ + \ IX\bigcirc \longrightarrow \sim\sim\sim M_{n+1}\!\!-\!\!\overset{+}{X}\!\!\bigcirc , A^-$$

The heterocyclic monomers that are the most sensitive to electrophilic active centers are of diverse nature:

Oxiranes   Oxetanes   Other ethers and acetals

but also aziridines, thiiranes, siloxanes, phosphazenes, and so on, all monomers whose polymerization leads to polymeric materials with various molecular structures and thus of different physical properties.

Most of the concepts concerning the structure of active species—aggregation, ionization, solvation, dissociation—which were described in the section on anionic polymerization, apply to cationic polymerizations; in particular, the more pronounced the ionic character of the species and the longer the interionic distance, the more prominent the reactivity of active centers.

Solvents that can be utilized in cationic polymerization must be inert with respect to strongly electrophilic active sites. They can play the role of diluent (aliphatic hydrocarbons) and that of solvating agents for electrophilic species (nitroparaffins) and/or of dissociating medium of ion pairs into free ions (dichloromethane: $\varepsilon_{CH_2Cl_2} = 8.93$ at $25°C$).

8.7.2. Initiation of Cationic Polymerizations

Numerous are the initiators that can be used in cationic polymerization, the monomer polymerizability determining their choice.

8.7.2.1. Protonic Acids (Brönsted Acids):

$$A^-, H^+ \longrightarrow A^- + H^+$$

These acids are all the more efficient as they are dissociated in the reaction medium.

More important than the pK_a in aqueous solution (of little interest), Table 8.16 gives the pK_a values of various protonic acids in acetic acid and acetonitrile. It can be noticed that acids which are reputed strong in aqueous solution are not dissociated in organic media. The most used Brönsted acids to initiate cationic polymerizations are:

Perchloric acid	$H-ClO_4$
Trifluoromethylsulfonic (triflic) acid	$H-SO_3-CF_3$
Trifluoroacetic acid	$H-O-OC-CF_3$
Hydroiodic acid	$H-I$

Depending upon the nature of the solvent used and, in particular, its basicity which represents its aptitude to trap protons, these initiators will be themselves more or less good proton donors. Among all the systems shown in Table 8.16, perchloric acid in acetonitrile solution is the best one.

Certain protonic acids add easily onto the monomer double bond, but when the associated counterion is more nucleophilic than the monomer, they form a covalent bond unable to propagate the reaction. Such a situation often occurs with hydracids:

$$HCl + H_2C=CHR \longrightarrow H-CH_2-HRC^+, Cl^- \longrightarrow H_3C-HRC-Cl$$

Table 8.16. pK_a values for some protonic acids in two different organic solvents

Protonic Acid	In Acetic Acid	In Acetonitrile
HSO_3CF_3	4.7	2.6
$HClO_4$	4.9	1.6
HBr	5.6	5.5
H_2SO_4	7.0	7.3
HCl	8.4	8.9
HSO_3CH_3	8.6	8.4
$HOOC-CF_3$	11.4	10.6
$HOOC-CCl_3$	12.2	12.7
$HOOC-CH_3$	12.8	22.5

In the same manner, protonic acids are generally capable of the following reaction with heterocycles,

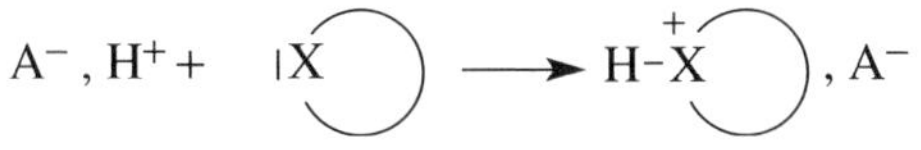

$$A^- , H^+ + \quad |X\bigcirc \longrightarrow H\text{-}\overset{+}{X}\bigcirc , A^-$$

but depending upon the relative nucleophilicity of the associated anion and the monomer, propagation may occur or not. Thus, HI initiates the polymerization of aziridines (three-membered cyclic amines)

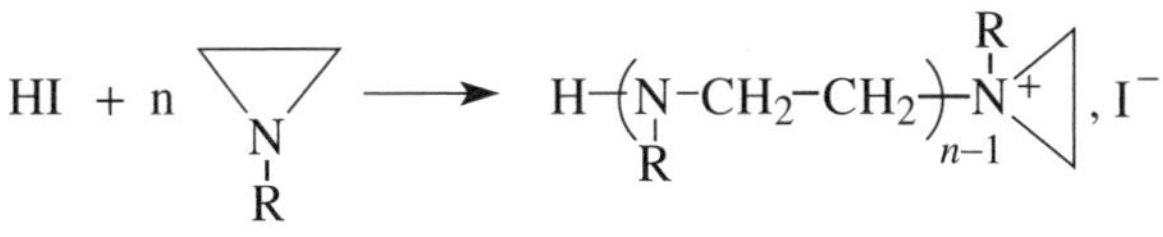

$$HI + n \underset{\underset{R}{|}}{\overset{}{\bigtriangleup}}_{N} \longrightarrow H\text{-}\!\!\left(\!N\text{-}CH_2\text{-}CH_2\!\right)_{n-1}\!\!\overset{\overset{R}{|}}{N^+}\!\!\bigtriangleup , I^-$$

but can only protonate oxiranes:

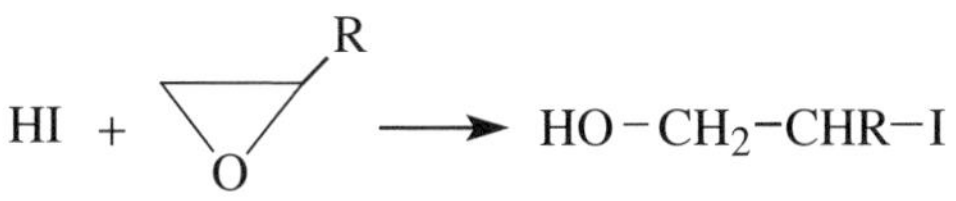

$$HI + \overset{R}{\underset{O}{\bigtriangleup}} \longrightarrow HO-CH_2-CHR-I$$

The protonic initiators that are the most used to polymerize heterocycles are trifluoromethylsulfonic ("triflic") and fluorosulfonic acids.

The kinetics of initiation of the polymerization of ethylenic monomers by protonic acids

$$A^-, H^+ + HC{=}CHR \longrightarrow H_3C\text{-}HRC^+, A^-$$

generally exhibits a first-order variation (expected) with respect to monomer and a second-order variation with respect to protonic acid.

This phenomenon is accounted for by a mechanism involving two acid molecules in the transition state which corresponds, for the case of HCl, to

$$
\begin{array}{c}
\text{H—Cl} \\
\text{H—Cl} \\
\text{C=C}
\end{array}
$$

The higher the rate constant of addition of acid molecule onto the monomer double bond (k_i), the greater the nucleophilicity of the monomer. Thus, the rate constant of initiation by trifluoromethylsulfonic acid at $0°C$ in dichloromethane (CH_2Cl_2) varies from $k_i = 10 \, L \cdot mol^{-1} \cdot s^{-1}$ for styrene, to $k_i = 10^3 \, L \cdot mol^{-1} \cdot s^{-1}$ for α-methylstyrene and $k_i = 5 \times 10^4 \cdot L \, mol^{-1} \cdot s^{-1}$ for p-methoxystyrene; these three monomers are ranked in the order of increasing nucleophilicity.

8.7.2.2. Lewis Acids. BF_3, $AlCl_3$, $TiCl_4$, $SnCl_4$, and $SbCl_5$ are the most generally used Lewis acids. In a few cases, it was shown that these Lewis acids can initiate polymerizations by themselves. For example, aluminum halides self-ionize from (generally) dimeric aggregates:

$$2AlCl_3 \rightleftharpoons (AlCl_3)_2 \rightleftharpoons AlCl_2^+, AlCl_4^-$$

and the cationic species ($AlCl_2^+$) formed adds to the double bond to generate the propagating carbocation:

$$AlCl_2^+, AlCl_4^- + H_2C=CHR \longrightarrow AlCl_2-CH_2-HRC^+, AlCl_4^-$$

Diiodine (I_2) is also able to initiate by itself the polymerization of certain vinyl monomers, and two mechanisms were proposed to account for this behavior. First is an addition of molecular iodine to the double bond

$$I_2 + H_2C=CHR \longrightarrow I-CH_2-CHR-I$$

followed by an ionization of a C–I bond under the effect of I_2 in excess, the latter playing the role of a co-initiator:

$$I-CH_2-CHR-I + I_2 \rightleftharpoons I-CH_2-{}^+CHR, I_3^-$$

The second mechanism proposes a self-ionization of I_2,

$$2I_2 \rightleftharpoons I^+, I_3^-$$

followed by an initiation according to a mechanism close to that described for $AlCl_3$.

Initiation by a Lewis acid can also occur by an electron transfer in analogy with the examples of anionic polymerization: for instance, 1,1-diphenylethylene, which cannot be polymerized for steric reasons, undergoes dimerization in the presence of $SbCl_5$ according to the mechanism shown hereafter:

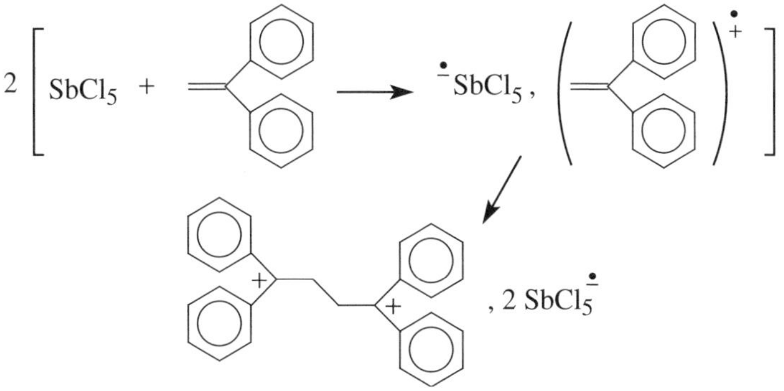

Several authors also propose that the initiation of the polymerization of dioxolane by BF_3 occurs without intervention of another molecule and thus leads to a zwitterion:

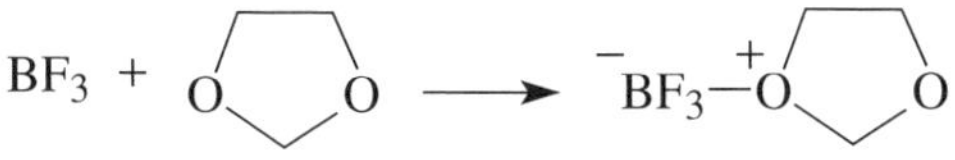

from which propagation occurs.

However, Lewis acids are more often active in the presence of either a weak acid or a *cationizing agent* (co-initiator). The reaction between the two components of the initiating system yields an extremely strong acid complex, the actual initiator is either the proton donor or the cationizing agent, and the Lewis acid serves as activator. The most common proton donors are water, alcohols, amines, and amides. For example, $TiCl_4/H_2O$ system is formed according to the scheme shown hereafter:

$$TiCl_4 + H_2O \longrightarrow TiCl_4OH^-, H^+$$

$$TiCl_4OH^-, H^+ + H_2C{=}CHR \longrightarrow H_3C{-}HRC^+, TiCl_4OH^-$$

There are many other systems that are excellent initiators for both unsaturated and heterocyclic monomers. Several reactions are shown below to demonstrate some of the possibilities offered by these systems. With a metal halide such as $MetX_n$ serving as Lewis acid, the various possibilities to generate a carbocationic initiator are:

$$MetX_n + RX \rightleftharpoons R^+, MetX_{n+1}{}^-$$

$$MetX_n + R{-}OR' \rightleftharpoons R^+, MetX_nOR'^-$$

$$MetX_n + R{-}O{-}CO{-}R' \rightleftharpoons R^+, MetX_nO{-}CO{-}R'^-$$

$$MetX_n + R{-}O{-}SO_2R' \rightleftharpoons R^+, MetX_nO{-}SO_2R'^-$$

Then, polymerization occurs by an electrophilic attack of the carbocation R^+ onto either the double bond or the heteroatom of a heterocyclic monomer.

8.7.2.3. Other Initiators. Compared to carbon, silicon is strongly electropositive and can thus be used in electrophilic reactions with strong nucleophiles serving as intermediates:

$$Me_3Si-O-SO_2-CF_3 + Me_2C=O \longrightarrow Me_3Si-O-Me_2C^+, {}^-SO_3-CF_3$$

then,

$$Me_3Si\text{-}O\text{-}Me_2C^+, {}^-SO_3\text{-}CF_3 + H_2C=CH\text{-}OR$$
$$\longrightarrow Me_3Si\text{-}O\text{-}Me_2C\text{-}CH_2\text{-}C^+H(OR), {}^-SO_3\text{-}CF_3$$

Cationic polymerizations can also be initiated by a photochemical process either in the presence of a photo-initiator or under the effect of ionizing radiations. These reactions are described in Section 8.3.

Advantage can be taken of the ionizing and dissociating effects produced by a solvent to activate inert molecules and initiate a polymerization. For instance, triphenylmethyl chloride in pure sulfuric acid solution undergoes an instantaneous ionization and produces a triphenylmethylium cation with a characteristic red color:

Strongly electrophilic species can also be generated by shifting the equilibrium of ionization through the precipitation of a salt, like in the following example:

$$R\text{-}CO\text{-}Cl + Ag^+, ClO_4^- \longrightarrow R\text{-}CO^+, ClO_4^- + AgCl$$

8.7.3. Propagation of Cationic Polymerizations

In the case of unsaturated monomers, the propagating active center is a carbocationic species (carbonium ion) reacting by electrophilic addition. To this carbocation is associated a negative counterion, and both may exist under the same ionic species (aggregates, ion pairs, free ions) as the carbanionic homologs. Thus, depending on the nature of monomers, the counterion and the solvent and depending on the temperature of the medium, growing chain ends may be more or less polarized, aggregated, ionized, solvated, or dissociated with a reactivity growing in the same order. The kinetics of polymerization are generally complex and reflect the multiple structures taken by the reactive species. The effect of temperature on the apparent reactivity of active species may sometimes result in overall negative (apparent) activation energies. As in anionic polymerization, the influence of temperature on the solvation of active centers and on the permittivity of the reaction medium and the fact that they vary in a different manner with the temperature are responsible for this unusual behavior.

When cationic polymerizations are initiated by γ radiations, propagation proceeds by means of free ions and it is then possible to evaluate the monomer polymerizability from determination of the corresponding rate constants of propagation. It can be observed that, contrary to anionic polymerizations whose rate constants of propagation of free ions are $\sim 10^4$ times higher than those measured for ion pairs, the same ratio of rate constants is only about 10 (or even less) in cationic polymerization. Such a difference is due to the faculty of certain solvents to solvate free cations and thus reduce their intrinsic reactivity.

The high reactivity of carbocationic species can be also responsible for rearrangement reactions during the propagation step of certain monomers. For instance, in the polymerization of 3-methylbut-1-ene, structural irregularities could be formed:

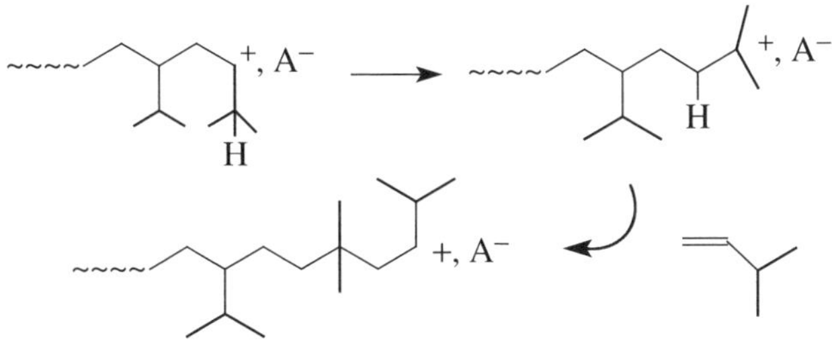

The rearrangement is due to a higher stability of the tertiary carbonium ion compared to that of the secondary carbonium ion formed after a propagation step. The frequency of these structural irregularities closely depends on the experimental conditions of the polymerization. Such a rearrangement is also observed in the cationic polymerization of β-pinene:

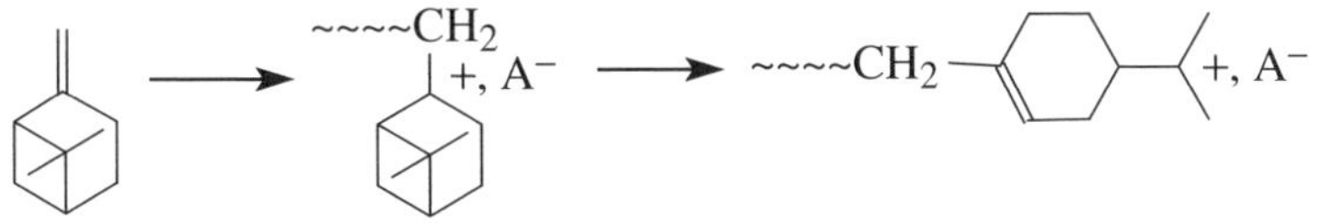

As for the propagation of heterocycles, it proceeds via "onium"-type cationic species, but actually involve transient carbonium ions:

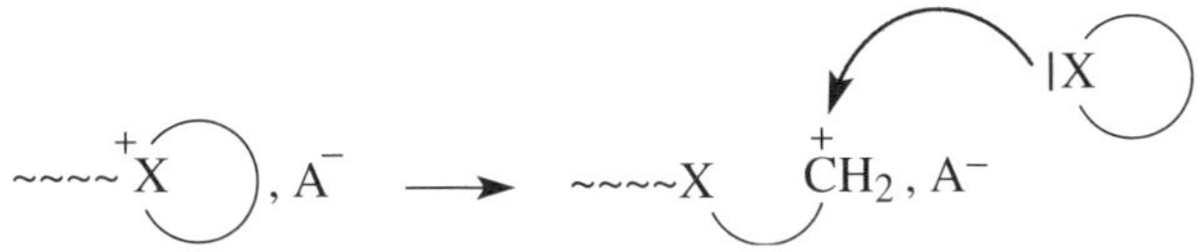

Due to their high energy level, the probability of existence of these carbocationic species is indeed very low compared to that of onium ions.

Table 8.17. Energy (in kJ·mol^{-1}) related to the strength of various heterocycles with variable number of links

Number of links	CH$_2$	O	S	N	O–O
3	115	117	77.8	96.1	—
4	109	—	79.0	—	—
5	25.5	28.0	4.1	—	30.5
6	0.4	—	9.2	—	12.1
7	25.5	—	—	—	—
8	40.5	—	—	—	—
9	52.5	—	—	—	—

In the cationic polymerization of heterocycles the propagation step is dominated by thermodynamic constraints and it can occur only if a negative enthalpy of polymerization resulting from the release of the cycle strain balances the variation of entropy. Heterocycles with low strain energy exhibit rather low ceiling temperatures. The values of molar ring strain energies for heterocyclic monomers of various natures and geometries are given in Table 8.17.

Certain heterocycles polymerize by very peculiar reaction pathways. For example, the cationic polymerization of lactams proceeds by activation of the monomer, as for their anionic polymerization:

$$NH\text{—}CO + H^+, A^- \longrightarrow NH\text{—}C\text{=}\overset{+}{O}H, A^-$$

then, the "activated" monomer reacts with another monomer molecule to give a dimer regenerating the initiator

$$NH\text{—}C\text{=}\overset{+}{O}H, A^- + NH\text{—}CO \longrightarrow H_2N \quad CO\text{—}N\text{—}CO + H^+, A^-$$

H$^+$,A$^-$ can again activate a monomer molecule that will form a trimer upon reaction with the NH$_2$ group of the dimer and regenerating H$^+$,A$^-$, and so on:

$$NH\text{—}C\text{=}\overset{+}{O}H, A^- + H_2N \quad CO\text{—}N\text{—}CO$$

$$\text{etc.} \quad H_2N \quad CO\text{—}NH \quad CO\text{—}N\text{—}CO + H^+, A^-$$

8.7.4. Transfer and Termination Reactions

They occur because of the high reactivity of cationic species and the necessity to form more stable species. In the case of transfer, the resulting species are, however, sufficiently reactive to reinitiate the polymerization whereas, in the case of termination, the species formed are totally inactive.

8.7.4.1. Transfer Reactions.

If the reinitiation is very slow, the transfer step slows down the entire kinetics of polymerization but such incidence would be negligible if the systems were to produce polymers with high degree of polymerization.

For polymerization of unsaturated monomers, two main types of transfer reactions can be distinguished; the first type occurs via proton β-elimination as shown hereafter for the polymerization of styrene:

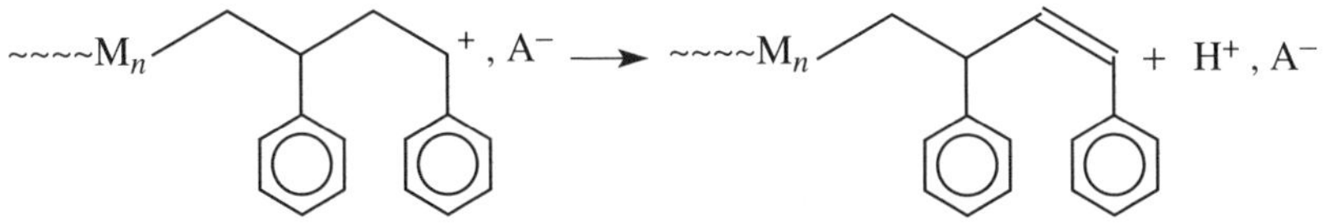

then

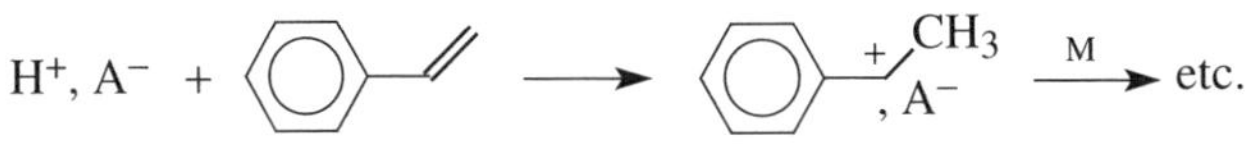

The elimination of H^+,A^- occurs with high probability whenever the acid released by transfer is of high stability—that is, the basicity of the associated anion (A^-) is strong. As already mentioned, the basicity of the solvent can influence the tendency of active centers to participate in such a reaction.

The second type of transfer reaction implies the presence of aromatic moieties that undergo Friedel–Crafts reactions. These aromatic rings can be carried by the monomer (styrene, α-methylstyrene, indene, coumarone, etc.) or be part of the solvent (toluene, etc.), the initiator or an impurity. These electrophilic substitution reactions also produce H^+,A^-. In the cationic polymerization of styrene, these Friedel–Crafts reactions can occur either intra- or intermacromolecularly and in the latter case, monomer or polymer sites will be involved. The various transfer reactions are shown hereafter:

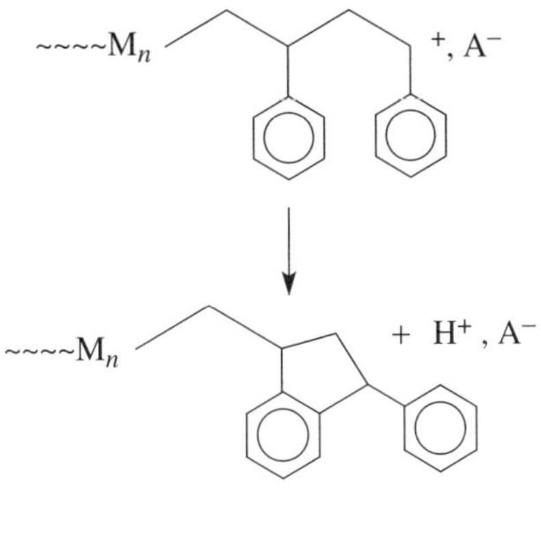

The spontaneous intramolecular reaction generates an indanylene group whereas the two intermolecular ones that follow correspond, for the first, to a transfer to monomer with formation of a *macromonomer* (macromolecule containing a poly-merizable group at one end):

and, for the second, to a transfer to another chain, resulting in an intermolecular coupling.

Depending upon the experimental conditions, one or another of these reactions can be favored or be absent. However, β-elimination is generally the dominating side reaction in the cationic polymerization of styrene.

Transfer reactions to polymer occur extensively in the cationic polymeriza-tion of heterocyclic monomers; they do this either intra- or intermacromolecularly through the attack of the electron-deficient α-carbon atom of the "onium" active center by a heteroatom of an unspecified monomeric unit, carrying an n electron pair:

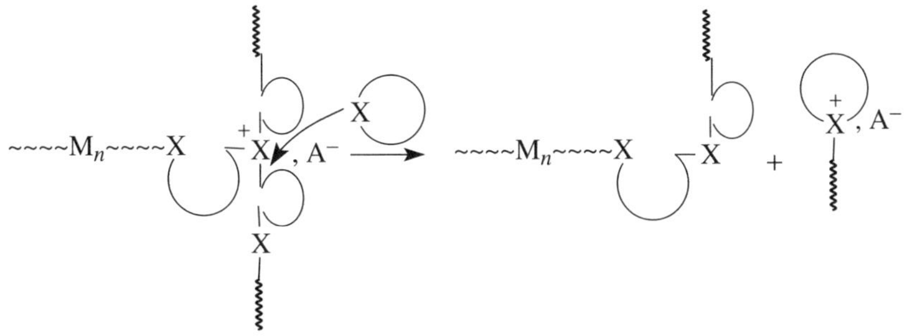

In contrast to the case of ethylenic monomers where transfer results in chain branching, intermolecular transfer reactions in the polymerization of heterocyclics correspond to exchange reactions between chains, with the number of chains remaining constant. Intramolecular transfer reactions—which is generally the case due to a higher probability of collision—yield inactive cycles which can possibly be the monomer ($n = 1$) and thus correspond to a depolymerization:

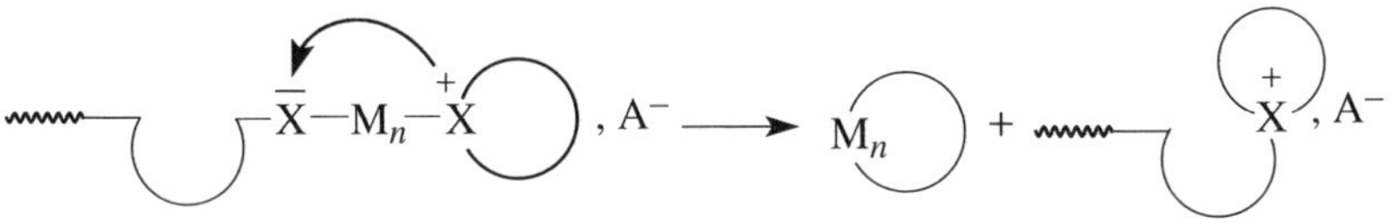

Such reactions are called *back-biting*. For example, in the cationic polymerization of ethylene oxide back-biting reactions produce 1,4-dioxane:

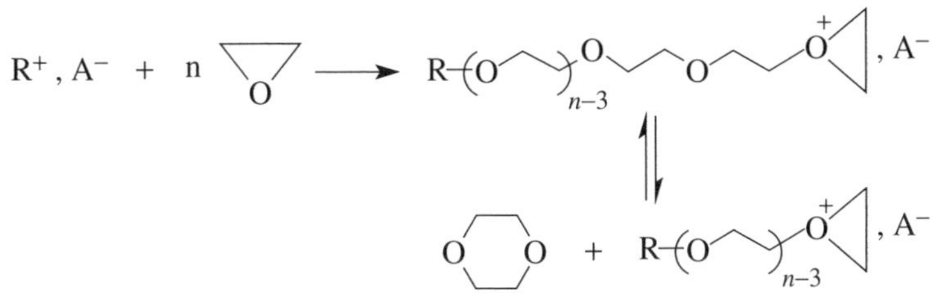

8.7.4.2. Termination Reactions. The reactions that result in the deactivation of **carbonium ions** are numerous. The latter indeed can rearrange and generate carbocationic species that are too stable (or too hindered) to propagate. Such a reaction generally occurs by transfer of hydride and generates bulky tertiary carbocations:

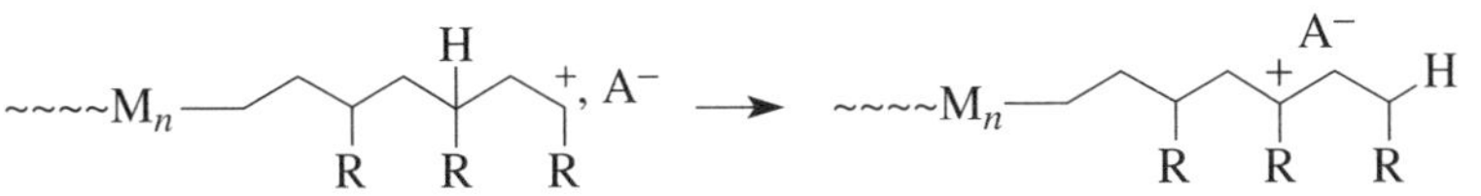

Termination reactions can also be due to so-called *"anion splitting,"* a reaction of the growing chain with the counterion resulting in an inactive covalent bond.

For example, the polymerization of styrene initiated by BF_3/H_2O undergoes such a reaction as shown below,

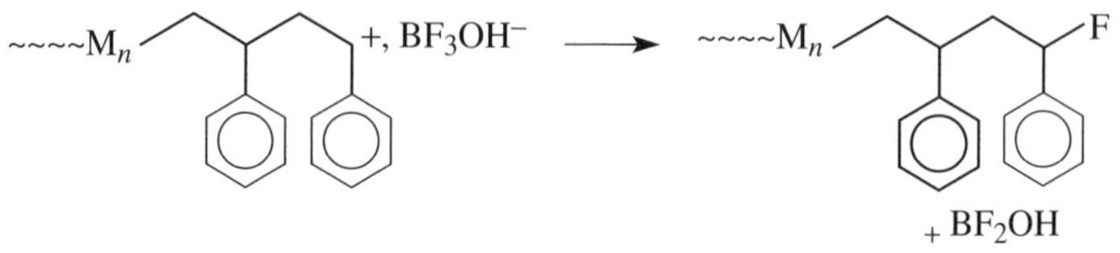

with BF_2OH being a too weak acid to reinitiate the polymerization process.

Certain terminations occur by deactivation due to the impurities present in the reaction medium. For instance, water is a particularly efficient co-initiator when it is used in conjunction with Lewis acids; but, if present in excess, it acts as Lewis base and deactivates carbocationic species:

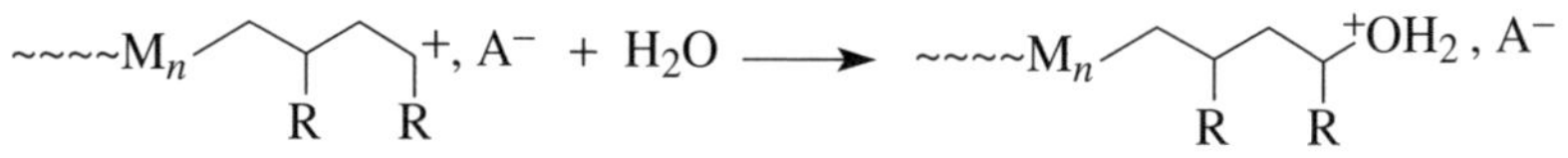

This behavior is encountered with all bases:

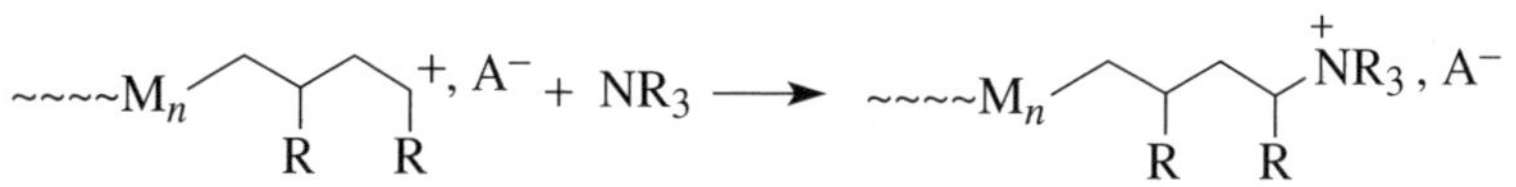

Added in low amounts, these bases can act as retarders.

With **heterocyclic monomers**, the termination can be due to the reaction of the growing site with the counterion A^-, with the formation of an inactive covalent bond:

Such a reaction occurs with monomers that are not nucleophilic enough compared to the counterion to prevent it. Other terminations encountered are similar to those mentioned for unsaturated monomers.

8.7.5. "Living" and/or "Controlled" Cationic Polymerizations

For many decades, polymer chemists strove to increase the reactivity of active centers responsible for chain propagation, expecting to improve both the polymerizability of certain monomers and the productivity of the corresponding processes. These efforts often resulted in an acceleration of the propagation but also in numerous side reactions—such as transfer and termination—involving these highly reactive species that precluded the possibility of controlling the polymerization process.

This trend was reversed after it was realized that lowering the intrinsic reactivity of propagating species could bring about many benefits. Inspired by the works in anionic polymerization by Szwarc, who discovered the features of "living" polymerizations, several teams have uncovered in the early 1980 the ways to control cationic polymerization. It was not yet a question of truly "living" polymerizations because transfer and termination reactions could not be completely avoided; but, under certain experimental conditions, it was possible to separate the initiation step from that of propagation and secure a satisfactory persistence of the propagating active centers.

The first "controlled" cationic polymerizations were achieved with **heterocycles** in spite of the tendency of these systems to undergo back-biting reactions. The "living" character could be retained only during the initial period of polymerization, and these systems were found useful only for monomers with a relatively fast rate of propagation compared to that of transfer, with the polymerizations being discontinued in the latter case before total monomer conversion. Tetrahydrofuran, when initiated by $\{R-CO-Cl/Ag^+, ClO_4^-\}$, polymerizes under controlled conditions. After generation of the primary active centers (see Section 8.7.2), initiation occurs:

$$R\text{-}CO^+,ClO_4^- + O\!\!\begin{array}{c}\fbox{}\end{array} \longrightarrow R\text{-}CO\!-\!\overset{+}{O}\begin{array}{c}\fbox{}\end{array},ClO_4^-$$

A controlled polymerization implies that the intramolecular attack of the growing oxonium ion by a heteroatom is—if not suppressed—difficult:

which is indeed the case with tetrahydrofuran.

Heterocycles often exhibit low ceiling temperature, which has an incidence on the kinetics of polymerization. Assuming a first-order kinetic law with respect to each reactive entity and a polymerization–depolymerization equilibrium occurring as follows,

$$\sim\!\sim\!\sim\!\sim\!M_n^* + M \underset{k_{-p}}{\overset{k_p}{\rightleftharpoons}} \sim\!\sim\!\sim\!\sim\!\sim\!M_{n+1}^*$$

one can write

$$R_p = -d[M]/dt = k_p[\sim\!\sim\!\sim\!M_n^*][M] - k_{-p}[\sim\!\sim\!\sim\!M_{n+1}^*]$$

As previously established k_{-p}/k_p is equal to $[M]_{equ}$ which corresponds to the concentration in monomer at equilibrium at a given temperature,

$$R_p = k_p[\sim\!\sim\!\sim\!M_n^*]\{[M] - [M]_{equ}\}$$

This relation shows that the rate of polymerization progressively slows down with the increase of the monomer conversion, and eventually tends to zero when the equilibrium monomer concentration is reached.

With certain cyclic ethers (dioxolane, oxiranes, etc.), the use of particular conditions results in a limitation of the back-biting reactions and a certain control of the polymerization. For instance, in polymerizations carried out in the presence of an alcohol and with a very low instantaneous monomer concentration, obtained by a slow addition of the monomer solution, almost the totality of electrophilic entities carried by the initiator reacts with the monomer to give protonic species ("activated" monomer); the latter then react with the nucleophilic sites that are the hydroxy groups of alcohols. Propagation occurs through nucleophilic attacks of hydroxyls onto activated monomer molecules:

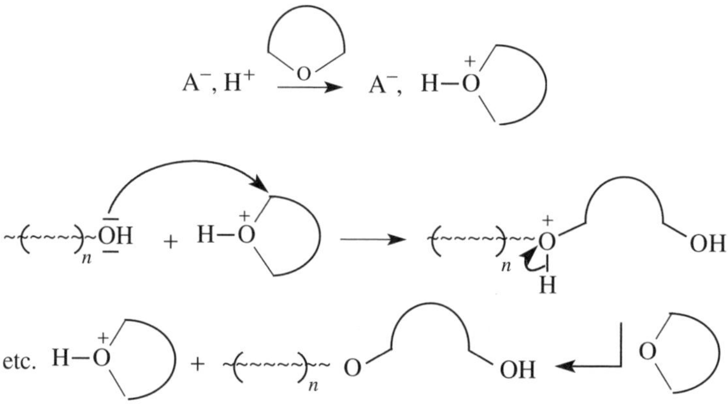

Under such conditions, the concentration in "free" monomer is very low, thus lowering the possibility of the side reactions.

With **vinyl and related monomers**, controlling their polymerizations was more difficult to achieve due to the very high reactivity of propagating carbocations. It was indeed necessary to take into account the effects of different structural factors: nature of monomer, nature of solvent, retrogradation of equilibrium between ion pairs and free ions, and nature of counterion. Monomers prone to polymerize under "living" conditions are also those whose corresponding carbocations are strongly stabilized:

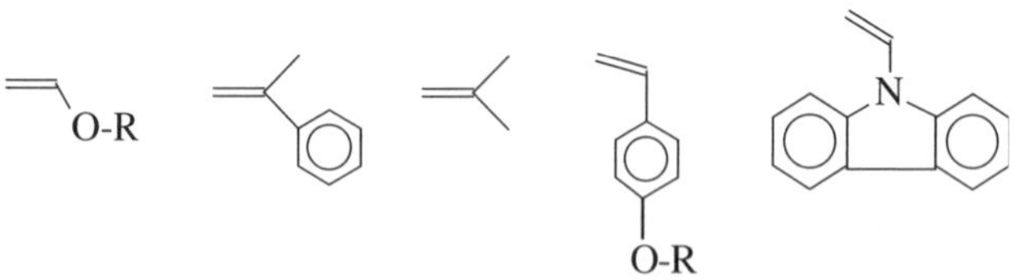

The polarity of solvents should not be too high in order to prevent the tendency of ion pairs to ionize and even to dissociate into free ions; aliphatic hydrocarbons meet the criterion of low-polarity medium; but due to the lack of solubility of the various components of the polymerization system in the latter solvents, the preference has generally been given to chlorinated solvents.

The retrogradation of the dissociation equilibrium is brought about by addition to the reaction medium of common ion salts whose cation is inert toward the polymerization system. As for the nature of the counterions, those that bring about a certain covalency of the active centers are preferred. Addition of a weak nucleophile is also an efficient means to curb the reactivity of carbocations. The following systems (monomer, initiating system, solvent) qualify, more or less perfectly, for the category of "living" polymerizations:

$$\text{O-R} \Big/ (\text{HI} + \text{I}_2 \longrightarrow \text{H}^+, \bar{\text{I}_3}) \Big/ \text{ toluene or hexane}$$

$$\Big/ (\text{O} + \text{O} - \overset{\text{O}}{\underset{}{<}} / \text{BCl}_3) \Big/ \text{ hexane} + \text{chloroform}$$

$$\text{O-R} / (\text{HI} + n\text{-Bu}_4\text{N}^+,\text{I}^-) / \text{ methylene chloride}$$

$$\Big/ \text{O} + \text{Cl} / \text{TiCl}_4 / \text{ dimethylsulfoxide or methylene chloride}$$

$$\Big/ \text{O} + \text{Cl} / \text{SnCl}_4 / n\text{-Bu}_4\text{N}^+, \text{Cl}^- / \text{ methylene chloride}$$

actually most of them could rather qualify for the category of "controlled" poly-merizations. These are some of the most illustrative systems, selected for their representativeness; the list of "living" cationic systems is richer than that.

8.7.6. Kinetics of Cationic Polymerizations

Given the multiplicity of initiation and termination reactions involved in cationic polymerizations and their strong effect on the overall kinetics of polymerization, it is difficult to propose a general kinetic equation that would account for all the possibilities of reaction rate constants and the associated equilibrium constants; each case is indeed a particular case.

However, the existence of a steady-state concentration in active species in con-ventional cationic polymerizations has allowed a kinetic treatment similar to the case of free radical polymerization; but, as in anionic polymerization, the rate constants are only apparent because several types of active species can be in equi-librium with each other, depending upon the experimental conditions used.

For "living" and/or "controlled" polymerizations, kinetic equations are similar to those established for anionic polymerizations (see Section 8.6.4).

8.8. COORDINATION POLYMERIZATION

8.8.1. General Characteristics

There are chain polymerizations whose propagation step entails the prior coordination of the monomer molecule (M) onto the vacant d orbitals of a transition metal atom (Met). This coordination occurs via either π electrons (for vinyl and related monomers) or n electrons (for heterocyclic monomers):

$$\sim\sim\sim\sim M_n \sim\sim\sim\sim \text{Met} + \text{M} \longrightarrow \sim\sim\sim\sim M_n \sim\sim\sim\sim \text{Met}$$
$$\sim\sim\sim\sim M_{n+1} \sim\sim\sim\sim \text{Met} \longleftarrow \quad \text{M}$$

This coordination not only influences the monomer polymerizability but can also determine the configuration of the inserted monomeric unit—in particular, the stereochemistry of its insertion.

The very first discoveries in the field of coordination polymerization date back to the early 1950s with the disclosures by Ziegler of a catalytic system allowing the linear polymerization of ethylene under a relatively low pressure and that of the so-called "Phillips" catalysis for the polymerization of the same monomer. Since then, many other systems have been put forward, developed, and progressively improved, so that coordination polymerization appears to date as one of the most powerful tools for designing and developing original polymeric materials. Its economic importance is already significant; in a few years, it could even surpass all the other methods of polymerization in view of the large variety of macromolecular structures to which it opens access.

It is beyond the scope of this textbook to describe comprehensively the various systems that have been discovered over the years because of their extreme diversity, even restricted to those used for industrial production. It is also not conceivable to make a presentation general enough to cover all the systems employed because each of them entails a specific reaction mechanism.

This presentation will thus be limited to the most representative and/or most promising systems.

8.8.2. "Ziegler–Natta" Catalysis

Strictly speaking, these systems should not enter in the category of catalysts since active species are consumed during polymerization. But the common terminology of "Ziegler–Natta" catalysis that is used to describe these systems is so widespread that it appears unrealistic to speak about initiators.

These systems were serendipitously discovered in the Ziegler laboratory (in Germany) in the course of a study of the oligomerization of ethylene in the presence of aluminum derivatives; the presence of colloidal nickel in an ill-cleaned reactor after a previous use had caused the formation of a strong proportion of butene, which was totally unexpected in the experimental conditions used. From nickel to other transition metals and from butene to linear polyethylene, Ziegler and co-workers were quick to bridge the gap.

Fully aware of the paramount importance of his discovery, Ziegler sought to optimize the catalytic composition, and it quickly appeared that the best system is the one comprised of a mixture of $TiCl_3$/$AlEt_3$ dispersed in a hydrocarbon solvent. These systems have been immensely improved since their discovery and, currently, the most efficient ones are supported catalysts utilized in association with $AlEt_3$ as cocatalyst and an "external" Lewis base as stereoregulating agent (for the polymerization of propene). The two essential components are $TiCl_4$ and $AlEt_3$; the role played by other components will be discussed after the presentation of the basic mechanisms. Active centers are generated through the reaction pathway shown below:

$$TiCl_4 + AlR_3 \longrightarrow TiCl_3R + AlR_2Cl$$

$$TiCl_4 + AlR_2Cl \longrightarrow TiCl_3R + AlRCl_2$$

$$TiCl_3R \longrightarrow TiCl_3 + R^{\bullet}$$

$$R^{\bullet} + R^{\bullet} \longrightarrow R\text{-}R$$

$$TiCl_3 + AlR_3 \text{ (or Al } R_2Cl) \longrightarrow TiCl_2R + AlR_2Cl \text{ (or } AlRCl_2)$$

The active species that actually initiate polymerization would be $TiCl_2R$, entities that are formed at the edges of $TiCl_3$ crystals.

Remark. Ti^{III}–type species ($TiCl_3$ and $TiCl_2R$) are generated from Ti^{IV} by reduction and carry a lone electron delocalized on the whole structure.

"Ziegler" catalysts were utilized by Natta to polymerize propylene and obtain isotactic polypropylene, which is presently one of the most important commodity polymers. The remarkable contribution of Natta to the mechanistic understanding and his utilization of these systems in propylene polymerization is the reason of his association with Ziegler in the designation of these catalysts.

A controversial debate about the detailed mechanism of the propagation reaction has lasted for years which is not totally settled to date. Both "monometallic" and "bimetallic" mechanisms featuring the sole Ti atom in the process of monomer insertion for the first and the association of Ti and Al for the second have had their proponents. After observing that at the very first moments polymerization occurs specifically at the $TiCl_3$ crystal edges, Rodriguez and Van Looy proposed the following reaction pathway, reconciling the two mechanisms:

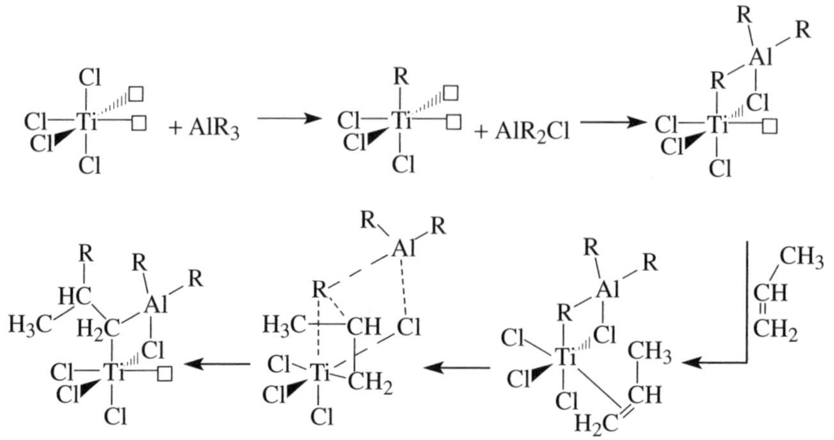

Even if the Rodriguez and Van Looy mechanism can account for most of the phenomena observed, the "monometallic" mechanism proposed by Cossee is today the generally accepted one as recent work seems to confirm its validity. Hereafter the isospecific polymerization of propene occurring according to the Cossee mechanism is represented:

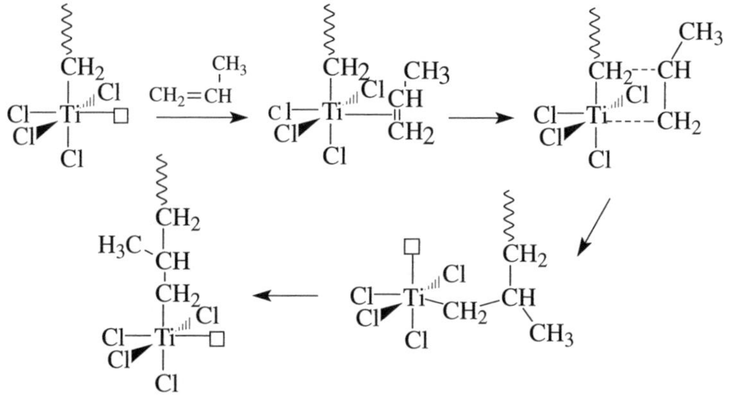

For an *isoregulating* process to occur, the propene molecule has to coordinate onto TiCl$_3$ crystal with a constant geometry. To address this issue, Cossee proposed that the coordination site tilts after each insertion, which is a weakness of this mechanism.

The oxidation of the titanium atoms is also a controversial matter. Indeed, if the main fraction of titanium atoms in $TiCl_3$ are at oxidation state III, it is not sure that it is the same for Ti atoms located at the crystal edges which could well be at oxidation state IV.

The **efficiency** of Ziegler–Natta systems—that is, the proportion of titanium atoms generating actual active centers—is relatively low ($\sim 30\%$) even after 50 years of improvements. In spite of that, the **activity**, which is the rate of formation of polymer per mass unit of titanium (or of catalytic system), is very high, due to the high **reactivity** of the propagating species:

$$\text{activity} = \text{constant} \cdot \text{efficiency} \cdot \text{reactivity} \cdot [M]$$

A significant increase in the efficiency could be obtained by co-milling Ti-based systems with **$MgCl_2$**. The role played by this additive in the efficiency of the catalyst is probably related to its geometry. Indeed, the ionic radius of Mg^{2+} cation (0.072 nm) is about that of Ti^{3+} (0.067 nm), favoring the co-crystallization of $MgCl_2$ with $TiCl_3$, thus causing defects in the crystal lattice and as many new edges in $TiCl_3$ crystal. The efficiency of the catalytic system would be increased in this way, with the formation of novel active sites at the crystal edges according to the Rodriguez and Van Looy mechanism.

Several authors also suggested that the presence of Mg^{2+} cation in the vicinity of C–Ti bond increases its reactivity.

Addition of suitable **"internal" Lewis bases** to the catalytic systems was found to cause a strong increase of the content in isotactic sequences in the polymerization of propene. Indeed, it is admitted that the crystal "defects" where the active sites lie exhibit various degrees of "Lewis acidities," depending upon their position on the crystal edge. The most acidic sites, which are also the nonstereospecific ones, would be "inhibited" upon addition of a weak Lewis base (ester, ether, amine, etc.), leaving unaffected the less acidic isospecific sites.

The mode of action of **"external" Lewis base**—so called because it is added to the catalytic system together with the organoaluminum cocatalyst—is more difficult to account for. This base generally consists of a dialkoxysilane molecule, and its role would be to inhibit certain nonspecific sites in addition to the "internal" Lewis base or transform some of them into isospecific ones by the steric effect. It is admitted that it coordinates simultaneously to a titanium atom and an aluminum atom via its two oxygen atoms, thus corroborating the Rodriguez and Van Looy mechanism.

Ziegler–Natta systems are prone to termination reactions; the latter can occur through reaction with impurities present in the reaction medium (O_2, H_2O, ROH, CO_2, etc.) or spontaneously:

$$\underset{R}{\sim\sim\sim CH\text{-}CH_2\text{-}Met} + H_2O \longrightarrow \underset{R}{\sim\sim\sim CH\text{-}CH_3} + Met\text{-}OH$$

$$\underset{R}{\sim\sim\sim CH\text{-}CH_2\text{-}Met} + CO_2 \longrightarrow \underset{R}{\sim\sim\sim CH\text{-}CH_2\text{-}CO\text{-}O\text{-}Met}$$

$$\underset{R}{\sim\sim\sim CH\text{-}CH_2\text{-}Met} + O_2 \longrightarrow \underset{R}{\sim\sim\sim CH\text{-}CH_2\text{-}O\text{-}O\text{-}Met}$$

$$\underset{R}{\sim\sim\sim CH\text{-}CH_2\text{-}Met} \longrightarrow \underset{R}{\sim\sim\sim C{=}CH_2} + H\text{-}Met$$

The last reaction is a termination only if the metal hydride formed is unable to reinitiate polymerization, which is often not the case as shown later.

Whatever the type of termination reactions, the k_p/k_t ratio is generally high and after consumption of impurities that are initially present in the reaction medium, many systems exhibit high persistence according to Rodriguez and Van Looy.

Because of the small number of truly active centers, the molar masses of the polymers obtained are very high; to control the sample molar mass, polymerizations are carried out in the presence of transfer agents and mostly H_2 is used; upon reaction with a macromolecular alkyltitanium, titanium hydride is produced that is able to reinitiate polymerization:

$$\underset{R}{\sim\sim\sim\sim CH\text{-}CH_2\text{-}Ti} + H_2 \longrightarrow \underset{R}{\sim\sim\sim\sim CH\text{-}CH_3} + HTi$$

then

$$HTi + H_2C{=}CH\text{-}R \longrightarrow R\text{-}CH_2\text{-}CH_2\text{-}Ti \quad etc.$$

Monomer concentrations and pressures are not very high in Ziegler–Natta polymerization (about 2–3 MPa), contrary to the free radical polymerization of ethylene which requires very high pressures. For obvious reasons, industrial producers of polyolefins focus their attention on the productivity of a catalytic system which corresponds to the mass of polymer produced (and not its rate of formation) per mass unit of transition metal, without mention of time. In Table 8.18 the typical components of a currently used Ziegler–Natta system for the industrial production of polypropylene are given.

The detailed kinetics of the polymerization process is extremely complex. Indeed, one has to take into account the existence of the two essential steps—that is, that of coordination and that of insertion—as well as the bursting of the $TiCl_3$ crystal lattice under the effect of the polymer produced. Several models exist that account for the first steps of the process and the overall kinetics.

Ziegler–Natta polymerization is used to produce high-density polyethylene (HDPE)—that is, exhibiting a high degree of crystallinity due to a low proportion

Table 8.18. Composition of a conventional Ziegler–Natta catalytic system used to polymerize propylene

Function of the Component	Nature of the Component	% Molar
Transition metal halide	$TiCl_4$	1.5
Aluminum alkyl	$AlEt_3$	75
Support	$MgCl_2$	20.5
Internal Lewis base	Ethyl phthalate	1.5
External Lewis base	Phenyltriethoxysilane	1.5

of structural defects (branches). It is also utilized to produce isotactic polypropene (i-PP), with degrees of isotacticity as high as 99%; due to the poor processability of the latter materials, for certain applications the degree of [mm] triads (degree of isotacticity) is generally limited to 96%.

The preparation of ethylene copolymers (or terpolymers) such as linear "low-density" polyethylenes (LLDPE), ethylene/propylene elastomers (EP), ethylene/propylene/diene terpolymers (EPDM) is also based on these catalytic systems. Stereoregular polyisoprene and polybutadiene elastomers are also obtained by this method of polymerization; the formation of 1,4-cis-polydienes requires the prior double coordination of the monomer onto the growing active center:

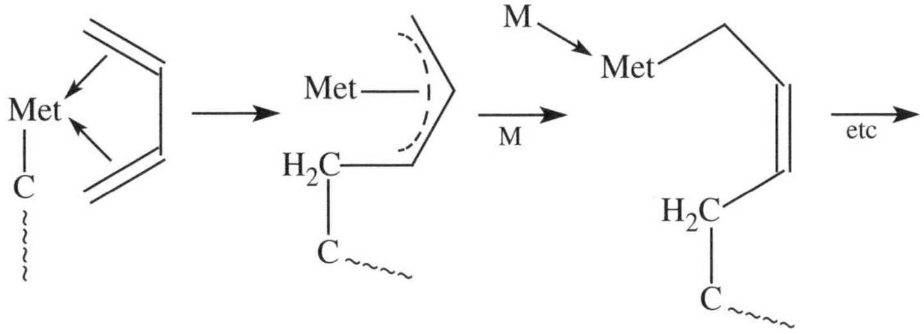

8.8.3. Metallocene-Based Systems

Metallocenes are complexes whose transition metal (mainly from IV.B group) is sandwiched between two aromatic rings, most generally of cyclopentadienyl (CP) type:

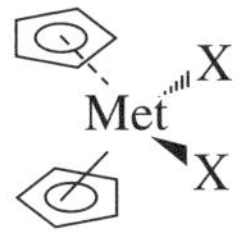

These metallocenes were tested a long time ago, either alone or together with an alkylaluminum derivative, to initiate olefins polymerization, but were found to be of low activity. In the beginning of the 1980s, Sinn and Kaminsky succeeded in activating them via the use of aluminoxanes and obtained extremely high productivities. Aluminoxanes are polymeric compounds resulting from controlled addition of water to an alkylaluminum; methylaluminoxane (MAO) can be represented by

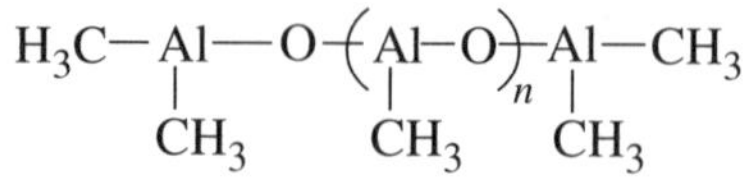

$$
H_3C-Al-O-(Al-O)_n-Al-CH_3 \\
\quad\quad | \quad\quad\quad | \quad\quad\quad | \\
\quad\quad CH_3 \quad\quad CH_3 \quad CH_3
$$

It contains systematically residual trimethylaluminum.

It is now well established that MAO plays a double role: in a first step, it methylates the metallocene by substituting a halogen atom and then "cationizes" the transition metal, permitting the coordination of an olefin monomer acting as a Lewis base. The reaction pathway of these two steps may be represented as follows:

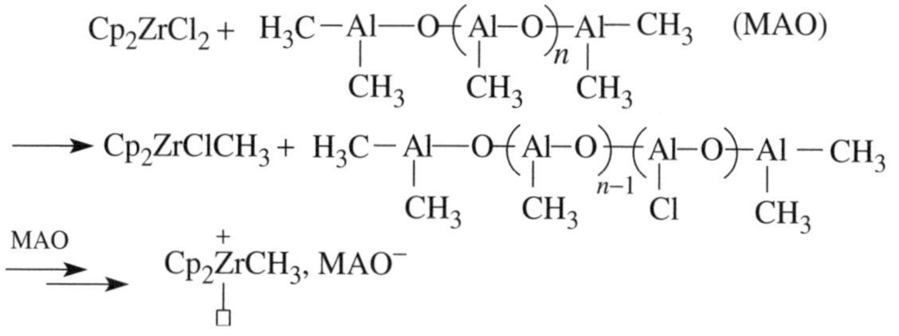

$$
Cp_2ZrCl_2 + H_3C-Al-O-(Al-O)_n-Al-CH_3 \quad (MAO) \\
\quad\quad\quad\quad\quad | \quad\quad\quad | \quad\quad | \\
\quad\quad\quad\quad\quad CH_3 \quad CH_3 \quad CH_3
$$

$$
\longrightarrow Cp_2ZrClCH_3 + H_3C-Al-O-(Al-O)_{n-1}-(Al-O)-Al-CH_3 \\
\quad\quad\quad\quad\quad\quad\quad\quad | \quad\quad\quad | \quad\quad\quad | \quad\quad | \\
\quad\quad\quad\quad\quad\quad\quad\quad CH_3 \quad CH_3 \quad Cl \quad CH_3
$$

$$
\xrightarrow{MAO} \quad Cp_2\overset{+}{Z}rCH_3, MAO^-
$$

The commonly admitted mechanism of olefin insertion is similar to that proposed by Cossee for Ziegler–Natta polymerization. It is shown hereafter in the case of the polymerization of propene initiated by a zirconocene possessing two indenyl ligands linked by an ethylene bridge flanking a zirconium atom:

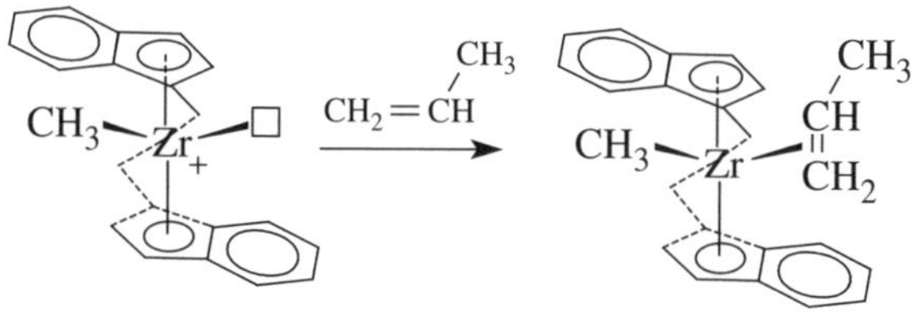

The coordination step is followed by insertion:

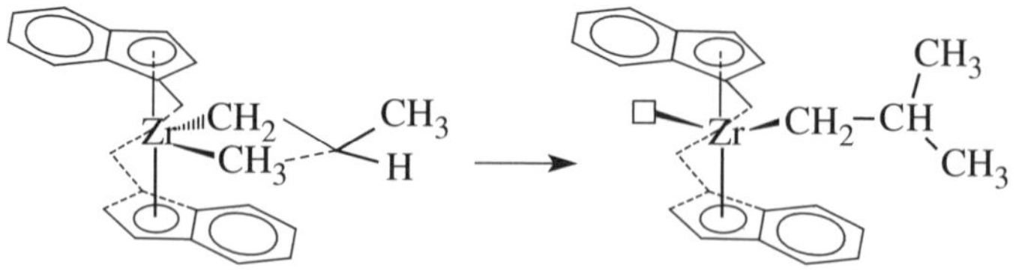

and so on.

Metallocene-based systems exhibit many advantages compared to Ziegler–Natta ones.

They are initially soluble in the reaction medium and thus have an efficiency close to unity which strongly increases their activity. Next, contrary to conventional Ziegler–Natta systems, they possess only one family of propagating species; for this reason, they are referred to as "monosites," meaning that all active centers exhibit the same characteristics and, in particular, the same reactivity. It is a very interesting property in the case of copolymerization for the "homogeneity" of the material formed. Lastly, their structure can be finely tuned by molecular engineering so as to obtain the targeted polymer. The size of the aromatic ligands, the substituents carried by the latter, and their covalent bridge are the principal structural parameters that can be adapted to achieve steric control while the monomer coordinates onto the transition metal; all these parameters determine the polymerizability and/or the stereochemical control of the process.

In addition to the homogeneity of the reaction medium and the presence of only one type of reactive species, the participation of *agostic* bonds between the metal atom and the hydrogen atom in α or β position to the polymer chain (the diagram hereafter gives an example of an α bond) in the coordination–insertion process is another essential feature of these catalytic systems. These agostic bonds can be described through the formalism of donor–acceptor interactions where the donor would be the C–H bond of the polymer and the acceptor would be the d^0-electron-deficient metal. Both the stabilization of certain chain conformations (and thus of catalytic site) and their high activity can be accounted for upon taking into account these interactions.

Vacant site   α–Agostic complex   Transition state

γ-Agostic complex

Mechanistic studies on the catalysis by metallocenes show that the tilting of the coordination site that is specific to the Cossee model is not essential to account for the isoregulation of the process. As a matter of fact, the steric hindrance around the vacant site and the symmetry of the catalytic site are the two parameters driving the stereospecificity. The metallocene shown below is well-suited to the copolymerization of cycloalkenes (by addition reaction) with ethylene and to the syndiospecific polymerization of propene,

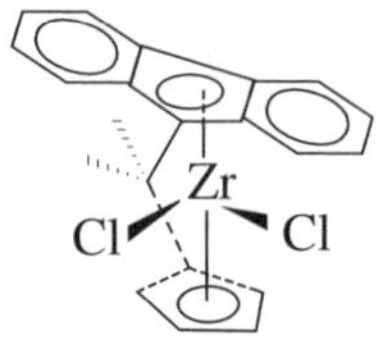

whereas the following is generally used to obtain a highly syndiotactic polystyrene:

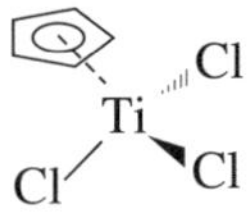

Some of these systems are tolerant of polar groups and are thus suitable for the homopolymerization or the copolymerization of polar monomers with olefins, thus adding to their versatility. The insertion of such polar monomers in the polymer chain is likely to induce new properties in the corresponding materials.

In the same manner, catalysts developed by Brookhart whose typical structure is shown hereafter,

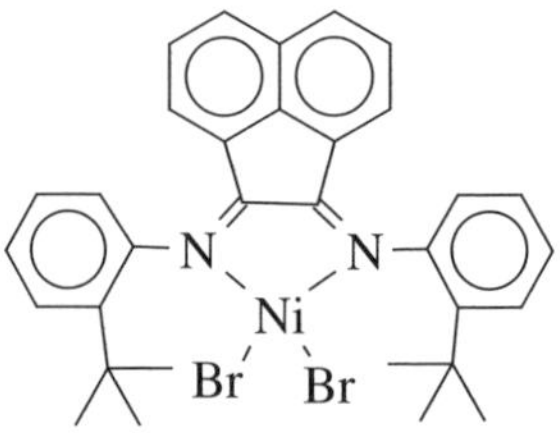

can be used to polymerize polar monomers with a certain control. They were also shown to afford highly branched polyolefins that could be functionalized by insertion of (meth)acrylic monomers at branch ends. The mechanism of action resembles that presented for metallocenes.

8.8.4. "Phillips"-Type Polymerization Catalysts

Although little is known about the actual mechanism of action of these catalysts, they are industrially utilized to produce high-density polyethylene (HDPE) in significant quantity. "Phillips" catalysts (so called because they were patented by Phillips Petroleum Cy) consisted of chromium oxide (CrO_3) deposited onto a silica support; the system is activated by air-drying and by a progressive heating up to 1000°C. CrO_3 is anchored on the support through the latter surface silanol functions, which generates chromate or bichromate (Cr^{VI}) species:

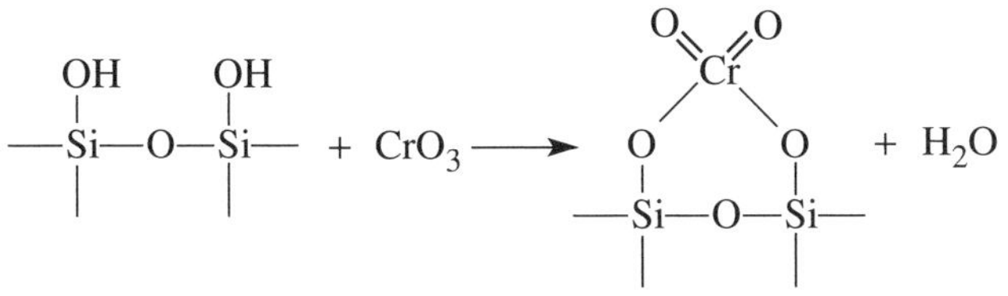

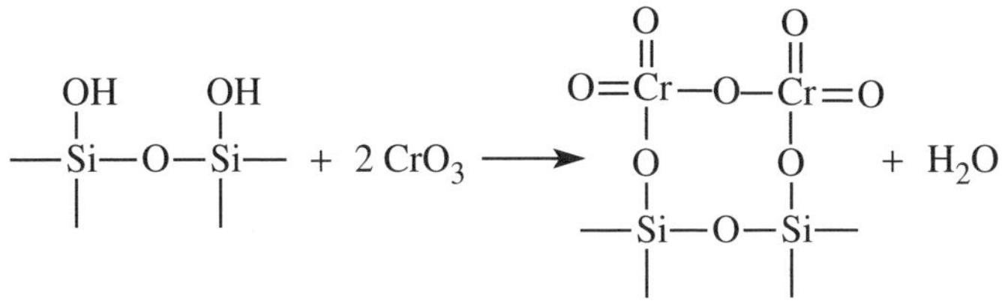

It is commonly admitted that the addition of ethylene reduces the chromium atom to form a Cr^{II}–carbon atom bond which would be the propagating active species. The use of other reducing agents improves the activity of these systems. These catalysts are only used for the polymerization of ethylene because the polymerization of higher olefins is not stereospecific. The originality of these systems lies in the occurrence of spontaneous transfer reaction via β-elimination of a hydride, causing the formation of chain-ended unsaturations. Contrary to Ziegler–Natta polymerization, the control of molar masses is not obtained by addition of a transfer agent but by varying the reaction temperature, which determines the rate of β-elimination.

In the same family of catalysts, one finds also MoO_3-based systems, on different supports (alumina, etc.).

8.8.5. Metathesis Polymerization

A metathesis reaction is catalyzed by transition metal derivatives and results in a transalkylidenation of olefinic hydrocarbons as shown hereafter:

$$2R-HC=CH_2 \underset{}{\overset{Cat}{\rightleftharpoons}} H_2C=CH_2 + RCH=CHR$$

Applied to cycloalkenes, this reaction affords polyunsaturated polymers called *polyalkenamers*:

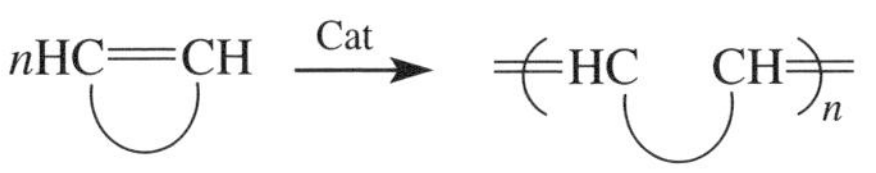

The reaction mechanism was first proposed by Chauvin (in 1970) and was experimentally corroborated a few years later.

> **Remark.** Chauvin was awarded the Nobel Prize in 2005 for this discovery. He shared this prize with R. Grubbs and R. Schrock, who intensely improved the efficiency and the specificity of initiators for metathesis ring-opening polymerization, with systems based on Ru and Mo respectively.

The mechanism proceeds via a coordination of the cycloalkene molecule onto the transition metal atom, followed by the formation of a metalacyclobutane whose reorganization generates an alkenamer unit. The active center, an alkylidenic species, is itself a ligand of the transition metal atom that also possesses a vacant coordination site: over a maximum number of n, $(n-1)$ ligands link to the transition metal atom:

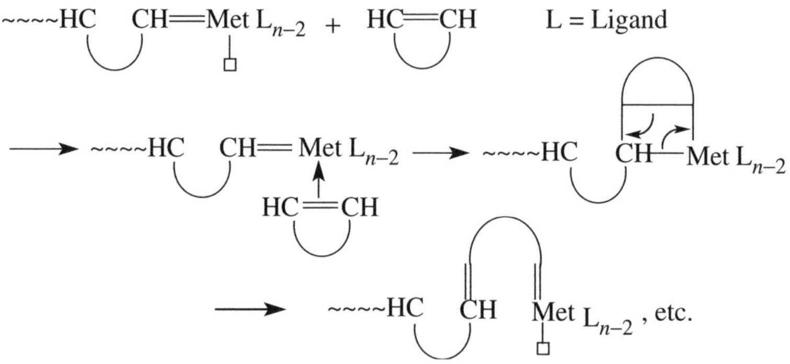

Polyalkenamers resulting from metathesis polymerization of cycloalkenes carry one unsaturation per monomeric unit. The basic rules of polymerizability apply to these monomers; in particular, it is imperative that the variation of the free energy of polymerization be negative, which means that a certain cyclic strain, even weak, must be released upon polymerization with $\Delta H_p < 0$.

Initiating systems that are used to induce a metathesis reaction can be classified into one of the three categories:

- Transition metal oxides or halides as

$$WO_3, MoO_3, Re_2O_7, RuCl_3, \text{etc.}$$

  in association with a photochemical activation. The mechanism of formation of the alkylidenic species by these systems is complex and is still pending;

- Binary systems composed of a transition metal halide and an organometallic compound; $WCl_6/Sn(CH_3)_4$ is one of the most used systems, with the alkylidenic species resulting from the three following reactions:

$$WCl_6 + (CH_3)_4Sn \longrightarrow (CH_3)_3SnCl + Cl_5WCH_3$$

$$Cl_5WCH_3 + (CH_3)_4Sn \longrightarrow Cl_4W(CH_3)_2 + (CH_3)_3SnCl$$

$$Cl_4W(CH_3)_2 \longrightarrow Cl_4W{=}CH_2 + CH_4$$

- Stable alkylidenic complexes among which some can bring about "living" polymerization under appropriate conditions:

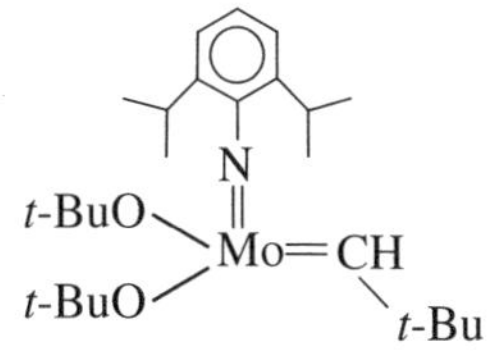

Schrock's complex 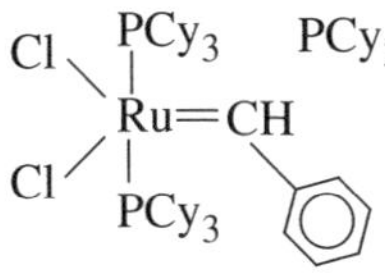   Grubbs' complex

It is worth stressing that polymerizations initiated by ruthenium-based derivatives are relatively insensitive to the presence of polar groups, unlike most of the coordinative initiating systems; these ruthenium-based complexes can thus be used to polymerize monomers carrying polar substituents and sometimes to carry out polymerizations in aqueous medium.

The main monomers that can be polymerized by metathesis polymerization are the following ones:

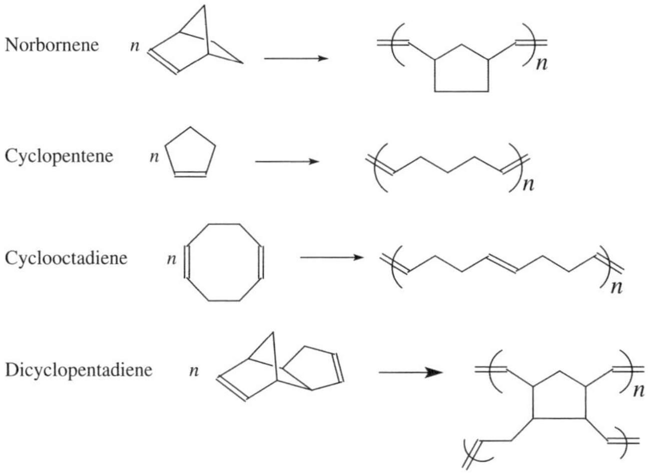

Because dicyclopentadiene is a tetravalent monomer, its polymerization results in the formation of a three-dimensional network.

When used with monomers such as **alkynes**, metathesis initiators afford poly-conjugated polymers. The reaction mechanism is similar to that of cycloalkene polymerization, with the reaction intermediate being here a metalacyclobutene:

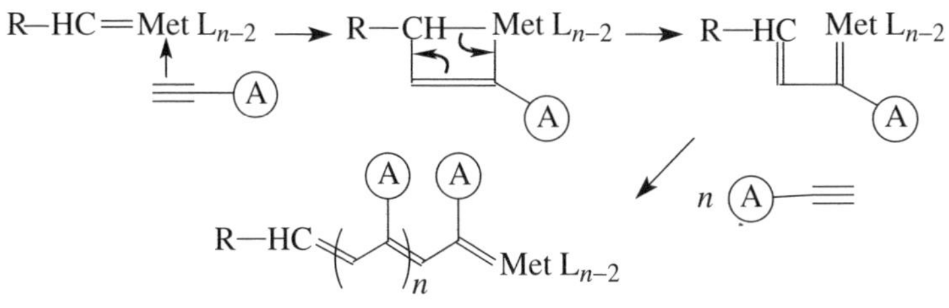

It is worth mentioning that such metathesis polymerization proceeds easily with monosubstituted alkynes but fails to efficiently occur with acetylene.

LITERATURE

G. C. Eastmond, A. Ledwith, S. Russo, and P. Sigwalt (Eds.), *Comprehensive Polymer Science*, Vols. 3 and 4: *Chain Polymerisation*, Pergamon Press, Oxford, 1989.

G. Moad and D. H. Solomon, *The Chemistry of Free Radical polymerization*, Pergamon Press, Oxford, 1995.

K. Matyjaszewski and T. P. Davis, *Handbook of Radical Polymerization*, Wiley-Interscience, New York, 2002.

R. D. Athey, Jr., *Emulsion Polymer Technology*, Marcel Dekker, New York, 1991.

M. Szwarc and M. Van Beylen, *Ionic polymerization and Living Polymers*, Chapman and Hall, New York, 1993.

H. L. Hsieh and R. P. Quirk, *Anionic Polymerization—Principles and Practical Applications*, Marcel Dekker, New York, 1996.

K. Matyjaszewski, (Ed.), *Cationic Polymerization—Mechanisms, Synthesis, and Applications*, Marcel Dekker, New York, 1996.

K. J. Ivin and J. C. Mol, *Olefin Metathesis and Metathesis Polymerization*, Academic Press, San Diego, 1997.

G. Odian, *Principles of Polymerization*, 4th Edition, Wiley, New York, 2004.

9

REACTIVITY AND CHEMICAL MODIFICATION OF POLYMERS

The chemical modification of polymers' molecular structure, their transformation, is a convenient means to get access to original polymers and to increase the variety of industrially available polymeric materials. Both **natural** precursors and **synthetic** ones can be used to obtain **artificial** polymers and generate structures that would be inaccessible by polymerization (or too expensive).

Organic polymers exhibit a reactivity that is not intrinsically different from that of simple homologous molecules; this principle, which was experimentally observed in the 1930s by Flory for a few examples, assumes a constant reactivity of the functions borne by the polymer regardless of its molar mass. Such a principle implies that only one functional group in the polymer chain is concerned by the reaction, and that it is soluble in the reaction medium and is as accessible as the same functional group carried by a simple molecule. These conditions, which were expressed by Alfrey, are seldom met simultaneously. For instance, due to the low mobility of the reagents in three-dimensional networks, the chemical modifications of inner segments are difficult to carry out though possible; obviously, the higher the cross-link density, the more significant this difficulty.

Thus, this chapter will be devoted to the presentation of typical reactions that are specific to linear polymers and to the modifications undergone by polymers' molecular structure under the effect of physical or chemical attacks. Because the reaction mechanisms are generally identical (or close) to those described for simple molecules, they will not be elaborated in detail in this chapter, except for a few cases.

Organic and Physical Chemistry of Polymers, by Yves Gnanou and Michel Fontanille
Copyright © 2008 John Wiley & Sons, Inc.

9.1. CHARACTERISTICS OF REACTIONS INVOLVING POLYMERS

Because each reaction is a particular case, it is difficult to establish general rules. Certain reactions on polymers can be strongly accelerated by the macromolecular nature of either the reagent or the substrate. Such behavior is generally accounted for by the presence of neighboring groups, or the influence of the chain stereochemistry, or the effect of electrostatic attractions or repulsions, and so on; but in most cases the apparent reactivity of polymers is definitely lower than that of simple model molecules, due to a reduced accessibility to the reaction sites.

The viscosity of polymers is very high in the condensed state, and only small molecules can freely move and react in the reaction medium. The variety of reagents (or substrates) that can be used under these conditions is thus limited, and most of the reactions involving polymers require the presence of a solvent. The lack of solubility of some of the polymers restricts the possibilities of modifying them.

Another specificity of polymers's modification is the statistical (generally random) and gradual character of the reactions. Each monomer unit of the chain may carry one or more reactive groups, and their transformation occurs as a function of their probability of collision with antagonist groups. For example, the esterification of cellulose, which is comprised of anhydroglucopyranose units with one primary and two secondary alcohols per monomer unit, exhibits a double randomness depending upon the hydroxyls esterified in a given monomer unit and the monomer unit modified along the chain. It is necessary to define a *degree of substitution* (DS), which corresponds to the average number of modified hydroxyl groups, the latter varying from zero to three in the case of cellulose derivatives. Only compounds corresponding to DS equal to zero and three are well-defined; those corresponding to intermediate values of DS exhibit a double randomness of their hydroxyl modification (Figure 9.1).

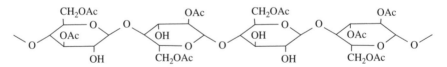

Figure 9.1. Tetrad of cellulose acetate corresponding to an average degree of substitution equal to two.

The reactions carried out on polymers enter into two main categories:

- The first one includes reactions that do not modify the dimensions of the macromolecular backbone even if they may weaken the latter. These reactions are generally intentional but may also occur in an aggressive environment.
- The second category includes reactions that cause either the breaking of the chains and are thus called degradations or, on the contrary, their cross-linking; in the latter case a one-dimensional polymer is transformed into a three-dimensional one with a considerable modification of its properties, in particular the mechanical ones.

The most representative and/or illustrative examples of polymer modification will be discussed in the next sections.

9.2. EFFECT OF THE MACROMOLECULAR NATURE OF POLYMERS ON THEIR REACTIVITY

9.2.1. Effect of Neighboring Groups

This effect is particularly discernible for reactions involving more than one functional group carried by the macromolecule considered. For instance, the proximity of carboxylic groups in poly(acrylic acid) accelerates the dehydration compared to that observed in simple homologous molecules:

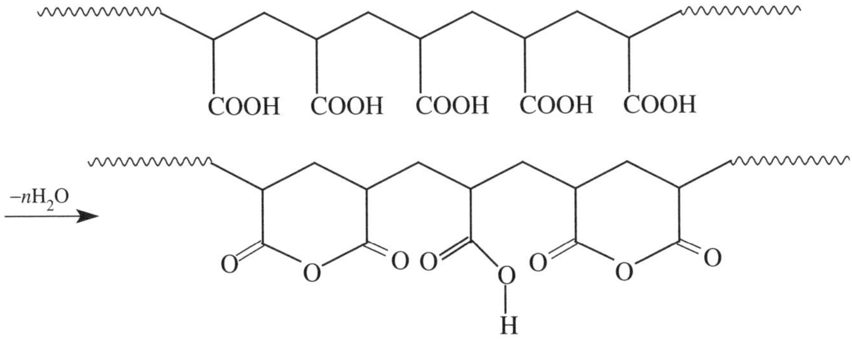

However, the formation of cyclic anhydrides leaves several carboxylic groups isolated; these groups have no option but to react intermolecularly to generate cross-links. A similar phenomenon is observed in the dechlorination reaction of poly(vinyl chloride) by zinc powder with the formation of cyclopropane cycles from 1,3-dichloro-type sequences (regular dyads):

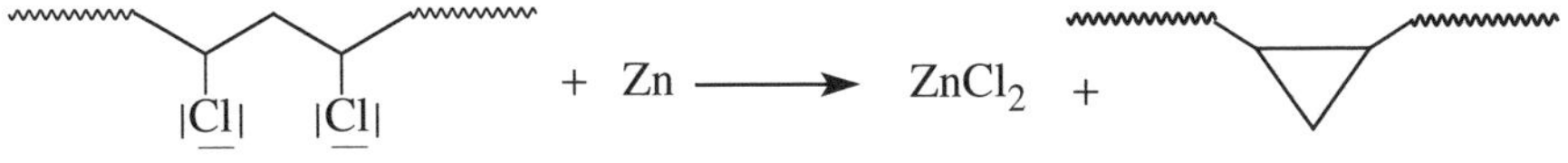

Larger cycles (cyclopentane, cycloheptane, etc.) can also be formed, although at a slow rate. Chlorine atoms that are isolated between two cycles find little possibility of reaction due to steric effects, with the yield of such chemical modification culminating at a maximum value of 86%; this value, which is indicative of the randomness of the process, is in close agreement with theoretical predictions. In the same polymer, the dechlorination of "tail-to-tail" dyads leads to the formation of double bonds instead of the cyclopropane groups:

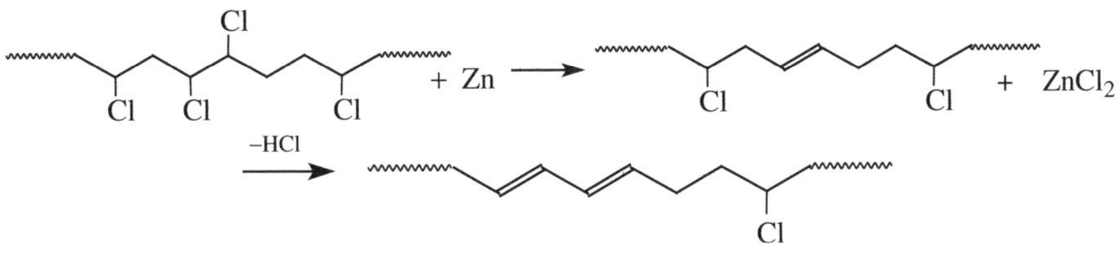

The presence of such double bonds favors the subsequent dehydrochlorination of other sites, in a reaction that can be regarded as self-catalytic.

Another important example of reaction involving two neighboring side groups is that of poly(vinyl alcohol) with formaldehyde:

A quantitative analysis of the influence of neighboring groups on the kinetics of reaction of functional groups carried by a macromolecule is very difficult to carry out. Indeed, in a reaction involving one type of functional group, each of such sites can react according to three different rate constants:

- k_1 for sites whose two immediate neighbors have not reacted yet
- k_2 for sites whose only one neighbor has reacted
- k_3 for sites whose two neighbors have already reacted

Mathematical solutions have been proposed under the assumption of a Markovian statistics for the chemical process, but the agreement with experimental data is far from excellent.

From a qualitative point of view, one can describe the behavior of certain systems such as those corresponding to the activation of a functional group A_1 by the product of reaction of a nearby functional group which was initially identical to A_1:

Generally, $k_3 > k_2 \gg k_1$; in this case, the reaction is slow at the beginning and evolves into a chain reaction, with each functional group being activated by the reaction of the neighboring functional group. If the k_3/k_1 ratio is much higher than the number of reactive functional groups per chain (i.e., the degree of polymerization), they react quickly as soon as one functional group has reacted. For non-disperse samples, the kinetics of chemical modification is first order, whereas for disperse ones the kinetics is likely to reflect the molar mass distribution. If the k_3/k_1 ratio is much lower than the number of reactive functional groups per chain, the effects due to the finite length of the chains can be neglected; in particular, if several sites of a same chain have reacted, their concentration is initially proportional to the reaction time. Since the rate of reaction is roughly proportional to the number of reacted functional groups, the extent of reaction is initially proportional to the square of the reaction time. A similar reasoning would apply for the deactivation of a given reactive site after chemical modification of a neighboring group.

Reactions involving electrically charged species are slowed down when the reactive species carry charges of the same sign and are accelerated in the presence of opposite charges. Such effects are more pronounced in polymers than in simple molecules.

A typical example of such reactions is given by the hydrolysis of pectins catalyzed by HO^- ions. Pectins are partially esterified derivatives of polygalacturonic acid that are frequently found in Nature. This polymer can be schematically represented as shown below:

In this case it has been shown that the second-order apparent rate constant of hydrolysis decreases rapidly as the hydrolysis reaction proceeds. Such behavior is interpreted as resulting from the mutual repulsion between the HO^- catalyst and the increasing number of negative charges arising from the hydrolysis of the polymer ester functional groups. Equilibria can also be affected by the effect of neighboring groups. As a slightly ionized polyacid, poly(acrylic acid) is characterized by an equilibrium constant of ionization defined as

$$K_{app} = \alpha^2/(1 - \alpha)$$

where α denotes the fraction of ionized carboxylic groups, which decreases when α increases. This peculiar variation can be interpreted by the existence of a free energy of ionization (ΔG^*_{el}), which represents the free energy of electrostatic activation required to increase by one unit the number of ionized functional groups carried by the chain. If K_0 is defined as the ionization constant of a model carboxylic acid carried by the chain, the relation between K_{app} and K_0 is given by

$$K_{app} = K_0 \exp(-\Delta G^*_{el}/kT)$$

Thus, for highly dilute solutions, values of K_0/K_{app} of 10^4 can be obtained. Addition of neutral electrolytes to the polyelectrolyte solution can considerably reduce the value of ΔG^*_{el} but can not cancel it.

9.2.2. Effect of Tacticity

It is interesting to reconsider the case of a poly(acrylic acid) dehydration reaction where two different steps can be distinguished for the atactic polymer; the first step is five to six times faster than the second one and obviously corresponds to dehydration of meso dyads (isotactic sequences). Hydrolysis of syndiotactic poly(acrylic anhydride) easily results in syndiotactic poly(acrylic acid); if the latter is treated with thionyl chloride so as to once again generate the initial anhydride, the reaction does not occur as expected:

H_2O | $SOCl_2$ Syndiotactic polymer

On the other hand, if the initial cyclopolymer is heated in presence of pyridine, isomerization of the syndiotactic form into the isotactic one is observed, probably by enolization of the position in α, whose hydrolysis gives rise to isotactic poly(acrylic acid); the latter can easily be transformed into polyanhydride by thionyl chloride:

H_2O | $SOCl_2$ Isotactic polymer

For certain reactions, the effect of tacticity plays differently as compared to the precedent case. For instance, the acetalization of poly(vinyl alcohol) is faster for racemo dyads (syndiotactic type) than for meso dyads (isotactic ones).

9.2.3. Influence of Conformation and Morphology

In an organized structure, conformation and morphology will play similar roles in bringing close functional groups that are away one from another along the macromolecular chain. A first example is related to the effect of the extent of the quaternization reaction of poly(4-vinylpyridine) on the conformation of the resulting chain:

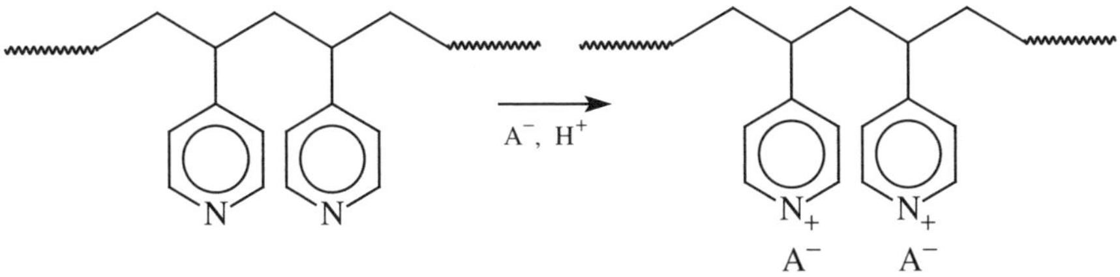

Due to the electrostatic repulsion between the pyridinium ions, the chains gradually unfold with the quaternization process, increasing in turn both its hydrodynamic volume and the viscosity of its solution. However, beyond a certain degree of quaternization, the solvating power of the medium toward the quaternized polymer decreases and the chains tend to shrink; thus, the viscosity of the corresponding solution initially increases and then decreases with the extent of the modification reaction.

The second example is drawn from the aminolysis of the ester side groups in poly(α-alkyl glutamate)s:

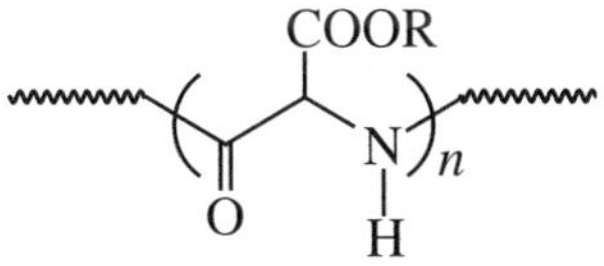

The reaction is relatively fast when the chains take the conformation of statistical coils, but is much slower in solvents (dimethylformamide, dimethylamine, etc.) favoring their helical conformation with strong molecular interactions (hydrogen bonds) stabilizing such a structure.

For reactions carried out in solid state, the polymer morphology plays a signifi-cant role because it determines the reagent accessibility to antagonist reactive sites. The modification of cellulose provides one such example.

Anhydroglucopyranose monomer units include one primary and two secondary alcohol functional groups:

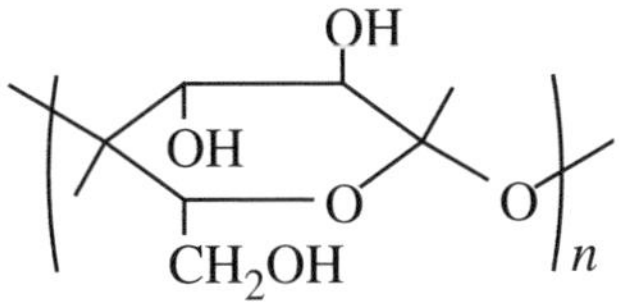

For their acetylation, acetic anhydride is used in a mixture of acetic and sulfuric acids. The reactivity of this acetylating reagent is closely dependent on the water content of cellulose; the higher the water content, the faster the esterification of cellulose. Through the formation of hydrogen bonds with hydroxyl groups, water is responsible for a modification of the morphology, forcing the amorphous zones of cellulose to swell. Chains part aside, thus facilitating the access of the reagent to their hydroxyl groups:

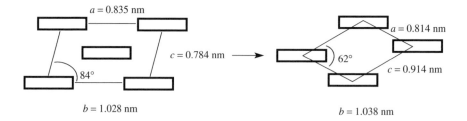

In contrast, water leaves the morphology of the crystalline regions of the sample unaffected and thus their reactivity unmodified. Elimination of water by drying leads to the regeneration of the hydrogen bonds between hydroxyl groups, which strongly reduces their reactivity.

The swelling of cellulose by a solvent is a phenomenon not restricted to water: Ethylamine ($CH_3-CH_2-NH_2$) swells not only the amorphous zones of cellulose but also crystalline ones. Actually, most polar organic solvents bring about the swelling in a proportion that depends on their polarity, molar volume, and especially their aptitude to break hydrogen bonds. Thus, methanol is an excellent swelling agent for cellulose unlike benzene and generally nonpolar solvents, but the latter can replace polar solvents with which they are miscible. In cellulose fibers, in particular, the substitution of methanol for water and then of benzene for methanol generates the so-called *inclusion celluloses* in which the hydrogen bonds are loosened by the presence of benzene, contributing to the significant increase in reactivity of the hydroxyl groups.

Mercerization of cellulose consists in treating fibers with an alkaline solution, in general with soda, before rinsing them abundantly with an acidic solution in order to subsequently get rid of the immobilized hydroxide groups. During this process, the average angle between the planes formed by the anhydroglucopyranose units undergoes a variation due to the rotation and translation of cellulosic chains, which results in a shortening of the fibers:

This modification of the crystalline structure of cellulose under the effect of soda increases its reactivity and is industrially exploited to chemically modify cellulose. The difference in reactivity between original and mercerized celluloses lies in the extent of hydration, with the mercerized cellulose fixing 100% more water than original cellulose. Such a mercerization process is also currently used to immobilize dyes molecules on cotton fibers.

As illustrated in the case of cellulose, the difference in reactivity between amorphous and crystalline zones is a general phenomenon: for instance, the treatment of a polyethylene film by concentrated nitric acid causes a selective degradative oxidation of its amorphous zones, which can be used to evaluate its degree of crystallinity.

9.3. CROSS-LINKING REACTIONS

The valence of a polymer chain carrying n v_i-*valent* functional groups is equal to nv_i. The reaction of such a polymer with a divalent antagonist molecule leads to its cross-linking through to the bridging of its chains:

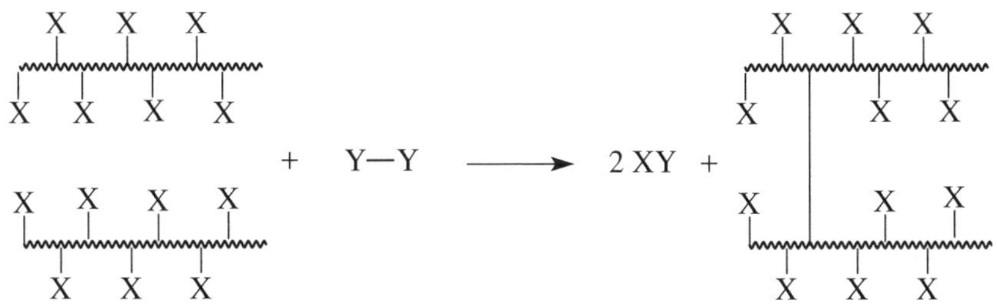

Such reactions may sometimes be spontaneous, but they are generally intentional and are used to modify the mechanical behavior of a sample by transforming one-dimensional chains into a three-dimensional network.

Chain cross-linking can be radio- or photoinduced, but most of the examples are chemically induced ones. Among these examples, the vulcanization of rubber has industrial significance.

> **Remark.** The term "vulcanization" is used only in the field of rubber technology. Cross-linking and sometimes curing are used in all other fields.

Natural rubber is a hydrocarbon polymer whose monomer units would be identical to those resulting from a perfectly regular polymerization of isoprene that would afford only 1,4 structures with a cis-configuration:

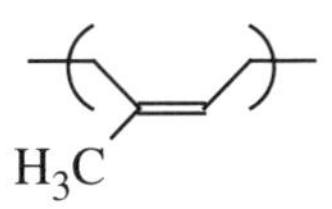

Several techniques of vulcanization can be used to carry out vulcanization; the oldest and still currently used in a large scale involves sulfur as a cross-linking agent in the presence of various additives. The comprehensive mechanism of such vulcanization of polyalkadienes by sulfur is still ill-known: many mechanisms have been proposed, but the most accepted one is ionic in nature as schematically represented below:

$$S_8 \xrightarrow{\Delta} S_x^{\delta+} ----- S_y^{\delta-} \quad \text{or} \quad S_x^+ + S_y^-$$

then

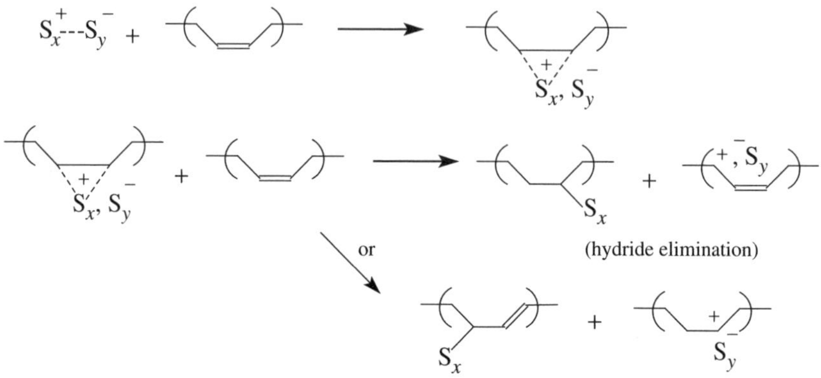

The carbocationic species generated along the chain would react with S_8 to generate a new cation,

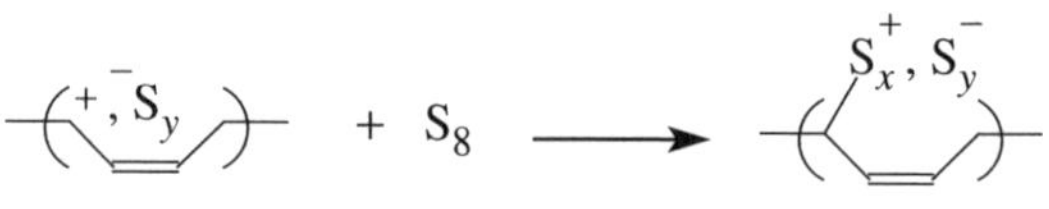

which in turn would attack a neighboring chain to produce cross-linking:

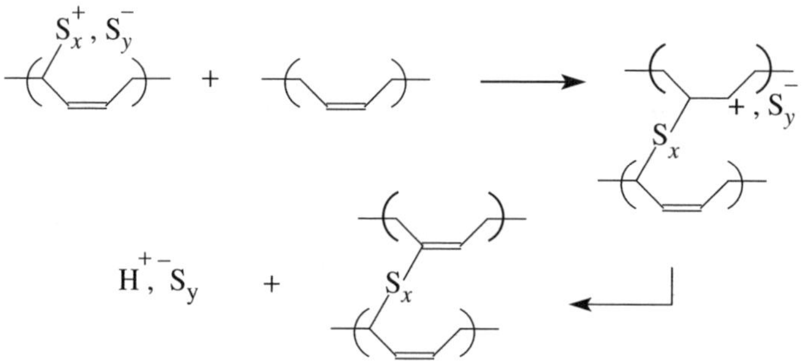

The released proton would then propagate the process by electrophilic addition. However, this mechanism is not universally accepted, and certain authors suggest a radical-based process.

Vulcanization by sulfur alone is of little help as it is too slow. The vulcanizing systems currently used contain accelerators like thiuram sulfides (R_2N-CS-S-S-CS-NR_2), zinc dithiocarbamates [(R_2N-CS-S)$_2$Zn], thiourea (H_2N-CS-NH_2), or zinc benzothiazoles, as well as activators such as ZnO, stearic acid, and so on, whose respective roles are still ill-elucidated.

The cross-linking of other types of polymers (polyethylene, polysiloxanes, etc.) is generally obtained by generating free radicals in the bulk of the material through the removal of hydrogen or addition on residual unsaturations, which results in interchain recombination:

9.4. DEGRADATION REACTIONS OF POLYMERS

9.4.1. Preliminary Considerations

Degradation of the side groups carried by polymer chains is not different from that observed for simple homologous molecules; only those reactions that cause the breaking of the macromolecular backbone will be considered in the following section.

Sometimes induced to obtain a controlled disappearance of polymer materials, the degradation of polymers is in most cases an undesirable phenomenon which limits their applications. The concept of durability had to be introduced to conform the characteristics of a polymer material to those required for the applications contemplated.

The degradation of polymers results from a supply of energy (chemical, thermal, photochemical, or radiochemical) to the macromolecule; when this energy is focused on a chemical bond and is higher than the corresponding bond energy, the rupture is effective. In most cases, the position of the broken bond in the chain is random.

9.4.2. Various Types of Degradation

The mechanisms of degradation reactions depend closely on the molecular structure of the polymer considered. The monomer units can be subject to attack by all sorts of reactive species, but only those resulting from the exposure to a natural environment are considered here. Thus, only hydrolytic, oxidizing, photochemical, and thermal degradations and any combination among them will be described below.

9.4.2.1. Hydrolytic degradation mainly concerns polymers possessing hetero-elements in their main chain (polyesters, polyamides, polyethers, etc.) and stems from the existence of equilibria governing the corresponding step-growth polymerization reactions. The rate of degradation depends not only on the nature of the substituents of the hydrolysable functional group but also on the temperature, the pH value, and the concentration of water in the material. This last parameter is extremely important: it closely depends on both the hydrophilicity of the material and the ambient hygrometry.

This dependence on the hydrophilicity is exploited in the controlled degradation of polyester-based biomaterials. For example, the copolymerization of glycolide and lactide

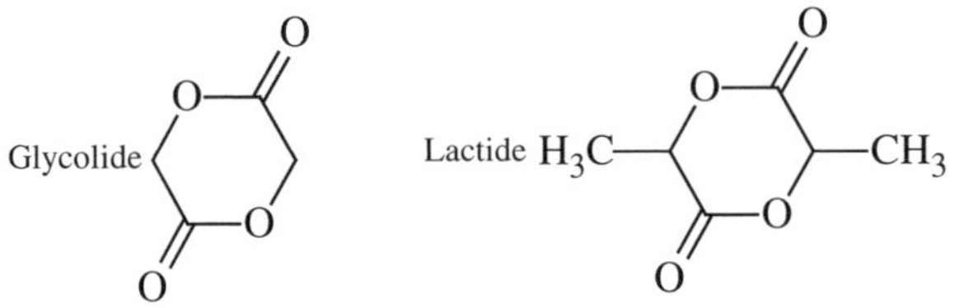

affords a copolyester whose hydrophilicity is determined by the respective propor-tion of the two comonomers; depending upon the application considered, the rate of hydrolysis can thus be known beforehand. In such polymers, the by-products formed by degradation are bioassimilable and the corresponding materials are called bioresorbable:

~~-(CO-CHR-O-CO-CHR-O)$_n$-~~ + 2nH$_2$O $\longrightarrow$ nHOOC-CHR-OH

The effect of pH can also be utilized to accelerate the hydrolytic process. For instance, polysaccharides can be rapidly degraded in acidic medium, with the degree of polymerization of the oligosaccharides formed depending on both the pH and the reaction time; for example, cellulose can be degraded down to glucose ($n = 1$):

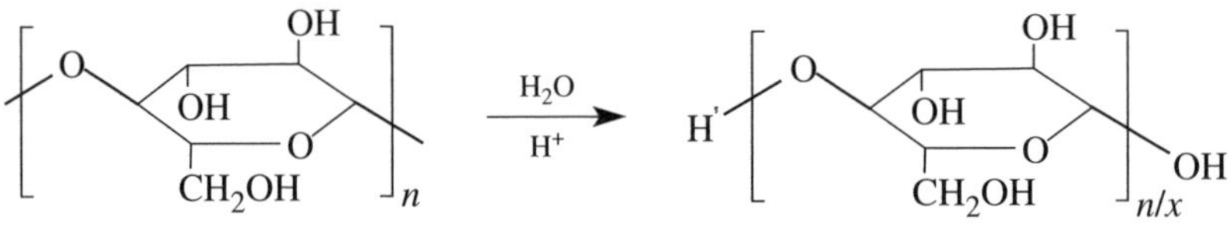

9.4.2.2. Degradation by Oxidation limits the long-term usage of several fam-ilies of polymers. Atmospheric oxygen and especially ozone (O$_3$) can indeed react with tertiary carbon atoms of CH bonds (PP, branching points of LDPE, etc.), and this occurs more easily at high temperature; one speaks of thermo-oxidizing degradation in this case.

The reaction mechanisms proposed to account for the experimental obser-vations are numerous. Each family of polymers is characterized by its own sensitivity to molecular oxygen. Hereafter, only one of them is presented which concerns the oxidizing decomposition of polypropene:

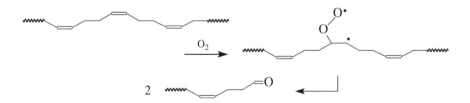

Actually, the kinetics of oxidation could not be accounted for by this succession of reactions, with the mechanism being much more complex as that shown above.

The degradation of polyalkadienes by oxidation is a well-known phenomenon that is accelerated by repeated mechanical stress. It is also a self-catalytic process that leads to the scission of the chains and at the same time to their cross-linking. Several mechanisms of oxidation involving molecular oxygen have been proposed, one of them being shown below:

Ozone (O_3), which mainly results from air pollution and is formed through the following mechanism,

$$NO_2 + h\upsilon \longrightarrow NO + O$$

$$O + O_2 \longrightarrow O_3$$

is also responsible for the degradation through cycloaddition:

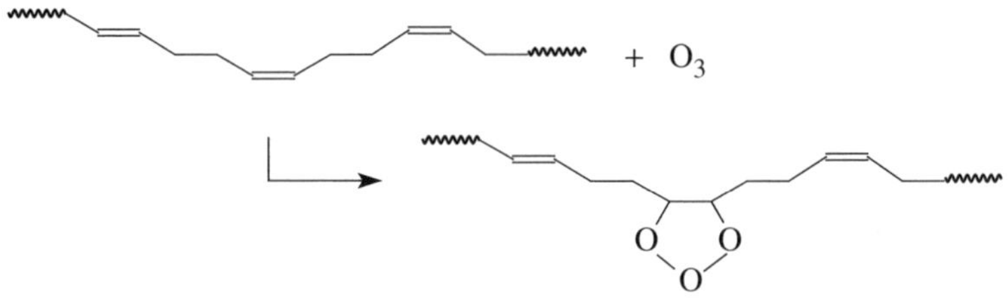

The trioxolane ring formed then decomposes into aldehyde (or ketone) and a biradical (or zwitterion):

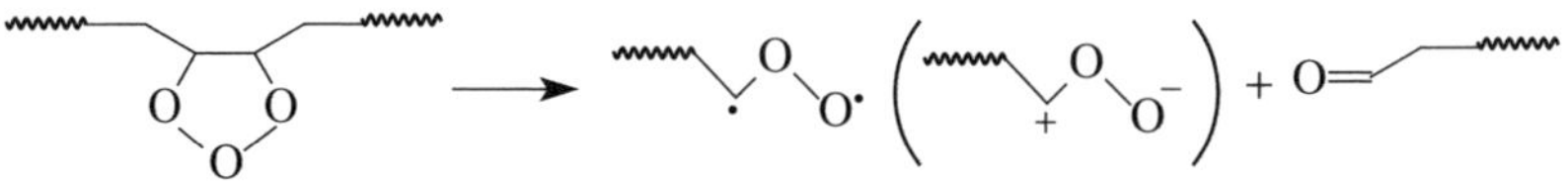

Such zwitterionic species then react to form either tetraoxanes or branched peroxides:

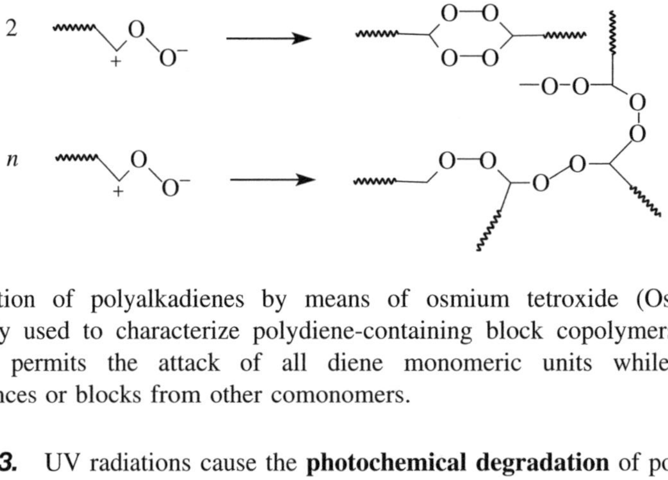

Oxidation of polyalkadienes by means of osmium tetroxide (OsO₄) is frequently used to characterize polydiene-containing block copolymers; this technique permits the attack of all diene monomeric units while preserving sequences or blocks from other comonomers.

9.4.2.3.   UV radiations cause the **photochemical degradation** of polymers containing chromophoric groups that absorb in the range of 200–290 nm. Such groups can be generated by oxidation ($-O-O-$, $R_1R_2C=O$, etc.), thermal degradation (giving, for example, $\sim\sim\sim\sim CH=CH-CH=CH-CH\sim\sim\sim\sim$ in the case of PVC), or the presence of various additives (phenols, transition metals, etc.), but also through copolymerization with photosensitive monomers. For instance, the copolymerization of ethylene with carbon monoxide affords a photodegradable polymer that is produced for this purpose:

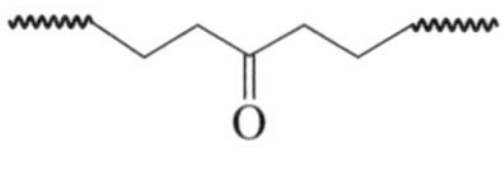

If Ch represents the chromophoric group, the following reactions can occur, where the exponents 1, 2, and 3 denote the singlet, doublet, and triplet states and the asterisk denote an excited state.

$$^1\text{Ch} + h\nu \longrightarrow {}^1\text{Ch}^* \tag{9.1}$$

$$^1\text{Ch}^* + h\nu \longrightarrow {}^1\text{Ch}^{**} \tag{9.2}$$

$$^1\text{Ch} + h\nu \longrightarrow {}^2\text{Ch}^* + e^- \tag{9.3}$$

$$^1\text{Ch}^* \longrightarrow {}^1\text{Ch}^* + \text{energy} \tag{9.4}$$

$$^1\text{Ch}^* \longrightarrow {}^3\text{Ch}^* + \text{energy} \tag{9.5}$$

$$^1\text{Ch}^{**} \longrightarrow {}^1\text{Ch}^* + \text{energy} \tag{9.6}$$

$$^1\text{Ch}^* \longrightarrow {}^1\text{Ch} + h\nu \tag{9.7}$$

$$^3\text{Ch}^* \longrightarrow {}^1\text{Ch} + h\nu \tag{9.8}$$

Equations (9.1) and (9.2) correspond to the absorption photons by the chromophore, (9.3) to its photoionisation, (9.4), (9.5) and (9.6) to the emission of a non-radiative energy, and finally (9.7) and (9.8) to a radiative emission.

In the case of ethylene/carbon monoxide copolymer, the degradation occurs generally with the triplet state which exhibits the maximum lifetime and corresponds to a free-radical carried by the carbonyl group. This radical evolves to give either

- an α-scission

(with all the subsequent reactions that implies) or
- a β-scission

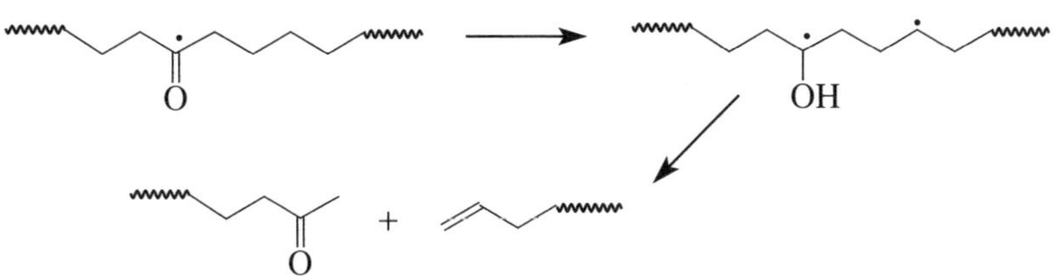

9.4.2.4. As for any organic molecule, the sensitivity to **thermal degradation** of a polymer is in close relationship with the energy of its bonds. Polymers are, however, less stable thermally than its homologous simple molecule. This is due to the disordered motion of the chains above the glass transition temperature and the energy associated with it, which can be concentrated on a particular bond of the macromolecular backbone.

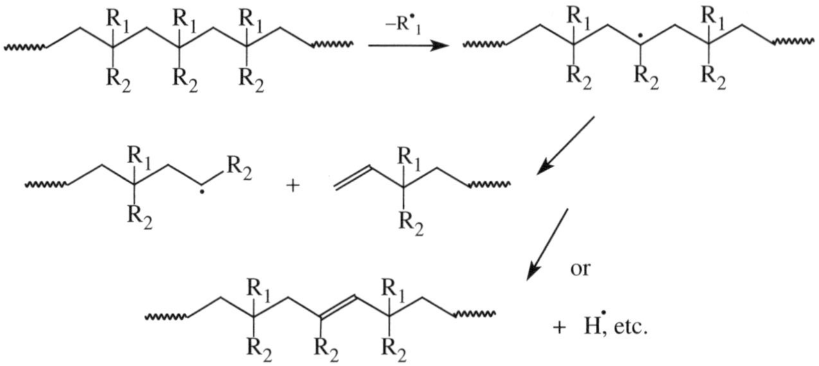

This additional energy supply applies only to monomer units that are sufficiently far away from chain ends and does not affect significantly the degradation of side groups. The latter can, however, cause the degradation of the main chain, when the bond linking the main chain to its substituents is weak, the process then corresponds to a chain reaction.

Hydrocarbon chains can also undergo a homolytic rupture of their carbon–carbon bonds that are particularly weakened by *head-to-head* sterically hindered irregular sequences:

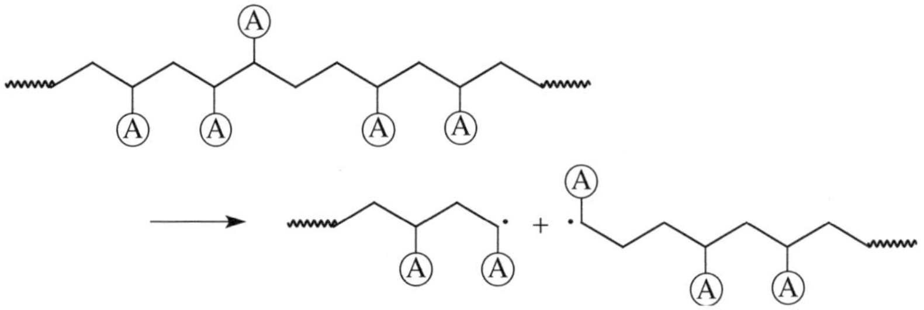

Two situations can occur, depending upon the temperature applied with respect to the polymer ceiling temperature (see Section 8.2.1). When this temperature is below its ceiling temperature, the free radicals generated undergo a rearrangement with stabilization of the species formed until disappearing by combination or disproportionation:

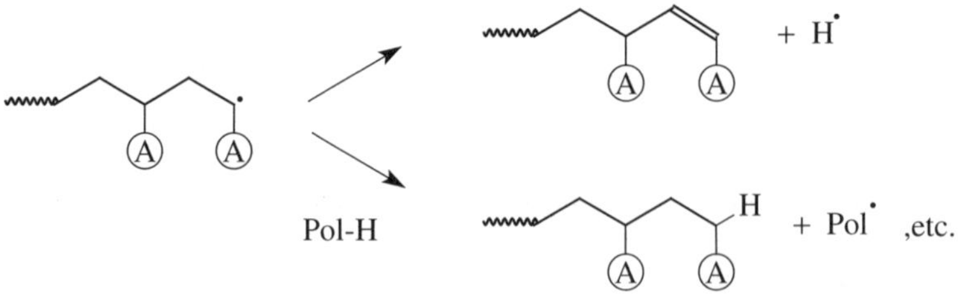

With chains containing hetero-elements in their backbone, each polymer is a particular case. For example, in the case of cellulose, the following degradation is observed above 180°C:

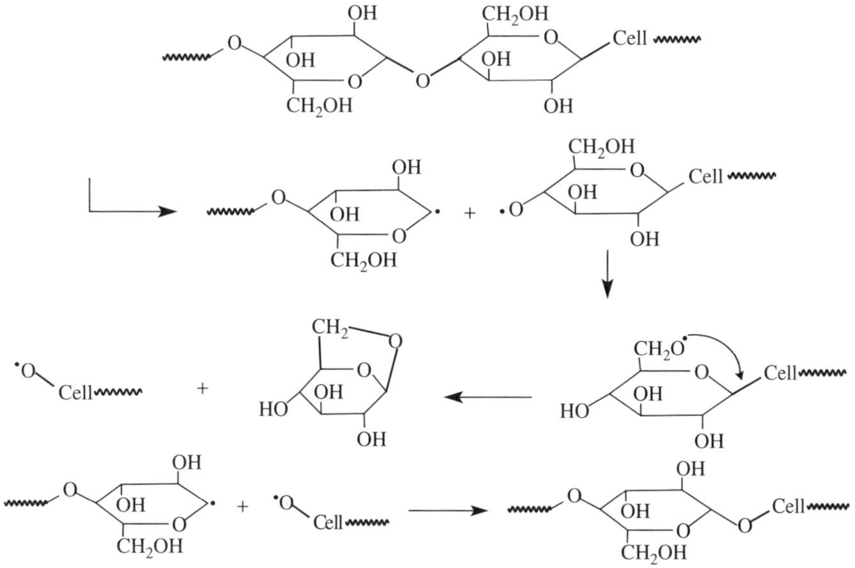

When the temperature applied approaches or exceeds the ceiling temperature, a homolytic rupture occurs that may result in the total depolymerization and regeneration of the monomer for several polymers. Methyl methacrylate is recovered in this way from waste of the corresponding polymer:

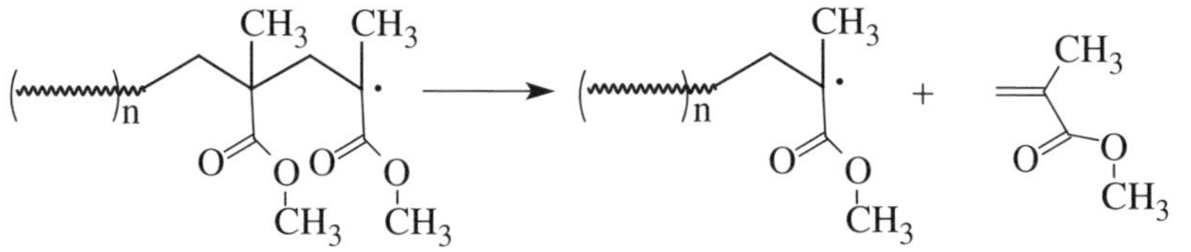

To get a better insight into the mechanism/type of degradation, thermogravimetric analysis is the most appropriate technique that is generally used in association with a technique of determination of molar masses. Figure 9.2 shows two typical behaviors.

From the knowledge of structural parameters that determine the polymer thermostability, the structure of the ideal thermostable polymer can be designed as follows:

- The interatomic chemical bonds should be strong.
- The chains should be rigid and generate strong molecular interactions in order to exhibit little mobility.

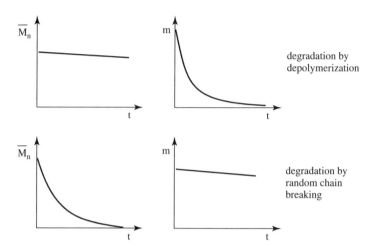

Figure 9.2. Variations of both molar mass and mass of polymer samples with time during degradation. (1) depolymerization; (2) random chain breaking.

For example, the structure of the polyimide that is shown below meets these criteria; it is approximately thermostable up to $450°C$:

9.4.2.5. Polymers are sensitive to **mechanical degradation** that occur through homolytic scissions, similarly to those caused by thermal degradation. The mastication of certain polymers in the molten state, in particular of natural rubber before addition of stabilizing additives, may cause a drastic reduction of their molar masses; in the case of polypropene, thermal and mechanical degradations both combine their effects to degrade it. The mechanical energy can also be provided by an ultrasonic generator.

9.4.2.6. When several sources of energy are combined to cause degradation, their synergism can be sometimes spectacular. For instance, degradations by **photo-oxidation** and **thermal oxidation** are particularly effective—in particular, in the degradation of polyethylene films.

It is frequent that additives that are incorporated in polymers to either generate or improve a given property impart an accelerating effect on the degradation process. The corresponding mechanisms are often complex but are not different from the basic phenomena described above.

9.5. STABILIZATION OF POLYMERS

The side effects due to degradation can be alleviated through a precise knowledge of the mechanism involved. For instance, the autocatalytic thermal dehydrochlorination of poly(vinyl chloride) and poly(vinylidene chloride) suggests that basic additives can well stabilize them and prevent their degradation by neutralizing the HCl gradually formed.

Bases utilized for this purpose can be either organic molecules (N,N'-diphenylurea, dihydropyridin, polyols, etc.) or salts or metallic oxides such as $3PbO \cdot PbSO_4 \cdot H_2O$ as well as barium, cadmium, calcium, zinc, lead carboxylates, or thiolates.

Instead of preventing the dehydrochlorination process, which generates colored conjugated sequences, it may be more appropriate to use additives that react with the chromophoric polyene formed and reduce the length of the conjugated sequences. A hypsochromic effect is observed in this case, which decreases the absorption in the visible range.

When the degradation of a polymer gives rise to free radicals, the addition of compounds that can trap and neutralize these radicals is the logical solution. For example, polyolefins are stabilized by addition of substituted phenols or polyphenols, which act as antioxidants:

The phenoxy radical formed are too stable to propagate the reaction of degradation. Carbon black is also an excellent antioxidant that is commonly used when its coloration is not a drawback for the application contemplated. It is systematically used to stabilize polyalkadienes as it contributes to the reinforcement of their mechanical properties in addition to its capacity to stabilize against heat and oxidation.

There are two categories of products that can protect against photodegradation:

- The first corresponds to UV radiation absorbers that possess a molecular structure enabling them to absorb sunlight up to $\lambda = 360$ nm. These compounds are thus used as screen that absorb photons and dissipate thermally the corresponding energy; derivatives of benzophenone are often used in this case. When the application contemplated permits, carbon black can also be utilized as a "total screen."

- The second category is that of free radicals traps. For example, hindered amines are excellent light-stabilizing compounds: their oxidation generates stable free radicals called nitroxide that are able to efficiently trap reactive free radicals. Very recently, such nitroxide radicals were used to prevent a polymer from degrading by a free radical mechanism as shown below:

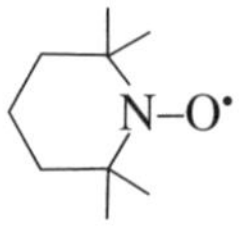

N–O˙ Tetramethylpiperidyloxyl (TEMPO)

~~~~~~pol˙ + TEMPO˙ ⟶ ~~~~~~pol-TEMPO

All the additives used should exhibit a very high compatibility with the polymer to stabilize in order to prevent their migration to the surface and their subsequent elimination; the durability of their effect depends on this factor.

LITERATURE

E. Maréchal, *Chemical Modification of Synthetic Polymers*, in *Comprehensive Polymer Science*, G. C. Eastmond, A. Ledwith, S. Russo, and P. Sigwalt. Pergamon Press., Oxford, p. 1, 1989.

J. C. Arthur, Jr., *Chemical Modification of Cellulose and its Derivatives* in *Comprehensive Polymer Science*, G. C. Eastmond, A. Ledwith, S. Russo, and P. Sigwalt (Eds.), Pergamon Press, Oxford, p. 49, 1989.

M. Lazár, T. Bleha, and J. Rychlý, *Chemical Reactions of Natural and Synthetic Polymers*, Ellis Horwood Ltd., Chichester, 1989.

N. A. Platé, A. D. Limanovich, and O. V. Noah, *Macromolecular Reactions*, Wiley, Chichester, 1995.

10

MACROMOLECULAR SYNTHESIS

10.1. INTRODUCTION

Chemical methods that are designed to afford polymers of controlled architecture/structure, corresponding to the expectation of the experimenter, are referred to as macromolecular synthesis. The whole set of these methods is also called *macromolecular engineering*, and in many aspects this domain of polymer chemistry is close to that of polymerization reactions and/or that of polymer chemical modification (see Chapters 7, 8, and 9).

The variety of potentially accessible macromolecular structures/architectures is endless, and only the most representative ones are described here. Recent developments in the field of "living" and/or "controlled" polymerizations further expand the possibilities of macromolecular engineering.

Arbitrarily, this chapter is divided into three parts devoted respectively to the synthesis of the following basic structures/architectures:

- End-functionalized polymers [i.e., α- and (α,ω-di) functionalized polymers], including macromonomers
- Block and graft copolymers
- Polymers with complex topology.

Organic and Physical Chemistry of Polymers, by Yves Gnanou and Michel Fontanille
Copyright © 2008 John Wiley & Sons, Inc.

10.2. END-FUNCTIONALIZATION OF POLYMER CHAINS (SYNTHESIS OF REACTIVE PRECURSORS)

Remark. It is important to point out the difference between *functionalized polymers*, which are polymers carrying functional groups, and *functional polymers*, which are those exhibiting a specific property and used for a particular application or a given function. Functionalized polymers may also be functional polymers.

When an accurate control of the structure targeted is not necessary, it is relatively easy to obtain functionalized chain ends. Widely used in industry, two methods of facile and straightforward functionalization are briefly described below.

The simplest method to obtain α,ω-difunctionalized polymers is to resort to step-growth polymerization and control the degree of polymerization of the formed polycondensates by means of the stoichiometry (r) of the initial reactants:

$$a\ \text{X--}\mathcal{A}\text{--X}\ +\ b\ \text{Y--}\mathcal{B}\text{--Y}\ \underset{a<b}{\overset{}{\rightleftharpoons}}\ 2a\ \text{XY}\ +\ \text{Y--}\mathcal{B}\text{--}(\mathcal{A}\text{--}\mathcal{B})_n\text{--Y}$$

The Carothers relation (Section 7.2.1) for nonstoichiometric conditions gives

$$\overline{X}_n = \frac{1+r}{r(1-2p)+1}$$

where $r = a/b$ is the stoichiometric imbalance and p is the extent of the reaction.

Alternatively, end-functionalization of growing chains can also be obtained *via* transfer in chain addition polymerization. For instance, dihydroxy polybutadiene telechelics (*i.e.* carrying one hydroxyl group at each chain-end) are industrially produced by free radical polymerization of butadiene initiated by hydrogen peroxide which simultaneously gives rise to transfer reactions:

$$\text{H}_2\text{O}_2\ \longrightarrow\ 2\ \text{HO}\cdot$$

$$\text{HO}^{\bullet}\ +\ n\ \diagup\hspace{-0.5em}=\hspace{-0.5em}\diagdown\ \longrightarrow\ \text{HO}\text{\~{}}\text{polybutadiene}\text{\~{}\~{}}^{\bullet}$$

$$\text{HO}\text{\~{}\~{}}\text{polybutadiene}\text{\~{}}\text{OH}\ +\ \text{HO}^{\bullet}\ \overset{\text{H}_2\text{O}_2}{\longleftarrow}$$

$$\text{etc.}\ \overset{n\ \diagup\hspace{-0.4em}=\hspace{-0.4em}\diagdown}{\longleftarrow}$$

As a matter of fact, the average functionality of polymers prepared under these conditions is slightly higher than 2, due to transfer reactions to polymer chains.

The "telomerization" of vinyl monomers is another example of chain end-functionalization by transfer. In the latter case, both propagation and transfer exhibit comparable rates, which affords α,ω-difunctionalized chains of low degree of polymerization also called *telomers*. For instance, the free radical polymerization of vinyl monomers in halogenated solvents produces α,ω-halogenated oligomers:

Experimental conditions are generally selected in such a way that n and n' are limited to few units. Some of these telomers can subsequently serve as precursors for polycondensation reactions.

To obtain better defined structures than the latter ones, it is advisable to rely on "living" and/or "controlled" polymerizations for the production of functionalized chains: the end-standing functions can be introduced at the initiation step upon selection of an appropriate initiator or at the end of the polymerization through a deactivating molecule carrying the desired functional group.

10.2.1. Functionalization Through Initiator

Due to the high reactivity of the propagating active centers, it is often neces-sary to protect the functional group carried by the initiator. For instance, hydroxyl functions in initiators for anionic polymerization require protection, and acetal func-tions are generally used to this end. Such acetal functional groups are introduced at the ends of polystyrene chains by means of an acetal-containing alkyllithium initiator that triggers "living" anionic polymerization; upon coupling, such "living" carbanionic polystyrene with dimethyldichlorosilane, α,ω-bisacetal chains can be indeed generated:

α,ω-Dihydroxyl chains, as well as other end functions (−C−I), can be subsequently obtained by chemical modification of these acetal functional groups.

By a similar process, macromonomers—that are chains carrying a polymerizable group at one of their ends—can be synthesized. The homopolymerization of such macromonomers affords "comb-like" polymers and their statistical copolymerization with simple monomers graft copolymers whose branches have all roughly the same size. The preparation of α-norbornenylpolystyrene is illustrated below:

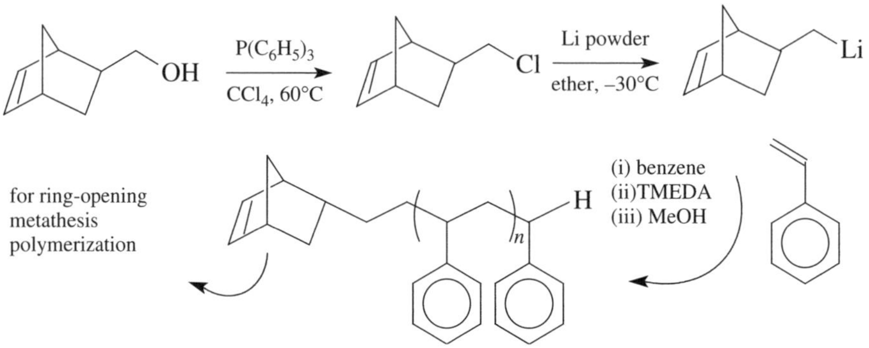

10.2.2. Functionalization by Deactivation of "Living" Chain Ends

This method is generally preferred to the preceding one because it allows one to synthesize end-functionalized chains with a variety of functional groups. One of the possible pathways is to react a "living" chain with a bifunctional deactivating molecule used in large excess to minimize the coupling of two growing chains. For instance:

$$\text{wwww PSLi} + \text{COCl}_2 \text{ (in excess)} \longrightarrow \text{wwww PS–COCl} + \text{LiCl}$$

However, the reaction of growing active centers with a deactivating molecule, serving as precursor of the functional group to be introduced, is the most commonly used method. The preparation of ω-hydroxy polybutadiene by deactivation of "living" polybutadiene carbanions by ethylene oxide is an illustration of this strategy:

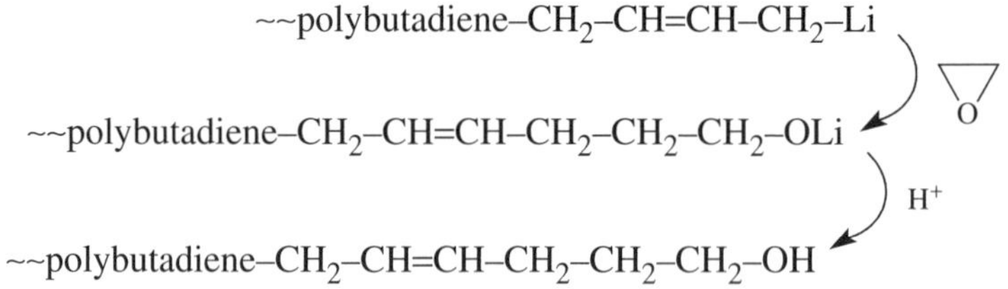

Macromonomers can be obtained by similar means, as illustrated below with the example of ω-methacryloyl poly(ethylene oxide) macromonomer:

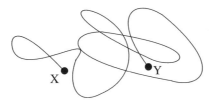

To obtain telechelic polymers—that are, those carrying a same reactive molecular group at each of their ends—the most direct method is to use bifunctional initiators. Examples of such initiators can be found for all "living" chain addition polymerizations; the example shown below illustrates the bifunctional initiation of "living" cationic polymerization of vinyl ethers:

Other examples of bifunctional initiators are described in Section 10.3.1, which is devoted to the synthesis of block copolymers.

> **Remark.** A distinction is generally made between ω-functionalized chains and α,ω-difunctionalized chains on the one hand and macromonomers on the other: the first are precursors that are used in polycondensation reactions, whereas the second serve as precursors for comb-type polymers and graft copolymers.

10.2.3. Hetero-functionalization

Hetero-telechelic polymers carry a different functional group at each of their end (X at one, Y at the other); these functional groups can possibly be antagonist and react with each other and bring about step-growth polymerization. Such polymers can be represented as

It is necessary to carry out each functionalization at different times of the synthesis to obtain such hetero-difunctionalized polymers—that is, for example, at the initiation step and then at the end of the polymerization—for the synthesis of α-amino,ω-hydroxy polycaprolactone:

$$n \left(\begin{array}{c} CO{-}O \\ (CH_2)_5 \end{array} \right)$$

Br–(CH$_2$)$_{12}$–O–AlEt$_2$ $\longrightarrow$ Br–(CH$_2$)$_{12}$–O–[CO–(CH$_2$)$_5$–O]$_n$–H

(i) NaN$_3$ | (ii) HCO–O$^-$, NH$_4^+$/ Pd (10%)

H$_2$N–(CH$_2$)$_{12}$–O–[CO–(CH$_2$)$_5$–O]$_n$–H

Whatever the nature of the end functional groups, these α- or α,ω-difunctionalized polymers are difficult to characterize by conventional methods of analysis due to the very low concentration of the said functional groups. The use of techniques of high sensitivity and precision such as mass spectrometry (MALDI-TOF) is helpful with respect to the identification of these functional groups.

10.3. BLOCK AND GRAFT COPOLYMERS

As shown in the chapter devoted to copolymerization, the distribution of monomer units in statistical copolymers is closely dictated by the concentration of the copolymerizing comonomers in the reaction medium and their reactivity ratios. Initially set by the experimenter, the relative concentrations of comonomers can be subject to an uneven variation throughout polymerization, depending upon the monomers'reactivity ratios, with the latter being primarily determined by the type of chain polymerization used. To obtain statistical copolymer chains of fairly constant composition, one generally resorts to continuous processes in which the composition of the reaction medium remains roughly constant throughout polymerization. Being detailed in Chapter 8, the synthesis of statistical copolymers will not be developed further here.

10.3.1. Synthesis of Block Copolymers

Owing to their propensity to self-organize in mesophases, block copolymers exhibit specific features and properties that are exploited at industrial level. The morphology of these mesophases depends primarily, among other parameters, on the nature of the comonomers and the relative length of the blocks, on the dispersion of molar masses, on the possible presence of residual homopolymers, and on the overall architecture of the copolymer (including more than two blocks, star block copolymers, etc.). It is thus essential to precisely control these structural parameters, and "living" and/or "controlled" polymerizations are particularly suitable for their synthesis.

There are three main synthetic strategies for the preparation of block copolymers:

- Sequential "living" polymerization of comonomers
- Polymerization of monomer B initiated by a macroinitiator polyA*
- Coupling of two polymer precursors through a covalent bond

10.3.1.1. Sequential "Living" and/or "Controlled" Polymerization of Two Monomers. The principle is schematically shown hereafter:

$$A^* \ + \ n_1M_1 \ \longrightarrow \ A\text{-}(M_1)^*_{n_1} \ \xrightarrow{\ n_2M_2\ } \ A\text{-}(M_1)_{n_1}\text{-}(M_2)^*_{n_2}$$

This method is industrially utilized to produce styrene–butadiene diblock copolymers (SB) by anionic polymerization. The corresponding triblock copolymer (SBS), which is a thermoplastic elastomer, is obtained by coupling the precedent "living" diblock copolymer using dimethyldichlorosilane:

$$\sim\!\sim\!\sim\!\sim\!\sim\text{-}S^-,Li^+ \ + \ nB \ \longrightarrow \ \sim\!\sim\!\sim\!\sim\!\sim\text{-}S\,(B)^-_n,\,Li^+$$

then

$$2\sim\!\sim\!\sim\!\sim\!\sim\text{-}S\,(B)^-_n,\,Li^+ \ + \ (CH_3)_2SiCl_2$$

$$\underbrace{\sim\!\sim\!\sim\!\sim\!\sim\text{-}S\text{-}(B)_n}_{\text{PS block}}\underset{\overset{|}{CH_3}}{\overset{\overset{CH_3}{|}}{Si}}\underbrace{\text{-}(B)_n\,S\sim\!\sim\!\sim\!\sim\!\sim}_{\text{PS block}}$$
$$\underbrace{\qquad\qquad\qquad}_{\text{Polybutadiene block}}$$

Remark. It has to be stressed that in the syntheses of block copolymers by anionic means, the first monomer (A) to be polymerized should not exhibit an electro-affinity higher than that of the second monomer (B) (see Section 8.6).

Triblock copolymers can be obtained in two steps from bifunctional initiators, as for the preparation of telechelic homopolymers. Such initiators are useful in the preparation of polystyrene-*block*-polydiene-*block*-polystyrene, which turns out to be excellent thermoplastic elastomers.

Such triblock copolymers can be derived by anionic polymerization using dilithium initiators. For the central polydiene block to exhibit a high content in 1,4-*cis* units, it is essential that the polymerization be carried out in hydrocarbon solvents (apolar medium) and in absence of any polar additive. Such dilithium initiators are generated from the reaction of butyllithium with an adequate precursor as shown below:

or

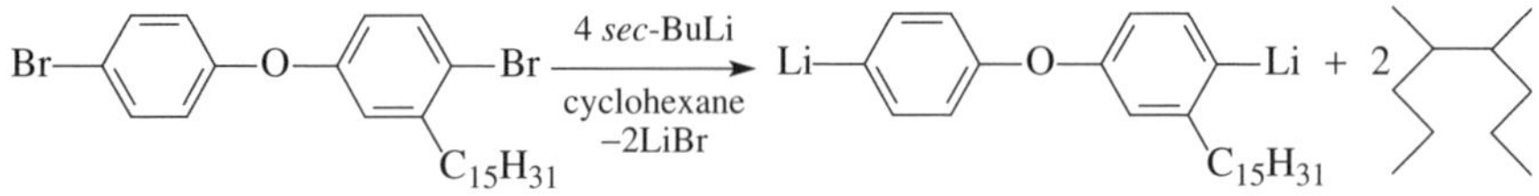

10.3.1.2. Polymerization of Monomer B Initiated by a Macroinitiator

PolyA*. This method differs from the preceding one by the absence of continuity in the growth of the two blocks. In this case the cross-over from polyA* to polyB* is obtained, after isolating poly A*, which is subsequently used as initiator for the polymerization of B. This method is useful whenever the two comonomers A and B are prone to polymerize by two different mechanisms; the active centers that are responsible for the polymerization of monomer (A) have to be transformed into reactive species that are appropriate for the initiation of the polymerization of monomer (B). For instance, the transformation of polystyryl carbanionic active centers into cationic sites was exploited to polymerize tetrahydrofuran and obtain PS-*block*-PTHF diblock copolymers as shown below:

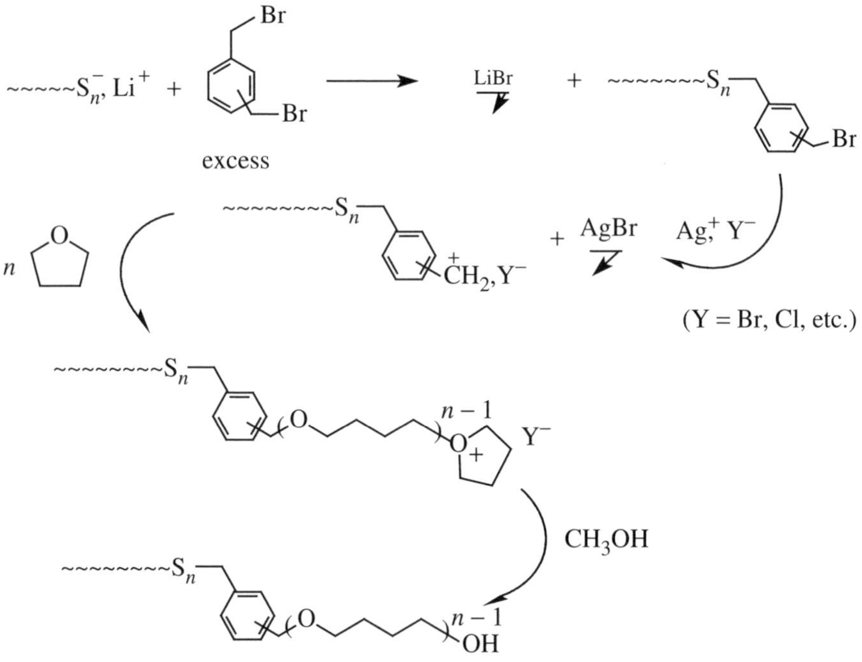

Another well-known example pertains to polysiloxane-*block*-polyamide diblock copolymers obtained by anionic copolymerization of ε-caprolactam from a poly-dimethylsiloxane (PDMS) macroinitiator; the latter is prepared by hydrosilylation of the unsaturated moiety carried by an acyllactam, using a α-hydrogenated PDMS:

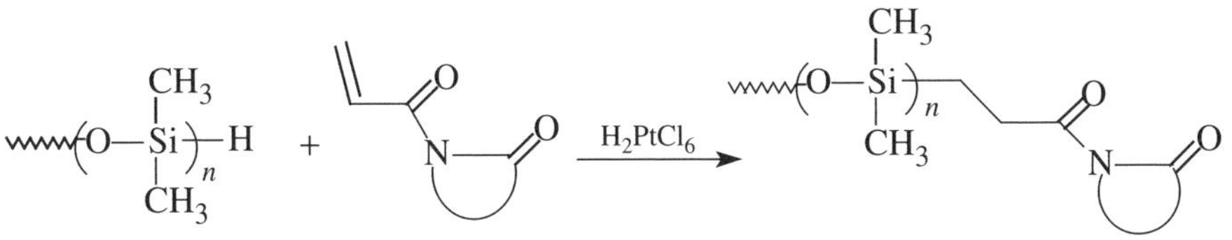

The polyamide (PA-6) block is grown by polymerization of ε-caprolactam in the presence of $NaAlH_2Et_2$ as catalyst through the so-called "activated monomer" mechanism:

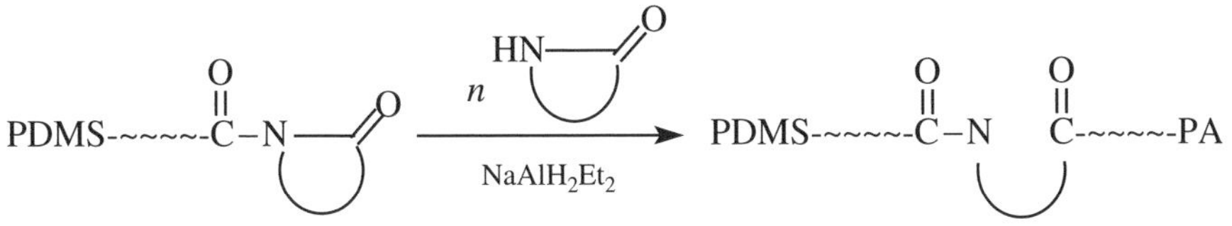

10.3.1.3. Covalent Coupling of Two Polymeric Precursors. This method requires selective, fast and complete coupling reaction to give satisfactory results. Due to their incompatibility, polymer chains of different nature tend to minimize their contacts and therefore the collisions between their antagonist reactive sites hardly occur. An example of this method is given by the synthesis of styrene-dimethylsiloxane block copolymers:

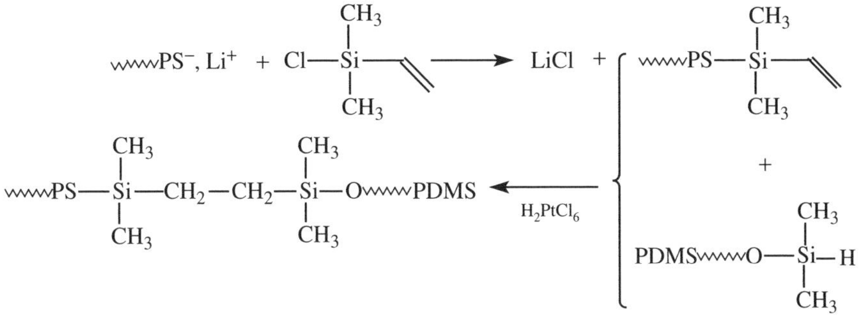

The synthesis of SBS triblock copolymer obtained by coupling of two "living" SB block copolymers by means of $Cl_2Si(CH_3)_2$ is another well-known example.

10.3.2. Synthesis of Graft Copolymers

Like block copolymers, graft copolymers also form mesophases when their backbone and grafts are incompatible; they are found in the same domains of applications as block copolymers, but they are generally easier to synthesize and are thus widely used.

Three general methods could be followed for the preparation of graft copolymers:

- Initiation of polymerization from a main chain carrying appropriate reactive sites (grafting from),
- Grafting of preformed chains onto the main chain (grafting onto),
- (Co)polymerization of macromonomers.

10.3.2.1. "Grafting from" Method. This method consists of taking advantage of the presence of reactive sites on a polyA backbone to initiate the polymerization of a monomer B and generate grafts polyB:

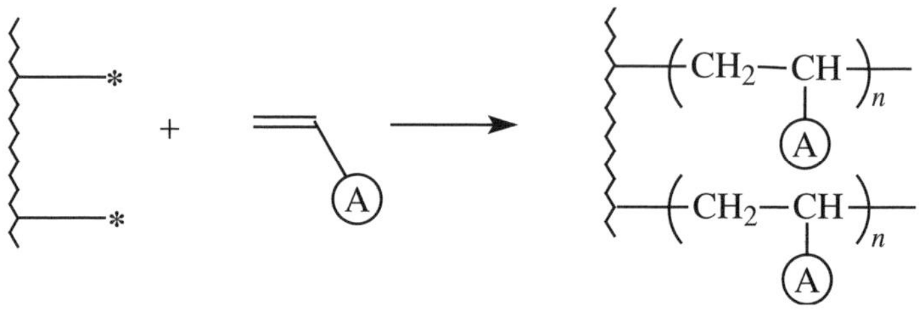

The example described below refers to nitroxide-mediated free radical polymerizations. Initially, a methacrylic monomer carrying an alkoxyamine group is copolymerized at moderate temperature with a comonomer (H$_2$C=CHA) to form the backbone. In a second step the grafts are grown by thermal activation of alkoxyamine groups, which produce, in addition to stable nitroxyl free radicals, reactive free radicals capable of initiating the polymerization of H$_2$C=CHB (see Section 8.5.11 for the mechanism).

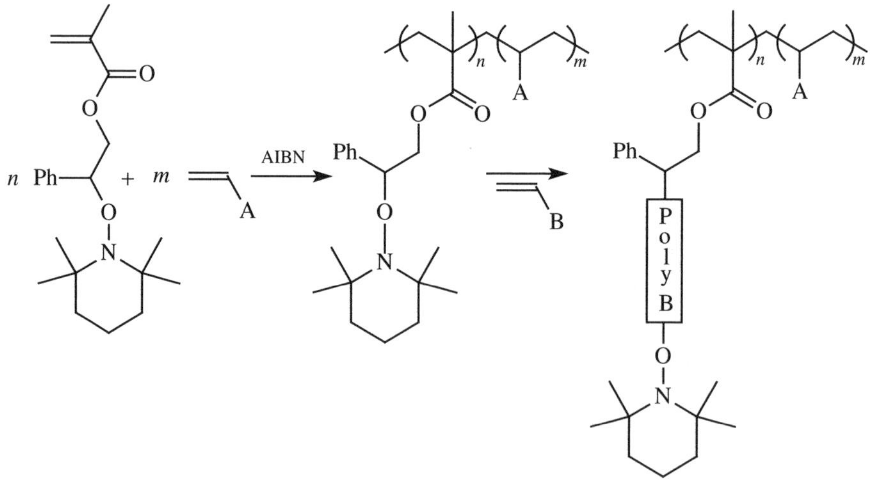

Active centers can also be generated on the backbone by chemical modification of its monomeric units:

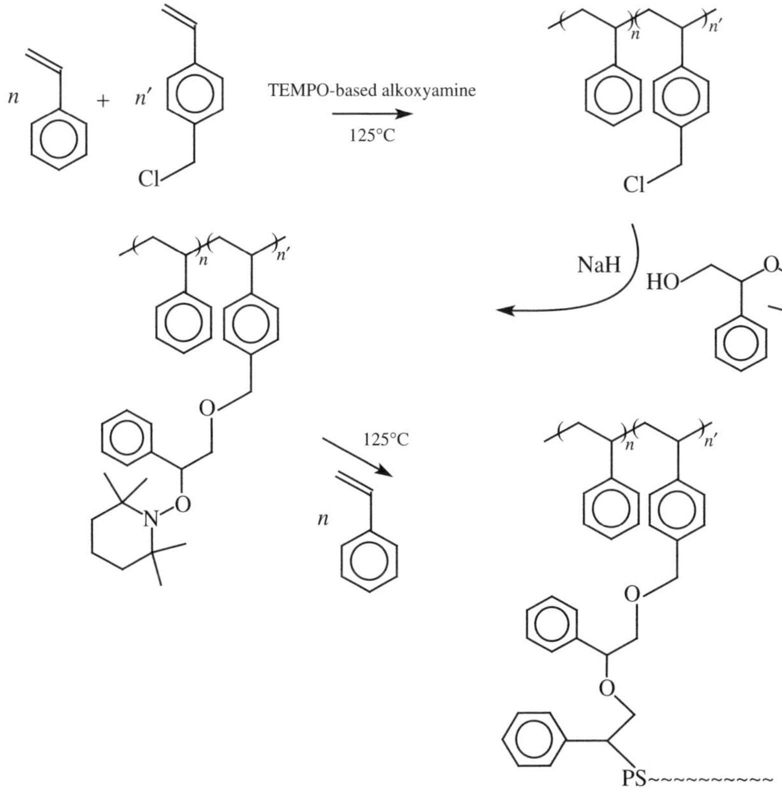

10.3.2.2. Grafting of Preformed Chains onto a Polymer Backbone.

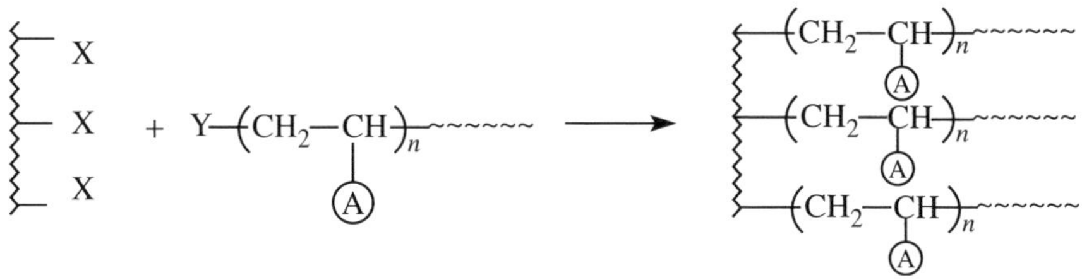

X and Y are antagonist functions whose mutual reaction brings about "grafting." The deactivation of "living" polystyrene chains prepared by anionic means on a PMMA backbone is a well-known example of the "grafting onto" method:

The proportion of grafts introduced in such graft copolymers is determined by the $[PS^-,Li^+]/[MMA]$ ratio, and the reaction is fast and total. Many other examples of graft copolymers have been described following the same principle.

10.3.2.3. Copolymerization of Macromonomers. "Comb-type" polymers can be obtained by homopolymerization of a macromonomer (see Section 10.2.1). In the latter case, all monomeric units of such a comb-type polymer carry a graft; in such structures the proportion of main-chain-forming monomers is very low. Moreover, main-chain crowding due to the grafts becomes even more rigid since it is long. Graft copolymers with varying graft density can also be obtained by copolymerization of a macromonomer with a conventional comonomer.

10.4. POLYMERS WITH COMPLEX TOPOLOGY

Interest in complex structures/architectures arises from the observation that the macroscopic properties of a material is largely affected by the chains'architecture and, in particular, by the presence and the location of branching points. It is not the aim of this section to give a comprehensive account of the synthesis of all possible architectures/topologies, but rather to describe the most important ones with respect to their applicability and their preparation by traditional methods of macromolecular synthesis.

10.4.1. Macrocycles

Mono- and polycyclic structures of macromolecular size are characterized by the absence of any chain end. Because the entropy of such architectures is lower than that of their linear counterparts, they exhibit specific conformational properties in solution as well as in the molten state.

The method generally utilized to obtain macrocyclic polymers is to react α,ω-difunctionalized prepolymers with a bivalent coupling reagent; it is a bimolecular coupling which has to be carried out in highly dilute solutions so as to favor the formation of macrocyclics at the expense of linear polycondensates. Its principle is schematically represented below:

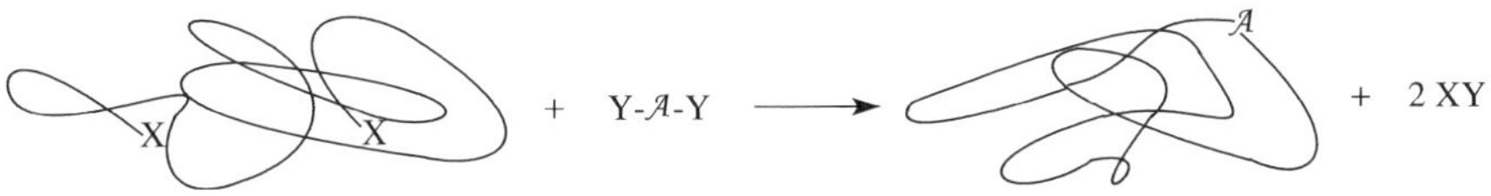

For instance, macrocyclic polybutadienes could be derived from living α,ω-dicarbanionic polybutadienes, which were reacted with 1,3-bis-(1-phenyl-ethenyl) benzene. As shown below, the coupling agent used here is also a precursor to a bifunctional initiator and the addition of ethylene oxide after cyclization permits the difunctionalization of the macrocyclics formed:

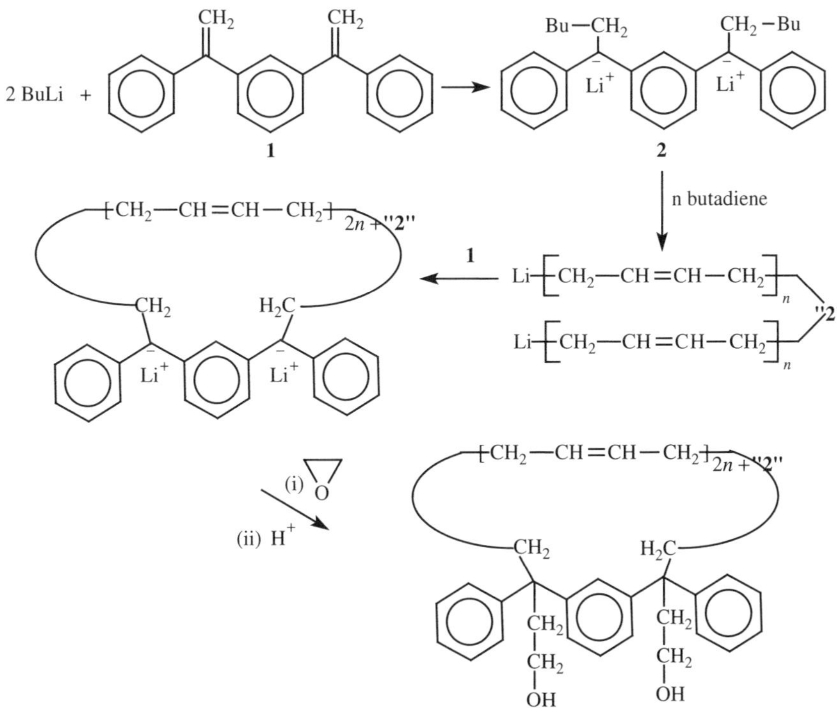

Whatever the experimental conditions utilized, a significant fraction of linear polymers resulting from intermolecular additions is always formed.

Monomolecular coupling affords higher yields in macrocyclics. The principle of such syntheses rests on the reaction between the chain ends of linear polymers carrying antagonist reactive sites. As shown in the example represented below, a vinyl ether is first polymerized cationically under "living" conditions, with the active centers formed being unable to react with the styrene-like double bond of the initial molecule. In a second step, upon addition of a strong Lewis acid, carbocations are created at the growing ends that can react with the styrenic unsaturation. Due to the high dilution of the medium, an instantaneous intramolecular cyclization occurs.

Following the same principle, it is possible to obtain macrotricyclics or eight-shaped bicyclics, and so on.

10.4.2. Star Polymers

Comb-like polymers (see Section 10.3.2), whose backbone is of small degree of polymerization, tend to adopt a conformation that reminds that of star polymers; as the degree of polymerization of their backbone increases, their conformation undergoes a transition toward a worm-like type. Star polymers including branches (if not isometric) of controlled size can be obtained through two methods.

10.4.2.1. "Convergent" Method. This method involves the coupling of mono-functional linear chains to a core agent fitted with antagonist functions. The number of the latter determines that of branches in the resulting stars. For example, four-arm polystyrene stars were obtained by reaction of polystyryllithium chains with a tetravalent molecule such as $SiCl_4$:

It is recommended to use a slight excess of monofunctional linear chains to obtain the expected four-arm stars in a quantitative yield.

The convergent method is particularly appropriate for anionic polymerization as plurifunctional electrophilic deactivators are available in large number.

10.4.2.2. "Divergent" Method. The divergent route involves the growth of the star arms -or branches- from a core-initiator whose functionality determines the resulting number of branches. Many examples can be found in the literature, all based on "living "and/or "controlled" polymerizations.

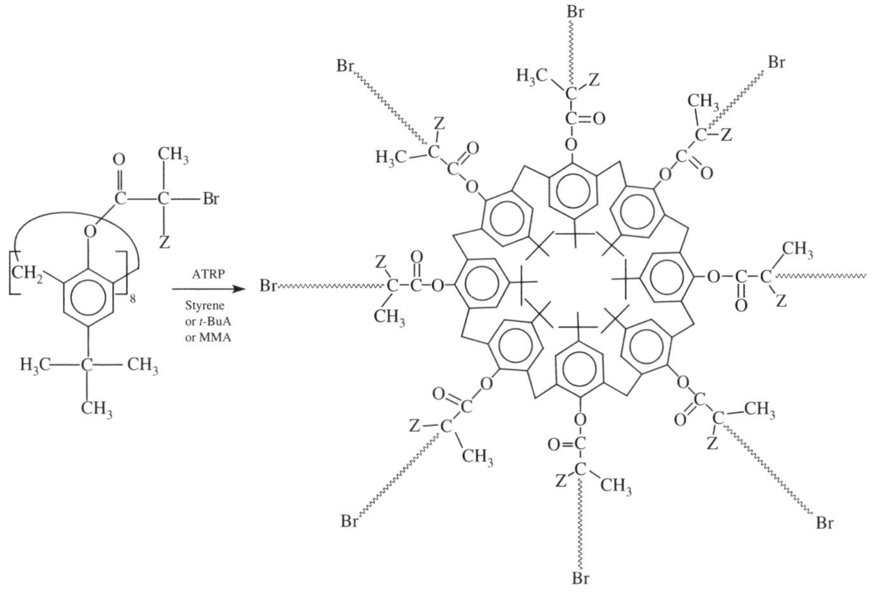

The example shown above illustrates the synthesis of eight-arm stars by Atom Transfer Radical Polymerization (ATRP). The initiator used derives from calixarenes, which are cyclic molecules whose degree of polymerization (4, 6, or 8) determines the resulting number of branches.

10.4.2.3. Combination of "Convergent" and "Divergent" Methods. Contrary to the stars described previously, those resulting from the combination of convergent and divergent methods are characterized by a large fluctuation of their functionality. Their core is obtained by copolymerization of a monofunctional linear chain with a diunsaturated monomer.

This method was used to prepare stars with branches of two different types *(miktoarm copolymers)*. For instance, the reaction of "living" carbanionic polystyrene chains (PS$^-$, K$^+$) with *p*-divinylbenzene (DVB = tetravalent molecule) gives rise to the formation of stars whose polyDVB core is fitted with a large number of carbanionic sites; the number of branches depends roughly on the [DVB]/[PS$^-$, K$^+$] ratio, but it cannot be controlled with full precision. It is possible to take advantage of the carbanionic species present at the core to initiate the polymerization of a different monomer—for instance, ethylene oxide:

$$n \ \text{wwwwPSwwww}^-, K^+ \ + \ x \ \text{(styrene)} \longrightarrow$$

10.4.3. Dendrimers

Dendrimers comprise a central core, a precise number of monomer units linked one to another by branching points, and an exact number of outer functional groups. Dendrimers are thus isometric objects with supposedly no fluctuation of their size or composition contrary to polymers produced by any conventional chain polymerization even "living" ones. With the samples of highest generations, it is difficult to precisely check the quality of the structure formed except for those obtained by the convergent method (cf. following section).

The synthesis of dendrimers is based on condensation reactions and requires generally successive protection-deprotection steps. After the growth of each generation, branching points are introduced at the dendrimer ends to multiply by a factor generally equal to 2 the number of outer functions from which the next generation will be grown. The molecular objects thus generated exhibit very a peculiar physicochemical behavior. Owing to their high number of branching points, dendrimers can be viewed as dense spherical objects whose interpenetration is hindered and viscosity particularly low.

Like the case of stars, there are two main methods of synthesis of dendrimers.

10.4.3.1. "Divergent" Method. This method consists of generating two identical functional groups from an initial one and then repeating the same operation several times. In the example shown below, four amino groups are first generated from a primary diamine, and the same sequence of reactions is repeated as many times as possible, the only limit being the space available to the functional groups of the generation to be grown:

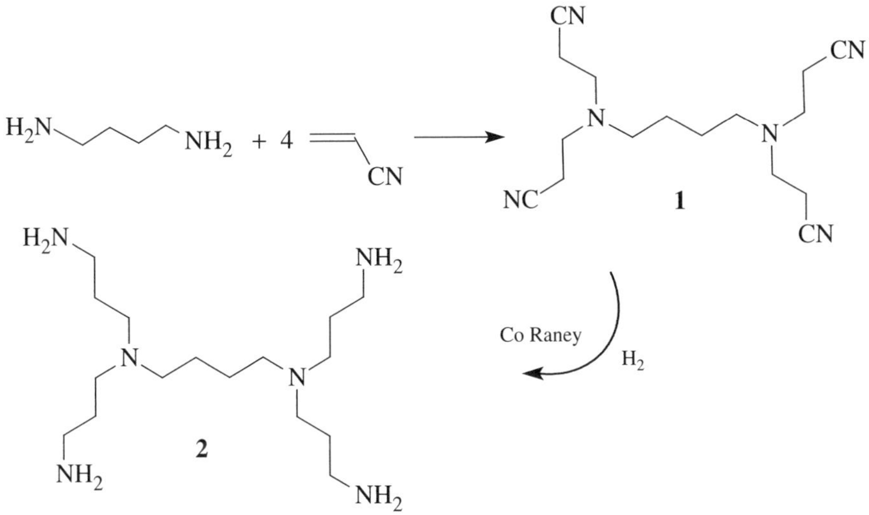

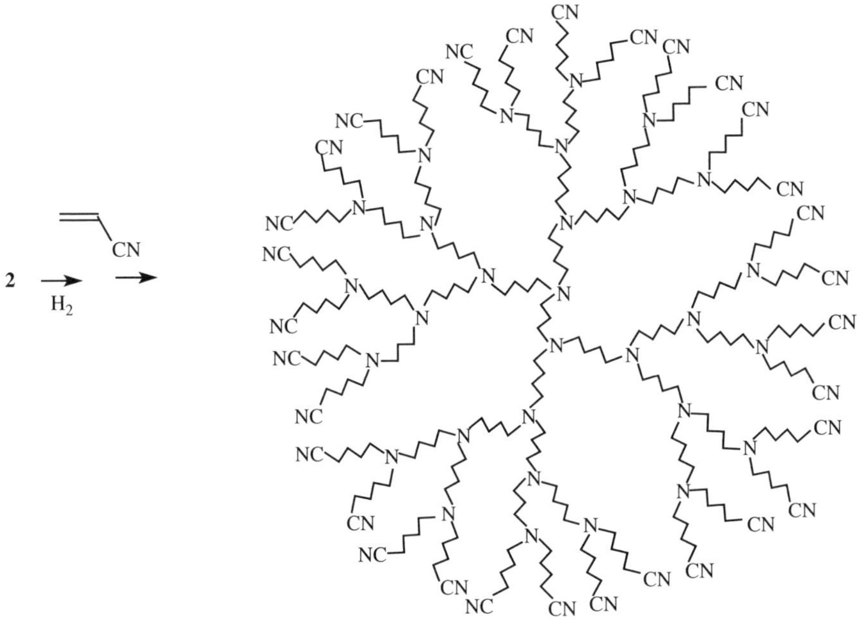

According to the authors who first described their synthesis, defect-free dendrimers of ninth generation could be obtained. Actually, the growing number of reactions occurring on each dendrimer at each new generation, as well as the absence of sufficiently powerful tools of analysis, cast some doubt about the perfection of the objects formed; if isometric objects up to the fourth or the fifth generation could be isolated free of defects, dendrimers of higher generations might comprise defects that could not be detected by analysis.

The divergent method was also used to produce dendrimers with macromolecular generations. Such dendritic polymers exhibit the same features as those of regular dendrimers, except that the generations connecting the branching points are of macromolecular size. The latter are obtained by "living" chain polymerization and are thus subject to a fluctuation—even minimal—of their size. In that they differ from the supposedly perfect regular dendrimers. Dendrimers with macromolecular generations exhibit properties that are peculiar and completely different from those of other architectures—in particular, star polymers. The size of these dendrimer-like polymers is not solely controlled by the number of generations, but also by the length of the branches of each generation. The example shown below illustrates the synthesis of dendrimers of poly(ethylene oxide). Dendrimers made of both polystyrene and poly(ethylene oxide) generations were also synthesized; they exhibit an amphiphilic character that was exploited for certain applications in aqueous medium.

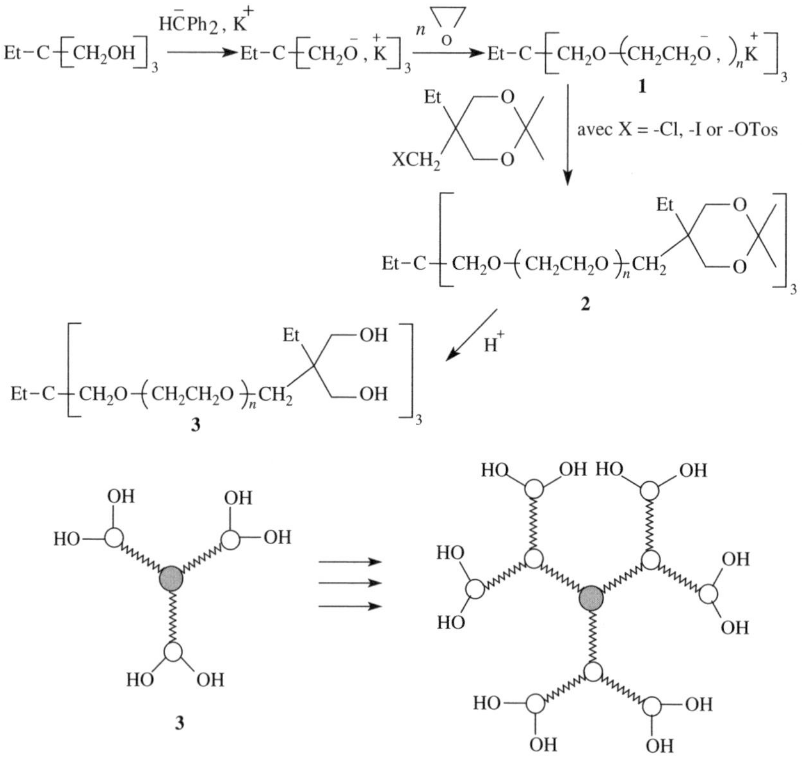

10.4.3.2. "Convergent" Method. This method consists of (a) the synthesis of the so-called *dendrons* — of more or less large size, depending on their generation — and (b) their coupling to a polyfunctional core. An example of this method is shown below.

The dendron, which hardly exceeds two or three generations, is prepared and isolated defect-free. This method is more appropriate than the divergent one to produce isometric samples. Because each dendron is of appreciable size, any flaw in the coupling reaction can be easily detected and the impurities easily removed.

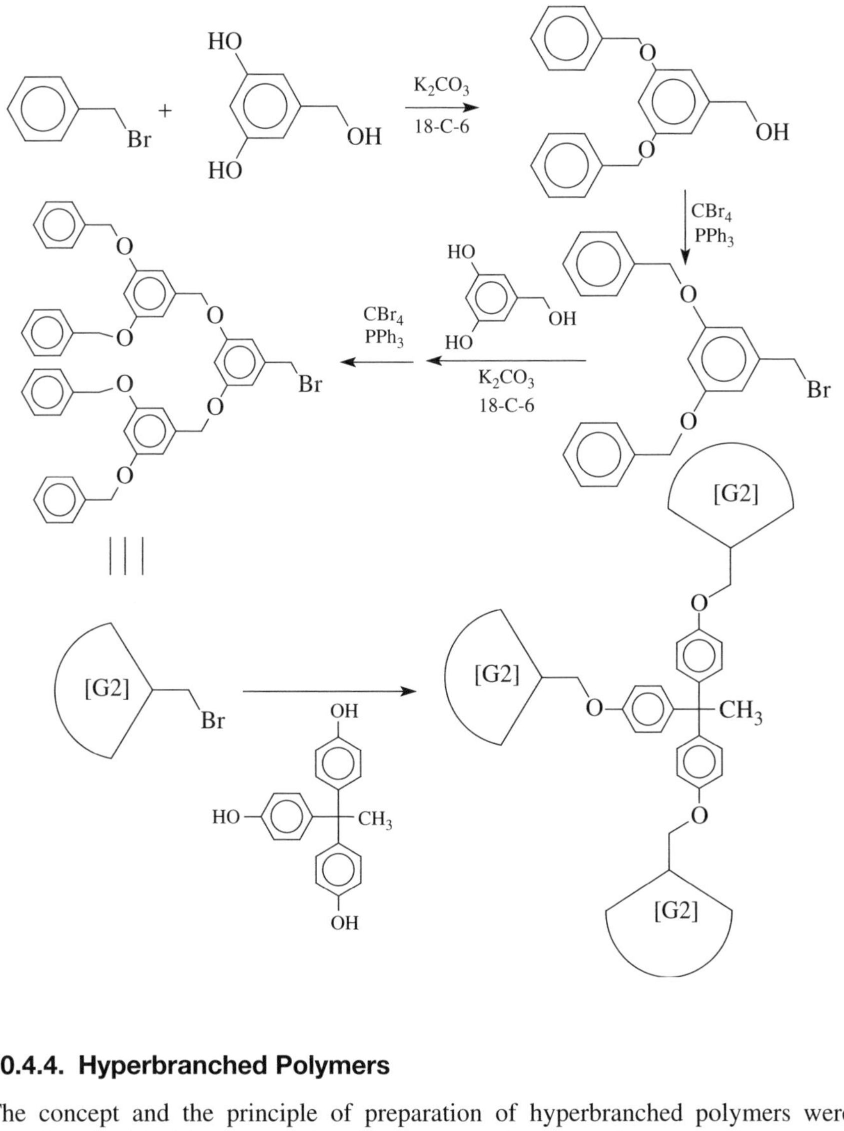

10.4.4. Hyperbranched Polymers

The concept and the principle of preparation of hyperbranched polymers were first described by Flory more than 40 years ago; however, they attract interest and attention only recently after the first dendrimers were described, appearing as substitutes for the latter. They share some common characteristics with dendrimers—for example, a globular shape and an inaptitude to entangle. However, hyperbranched polymers and regular dendrimers differ in one major feature, which is the large distribution in molar mass of the first compared to the perfect isometry of the second.

As shown by Flory, hyperbranched polymers can be obtained from self-condensing monomers carrying one functional group (X) and antagonist functional groups (Y). The following scheme illustrates such a hyperbranched polymer resulting from the self-condensation of the XY_2 monomer

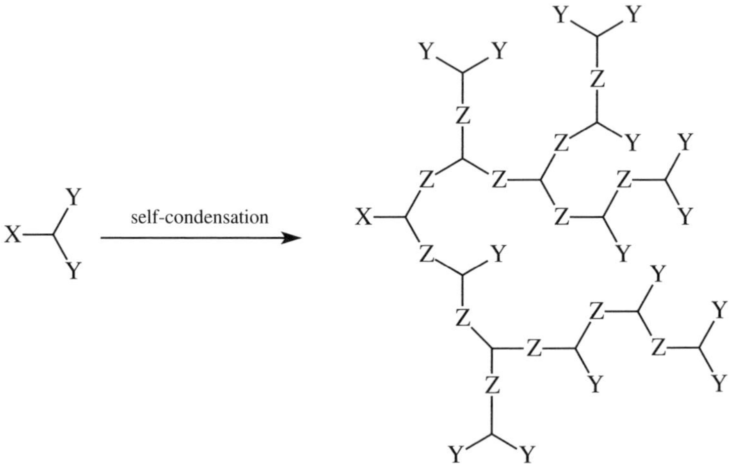

where X can well be a functional group such as $-C-OH$, Y an antagonist function such as $-CO-OH$, and Z an ester function. In addition to the focal point (X), three types of units coexist in such a hyperbranched architecture:

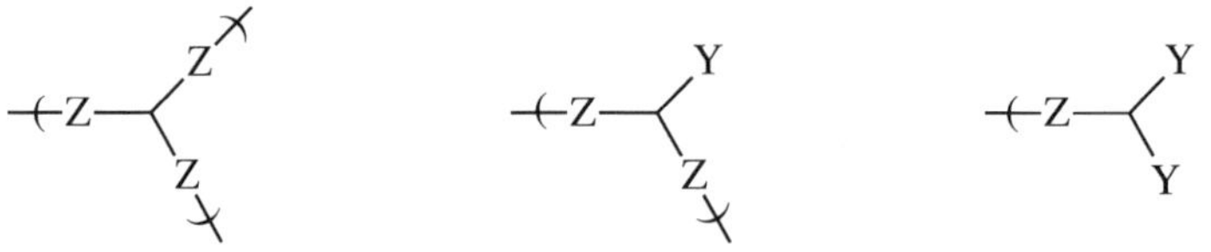

| Dendritic unit | Linear unit | Terminal unit |

The "dendritic units" are those whose both Y functional groups have reacted with an X functional group; on the contrary, those whose both Y functional groups are left unreacted are called "terminal units." Those whose only one functional group Y has reacted are called "linear units." To determine the branched character of the resulting condensation polymer, the notion of "degree of branching" (DB) was defined as

$$DB = \frac{\sum D + \sum T}{\sum D + \sum L + \sum T}$$

For a perfect dendrimer, DB is equal to 1; and, by contrast, for a linear polycondensate, DB is close to 0. For XY_2-type monomers, the theory predicts a degree of branching of the resulting polymer equal to 0.5, assuming the isoreactivity of all Y functional groups. When the second Y functional group of a linear unit exhibits an enhanced reactivity compared to that of the first one, the degree of branching approaches a value equal to unity. In the opposite case—that is, whenever the second Y functional group of a linear unit exhibits a lower reactivity—the polymers obtained are characterized by degrees of branching lower than 0.5. A large number of polymers (polyethers, polyesters, polyurethanes, polycarbonates, etc.) have been prepared by this method from XY_n-type monomers with a wide variety of antagonist functional groups.

The easy synthesis and access to such hyperbranched polymers—compared to dendrimers which are difficult to prepare—explain the current interest in this kind of structures.

It is also possible to generate hyperbranched polymers from vinyl monomers that undergo chain polymerizations. In addition to a polymerizable unsaturation, these monomers also contain a site that can initiate polymerization after activation:

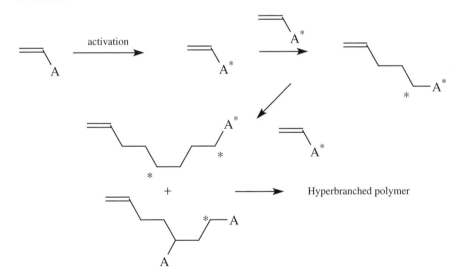

A monomer such as p-(1-chloroethyl)styrene lends itself to this type of reaction when activated either cationically or by radical means through halogen transfer:

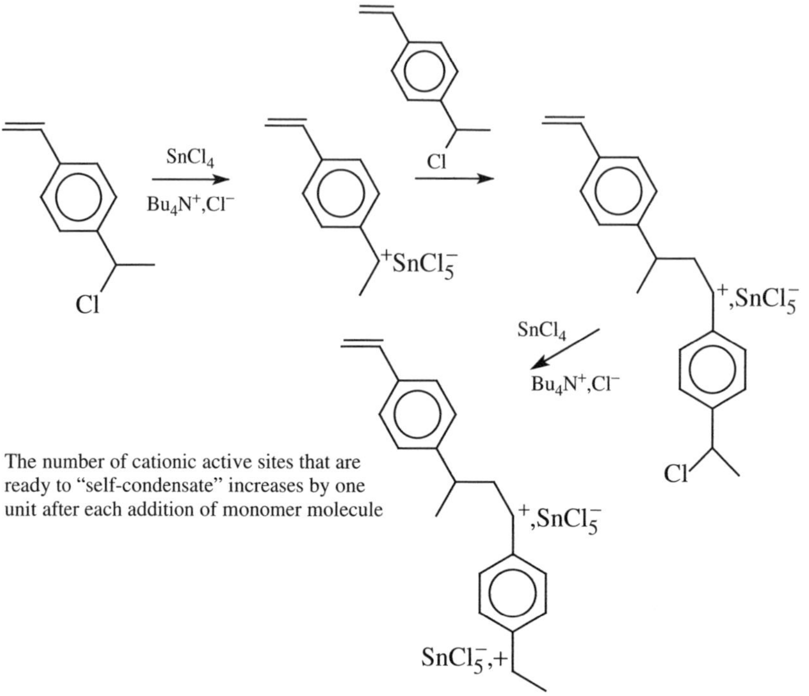
SnCl4
Bu4N+,Cl−
Cl
+SnCl5−
+,SnCl5−
The number of cationic active sites that are ready to “self-condensate” increases by one unit after each addition of monomer molecule
SnCl5−,+

Similarly to dendrimers with macromolecular generations, hyperbranched polymers comprising true macromolecular chains between their branching points have also been synthesized. They were grown upon repeating generation after generation the reaction of ω-functionalized (or “living”) chains with a polymer backbone carrying reactive sites. Molecular objects with very high molar mass ($>100 \times 10^6$ g·mol^{-1}) and low hydrodynamic volume were obtained by this method. The example shown below refers to the deactivation of polystyryllithium chains (obtained by anionic polymerization) on short blocks of poly(chloroethylvinylether) (obtained by cationic polymerization):

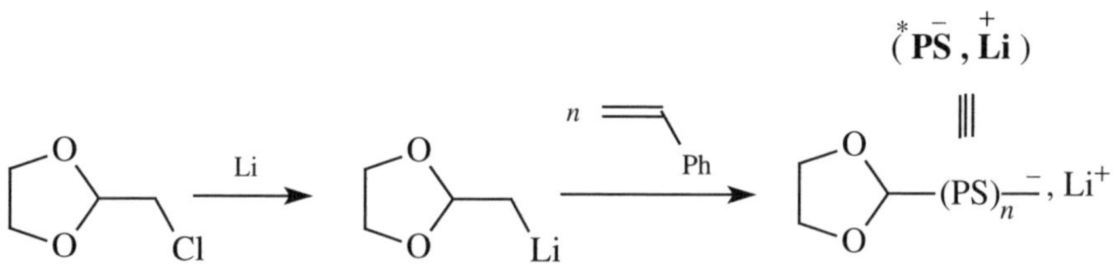
(PS, Li)
O
Cl
Li
n
Ph
(PS)n−, Li+

The comb-type polymers obtained in the first instance carry a functional group at each end of their side grafts that serves to initiate the cationic polymerization of chloroethylvinylether (CEVE), allowing the deactivation of a further generation of "living" polystyryllithium chains. Repeated several times, this gives

leading to

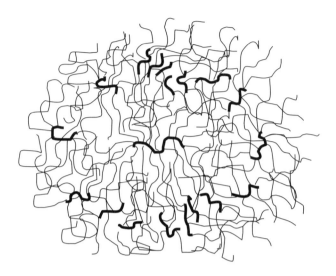

Under stoichiometric conditions {[PS$^-$,Li$^+$] and [CEVE units] = 1}, the relative composition [S]/[CEVE] is determined by the degree of polymerization of PS$^-$,Li$^+$. Such molecular objects thus mainly consist of polystyrene.

LITERATURE

E. J. Goethals (Ed.), *Telechelic Polymers: Synthesis and Applications*, CRC Press, Boca Raton, FL, 1989.

Y. Gnanou, Macromonomers: synthesis, polymerization and utilization. In: *The Polymeric Materials Encyclopedia* J. C. Salamone (Ed.), CRC Press, Boca Raton, FL, 1996.

Y. Gnanou, Tailor-made polymers. In: *The Polymeric Materials Encyclopedia*, J. C. Salamone (Ed.), CRC Press, Boca Raton, FL, 1996.

A. Hult, M. Johanson, E. Malmström, Hyperbranched polymers, *Adv. Polym. Sci.* 143, 1 (1999).

G. Hawker, Dendritic and hyperbranched macromolecules: Precisely controlled macromolecular architectures, *Adv. Polym. Sci.* 147, 113 (1999).

W. J. Mijs, *New Methods for Polymer Synthesis*, Plenum Press, New York, 1992.

K. Mishra (Ed.), *Macromolecular Design: Concepts and Practice*, PFI, New York, 1994.

M. Lazzari, G. Liu, and S. Lecommandoux (Ed.), *Block Copolymers in Nanoscience*, Wiley VCH, Weinheim, 2006.

K. Matyjaszewski, Y. Gnanou, and L. Leibler (Eds.), *Macromolecular Engineering*, Wiley-Interscience, New York, 2007.

11

THERMOMECHANICAL PROPERTIES OF POLYMERS

Thermal properties that directly affect the mechanical characteristics of polymeric materials are also referred to as thermomechanical properties. They are closely related to the morphological structure taken by macromolecular systems but are little affected by their dimensionality—at least when the cross-linking density is low. Accordingly, only linear and related polymers will be subsequently considered.

11.1. GENERAL CHARACTERISTICS

Because thermomechanical properties are specific to a given structural state, it is useful to point out that solid polymeric materials exist under one of the following three physical states, each one being characterized by a specific morphology:

- The crystalline state, which corresponds to an almost perfect ordering of macromolecular entities and appears in the form of small-size single crystals,
- The amorphous state, which is a disordered entanglement of polymer chains and, finally,
- The semicrystalline state, which comprises the two preceding states in varying proportions measured by the degree of crystallinity.

Each one of these states exhibits specific thermomechanical properties and responses, with the semicrystalline state combining the properties of both the amorphous and of crystalline states at the macroscopic level.

Organic and Physical Chemistry of Polymers, by Yves Gnanou and Michel Fontanille
Copyright © 2008 John Wiley & Sons, Inc.

When comparing macromolecular systems with simple molecules, one may, *a priori*, think that the thermomechanical properties of polymers would be easier to describe since matter exists only in two physical states (solid and liquid) in the polymer case instead of three for simple molecules. Indeed, the transition from the liquid to the gaseous state does not exist for polymers because the multiplicity of their molecular interactions prevents their vaporization at temperatures lower than that of their degradation.

For low-molar-mass molecules, the transitions between the three physical states are governed by thermodynamic equilibria and are associated with a transition temperature that is perfectly defined for a given pressure. These transitions are first-order thermodynamic transitions during which primary thermodynamic functions (P) such as specific heat, specific volume, and so on, undergo an abrupt change at the transition temperature (Figure 11.1a, b).

For certain low-molar-mass molecules, transitions referred to as "second-order" transitions are observed; in the latter case a change of the $P = f(T)$ slope is observed (Figure 11.1c, d).

In the case of polymers, the situation is more complex. Indeed, phase transitions in polymers do not occur at a well-defined temperature as in the case of low-molar-mass molecules. The dispersity of macromolecular systems causes their transition to widen over a range of temperature that depends on the degree of dispersion of the structures. In addition, in a large number of cases, solid-state polymers are completely amorphous (glassy state) and do not undergo first-order but only

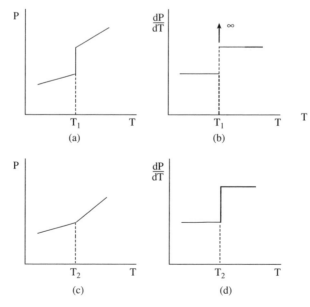

Figure 11.1. Variation of a primary thermodynamic function (P) as a function of the temperature. (a, b) Systems constituted by crystallizable simple molecules. (c, d) Amorphous macromolecular systems. T1 and T2 are transition temperatures.

second-order transitions. Finally, there is a transition peculiar to polymers that reflects the dependence of their mechanical properties on their molecular dimensions as illustrated in the example below:

A rigid sample of poly(methyl methacrylate) (PMMA) of $\overline{M}_w \sim 10^5 \mathrm{g \cdot mol}^{-1}$ softens and becomes deformable when heated above $100°C$, indicating that it has undergone a transition from the glassy state into the rubbery one. Beyond $150°C$, this same PMMA sample flows as a viscous liquid when subjected to an elongational stress. If this experiment is repeated with a sample of $\overline{M}_w > 2 \times 10^6 \mathrm{g \cdot mol}^{-1}$, the transition from the rubbery state to that of a viscous liquid is not observed, with the sample retaining its rubbery character whatever the temperature applied. The origin of such a difference in the behavior of the two samples is discussed in Section 12.4.1 and accounted for through the concept of critical molar mass between entanglements.

From the technological point of view, the two transitions observed in amorphous polymers have a considerable importance. Indeed, the transition from the glassy to the rubbery state, also called glass transition, determines the minimum service temperature for an elastomer and also the maximum service temperature for an amorphous glassy polymer (PS, PMMA, PVC, etc.). On the other hand, the transition from the rubbery state to a viscous liquid sets the minimum temperature at which polymeric materials can be processed. Plastic or viscous behaviors, which correspond to the partial or total irreversibility of deformations in response to a mechanical stress, are imperative for the usual techniques of polymer processing (thermoforming, extrusion, injection molding, etc.) to be applied. Figure 11.2 schematically illustrates the existence of these various states as a function of the temperature and the length or the molar mass of macromolecular chains.

In the case of polymer single crystals, the situation is not very different from that of regular crystals; these entities undergo fusion or melting at a rather well-defined temperature because the dimensions of the crystalline domains are relatively large. The only structural "defects" are those corresponding to chain folding, hairpin turns, loops, and chain ends.

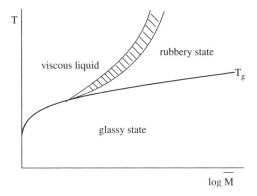

Figure 11.2. Representation of the effect of temperature and average molar mass on the physical state of an amorphous polymer.

The behavior of semicrystalline polymers is more difficult to describe. Indeed, they not only contain amorphous and crystalline zones but also exhibit a more or less ideal chain packing depending on the molecular irregularities present and the crystallization conditions. On the other hand, depending upon the degree of crystallinity (X) of the sample considered, its mechanical properties vary with the service temperature (T_s) chosen and with the value of the latter with respect to that of the glass transition (for low values of X) or to that of melting (for high values of X). Indeed it has to be stressed that for $T_g < T_s < T_m$, the material is hard but deformable (ductile), whereas for $T_s < T_g$ the material is hard and breakable.

11.2. GLASS TRANSITION

11.2.1. The Phenomenon

A completely amorphous polymer ($X = 0$) is rigid and breakable at $T < T_g$ and becomes viscous or elastomeric—depending upon its molar mass or the duration of the stress applied—at $T > T_g$. The origin of this phenomenon is still a matter of discussions, but the following points are beyond any controversy:

- The highly reversible extensibility (rubber elasticity) of polymer chains is related to their ability to coil and uncoil. The extension of an elastomer corresponds to a negative variation of its entropy (see Section 12.1). The examination of X-ray diffraction patterns of a stretched elastomer and of an unstretched one confirms this statement.

- The motion of a limited number of subchains is enough to impart a rubbery behavior to a polymer sample as observed in networks above their glass transition temperature. However, for very dense networks, the movement of chain segments is limited and the phenomenon is difficult to detect. Conversely, for polymers of short size, the motion of the whole chain can result in irreversible phenomena such as their creep and relaxation (see Chapter 12).

- In the glassy state ($T < T_g$) the movement of subchains are not completely absent; they are only slowed down, especially in the absence of an external mechanical stress.

- For polymers of very high molar mass, the glass transition temperature is independent of their molar mass; thus, the glass transition temperature of vulcanized elastomers is close to that of the corresponding linear polymers.

- Finally, for compounds of intermediate molar mass, a second transition is observed at $T = T_{\text{flui}}$, which likely corresponds to the motion of the whole macromolecule and is generally identified with the glass transition temperature (at $T = T_g$) of polymers of low degree of polymerization.

Remarks   The glass transition phenomenon is not restricted to polymers. Due to a molecular structure that lacks symmetry, certain low molar mass molecules cannot crystallize quickly and can thus be cooled below their melting point, leading to amorphous systems that undergo the said phenomenon.

From these factual observations, various theories were elaborated which, considered all together, can practically account for all aspects of the observed phenomena. Depending upon the experimental conditions, one or another among these theories may appear more suited to describe the behavior observed.

11.2.2. The Free Volume Theory. In spite of the fact that polymers are denser than homologs of low molar mass, the apparent volume they fill in the condensed state is not entirely occupied by their constitutive atoms. For instance, it is observed that the volume occupied by a solution of a polymer is lower than the sum of volumes of each component considered separately. This "unoccupied" volume in the condensed state is referred to as *free volume*; it results from the impossibility of molecular groups to take any position available in the vacant space due to restrictions imposed by valence angles. Whatever the temperature of the system and, in particular, for $T = 0\,\mathrm{K}$, the following relation can be written

$$V_{T0} = V_0 + V_{f0}$$

where V_{T0} is the apparent volume of the sample, V_0 is the actually occupied volume, and V_{f0} is the free volume.

Raising the temperature of the system results in, a dilation that, strictly speaking, cannot be that of the constitutive atoms, but corresponds to their vibration due to thermal energy

$$V_T = V_0 + V_{f0} + (dV/dT)_g T$$

where $(dV/dT)_g$ is the volume expansion coefficient in the glassy state.

Whatever the temperature considered, the system is said to be in the glassy state when its free volume fraction is constant, a phenomenon observed for all polymers. Figure 11.3, which represents the variation of the volume of a completely amorphous polymer as a function of its temperature, illustrates this phenomenon. The difference between experimental values of $V = f(T)$ extrapolated to $T = 0\,\mathrm{K}$ corresponds to the "incompressible" free volume, namely:

$$V_{f0} = V_{T0g} - V_{T0\ \mathrm{liq}}$$

At the glass transition temperature (T_g), the free volume is large enough to allow the chains to move under a stress, and at this precise temperature the total volume can be expressed as

$$V_{Tg} = V_0 + V_{f0} + (dV/dT)_g T_g$$

Above the glass transition temperature $(T > T_g)$, the volume expansion coefficient of a polymer is not that prevailing in the glassy state but that of the liquid or rubbery state, depending upon the range of molar masses considered:

$$V_T = V_{Tg} + (dV/dT)_{\mathrm{liq}}(T - T_g)$$

This overall behavior is schematically represented in Figure 11.3.

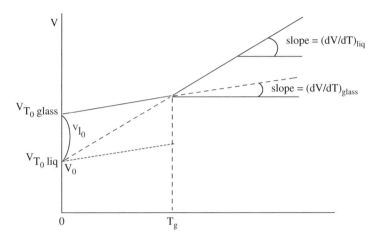

Figure 11.3. Variation of the volume V of an amorphous polymer sample versus temperature.

According to the free-volume theory, at the glass transition temperature all polymers have practically the same fraction of free volume (f_{Tg}):

$$f_{Tg} = [V_{f0} + (dV/dT)_g T_g]/V_{Tg} = 0.025$$

and its value, which was obtained from measurements over a whole range of polymers different in nature and molar masses, is a quasi-universal constant. Above T_g, the fraction of free volumes (f_T) can be written as

$$f_T = f_{Tg} + [(dV/dT)_{liq} - (dV/dT)_g](T - T_g) = f_{Tg} + (\alpha_{liq} - \alpha_g)(T - T_g)$$

As compared to the "incompressible" volume, there exists an excess of free volume which makes the motion of chains all the more easy as the free volume increases.

The difference between the volume expansion coefficients in the liquid and the glassy state for all polymers is approximately equal to

$$\Delta\alpha = (\alpha_{liq} - \alpha_g) \sim 5 \times 10^{-4}\, K^{-1}$$

In a study devoted to the viscosity of noncrystallizable n-alkanes, Doolittle established that the viscosity of these compounds varies with the temperature according to an Arrhenius-type law:

$$\eta_T = A \exp[B V_0/(V_T - V_0)$$

corresponding to

$$\ln \eta_T = \ln A + B V_0/(V_T - V_0)$$

Since

$$V_0/(V_T - V_0) \simeq V_T/(V_T - V_0) = V_T/V_{\text{free at } T} \simeq 1/f_T$$

one has

$$\ln \eta_T = \ln A + \frac{B}{f_T}$$

If the viscosity of amorphous polymers is measured at two different temperatures T and T_g, the above expression becomes

$$\ln(\eta_T/\eta_{Tg}) = -B(1/f_T - 1/f_{Tg})$$

which corresponds to

$$\ln(\eta_T/\eta_{Tg}) = \frac{-B\Delta\alpha(T - T_g)}{f_{Tg}[f_{Tg} + \Delta\alpha(T - T_g)]}$$

$$\ln(\eta_T/\eta_{Tg}) = \frac{-(B/f_{Tg})(T - T_g)}{f_{Tg}/\Delta\alpha + T - T_g}$$

If one poses

$$B/f_{Tg} = C_1 \quad \text{and} \quad f_{Tg}/\Delta\alpha = C_2$$

one has

$$\ln(\eta_T/\eta_{Tg}) = \frac{-C_1(T - T_g)}{C_2 + T - T_g}$$

This relationship is analogous to the empirical one established by Williams, Landel, and Ferry (WLF equation) which serves to relate the dielectric and mechanical relaxation times measured at a temperature T with those measured at the reference temperature (here T_g).

Introducing a_T, the *shift factor*, as the ratio of the relaxation time (τ_T) at T to that (τ_{Tg}) at T_g, the WLF equation can be written as

$$\ln a_T = \frac{-D_1(T - T_g)}{D_2 + T - T_g}$$

The viscosity of a polymer in the liquid state can thus be related to its relaxation time:

$$a_T = \tau_T/\tau_{Tg} \simeq \eta_T/\eta_{Tg}$$

From the measurement of the relaxation times determined at various temperatures, the values of D_1 and D_2 and hence the values of f_{Tg} and $\Delta\alpha$ can be deduced using the above expression.

11.2.3. The Kinetic Theory

This theory is not independent of the preceding one since it considers that the amorphous matter contains a series of "holes" of molecular size corresponding to the free volume, whose motion determines the capacity of polymer chains to move.

This theory is based on the measurement of the variation of the specific volume with temperature and time. Indeed, certain authors noted that, depending upon the cooling rate, the values measured can be appreciably different for a given polymer. They observed (Figure 11.4) that the faster the cooling process, the higher the temperature at which the change of the slope occurs. In other words, if the temperature of glass transition is taken at the change of the slope, it appears to vary with the heating rate of the system.

From this observation, it was assumed that the kinetics of contraction (or expansion) of volume which pertains to the excess free volume (thus beyond T_g) is proportional to that excess free volume. It is a first-order kinetics which can be expressed as

$$dV/dT = c^{st}(V_t - V_\infty) = -1/\tau(V_t - V_\infty)$$

or

$$(V_t - V_\infty) = (V_0 - V_\infty)\exp^{(-t/\tau)}$$

In this relation, V_t denotes the volume of the sample at time t and V is its volume at infinite time (*i.e.*, at equilibrium). The rate constant of the process is expressed in t^{-1} and in the form of $1/\tau$, with τ being the relaxation time of the system—that is, the time required for a certain fraction of volume (conventional) to reach its equilibrium state.

According to Kovacs, the first-order kinetics of the process is not perfectly observed as its rate constant depends itself on the temperature and, likely in more complex cases, on the time (which can be expressed as $\tau = at + b$). Then, one has

$$\tau = \tau_0 \exp(E_a/RT)$$

which means that the higher the temperature, the faster the relaxation. The "actual" glass transition temperature could thus only be reached at infinite relaxation times and thus measured at an infinitely slow rate.

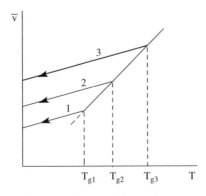

Figure 11.4. The effect of cooling rate (1: slow, 3: fast, 2: intermediate) on the variation of the mass volume of an amorphous polymer according to the temperature.

11.2.4. Thermodynamic Theory

Proposed by Gibbs and DiMarzio, this theory considers that the glass transition temperature of a given polymer corresponds to the value at which a given chain exhibits only one unique conformation.

From the calculation of the variation of entropy of a chain (see Chapter 6) as a function of the temperature, T_g is defined as the temperature at which the conformational entropy is equal to zero. The diagram in Figure 11.5 schematically illustrates the corresponding variation. Values of T_g determined by this method are approximately lower by 55°C than those commonly used.

For polymer networks, the conformational entropy closely depends on the cross-linking density; according to the thermodynamic theory, the dimensionality of the polymer system should affect the value of T_g which is indeed observed on highly cross-linked systems, confirming the predictions of this theory.

11.2.5. Structural Parameters Affecting the Glass Transition Temperature

External factors affecting T_g, such as the cooling rate and the frequency of the mechanical stress, depend on the service conditions of the material considered. With respect to the structural or compositional parameters, they can be modified at will by the chemist or the formulator in order to adjust T_g and thus meet the requirements of a given application.

In the preceding paragraph the dimensionality was mentioned as affecting the value of T_g; the effect is actually not very pronounced and is limited to highly cross-linked materials; this observation is logical since the glass transition phenomenon is related to the motion of relatively short sequences.

11.2.5.1. Effect of Molar Masses. The proportion of free volume introduced by chain ends decreases as the molar mass increases. Experimentally, it is thus logical to observe an increase of T_g with M. However, only empirical relationships

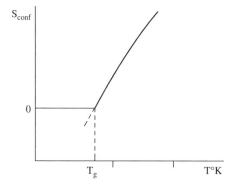

Figure 11.5. Variation of the conformational entropy of a polymer chain as the function of temperature.

could be established; for example, in the case of polystyrene, the generally used relationship is as follows:

$$T_g = 373\,\text{K} - \frac{1.8 \times 10^5}{\overline{M}_n}$$

The variation of T_g with M thus becomes undetectable for molar masses higher than $10^5\,\text{g}\cdot\text{mol}^{-1}$.

11.2.5.2. Effect of the Polymer Molecular Structure.
This effect is multiple since it can be related to the topology of the monomer unit or to the tacticity of the monomer sequence as well as to the extent of molecular interactions.

The **cohesive properties** thus play an important role in the glass transition phenomenon. Indeed, molecular interactions decrease the mobility of the chains which, according to the kinetic theory, must contribute to an increase of T_g and can be beneficial for certain applications. The comparison of the glass transition temperature of polypropylene$-[CH_2-CH(CH_3)]_{n-}$, whose T_g is $-10°C$, with that of poly(vinyl alcohol)$-[CH_2-CHOH]_{n-}$, whose T_g is $+70°C$, gives an idea of the importance of interactions.

The effect of **main chain substituents** is more complex. The size of certain atoms or the geometry of some substituents may contribute to stiffen the chain backbone and cause an increase of T_g. For instance, the atomic radius of fluorine atoms in polytetrafluoroethylene (PTFE) imposes a conformational structure that is mirrored in a higher glass transition temperature for this polymer ($T_{gPTFE} = +125°C$) than that of polyethylene ($T_{gPE} \sim -120°C$). Conversely, the grafting of side chains or branches of increasing size onto polymer backbones will distance one from another, which generates free volume and may in turn induce a decrease of T_g. Each additional branch indeed introduces a chain end that contributes to the free volume of the system. However, very long linear side chains can "consume" free volume by folding up regularly, which may stiffen the backbone and thus raise T_g. Table 11.1 gives the values of T_g for a series of polyacrylates, illustrating the effects due to the parameters previously discussed.

The effect of **tacticity** can be considerable for certain types of polymers. For instance, it has little impact on polyacrylates but is pronounced for polymethacrylates. Thus, for PMMA, $T_{g,\text{iso}} = +45°C$ whereas $T_{g,\text{syndio}} = +150°C$. It could indeed be established for 1,1-disubstituted ethylenic polymers that such a large difference in T_gs is due to a larger rotational energy of syndiotactic polymers compared to that of isotactic ones.

A similar behavior is observed for polymers carrying intrachain ethylenic unsaturations. The phenomenon is observed, in particular, for natural 1,4-polydienes: 1,4-*cis*-polyisoprene (natural rubber) exhibits a T_g equal to $-72°C$, whereas that of 1,4-*trans*-polyisoprene (gutta-percha) is $-58°C$. Such a difference originates from the configurations of the two polymers which can also be correlated with their corresponding stiffness.

Table 11.1. Glass transition temperature (T_g) for some atactic poly[alkyl (and other) acrylates]

R	T_g (°C)	R	T_g (°C)
$-CH_3$	+10	$-CH_2-CH-(CH_2)_3CH_3$ $\quad\quad\;\; CH_2$ $\quad\quad\;\; CH_3$	−50
$-CH_2-CH_3$	−24	$-(CH_2)_{11}CH_3$	−3
$-(CH_2)_2CH_3$	−37	$-(CH_2)_{15}CH_3$	+35
$-(CH_2)_3CH_3$	−54	⬡ (cyclohexyl)	+19
$-CH_2-CH{\large<}^{CH_3}_{CH_3}$	−24	⬡ (phenyl)	+6
$-(CH_2)_5CH_3$	−57	$-$⬡$-CH_3$ (p-tolyl)	+43
$-(CH_2)_7CH_3$	−65	$-$⬡ pentachlorophenyl (Cl, Cl, Cl, Cl, Cl)	+147

One convenient means to manipulate the glass transition temperature of polymers is through **copolymerization**. Whatever the phenomenological interpretation given to this effect (additivity of free volumes or combined effects due to the chain stiffness and to molecular interactions), empirical relations can be used to evaluate the T_g of random copolymers. The relationship

$$1/T_{g1,2} = (w_1/T_{g1}) + (w_2/T_{g2})$$

is sometimes used, where T_{g1}, T_{g2}, and $T_{g1,2}$ are the glass transition temperatures of the homopolymers and the copolymer respectively, with w_1 and w_2 being the mass fractions of the two types of monomeric units forming the copolymer; this relationship gives satisfactory results when T_{g1} and T_{g2} are not too different. In other cases, it is preferable to use the Gordon–Taylor relationship

$$(T_{g1,2} - T_{g1})w_1 = K(T_{g2} - T_{g1,2})$$

where the constant K for a given pair of comonomers corresponds to

$$K = (\alpha_{\text{liq},2} - \alpha_{g,2})/(\alpha_{\text{liq}1} - \alpha_{g,1})$$

where the various α coefficients of this equation represent the coefficients of volume expansion of homopolymers 1 and 2 in the liquid and the glassy states.

Actually, this relationship is not completely an empirical one since it can also be established using the free volume theory. A calculation based on this theory is presented below to evaluate the effect of external plasticization.

11.2.5.3. Effect of Plasticizers.

The phenomenon of plasticization results from the addition of a diluent (called *plasticizer*) to a polymer in which it is miscible in all proportions so as to lower its glass transition temperature. This diluent introduces free volume in the material and, like any solvent, promotes polymer–diluent interactions at the expense of polymer–polymer interactions. This phenomenon of plasticization results in the decrease of T_g. In the case of poly(vinyl chloride) (PVC), it is a process of great economic importance because most of the PVC annually produced and used is plasticized.

There are requirements to fulfill for a good plasticization, and it turns out that PVC is well-suited to these requirements. Among these, the following ones are essential: the polymer–plasticizer miscibility, which requires that the Flory–Huggins parameter (χ_{12}) be as low as possible (negative), a plasticizer of low vapor pressure (and then a rather high molar mass) to secure the polymer plasticization over a long period, and, finally, all the toxicological, economic, etc, requirements determining the feasibility of the process.

Using the free volume theory, the value of T_g for the plasticized polymer can be calculated as a function on the composition of the blend. If one assumes, on the one hand, the additivity of the polymer and the plasticizer (diluent: d) and, on the other, a common critical value (f) for the fraction of free volume for the two components and from their blend as well, then the fraction of free volume of a pure component at a given temperature T is equal to

$$f = f_g + (\alpha_{\text{liq}} - \alpha_g)(T - T_g)$$

The fraction of free volume introduced by the polymer in the blend at the glass transition temperature (T_{gm}) of the blend is expressed as

$$f_p = [f_{\text{gp}} + (\alpha_{\text{liq}} - \alpha_g)(T_{\text{gm}} - T_{\text{gp}})]\Phi_p$$

where Φ_p is the volume fraction of polymer.

For the plasticizer, one has

$$f_d = [f_{\text{gd}} + \alpha_d(T_{\text{gm}} - T_{\text{gd}})]\Phi_d$$

where Φ_d, the volume fraction of the plasticizer, is equal to $(1 - \Phi_p)$. It is thus possible to deduce the relationship

$$T_{gm} = \{T_{gp}(\alpha_{liq} - \alpha_g)\Phi_p + T_{gd}\alpha_d(1 - \Phi_p)\}/\{\Phi_p(\alpha_{liq} - \alpha_g) + (1 - \Phi_p)\alpha_d\}$$

A particularly well-known blend is that of poly(vinyl chloride) with dioctyl phthalate; the plasticizer extent can reach a value as high as 50% and the T_g can be lowered by as much as 100°C.

Flory proposed a constant value of $(\alpha_{liq} - \alpha_g)$, equal to $4.8 \times 10^{-4}\,\mathrm{K}^{-1}$ for all polymers. Actually, this value is not universal; but, whatever the polymer considered, the measured ones generally fall in that range.

11.2.6. Determination of the Glass Transition Temperature

The monitoring of any property that is sensitive to the glass transition phenomenon can be used to measure T_g; however, the variations of these properties are generally very difficult to detect, and only techniques based on dilatometric, calorimetric, or mechanical measurements are used.

The variation of the mechanical properties near the glass transition is discussed in Chapter 12, and techniques developed to this end such as dynamic mechanical analysis (DMA) are usually employed. Figure 11.3 illustrates the shape taken by the curve that reflects the variation of the volume of a polymer sample as a function of the temperature: this technique is sometimes used; but in spite of the simplicity of its principle, it is delicate to handle. The calorimetric technique is preferred, especially differential scanning calorimetry (DSC). The typical shape taken by a thermogram for a noncrystallizable polymer is shown in Figure 11.6.

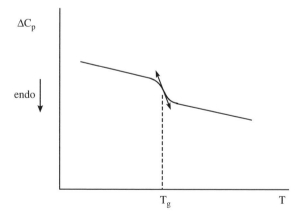

Figure 11.6. Curve of variation of the heat capacity (c_p) of an amorphous polymer as a function of the temperature (T_g is measured at inflection point).

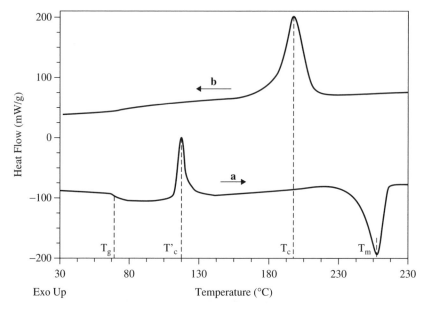

Figure 11.7. DSC thermograms of a quenched poly(ethylene terephthalate) (PET) sample: (a) upon increasing temperature; (b) upon cooling from the melt.

For a semicrystalline polymer, the lower the degree of crystallinity, the more discernable the signal, as shown in Figure 11.7; the thermogram of a quenched PET (carried out upon increasing temperature) and that of a partially crystalline PET (carried out upon decreasing temperature) illustrate it clearly.

11.3. MELTING OF SEMICRYSTALLINE POLYMERS

11.3.1. General Scope

The transformation of crystalline zones of a semicrystalline polymer into a viscous liquid state a first-order thermodynamic transition.

The free energy of melting

$$\Delta G_m = \Delta H_m - T_m \Delta S_m$$

is independent of the molar mass beyond a certain degree of polymerization (X). To account for this phenomenon, one considers that both ΔH_m and ΔS_m result from the addition of two independent terms; one (ΔH_0 or $T\Delta S_0$) would be specific to the structure of the monomer unit and thus be invariant with respect to X; the other one ($X\Delta H_1$ or $X\Delta S_1$) would be not only specific to the monomer unit but also proportional to the degree of polymerization:

$$\Delta G_m = (\Delta H_0 + X\Delta H_1) - T_m(\Delta S_0 + X\Delta S_1)$$

At equilibrium, $\Delta G_m = 0$ and thus

$$T_m = \frac{\Delta H_m}{\Delta S_m} = \frac{\Delta H_o + X \Delta H_1}{\Delta S_0 + X \Delta S_1}$$

For very high values of X: $T_m \simeq \dfrac{\Delta H_1}{\Delta S_1}$

Remarks

(a) Polymer single crystals are generally too small to permit studies of their melting.

(b) From the thermodynamic point of view, polymer melting cannot be regarded as an equilibrium phenomenon. Owing to the disentanglement of polymer chains in the molten state, crystallization occurs at temperatures lower than T_m.

11.3.2. Factors Determining the Melting Point

As previously mentioned, melting temperatures increase with the molar mass of the sample in the range of low $\overline{X}_n s$, until reaching a plateau for samples of higher degrees of polymerization. Such a pattern reflects the role played by chain ends in low degree of polymerization samples, similar to the effect of structural or compositional irregularities (various impurities, branching points, additives, etc.) (see next page) in samples of higher degree of polymerization.

The molecular structure of constitutive units is obviously an essential factor determining T_m for chains of long length and free from structural irregularities. The enthalpy term (ΔH_m), which corresponds to the heat of melting, is conditioned by the extent of molecular interactions (their nature, their distance, and consequently their energy). Table 11.2 gives the thermodynamic parameters and the melting points of some important polymers with varying cohesion and chain rigidity.

It is important to emphasize that the values of the enthalpy of melting per monomer unit are not very different from one polymer to another of a homologous series when the number of strongly cohesive groups is the same as in the case of poly(ethylene terephtalate) and poly(butylene terephtalate). On the other hand, it should be emphasized that enthalpies per volume unit can be very different. One can also note the difference between the values of ΔH_m for polycaprolactone ($143 \, J \cdot g^{-1}$) and of polycaprolactam ($160 \, J \cdot g^{-1}$) due to the cohesive nature of amide functional groups (H bonds).

Table 11.2 gives the values of the entropy of melting for the same polymers; they show that the variation of this term, which is an indication of the degree of freedom experienced by the polymer during melting, depends closely on its molecular structure and in particular on its backbone. For instance, the high melting temperature of polytetrafluoroethylene (PTFE) ($T_m = 327°C$) is primarily due to the rigidity of its backbone even in the molten state, which is mirrored in a low value of the entropy of melting.

Table 11.2. Thermodynamic parameters related to the melting of some synthetic polymers.[a]

Polymer	ΔH_u (J·mol^{-1})	ΔS_u (J·K^{-1}·mol^{-1})	T_m^0 (K)
Polyethylene $(-CH_2-)_n$	4,110	9.9	415
Polytetrafluoroethylene $(-CF_2-)_n$	3,420	5.7	600
Isotactic polypropylene $[-CH_2-CH(CH_3)-]_n$	8,200	17.6	461
Isotactic poly(4-methylpent-1-ene). $[-CH_2-CH(C_4H_9)-]_n$	5,300	10.1	523
Polyisobutene $[-CH_2-C(CH_3)_2-]_n$	12,000	37.8	317
1,4-cis-Polybutadiene $(-CH_2-CH=CH-CH_2-)_n$	9,205	33.7	273
1,4-cis-Polyisoprene $[-CH_2-C(CH_3)=CH-CH_2-)_n$	8,700	28.9	301
Poly(oxymethylene) $(-CH_2-O-)_n$	9,800	21.4	457
Poly(ethylene oxide) $(-CH_2-CH_2-O-)_n$	8,700	24.6	353
Polycaprolactone $[-(CH_2)_5-CO-O-]_n$	16,300	48.3	337
Poly(ethylene terephthalate) $[-CO-(C_6H_4)-CO-O-(CH_2)_2-O-]_n$	26,150	48.9	535
Poly(butylene terephthalate) $[-CO-(C_6H_4)-CO-O-(CH_2)_4-O-]_n$	31,700	61.8	518
Polycaprolactam PA-6 $[-NH-(CH_2)_5-CO-]_n$	17,950	35.7	502
Poly(hexamethylene adipamide) $[-NH-(CH_2)_6-NH-CO-(CH_2)_4-CO-]_n$	43,370	80	542
Poly(decamethylene sebacamide) $[-NH-(CH_2)_{10}-NH-CO-(CH_2)_8-CO-]_n$	72,000	147.2	489

[a]The melting point is given for the perfect crystal (values of ΔH_u and ΔS_u are given for one mole of repeating unit).

However, except for a few polymers such as PTFE, which has just been mentioned, the values of ΔS_m relative to chain bonds are not very different; this means that the melting point of most of the polymers is mainly determined by their cohesive energy density.

The difference is considerable between ideal systems such as single crystals and semicrystalline polymers. Indeed, the presence of impurities (diluent), of structural defects, or even junctions and loops due to chain entanglement considerably lowers the regularity of crystalline assemblies. This results in a weakening of molecular interactions and hence in a depression of melting points. Flory proposed a theory based on thermodynamic considerations that relates the depression of melting points to the mole fraction of impurities. Considering that the chemical potentials of repeating units are equal in the crystalline phase (μ_u^c) and in the viscous "liquid" phase (μ_u^l) at the melting temperature (T_m), we have

$$\mu_u^l = \mu_u^c$$

If the polymer is "pure", we obtain

$$\mu_u^l \equiv \mu_u^0$$

where μ_u^0 denotes the chemical potential of the repeating unit in the standard state, the melting temperature of the corresponding "pure" polymer corresponding to T_m^0. If an impurity is present in the system, one has

$$\mu_u^l < \mu_u^0$$

To reach thermodynamic equilibrium, which means $\mu_u^l = \mu_u^c$, the melting temperature should thus be lowered. The difference in the chemical potentials of repeating units in the standard state and in the new crystalline state is equal to the variation of free energy:

$$\mu_u^0 - \mu_u^c = \Delta G_{mu} = \Delta H_{mu} - T_m \Delta S_{mu}$$

Considering that

$$T_m^0 = \frac{\Delta H_{mu}^0}{\Delta S_{mu}^0}$$

and given the basic thermodynamic relationship for mixing, one arrives at

$$1/T_m - 1/T_m^0 = (R/\Delta H_u)(V_u/V_1)(\Phi_1 - \chi_{12}\Phi_1^2)$$

where V_u and V_1 are the molar volumes of the repeating unit and of the "impurity," Φ_1 being the volume fraction of this impurity and χ_{12} the Flory–Huggins interaction parameter corresponding to the system. For an ideal solution, $\chi_{12} = 0$ and the following relationship is obtained:

$$1/T_m - 1/T_m^0 = -(R/\Delta H_2)ln N_2$$

where ΔH_2 is the enthalpy of melting of the polymer and N_2 is its molar fraction. This relationship can be applied to evaluate the depression of the melting point due to the presence of a small content in comonomer units in a semicrystalline statistical copolymer:

$$1/T_m - 1/T_m^0 = (R/\Delta H_u) \ln N_A$$

where N_A denotes the molar fraction of crystallizable comonomeric units present in majority. According to the same principle, one can determine the effect of chain ends on T_m, which gives for a Gaussian distribution of chain lengths

$$1/T_m - 1/T_m^0 = (R/\Delta H_u)(2/\overline{X}_n)$$

where $\overline{X}_n$ is the number average degree of polymerization.

11.4. CRYSTALLIZATION OF CRYSTALLIZABLE POLYMERS

11.4.1. The Phenomenon

From a thermodynamic point of view and at equilibrium, the phenomenon of crystallization is the opposite to that of melting, but the macromolecular nature of polymers makes it more complex than for low-molar-mass species. Indeed, the kinetics plays a key role in the case of polymers, with their crystallization being conditioned by the need of chains to disentangle before regular arrangement; for instance, certain polymers of regular molecular structure and hence potentially suitable for crystallization cannot crystallize at all due to their inability to disentangle.

Single crystals are obtained from dilute solutions upon decreasing the temperature of the medium or evaporating a solvent of low solvating power. The presence of a diluent decreases the viscosity of the medium and thus increases the polymer crystallizability. The formation of spherulitic structures is favored in the case of concentrated solutions or in bulk with the possibility of amorphous zones of the spherulites to be filled with solvent molecules.

Crystallization from the melt has been studied to a larger extent than that of single crystals because it occurs regularly when processing polymer melts. Barring the medium viscosity there is no major difference between the two conditions of polymer crystallization except when otherwise mentioned, only crystallization from the melt will be discussed here.

Crystallization is not symmetrical to melting, as seen in the thermogram of Figure 11.7. Indeed, when polymers crystallize from the melt, they usually *supercool (supercooling effect)* to lesser or greater extents depending upon the rate of cooling. For polymers crystallizing slowly, it is even possible to prevent them from crystallizing by a fast cooling *(quenching)* and thus to retain them in a completely amorphous state. For instance, poly(ethylene terephthalate) (PET) affords two definitely different materials: one amorphous and the other one semicrystalline, depending upon the rate of cooling.

Thermogram (b) shown in Figure 11.7 is that of a quenched sample of PET heated from low to high temperatures. A first "endothermic" signal (T_g) is seen corresponding to the glass transition at $70°C$; then an exothermic event at T_c' occurs, reflecting a first-order transition due to the chain crystallization upon heating (the maximum rate of crystallization for such chains occurring at $130°C$) and finally, an endothermic event at $T_m = 264°C$ indicative of the melting of the previously crystallized phase.

Thermogram (a) of Figure 11.7 shows the signals recorded upon cooling the same sample from the melt. The area under the exothermic peak corresponds to the heat of crystallization (at T_c) and is smaller than that of the endothermic signal that reflects the heat of melting (a). This means that the rate of cooling of this PET sample was too fast (but lower than the rate of quenching) to allow the chain to crystallize to the same extent as that observed upon heating.

The observation by optical microscopy showed that crystallization is initiated by relatively few nuclei randomly distributed in the bulk of the material. The phenomenon of crystallization then proceeds in two essential steps, namely, *nucleation* (also called *germination*) and growth into spherulites.

11.4.1.1. Nucleation.
The **nucleation** generally occurs through two competing and simultaneous mechanisms whose relative importance depends on both the purity of the system and the experimental conditions of crystallization.

In the case of *homogeneous nucleation*, thermal motion causes subchains or segments of the crystallizing polymer to sporadically cluster within a homogeneous phase and form unstable embryons that may grow into stable nuclei. This occurs when the temperature (T) of the sample considered is markedly lower than its melting point (T_m) and when one sequence in the crystallizable chains adopts a conformation of lower energy than that of others in the polymer melt at the same temperature. The permanent chain motion induces the regular arrangement of these short sequences, and the structure thus generated becomes stable when

$$\Delta G_c = \Delta H_c + T \Delta S_c < 0$$

The phenomenon of supercooling arises from the additional free energy required to align these segments; this energy comes mainly from the surface rather than from the bulk of the microcrystals. Thus,

$$G_{\text{crystal}} = G_{\text{bulk}} + \sum_i s_i G_i$$

where s_i denotes the surface area of a crystal face, G_i denotes the free energy of specific surface, and thus $\sum_i s_i G_i$ denotes the total surface free energy of the crystal.

When crystallization occurs, we have

$$\Delta G_c = \Delta G_\infty + \Delta \sum s_i G_i$$

where ΔG_c is the specific free energy of crystal formation, ΔG_∞ is the specific free energy of formation of a hypothetical crystal having infinite size, and $\Delta \sum s_i G_i$ is the specific free energy of formation of i faces of the crystal. This last term is always positive; and when ΔG_c is plotted versus the size of the crystal, one obtains

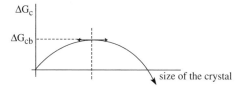

As shown above, up to a certain size, crystal growth is thermodynamically not favorable and hence less probable. This situation corresponds to the subcritical state of the nucleus. The barrier free energy (ΔG_{cb}) is crossed only by a very small fraction of constitutive units, which accounts for the small number of nuclei that sporadically form. After crossing ΔG_{cb}, ΔG_c decreases, making the crystal growth increasingly probable (supercritical state); and when ΔG_c becomes negative, the nucleus is thermodynamically stable, paving the way for the second growth step. The critical size of the nucleus is estimated to be within the range of $10 \, \text{nm}^3$.

Heterogeneous nucleation is considered by many as the only process of generation of nuclei. The following experimental facts support such a statement:

- Microscopic observation of the crystallization shows that the nuclei are almost always located on the same sites after successive meltings and crystallizations;
- The crystallization of a very pure polymer in the form of droplets dispersed in an inert liquid may lead to strong supercooling with crystallization temperatures that may be $100°C$ lower than those of other sites. Due to the lack of connectivity between crystallizable domains, certain droplets remain amorphous, a phenomenon interpreted as resulting from the absence of heterogeneities in these elements.

These heterogeneities are extraneous nuclei of small size (2–10 nm) that may be dust, walls of the container, catalytic residues, residual nuclei present in the polymer melt, or purposely introduced nucleating agents. From a thermodynamic standpoint, the barrier free energy for heterogeneous nucleation is definitely lower than that of a sporadic nucleation because the free energy of interface with a given heterogeneity is low and hence does not oppose the growth of other faces of the crystal.

11.4.1.2. Growth of Spherulites. The **growth of spherulites** from the melt or highly concentrated solutions occurs in two steps. First, the nucleus induces the growth of lamellae- or pyramid-shaped single crystals (in the case of a pyramid *via* a screw-type dislocation) in a perpendicular direction. Then, as these growing lamellae diverge and fan outward in a splaying motion, new lamellar structures are initiated and, by repeated splaying, spherulites of spherical shape are eventually formed that pervade the entire sample. Structural defects and impurities present in small proportions are rejected from lamellae in a direction perpendicular to their growth and lie in the amorphous part; their rate of diffusion in the melt determines the size of the crystalline structures. Because chains can partly lie into two different lamellae, the latter can be tied one to another within the spherulite.

The study of the rate of crystallization (R_c) as a function of the temperature shows that $\log R_c$ varies linearly with $1/\Delta T_c$ and that secondary nucleation (new chain segments adding to the broad plane constituted by the growing lamellae) exerts a prominent effect on the overall kinetics of the process.

11.4.2. Elements of Crystallization Kinetics

The kinetics of crystallization from melts was studied by Avrami and was adapted later on to the case of polymers. Due to the economic repercussions of the crystallization phenomenon and of its kinetics, many studies were conducted to simplify or refine Avrami's treatment. With respect to nucleation and to its kinetics, the temperature is obviously a parameter of paramount importance. For a sporadic nucleation, it was established that the rate N, which corresponds to the number of nuclei generated per unit time and volume, is given by

$$N = N_0 \exp\left(\frac{\Delta E_D}{k(T - T_g)} \times \frac{\Delta G^*}{k(T_m - T)}\right)$$

In this relation, N_0 is the number of crystallizable elements that can give rise to nuclei (entities of size higher than the critical one), E_D is the energy of activation of diffusion of the crystallizable matter through phase boundary, and, finally, ΔG^* is the free energy of crystallization of a nucleus that has reached the critical size. These two terms vary in opposite directions depending on the temperature considered.

The $\Delta G^*/kT = f(T)$ plot represented below shows that ΔG^* increases considerably close to the melting point and must be unfavorable for nucleation as the temperature comes close to T_m:

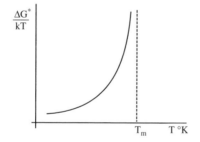

With regard to ΔE_D, it is different since the mobility of the crystallizable entities grows with temperature (similar to the WLF equation). On the other hand, close to T_g, the viscosity of the medium is very high and the diffusion is prevented, if not blocked:

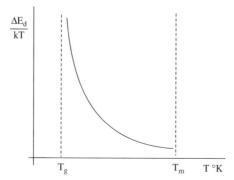

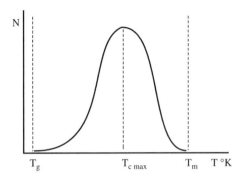

Figure 11.8. Curve giving the shape of the variation of the overall rate of crystallization (N), *versus* temperature.

By combination of the two aforementioned curves, the shape of the overall effect of temperature on the rate of the secondary nucleation (i.e., that of crystallization) can be drawn (Figure 11.8). The temperature of the maximum rate of nucleation is thus below the melting point and clearly above T_g, as shown in the thermograms of Figure 11.7.

These thermograms point to the effect of the degree of crystallinity on the intensity of the signal corresponding to T_g; indeed, row (a) of the thermogram is that of an initially amorphous polymer whose exothermic signal, which lies between 400 and 470 K, reflects the phenomenon of crystallization before melting. The rate of crystallization is seen maximum at 433 K upon heating and at 480 K upon cooling; this is due to the uneven influence of the rates of heating and cooling on the rates of nucleation and crystal growth, whose energies of activation are different.

The experimental study of the kinetics of nucleation is very delicate because both this step and the spherulite growth are superimposed one on another. The studies are generally performed at temperatures slightly lower than T_m (i.e., in a range of temperature where the $\Delta G^*/kT$ term dominates); two types of situations can be distinguished: the first ones correspond to an instantaneous nucleation as soon as the appropriate temperature is reached, which corresponds to a process independent of time. This situation is also seen when extraneous nuclei (mineral microcrystals, etc.) are added to the sample in quantity larger than that formed in a sporadic nucleation. The second type of situation corresponds to a direct first-order dependence of the nucleation on time, described by the relation

$$N = K_g t$$

where K_g is the rate constant of nucleation. This relationship does not take into account the quantity of crystallizable matter, and therefore it applies only to the early steps of the process before most of the crystallizable matter is consumed through spherulitic growth.

Concerning the latter step, the radius r of a spherulite is observed to increase in a linear manner with time,

$$r = Ct$$

where C is the growth constant, provided that $T_c \gg T_g$. The above relation implies that the kinetics of growth is not controlled by diffusion phenomena (one would have then $r = C't^{1/2}$) whatever its nature (whether this diffusion concerns the crystalline matter or the energy released by crystallization). This relation also accounts for the shape of the patterns shown in Figures 5.31 and 5.32—in particular, the existence of a straight border between spherulites at the end of the crystallization process.

Most kinetic studies of isothermal crystallization rely on dilatometry and on the Avrami equation, relating the variation of the total density of the system (ρ) to the duration of the phenomenon. Measurements are more precise when the difference between the density of the crystalline phase (ρ_C) and that of the liquid phase (ρ_L) is large.

If m_0 is the initial mass of crystallizable polymer and if the nucleation process occurs sporadically, then it can be written

$$dN = K_g m_0 dt_x / \rho_L$$

where dN is the number of nuclei formed during the time interval dt_x.

If dm_s is the mass of polymer crystallized during time $(t - t_x)$ and which has nucleated during the time interval dt_x at the time t_x, then one has

$$dm_s = K_g m_0 dt_x / \rho_L \frac{4}{3} \pi C^3 (t - t_x)^3 \rho_s$$

The total mass of matter converted into spherulites at time t is thus equal to

$$m_S = \int_{x=0}^{x=t} \frac{4\pi K_g C^3 \rho_s (t - t_x)^3}{3\rho_L} dt_x \qquad (11.1)$$

which corresponds to

$$\frac{m_s}{m_0} = \frac{\pi K_g C^3 \rho_s t^4}{3\rho_L} \qquad (11.2)$$

and if m_L indicates the mass of residual liquid polymer, then

$$\frac{m_L}{m_0} = \frac{1 - \pi K_g C^3 \rho_s t^4}{3\rho_L} \qquad (11.3)$$

This relation is referred to as **Avrami equation**, and the time exponent (n) is called the **Avrami exponent**. Here it is equal to 4 as the nucleation is assumed to be directly proportional to time and the growth of spherulites is three-dimensional.

Thus, knowledge of the Avrami exponent gives information about both the nucleation and the spherulite growth.

The above relations are valid only for the early stages of crystallization, as they do not take into account the consumption of the crystallizable matter. For crystallizations reaching an advanced stage, one has to imperatively take into account that the crystalline growth stops as soon as spherulites come into contact. If m_s is the mass of crystallizable polymer that has germinated at time t_x during the time interval dx and has grown at the time t, where m_s' is this same quantity without the spherulites coming into contact, then it can be written

$$\frac{dm_s}{dm_s'} = 1 - \frac{m_s}{m_0}$$

According to relation (11.2), one has

$$dm_s' = \frac{4\pi m_0 K_g C^3 \rho_s t_x^3 dt_x}{3\rho_L} = \frac{dm_s}{1 - m_s/m_0}$$

and after integration from $t_x = 0$ to $t_x = t$

$$\ln\left(\frac{m_0 - m_s}{m_0}\right) = \frac{-\pi K_0 C^3 \rho_s t^4}{3\rho_L}$$

Posing

$$\frac{\pi K_g C^3 \rho_s}{3\rho_L} \quad \text{as equal to } z \text{ the rate constant of crystallization,}$$

this gives

$$m_L/m_0 = \exp(-zt^4) \tag{11.4}$$

whose expansion of the two first terms into series gives equation (11.3).

The value of n can be obtained from dilatometry measurements by relating the extent of crystallization, namely $(M_0 - M_L)/M_0$ to the polymer concentration. The total volume of polymer partially solidified at time t is given by

$$V_t = \frac{m_L}{\rho_L} + \frac{m_0 - m_L}{\rho_S} = \frac{m_0}{\rho_S} + \frac{m_L}{\rho_L} - \frac{m_L}{\rho_S} = \frac{m_0}{\rho_S} + m_L\left(\frac{1}{\rho_L} - \frac{1}{\rho_S}\right)$$

Since $M_0 = \text{constant}$, we have

$$V_\infty = m_0/\rho_S.$$

In the same way,

$$V_0 = \frac{m_0}{\rho_L}$$

$$V_t = V_\infty + m_L(V_0 - V_\infty)/m_0$$

Table 11.3. Values of the Avrami exponent for different types of polymer crystallization

Morphology of Crystalline Zones	Type of Nucleation	Avrami's Exponent
Spherulites	Sporadic	$3 + 1 = 4$
Spherulites	Instantaneous	$3 + 0 = 3$
Discs	Sporadic	$2 + 1 = 3$
Discs	Instantaneous	$2 + 0 = 2$
Rods	Sporadic	$1 + 1 = 2$
Rods	Instantaneous	$1 + 0 = 1$

which corresponds to

$$m_L/m_0 = (V_t - V_\infty)/(V_0 - V_\infty)$$

Since heights can be considered directly proportional to volumes in dilatometry measurements, one has

$$\frac{m_L}{m_0} = \frac{H_t - H_\infty}{H_0 - H_\infty} = \exp(-zt^4)$$

in the case of an Avrami exponent equal to 4 or, more generally,

$$\frac{H_t - H_\infty}{H_0 - H_\infty} = \exp(-zt^n)\,(\text{with } n = \text{Avrami exponent})$$

The plotting of $\log\{-\log[(H_t - H_\infty)/(H_0 - H_\infty)]\}$ against $\log T$ gives access to the experimental value of n (from the slope) and to z (from the ordinate at the origin). However, the knowledge of C and K_g requires separate determinations. The theoretical values of n for the various crystallization processes and the morphology of the crystalline zones are given in Table 11.3.

Actually, it is very rare to obtain integer values of the Avrami exponent, which suggests the existence of parallel and/or competing processes of nucleation and growth of the crystalline zones. The treatment of experimental data using the Avrami model has thus less significance. Other, more complex models, which are better adapted to the case of polymers, were proposed—in particular, those that take into account the necessary disentanglement of the chains before crystallization.

LITERATURE

L. H. Sperling, *Introduction to Physical Polymer Science*, 2nd edition, Wiley, New York, 2002.

K. Mezghani and P. J. Phillips, *Crystallization Kinetics of Polymers, in Physical Properties of Polymers Handbook*, J. Mark (Ed.), AIP Press, Woodbury, NY 1996.

D. W. Van Krevelen, *Properties of Polymers*, Elsevier, Amsterdam, 1990.

L. Mandelkern, *Crystallization and Melting in Understanding Polymer Science*, Vol. 2, Pergamon Press, Elmsford, NY, 1989.

B. Wunderlich, *Macromolecular Physics*, Vols. 1, 2, and 3 Academic Press, New York, 1973, 1976, and 1980.

U. W. Gedde, *Polymer Physics*, Chapman & Hall, London, 1995.

G. Strobl, *The Physics of Polymers*, 2nd edition, Springer, Berlin, 1997.

C. W. Macosko, *Rheology: Principles, Measurements and Applications*, Wiley-VCH, New York, 1994.

12

MECHANICAL PROPERTIES OF POLYMERS

12.1. ORIGIN OF ELASTICITY IN POLYMERS

12.1.1. Preliminary Definitions

Generally, a body subjected to an external force undergoes a deformation. If the deformed body recovers instantaneously its initial dimensions upon removal of the force, the deformation undergone is said to be reversible. In physics, the reversibility of the deformation is called *elasticity*, and any material has a limit of deformation below which it is elastic.

The behavior of an ideal elastic material is independent of time. In response to a mechanical stress σ—which is the ratio of the force (F) to the cross-sectional area of the tested specimen (S)—and whatever the rate applied, the body undergoes a deformation proportional to the stress. In the case of a monoaxial stretching, tensile stress and tensile strain are related by Hooke's law (linear part of stress–strain curves; see, for example, Figure 12.15).

$$\sigma_{11} = E\varepsilon^* \tag{12.1}$$

where E is Young's modulus and ε is the relative strain expressed by

$$\varepsilon = (L - L_0)/L_0 = \Delta L/L_0 \tag{12.2}$$

*In the notation σ_{ij}, the first number indicates the direction perpendicular to the plane on which a constraint is applied and the second number determines the direction of the constraint.

Organic and Physical Chemistry of Polymers, by Yves Gnanou and Michel Fontanille
Copyright © 2008 John Wiley & Sons, Inc.

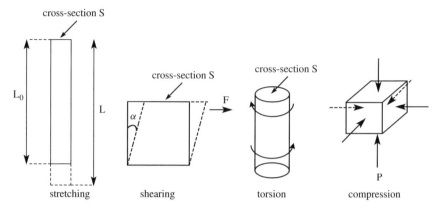

Figure 12.1. Commonly observed modes of stress.

(refer to Figure 12.1). Materials which obey Hooke's law are said to be *Hookean*, and the maximum stretching at which this law still applies is called the *elastic limit*.

The other modes of stress—shearing, torsion, compression—also involve a proportional modification of the shape or dimensions of elastic material but the corresponding modulus, that is the ratio of each type of stress to the corresponding deformation will be different from E; thus, the shear modulus (G) is the ratio of shear stress (σ_{21}) to the corresponding deformation (γ), while the bulk modulus (K) is the ratio of the hydrostatic pressure (p) to the volumic dilation ($\Delta V/V_0$).

In Hookean materials, the three moduli (E, G, K) are interrelated by the Poisson ratio (q), which is a measure of the lateral contraction undergone by a sample during a stretching experiment:

$$q = \frac{1}{2}\left(1 - \frac{1}{V}\right)\left(\frac{\partial V}{\partial \varepsilon}\right) \tag{12.3}$$

V is the volume of the sample and q can take values between 0 and 1/2. If the stretching is not accompanied by a significant change of volume, $\partial V/\partial \varepsilon$ is then equal to 0 and therefore q is equal to 0.5; this value is characteristic of ideal elastomers. If the lateral contraction is totally absent, q is then equal to 0; in the case of steel, q is equal to 0.2; and in the case of quartz, the value of q is equal to 0.07.

In the region of elastic deformation, the relation between the various moduli is written as

$$E = 2G(1 + q) = 3K(1 - 2q) \tag{12.4}$$

The ratio E/G is roughly equal to 2.5 for rigid solids and 3 for elastomers.

Polymers behave as ideal elastic materials only for relatively short periods of time because they are basically *viscoelastic* bodies. It is in this context that the elasticity of polymers is described below.

12.1.2. Two Types of Elasticity

Although samples of rubber and of steel can be regarded as Hookean (within the limit of small deformations), they do not undergo the same phenomena during deformation; upon stretching, a steel sample experiences a cooling phenomenon while that of rubber a heating one.

This means that the actual mechanisms of deformation are different in these two materials when a stress is applied. The deformation of a steel sample causes an *affine** displacement of the iron atoms compared to their equilibrium position; since the energy required to perform this work is provided by the system, steel cools; such elasticity is said to be of enthalpic origin. On the contrary, the deformation of rubber induces an orientation of the polymer in the direction of the stress generating extramolecular interactions that cause the phenomenon of heating. Because such a stretching also reduces the number of possible conformations of the polymer segments—without modifying the valence angles—such elasticity is said to be of entropic origin.

12.1.2.1. Elasticity in Rigid Polymers.
Rigid polymers, those amorphous or strongly oriented, are characterized by an elasticity of enthalpic origin. Depending upon the degree of orientation of their chains, their Young modulus (also called tensile modulus or elastic modulus, E) can vary in significant proportions; because the chain elements are randomly dispersed in an amorphous polymer sample, the stretching forces only pull those which are oriented in the direction of deformation. As a result, the higher the proportion of chain segments oriented in the direction of stretching, the higher the tensile modulus of the sample. Thus, ultra-stretched polymer fibers or liquid-crystalline polymers exhibit a particularly high elastic modulus (E). Because the chain orientation is never perfect, the tensile modulus of such samples is lower than that of segments that would be perfectly oriented.

To determine such a modulus, the measurement of the deformation is not made at the macroscopic level but rather at the molecular one using Raman spectroscopy and/or X-ray diffractometry. These experimental methods give access to the "absolute modulus": from the modifications induced by the applied stress in the Bragg reflection—and therefore in the position of atoms—, one can indeed evaluate Young's modulus in the direction of stress—and even in the perpendicular direction—thanks to the Hooke law. The values of moduli obtained in this manner are remarkably high. Diamond, which is exclusively constituted of carbon–carbon bonds, has a tensile modulus equal to 1160 GPa in the [110] direction for a cross-section of 0.049 nm^2. In comparison, polyethylene chains which also consist of C–C bonds substituted by hydrogen atoms and whose cross-section is 0.180 nm^2, should exhibit a tensile modulus of about 310 GPa [*i.e.,* 1160 (0.049/0.180)]. This value corresponds almost ideally to the absolute modulus of polyethylene fibers determined at the molecular level by X-ray diffractometry. In contrast, the tensile modulus obtained from a macroscopic measurement of the deformation represents

Affine qualifies a property or a deformation in a plane or in space that does not vary by means of a linear correspondence.

only one-third of this absolute value (130 GPa); this means that chain orientation is far from perfect even in a highly stretched sample!

Treloar proposed a model that allows the calculation of the theoretical tensile modulus for hydrocarbon chains. In this model, chains are all assumed to locally adopt transoidal conformations, and its purpose is to calculate the stretching of chains consisting of n C–C bonds of length L. At rest, the contour length of such a chain can be written as

$$r_{cont} = nL \sin(\theta/2)$$

where $\theta = 109°28'$ represents the valence angle. Expressing r_{cont} as a function of α, the complementary angle of θ ($\alpha = 180° - \theta$), one obtains

$$r_{cont} = nL \cos(\alpha/2) \qquad \text{or also } r_{cont} = nL \cos \beta$$

where $\beta = \alpha/2$. A force applied to such a chain contributes to an increase of its bond length by an increment $\Delta\beta$ in β. Therefore, the chain is stretched to an extent

$$\Delta r_{cont} = \Delta[nL \cos \beta] = n[\Delta L \cos \beta - L\Delta\beta \sin \beta] \qquad (12.5)$$

As for ΔL and $\Delta\beta$, they can be calculated as follows: in the direction corresponding to the chain axis, the component of the force (F) applied is $F \cos \beta$. The C–C bond is stretched to an extent $\Delta L = F \cos \beta/K_b$, where K_b is the constant that can be obtained by Raman or infrared spectroscopy. As for the valence angle, its increase $\Delta\theta$ is equal to the ratio of the angular moment exerted on each valence angle to the corresponding stress (K_θ). Since the moment M corresponds to $M = 1/2 \, FL \sin \beta$ and $\Delta\theta$ to $-2\Delta\beta$, one can easily deduce the expression for $\Delta\beta$ and then that for the theoretical Young modulus (E_{th}):

$$\Delta\beta = -FL \sin \beta/4K_\theta$$

In Hooke's law ($\sigma = E\varepsilon$), where the stress is the ratio of the force F to the cross-section (S_c) of the chain and β is the half of the complementary angle to θ, E_{th} is deduced as follows:

$$E_{th} = \frac{L \sin(\theta/2)}{S_c} \left[\frac{\sin^2(\theta/2)}{K_b} + \frac{L^2 \cos^2(\theta/2)}{4K_\theta} \right] \qquad (12.6)$$

This expression shows that the smaller the cross-section of the chain, the higher E_{th}; it applies to any chain with local transoidal conformation irrespective of its nature. For instance, the theoretical modulus obtained from this expression gives a value of 180 GPa for fibers of poly(p-phenyleneterephtalamide), which is hardly different from the "absolute modulus" measured at the molecular level (200 GPa) and even from the tensile modulus (132 GPa) determined from a macroscopic sample. For polyethylene chains, the calculated value of 340 GPa is in agreement with the "absolute modulus" value of 325 GPa.

12.1.2.2. Elasticity of Rubber. Rubber and more generally elastomers are characterized by their capability to undergo large deformations before breaking while exhibiting a Hookean behavior. They simultaneously resemble liquids by their ability to undergo deformations and resemble solids by their capacity to recover their initial dimensions.

They are typically slightly cross-linked long polymer chains whose glass transition temperature is lower than that of their use. The necessity of long chains can be easily understood: as the deformation of the sample merely implies a rearrangement of the chains, this deformation can be larger as the chains exhibit a larger number of possible conformations; that is, all the chains are long. The presence of junctions which can be occasional (entanglements) or permanent (cross-links), is essential for the restoration of the initial dimensions of the sample after the removal of the stress; indeed, except for very long chains, non-cross-linked polymers would dissipate the orientation induced by the deformation by acquiring different conformations. Experimentally, one can induce an elastomeric behavior for any polymer, provided that the temperature of the experiment is properly selected or a plasticizer is added.

Gough (contemporary to Dalton) in 1805, then Joule (in collaboration with Kelvin) in 1859, who was working on vulcanized rubber, observed that the stretching of such a sample results in a rise of the temperature as in the case of gases. In the meantime they observed that the same sample maintained under constant load will contract as the temperature is raised. They concluded, on the one hand, that $(\partial L/\partial T)_{p,F}$ is negative and, on the other, that $(\partial F/\partial T)_{p,L}$ is positive for rubbery materials. Then it was not before 1932 (*i.e.,* a few years after Staudinger had shown the polymeric nature of rubber) that a relation could be established between the behavior of rubbers as described previously and the origin of their elasticity. The stretching of an elastomeric sample—and consequently of the corresponding chains—orients the latter and thus contributes to decrease their entropy of conformation.

12.2. ELASTIC BEHAVIOR OF ELASTOMERS

12.2.1. Thermodynamic Relationships

According to the first law of thermodynamics, the variation of the internal energy $(d\mathcal{E})$ of an elastic body is the sum of the heat absorbed (dQ) and the work $(-dW)$ done by this body on the surroundings:

$$d\mathcal{E} = dQ - dW \qquad (12.7)$$

If the process to which the sample is subjected is reversible, dQ is equal to TdS (S being the entropy of the elastic body). As for dW, it includes an elastic contribution FdL, in addition to the term due to the external pressure (pdV). One can thus write

$$dW = pdV - FdL$$

where F denotes the drawing force being exerted on the sample. By taking into account the two preceding expressions, one obtains

$$dE = TdS - pdV + FdL$$

Expressing the Gibbs free energy

$$G = H - TS = E + pV - TS$$

one obtains, after derivation,

$$dG = dE + pdV + Vdp - TdS - SdT$$

By introducing the expression corresponding to dE, one obtains

$$dG = Vdp - SdT + FdL$$

This equation expresses the differential of the free energy in terms of the differentials of the experimentally accessible independent variables p, T, and L.

Since G is a variable of state and forms exact differentials, an alternative expression for dG is

$$dG = \left(\frac{\partial G}{\partial p}\right)_{T,L} dp + \left(\frac{\partial G}{\partial T}\right)_{p,l} dT + \left(\frac{\partial G}{\partial L}\right)_{p,T} dL \qquad (12.8)$$

It follows from equation (12.8) that

$$F = \left(\frac{\partial G}{\partial L}\right)_{p,T}$$

Taking into account the definition of G, the force F can thus be expressed in the following form:

$$F = \left(\frac{\partial H}{\partial L}\right)_{p,T} - T \left(\frac{\partial S}{\partial L}\right)_{p,T} \qquad (12.9)$$

This expression is sometimes called the equation of state for elastomers, in analogy with the thermodynamic equation of state, that is,

$$p = \left(\frac{\partial E}{\partial V}\right)_{T} - T \left(\frac{\partial S}{\partial V}\right)_{T} \qquad (12.10)$$

in which pressure and force, on the one hand, and volume and length, on the other hand, play parallel roles. Equation (12.9) shows that the force required to stretch an elastomer sample includes two strain-dependent contributions, one enthalpic and the other one entropic. Similarly, the pressure of a system also includes two contributions varying with the volume. Noting that the internal energy of a perfect gas is independent of the volume (no molecular interactions between molecules), which means that $(\partial \mathcal{E}/\partial V)_T = 0$ and thus that $p = T(\partial S/\partial V)_T$, the analogy between gas and elastomers leads to the outcome that $(\partial H/\partial L)_{p,T}$ is equal to 0 for an ideal elastomer. Under these conditions,

$$F = -T \left(\frac{\partial S}{\partial L} \right)_{p,T} \tag{12.11}$$

As a matter of fact, interactions between molecules can be modified upon stretching the sample and hence the chains, so that $(\partial H/\partial L)_{p,T}$ cannot be strictly taken equal to 0. To evaluate the respective contributions of the enthalpy and entropy terms, use can be made of the rules of partial differentials:

$$\frac{\partial}{\partial T} \left(\frac{\partial G}{\partial L} \right)_{T,p} = \frac{\partial}{\partial L} \left(\frac{\partial G}{\partial T} \right)_{p,L}$$

which gives

$$-\left(\frac{\partial S}{\partial L} \right)_{T,p} = \left(\frac{\partial F}{\partial T} \right)_{p,L} \tag{12.12}$$

This relation shows that if the second term is positive, an increase of the length of the sample results in a reduction of its entropy. Thus, equation (12.9) can be written in the form

$$F = \left(\frac{\partial H}{\partial L} \right)_{p,T} + T \left(\frac{\partial F}{\partial T} \right)_{p,L} \tag{12.13}$$

This expression reflects not only the variations of energy and entropy induced by the chain stretching and orientation (which is sought by the experimenter), but also the variations caused by the change of the sample volume.

Similar expressions can be written as a function of the Helmholtz free energy A ($A = \mathcal{E} - TS$):

$$dA = -pdV - SdT + FdL$$

and thus

$$F = \left(\frac{\partial A}{\partial L} \right)_{T,V} = \left(\frac{\partial \mathcal{E}}{\partial L} \right)_{T,V} - T \left(\frac{\partial S}{\partial L} \right)_{T,V} \tag{12.14}$$

In the latter expression, $-(\partial S/\partial L)_{T,V}$ can be replaced by $(\partial F/\partial T)_{V,L}$ using equation (12.12), which gives

$$F = \left(\frac{\partial \mathcal{E}}{\partial L} \right)_{T,V} - T \left(\frac{\partial F}{\partial T} \right)_{V,L}$$

This expression is difficult to use in this form because the measurement of $(\partial F/\partial T)_{V,L}$ would necessarily induce an increase of the volume of the sample due to thermal expansion and would therefore require the application of a hydrostatic pressure to compensate. Such an experiment is thus not practical. Because Flory showed that $(\partial S/\partial L)_{T,V}$ corresponds to $(\partial F/\partial T)_{p,\lambda}$ where λ denotes the relative elongation L/L_0, F can accordingly be expressed as follows:

$$F = \left(\frac{\partial E}{\partial L}\right)_{T,V} + T \left(\frac{\partial F}{\partial T}\right)_{p,\lambda} \tag{12.15}$$

The latter equation is useful because it allows one to determine the variation of entropy and internal energy due to the chain orientation at constant volume from the measurement of F as a function of the temperature (L_0 should be measured at each temperature) at constant pressure and elongation. The experimenter then observes that the contribution of internal energy is negligible compared to that of entropy.

12.2.2. Statistical Theory of Rubber Elasticity

As indicated previously, an elastomer can be identified with an assembly of N_υ chains connected through N_μ randomly distributed cross-links that are separated from one another by a quadratic average end-to-end distance ($\langle r^2 \rangle_0$) satisfying a Gaussian distribution function $P(n, r)$ (see Chapter 5). In the following treatment, the network is considered ideal, without dangling chains and entanglements.

$$P(n, r) = [3/(2\pi\langle r^2 \rangle_0)]^{3/2} \exp[-3\langle r^2 \rangle/2\langle r^2 \rangle_0] \tag{12.16}$$

On the one hand, this function can be used to calculate the probability of finding a given chain end of the network in the spherical envelope of radius r and thickness dr, with the other end corresponding to the origin; on the other hand, the function can be used to calculate the entropy of the same chain:

$$S = k \ln P(n, r)$$

where k is Boltzmann's constant.

Thus the free energy $G_i(r)$ of this chain is written as

$$G_i(r) = H - kT \ln[3/(2\pi\langle r^2 \rangle_0)]^{3/2} + kT[3\langle r^2 \rangle/2\langle r^2 \rangle_0] \tag{12.17}$$

After regrouping the first two terms in the constant $C(T)$ depending on temperature, one obtains for an assembly of N_υ elastic chains

$$G(r) = C(T) + N_\upsilon kT[3\langle r^2 \rangle/2\langle r^2 \rangle_0]$$

The stretching of an elastomeric network changes its Gibbs free energy (ΔG_{elas}) in two ways: on the one hand, by inducing conformational changes within each elastic chain of the network [$G(r) - G(r0)$] and, on the other hand, by modifying the spatial distribution of the cross-links. A term corresponding to the dispersion

(ΔG_{disp}) and similar to that found in the case of gases where there is also volume expansion from V_0 to V has to be introduced:

$$\Delta G_{disp} = N_\mu kT \ln(V/V_0)$$

In this relation N_μ is the number of v-valent cross-links and can be written as

$$N_\mu = 2N_v/v \tag{12.18}$$

for an ideal network.

Under these conditions, ΔG_{elast} is expressed as

$$\Delta G_{elast} = G(r) - G(r)_0 + \Delta G_{disp} \tag{12.19}$$

Initially, when the network is at rest $\langle r^2 \rangle = \langle r^2 \rangle_0$, which gives for ΔG_{elast}

$$\Delta G_{elast} = N_v kT[3\langle r^2 \rangle/2\langle r^2 \rangle_0] - N_\mu kT(3/2) - N_\mu kT \ln(V/V_0)$$

an expression that can be simplified into

$$\Delta G_{elast} = (3N_v/2)kT[(\langle r^2 \rangle/\langle r^2 \rangle_0) - 1] - (2N_v/v)kT \ln(V/V_0) \tag{12.20}$$

Unless resorting to coherent neutron scattering, the experimenter does not have direct access to information contained in this expression, that is, to the variation of the dimension of elastic chains as a result of a macroscopic deformation. Models that express the impact of a stress applied at a macroscopic scale—and thus of the deformation undergone—on the molecular level have been proposed.

12.2.3. "Affine" and "Phantom" Models

The two models postulate an affine displacement of the positions occupied by the cross-links of the network resulting from a deformation, but differ about the movements undergone by these cross-links. For the Flory–Rehner affine model, cross-links move proportionally to the macroscopic deformation and remain in a given position of space at constant deformation. In the James–Guth "phantom" model, cross-links are assumed to freely move or fluctuate around an average position corresponding to the affine deformation. The amplitude of such fluctuations is independent of the deformation but depends on the valence of the cross-links and the length of elastic chains:

$$\langle \Delta r^2 \rangle = (2/v)\langle r^2 \rangle_0 \tag{12.21}$$

Actually these two models correspond to two extreme situations. The affine model is well appropriate to describe the case of networks made of short elastic chains—in this case, the fluctuation of cross-links is hindered by the presence of adjacent chains—whereas the "phantom" model is better suited to networks comprising long elastic chains. In the affine model, the macroscopic elongation λ_x in the direction x is translated in a corresponding deformation at the molecular level:

$$\lambda_x^2 = \langle x^2 \rangle / \langle x^2 \rangle_0 \tag{12.22}$$

x_0, y_0, z_0 being the coordinates of the vector r_0 connecting the two ends of a given elastic chain at rest, and x, y, z those of the same stretched chain. Since the sample considered is isotropic in the initial state, one can write

$$\sum_i^{N_v} x_0^2 = \sum_i^{N_v} y_0^2 = \sum_i^{N_v} z_0^2 = \frac{1}{3} \sum_i^{N_v} r_0^2 = \frac{N_v \langle r^2 \rangle_0}{3} \tag{12.23}$$

where N_v denotes the total number of chains present in the sample.

Due to the isotropic character, expression (12.22) can be rewritten in the form

$$\lambda_x{}^2 = \langle x^2 \rangle / (\langle r^2 \rangle_0 / 3)$$

In the three directions of space, this gives

$$\langle r^2 \rangle = \langle x^2 \rangle + \langle y^2 \rangle + \langle z^2 \rangle = (\lambda_x{}^2 + \lambda_y{}^2 + \lambda_z{}^2)(\langle r^2 \rangle_0 / 3)$$

Because the volume of the sample before deformation is equal to

$$V_0 = L_0^3 \quad \text{with} \quad L_0 = L_{x,0} = L_{y,0} = L_{z,0},$$

one obtains after deformation

$$V = L_x L_y L_z = \lambda_x \lambda_y \lambda_z L_0^3 = \lambda_x \lambda_y \lambda_z V_0$$

Taking into account these two last expressions for $\langle r^2 \rangle / \langle r^2 \rangle_0$ and V/V_0, the expression for ΔG_{elas} (12.20) can be reexpressed as

$$\Delta G_{elas} = N_v (kT/2)\{[\lambda_x^2 + \lambda_y^2 + \lambda_z^2 - 3] - (4/v)\ln(\lambda_x \lambda_y \lambda_z)\} \tag{12.24}$$

The variation of the Gibbs free energy due to the deformation of the network thus depends on the number of elastic chains, the valence of its cross-links, the elongation λ, and the temperature, but not on the chemical nature of the network.

12.2.4. Uniaxial Stretching of Elastomers

In an uniaxial stretching experiment in the direction x, elongation is expressed as

$$\lambda_x = L/L_0 \tag{12.25}$$

Assuming that such stretching occurs without change in volume, one can deduce

$$\lambda_y = \lambda_z = (1/\lambda_x)^{1/2} \qquad (12.26)$$

and in the opposite case,

$$\lambda_y = \lambda_z = [(V/V_0)(1/\lambda_x)]^{1/2} \qquad (12.27)$$

The expression for the variation of the free energy resulting from a deformation then simplifies to give

$$\Delta G_{elas} = N_\upsilon(kT/2)\{[\lambda^2 + 2(V/V_0)\lambda^{-1} - 3] - (4/\upsilon)\ln(V/V_0)\} \qquad (12.28)$$

By definition, the force F responsible for the elongation of a sample is written as

$$F = \partial(\Delta G_{elas}/\partial L)_{T,V} = \partial(\Delta G_{elas}/\partial\lambda)_{T,V}/L_0$$

that is,

$$F = N_\upsilon(kT/L_0)[\lambda - (V/V_0)\lambda^{-2}] \qquad (12.29)$$

Substituting the stress ($\sigma_{11} = F/A_0$ with $V_0 = A_0L_0$) for the force and introducing molar concentrations with $k = R/\mathcal{N}a$, one obtains the expression below for the applied stress at constant volume,

$$\sigma_{11} = \upsilon RT(\lambda - \lambda^{-2}) \qquad (12.30)$$

where υ is the molar concentration of elastic chains per unit volume, which is equal to $N_\upsilon/\mathcal{N}_a V_0$.

Using a slightly different calculation for the variation of the free energy (ΔG_{elast}), the phantom model leads to a very simple yet slightly modified expression:

$$\sigma_{11} = (\upsilon - \mu)RT(\lambda - \lambda^{-2}) \qquad (12.31)$$

where μ denotes the molar concentration of cross-links per unit volume.

The swelling of a network by a solvent present in large excess—and hence the network deformation—can be treated in the same manner as indicated previously. Such a swelling occurs in an isotropic manner, with the volume of the network changing from V_0 to V. The volume fraction of polymer is then equal to $\Phi_2 = V_0/V$. If λ_x, λ_y, λ_z denote the variations in dimensions induced simultaneously by the swelling and the deformations resulting from application of an uniaxial stretching, then one deduces that $\lambda_x\lambda_y\lambda_z = 1/\Phi_2$. In other words, λ_x can be expressed as the product of $L_{0,s}/L_{0,d}$ and λ, the deformation due uniquely to the stress applied to the sample ($L_{0,s}$ and $L_{0,d}$ being lengths of the unstretched sample in swollen and dry states, respectively). Since $L_{0,s}/L_{0,d}$ is equal to $(V/V_0)^{1/3}$, one obtains

$$\lambda_x = \lambda(V/V_0)^{1/3} = \lambda/\Phi_2^{1/3} \qquad (12.32)$$

From the following equality $\lambda_x\lambda_y\lambda_z = 1/\Phi_2$, it can be deduced that

$$\lambda_y = \lambda_z = (\Phi_2\lambda_x)^{-1/2} = 1/\lambda^{1/2}\Phi_2^{1/3} \tag{12.33}$$

After introducing the terms corresponding to λ_x, λ_y, and λ_z in the relation (12.28) giving ΔG_{elas}, and expressing the derivative of ΔG_{elas} with respect to the length L $(\partial(\Delta G_{elas}/\partial L)_{T,V})$, the expression of the force F and hence of the stress σ_{11} for a swollen network now becomes

$$\sigma_{11} = RT\upsilon\Phi_2^{1/3}(\lambda - \lambda^{-2}) \tag{12.34}$$

12.2.5. Real Behavior of an Elastomer

In practice, none of the two models—"affine" and "phantom"—accounts satisfactorily for the behavior of elastomers in the entire spectrum of the strains. The affine model is generally appropriate for small deformations, under conditions of limited motion of cross-links due to the presence of neighboring cross-links and of the entanglements. For larger deformations, when chains are disentangled, the experimental behavior is better described by the "phantom" model. Thus, the tensile modulus tends to decrease with the applied deformation and gradually comes close to that predicted by the "phantom" model (Figure 12.2). Mooney and Rivlin took into account this nearly general behavior in elastomers and thus proposed a semiempirical model. At rest, the network is considered isotropic and incompressible and is assumed to behave like a Hookean material upon shearing. The authors proposed the following expression for the stress:

$$\sigma_{11} = [2C_1 + (2C_2/\lambda)](\lambda - \lambda^{-2})$$

where C_1 and C_2 are coefficients that vary with the material under consideration and its molecular characteristics. Contrary to molecular models, the stress is supposed to vary with the deformation applied.

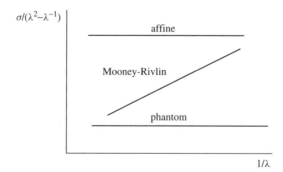

Figure 12.2. Variation of the reduced stress versus the reverse of elongation for various models of elastomeric networks.

The coefficient C_2 takes higher values when the network contains entanglements and the applied deformation is small. On the other hand, C_2 is close to zero in the case of highly swollen networks ($\Phi \leq 0.2$). As for the coefficient C_1, it can be identified with the RT_υ term of statistical models.

12.2.6. Shearing of the Networks

Contrary to an uniaxial stress, shearing implies deformations in two opposite directions (x, y) of space, with the third (z) being preserved from any change. Thus, a sample subjected to shearing undergoes a stretching λ_x in direction x and a compression $\lambda_y = 1/\lambda_x$ in the other direction (y), with its volume $(\lambda_x \lambda_y \lambda_z = 1)$ remaining constant.

A strain due to shearing is defined as

$$\gamma = \lambda_x - \lambda_y = \lambda - \lambda^{-1} \tag{12.35}$$

In such a context the second term of expression (12.24) is equal to 0 and the first term $(\lambda_x{}^2 + \lambda_y{}^2 + \lambda_z{}^2 - 3)$ becomes

$$(\lambda^2 + \lambda^{-2} - 2) = (\lambda - \lambda^{-1})^2 \tag{12.36}$$

Thus the variation of the free energy resulting from shearing is written as

$$\Delta G_{\text{elas}} = N_\upsilon(kT/2)(\lambda - \lambda^{-1})^2 = N_\upsilon(kT/2)\gamma^2 \tag{12.37}$$

The calculation of the derivative of ΔG_{elas} with respect to γ gives

$$Q = (\partial \Delta G_{\text{elas}}/\partial \gamma)_{T,V} = N_\upsilon kT(\lambda - \lambda^{-1}) = N_\upsilon kT\gamma \tag{12.38}$$

The shear stress (σ_{12}, denoted by $\mathcal{T}$ in order to avoid any confusion with tensile stress σ_{11}) applies not to a surface as in the case of the tensile stress, but to the entire volume (V_0) so that

$$\mathcal{T} = Q/V_0 \tag{12.39}$$

Substituting the molar concentration of elastic chains (υ) for their number (N_υ) $(k = R/\mathcal{N}_a$ and $\upsilon = N_\upsilon/\mathcal{N}_a V_0)$ results in

$$\mathcal{T} = RT\upsilon\gamma \tag{12.40}$$

The shearing modulus $(\mathcal{G})$ corresponds to $RT\upsilon$. This term is also found in the expression of the tensile stress but with a different strain component [$\sigma_{11} = \upsilon RT(\lambda - \lambda^{-2})$]. To express the latter relationship in the form of Hooke's law ($\sigma_{11} = E\varepsilon$), one can observe that $\varepsilon = \lambda - 1$ and hence $\lambda - \lambda^{-2} = 1 + \varepsilon - (1 + \varepsilon)^{-2} \approx 3\varepsilon$. Thus the Young modulus (E) corresponds to

$$E = 3\upsilon RT$$

The relation between the two moduli, G and E, is expressed as

$$E = 3G$$

in agreement with expression (12.4), which was established on the basis of a deformation with constant volume, with a Poisson ratio equal to 1/2.

12.3. VISCOELASTICITY OF POLYMERS

12.3.1. A Specific Property

Beyond short periods of time, a polymer cannot be regarded as a purely elastic material for the ratio of the stress applied to the strain undergone does not remain constant with time. Polymers indeed behave simultaneously like Hookean objects and like Newtonian (or non-Newtonian) viscous liquids. The matter in the latter case is not elastic and does not strain but flows under mechanical forces:

$$\mathcal{T} = \eta\dot{\gamma} \tag{12.41}$$

where $\dot{\gamma}$ denotes the rate of shearing and η the viscosity of this liquid.

A body exhibiting simultaneously elastic properties (which are independent of time) and a viscous behavior (which depends by definition on time) is called *viscoelastic* (Figure 12.3). Depending on the time scale of the experiment, either the elastic or the viscous component dominates.

Viscoelasticity is typical of polymers; it can be characterized through three types of experiments, creep, stress relaxation, and dynamic mechanical analyses.

In a **creep** experiment, a body is subjected to a constant stress (σ_0) under isothermal conditions and the variation of its dimensions is followed as a function of time. After a rapid application of the stress to the sample—whatever the nature of this

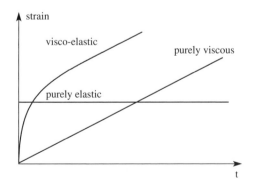

Figure 12.3. Typical behavior of viscoelastic materials.

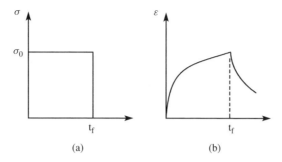

(a)   (b)

Figure 12.4. Principle of a creep experiment: (a) Application of constant stress (σ_0) between t_0 and t_f, (b) Measurement of the deformation as a function of time.

stress, uniaxial or caused by shearing—, the strain $[\varepsilon(t)]$ is recorded as a function of time (Figure 12.4). One can then define the elongational compliance $[D(t)]$ and the shearing compliance $[J(t)]$ of this sample in the following manner:

$$D(t) = \varepsilon(t)/\sigma_{11} \quad \text{and} \quad J(t) = \gamma(t)/T \tag{12.42}$$

In a **stress relaxation** test, the sample is subjected to a sudden deformation (ε_0) that is maintained constant and the variation of the stress—$\sigma_{11}(t)$ or $T(t)$—is followed as a function of time (Figure 12.5).

Then the tensile $[E(t)]$ or shearing relaxation modulus $[G(t)]$ can be defined as

$$E(t) = \sigma_{11}(t)/\varepsilon_0 \tag{12.43}$$

$$G(t) = T(t)/\gamma_0 \tag{12.44}$$

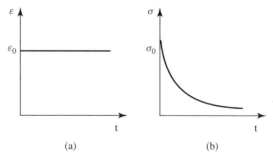

(a)   (b)

Figure 12.5. Principle of a stress relaxation experiment: (a) Application of a constant deformation (ε_0), (b) Variation of the stress as a function of time.

12.3.2. Dynamic Mechanical Analysis

A complete description of the viscoelastic behavior of a polymer through creep and relaxation tests would require monitoring over long periods of time. This limitation can be overcome through dynamic experiments. The latter involve stress and strain that vary periodically in a sinusoidal manner. The sinusoidal oscillation of frequency (υ) corresponds to cycles/s or ω ($= 2\pi\upsilon$) corresponds to rad/s. Practically, a sinusoidal experiment at frequency υ is equivalent to creep or relaxation experiments at a time $t = 1/\omega$.

When a dynamic stress is applied, the latter is directly proportional to the strain only in the limit of small deformations; stress and strain then vary sinusoidally and, in certain cases, completely in phase. When submitted at sufficiently high frequencies, a polymer network also behaves in an exclusively elastic manner within the limit of small deformations.

In contrast, stress and strain can be $90°$ out of phase when sufficiently low frequencies are used, a situation that is characteristic of liquid bodies. At intermediate frequencies, the phase difference between stress and strain is less pronounced. Sinusoidal variations of the stress can be represented as a rotating vector (0A) (Figure 12.6) whose projection (0B) on the vertical axis corresponds to the stress applied at a given time in that direction.

In such a representation, the vector 0A rotates at frequency ω, which is that of the sinusoidal stress, with the direction 0A thus corresponding to that of the maximum stress. The cycle of deformation undergone by the sample is symbolized by the vector 0C, which rotates at the same frequency (ω); its projection 0D on the vertical axis denotes the deformation of the sample. The strain thus lags behind the stress by the phase angle δ, also called the *loss angle*.

In such a diagram, the strain vector can be divided into two components: the first, 0E, in-phase with the stress and the second, 0F, out-of-phase. The projection of 0E on the vertical axis (0H) reflects the strain in-phase with the stress, and the projection of 0F on the same axis (0I) corresponds to the strain out-of-phase ($90°$) with the stress in that direction. If a sinusoidal stress is applied by uniaxial shearing, this stress will be defined as a function of time by the following relation:

$$\mathcal{T}(t) = \mathcal{T}_0 \sin \omega t \tag{12.45}$$

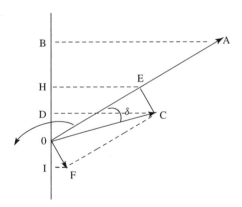

Figure 12.6. Stress vector and strain vector split into two components in a dynamic experiment.

T_0 denotes the maximum amplitude of this stress and ω denotes its frequency. For purely elastic "Hookean" bodies with no energy dissipated, the strain is written as

$$\gamma_t = \gamma_0 \sin \omega t$$

where γ_0 is the maximum amplitude of strain.

For real viscoelastic materials, shear deformations trail the applied stress, with some energy being dissipated in viscous resistance (Figure 12.7) which gives the relation

$$\gamma_t = \gamma_0 \sin(\omega t - \delta) \tag{12.46}$$

As shown earlier, the stress includes two components: one in-phase with the strain and the other one in-advance with respect to the latter; the in-phase component is expressed as

$$T'_{(\omega)} = T_0 \cos \delta \tag{12.47}$$

and the one in-advance is written as

$$T''_{(\omega)} = T_0 \sin \delta \tag{12.48}$$

Similarly, the modulus can be expressed through two components: one in-phase with the direction of deformation (0C) and the other one out-of-phase in advance of $90°$. From these elements, the storage modulus can be defined as the ratio of the in-phase component of the stress to the maximum amplitude of the strain. This storage modulus $G'(\omega)$ can be written as

$$G'(\omega) = T'(\omega)/\gamma_0 = (T_0/\gamma_0) \cos \delta$$

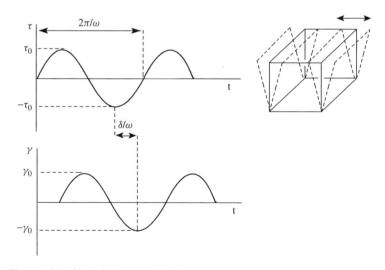

Figure 12.7. Sinusoidal stress (τ) and the corresponding out-of-phase strain (γ).

The ratio of the out-of-phase (in-advance of $90°$) component of the stress to the strain corresponds to the loss modulus $G''(\omega)$:

$$G''(\omega) = T''(\omega)/\gamma_0 = (T_0/\gamma_0) \sin \delta$$

Likewise, one can also define the *instantaneous compliance* or *storage compliance* $J'(\omega)$; this corresponds to the ratio of the in-phase strain component to the stress; the loss compliance $[J''(\omega)]$ then corresponds to the ratio of the out-of-phase component of the strain lagging behind the stress to the latter. $G'(\omega)$ and $J'(\omega)$ reflect the propensity of a sample to retain a supplied mechanical energy and to restore it in the form of an elastic strain; $G''(\omega)$ and $J''(\omega)$, on the other hand, reflect the loss of this same energy due to viscous dissipation (flow).

Under the conditions of a brief shearing of low amplitude, the storage modulus $[G'(\omega)]$ is equivalent to the previously defined shearing modulus (G). At a given temperature and frequency, a sample can thus be identified by characteristics such as $G'(\omega)$, $G''(\omega)$, $J'(\omega)$, $J''(\omega)$, and $\tan \delta$. For the sake of didactics, two different representations—rotating vector (Figure 12.6) and sinusoidal shear stress (Figure. 12.7)—have been used above to describe the behavior of a system submitted to a dynamic experiment; in practice, the experimenter prefers to resort to complex variables to express characteristics such as the dynamic mechanical modulus. Thus, for a sinusoidal shearing, one can define a complex modulus $[G^*(\omega)]$ whose real component is the storage modulus (G') and the imaginary one is the loss modulus (G''):

$$G^*(\omega) = G'(\omega) + i G''(\omega) \tag{12.49}$$

In a similar manner, the *complex compliance* $J^*(\omega)$ is the sum of two components: one of these is real, corresponding to the storage component $[J'(\omega)]$, while the other is imaginary, corresponding to the loss compliance:

$$J^*(\omega) = J'(\omega) - i J''(\omega) = 1/G^* = 1/[G'(\omega) + i G''(\omega)] \tag{12.50}$$

Using complex notations, the complex shear modulus (G^*) becomes

$$|G^*| = [(G')^2 + (G'')^2]^{1/2} \tag{12.51}$$

which gives

$$\tan \delta = G''(\omega)/G'(\omega) = J''(\omega)/J'(\omega) \tag{12.52}$$

$\tan \delta$ is thus a measure of the ratio of the energy dissipated by the sample in the form of heat to the energy retained and subsequently restored during one cycle of sinusoidal deformation. The heat produced results from the chains' flow and from their friction; it mirrors the damping capacity of the sample. The energy dissipated by cycle is written as $\pi \gamma^2 G''$.

Dissipative phenomena can also be accounted for through the viscosity, which corresponds to the ratio of a stress to the rate of deformation—in fact the shearing—and can also be expressed in a complex manner. The complex viscosity (η^*) is written as follows:

$$\eta^*(\omega) = G^*(\omega)/i\omega = [(G''(\omega)/\omega) - i(G'(\omega)/\omega)] = \eta'(\omega) - i\eta''(\omega) \quad (12.53)$$

From this it follows that $\eta'(\omega)$ is equal to $G''(\omega)/\omega$ and $\eta''(\omega)$ is equal to $G'(\omega)/\omega$. Like G'', $\eta'(\omega)$ (also called *dynamic viscosity*) reflects the dissipation of energy and is equivalent to η_0 (viscosity in continuous flow) within the range of low-frequency shearing.

12.3.3. Linear Viscoelasticity

Linear viscoelasticity is characterized at a given temperature and frequency by a linear dependence of stress relaxation on strain even if they are not necessarily in-phase. It can be accounted for by combining both the Hooke and Newton laws which stipulate a linear variation of the stress with the strain for the first and of the stress with the rate of deformation for the second.

This proportionality between the cause and the effect can be expressed in a more general manner by the Boltzmann superposition principle. The effect A produced by the sum of several causes or loads is equivalent to the sum of the effects produced by each of these causes taken separately:

$$A = \sum_i k_i C_i = \sum_i a_i \quad \text{with} \quad a_i = k_i C_i \quad (12.54)$$

In practical terms, the assumption of linear viscoelasticity implies that mechanical behaviors observed under different conditions are interdependent; in other words, the behavior of a material in a given circumstance or condition can be predicted from the behavior observed in a completely different circumstance or condition.

In addition to the Boltzmann superposition principle, the second consequence of linear viscoelasticity is the **time–temperature equivalence**, which will be described in greater detail later on. This equivalence implies that functions such as $\sigma = f(\varepsilon)$, but also moduli, behave at constant temperature and various extensional rates similarly to analogues that are measured at constant extensional rates and various temperatures. Time- and temperature-dependent variables such as the tensile and shear moduli (E, G) and the tensile and shear compliance (D, J) can be transformed from $E = f(t)$ into $E = f(T)$ and vice versa, in the limit of small deformations and homogeneous, isotropic, and amorphous samples. These principles are indeed not valid when the sample is anisotropic or is largely strained.

12.3.4. Boltzmann Superposition Principle

The Boltzmann superposition principle states that:

1. The stress or strain status of a sample at a given instant mirrors all the stresses or strains it has undergone.

2. Each new stress contributes independently to the final deformation, which thus corresponds to the algebric sum of the various loads applied. In other words, a deformation or a recovery caused by an additional load or its removal is independent of previous loadings or unloadings.

In a tensile experiment, if the initial constraint $\sigma_{11,0}$ is followed by a second one $\sigma_{11,1}$ at time t_1, the deformation resulting only from $\sigma_{11,1}$ is written as

$$\varepsilon(t) = \varepsilon_{11,1} D(t - t_1) \tag{12.55}$$

with the total strain being expressed as

$$\varepsilon(t) = \sigma_{11,0} D(t) + \sigma_{11,1} D(t - t_1)$$

In the general case, the following relationship is obtained:

$$\varepsilon(t) = \sigma_{11,0} D(t) + \sum_{i=1}^{n} \sigma_{11,i} D(t - t_i) \tag{12.56}$$

In the particular case where the initial constraint and thus the stress are stopped at time t_1 (creep experiment Figure 12.8), $\varepsilon(t)$ becomes

$$\varepsilon(t) = \sigma_{11,0}[D(t) - D(t - t_1)] \tag{12.57}$$

The stress $\sigma_{11,0}$ applied at time $t = 0$ results in a deformation equal to ε_1 at time t_1. Discontinuing this stress at time t_1 causes the initial deformation (ε_1) to decrease to a value equal to ε at t_2. The difference in deformation $\Delta\varepsilon_1$ is expressed as

$$\varepsilon = \varepsilon_1 - \Delta\varepsilon_1$$

If the stress is not discontinued at t_1, the deformation grows steadily to a value equal to ε_2 at t_2 and the difference $\Delta\varepsilon_2$ then corresponds to

$$\varepsilon = \varepsilon_2 - \Delta\varepsilon_2$$

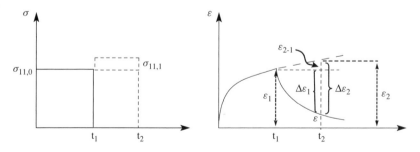

Figure 12.8. Illustration of the Boltzmann superposition principle for a creep experiment.

$\Delta\varepsilon_2$ can also be written as

$$\Delta\varepsilon_2 = \varepsilon_2 - \varepsilon_1 + \Delta\varepsilon_1 = \Delta\varepsilon_1 + \varepsilon_{2-1}$$

a relation in which ε_{2-1} denotes the additional deformation—with respect to ε_1—that this sample would undergo at time t_2, if a putative constraint $\sigma_{11,1}$ is applied in addition to $\sigma_{11,0}$ at time t_1.

In other words, the deformation caused by an additional load or stress is independent of the preceding ones. By adopting a similar reasoning for the case of a relaxation experiment, one obtains the following relation for the stress versus time:

$$\sigma_{11}(t) = \varepsilon_0 E(t) + \sum_{i=1}^{n} \varepsilon_i E(t - t_i) \qquad (12.58)$$

12.3.5. Empirical Analogical Models

As indicated previously, the behavior of a polymer satisfying the criterion of linear viscoelasticity and hence the Boltzmann principle can be described as a linear combination of Newtonian viscous and Hookean elastic behaviors.

To illustrate this linear combination, various analogical models with no relationship with the molecular nature of the phenomena were proposed, identifying a polymer with a combination of springs and dashpots. The spring can be strained without inertia and thus reflects a purely elastic mechanical behavior whereas dashpots, which are pistons moving in cylinders filled with a viscous liquid, cannot respond instantaneously to a stress. These two elements were thus associated under various combinations to simulate the response of a viscoelastic body to a mechanical stress.

12.3.5.1. The Maxwell Model. In this model the spring and the dashpot are connected linearly in series:

As was previously mentioned, stress (σ) and strain (ε) undergone by a spring can be related by the Hooke law:

$$\sigma_r = E\varepsilon_r \qquad (12.59)$$

In a dashpot, stress and strain obey the Newton law:

$$\sigma_a = \eta \frac{d\varepsilon_a}{dt}$$

In this model, the total stress (σ) felt by the spring and by the dashpot is identical:

$$\sigma = \sigma_r = \sigma_a$$

On the other hand, the total deformation is additive:

$$\varepsilon = \varepsilon_r + \varepsilon_a$$

Under these conditions, relation (12.59) can be written as

$$\frac{d\sigma_r}{dt} = E\frac{d\varepsilon_r}{dt} = \frac{d\sigma}{dt} \tag{12.60}$$

in the same way

$$\frac{d\varepsilon}{dt} = \frac{1}{E}\frac{d\sigma}{dt} + \frac{\sigma}{\eta} \tag{12.61}$$

In the case of relaxation or creep experiments, this differential equation can be easily solved. For a relaxation experiment the deformation is constant by definition:

$$\varepsilon = \varepsilon_0 \quad \text{and} \quad d\varepsilon/dt = 0$$

Relation (12.61) becomes

$$\frac{d\sigma}{\sigma} = -\frac{E}{\eta}dt \tag{12.62}$$

which gives, after integration, for $\sigma = \sigma_0$ at $t = t_0$

$$\sigma(t) = \sigma_{11,0} \exp\left(-\frac{E}{\eta}t\right) \tag{12.63}$$

Under these conditions the relaxation modulus, $E(t)$ ($\equiv \sigma(t)/\varepsilon_0$) is written as

$$E(t) = \frac{\sigma_{11,0}}{\varepsilon_0} \exp\left(-\frac{E}{\eta}t\right) \tag{12.64}$$

After noting that the ratio $\eta/E = t_{\text{relax}}$ has the dimensions of time and $\sigma_{11,0}/\varepsilon_0$ corresponds to E, one obtains

$$E(t) = E \exp\left(-\frac{t}{t_{\text{relax}}}\right) \tag{12.65}$$

and t_{relax} is called the *relaxation time*.

For very short times, the Maxwell model predicts that the material behaves essentially like a spring for a relaxation experiment. Over long times, it actually

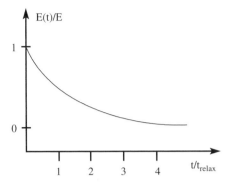

Figure 12.9. Behavior of a material agreeing with Maxwell model for a relaxation test.

behaves only like a dashpot, where the stress decreases exponentially with time in a sample subjected to a constant strain ε_0. For a time comparable to t_{relax}, which is the time required to reduce the stress by a factor $(1/e)$ (i.e., by 36.8%) with respect to initial one, the model predicts a response including simultaneously the contribution of both the spring and the dashpot (see Figure 12.9).

The treatment of a creep experiment by the same Maxwell model leads to the following equations. In such an experiment, the stress is constant: $\sigma = \sigma_{11,0}$, $d\sigma/dt = 0$ and thus the relation (12.61) can be written as

$$\frac{d\varepsilon}{dt} = \frac{\sigma_{11,0}}{\eta} \tag{12.66}$$

which, upon integration for the strain, gives

$$\varepsilon(t) = \frac{\sigma_{11,0}}{\eta}t + \varepsilon_0 \tag{12.67}$$

The creep compliance is thus expressed as follows:

$$D(t) = \frac{t}{\eta} + \frac{1}{E} \tag{12.68}$$

Upon removal of the load (or stress) at time t, a sample corresponding to the Maxwell model will retract by a value equal to its elastic contribution $(\varepsilon_0 = \sigma_{11,0}/E)$, but will be permanently strained by a value $\varepsilon(t) = (\sigma_{11,0}/\eta)t$. In a creep experiment, such a sample behaves at the onset like an elastic solid and then like a viscous liquid; thus it exhibits the characteristics of a viscoelastic liquid. However, this Maxwell model also predicts a linear deformation as a function of time when subjected to a constant stress which is not realistic, because no such example could be found in the field of polymers (see Figure 12.10).

12.3.5.2. The Voigt Model. It consists of a spring and a dashpot connected in parallel:

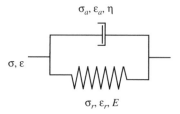

$$\sigma_a, \varepsilon_a, \eta$$

$$\sigma, \varepsilon$$

$$\sigma_r, \varepsilon_r, E$$

In the latter case, the deformation undergone by the spring and the dashpot are identical: $\varepsilon = \varepsilon_r = \varepsilon_a$. On the other hand, the total stress is the sum of those felt by the spring and the dashpot ($\sigma = \sigma_r + \sigma_a$).

In the context of Hookean solids and Newtonian liquids, the Voigt model thus predicts

$$\sigma = E\varepsilon + \eta \frac{d\varepsilon}{dt} \tag{12.69}$$

and consequently

$$\frac{d\varepsilon}{dt} = \frac{\sigma}{\eta} - \frac{E\varepsilon}{\eta} \tag{12.70}$$

This model is primarily used to describe the viscoelastic behavior in creep experiments. On the other hand, it cannot be used for relaxation experiments since it is unable to account for instantaneous deformation (stress should be infinite). According to the Voigt model, the relaxation modulus thus reduces to

$$E(t) = E \tag{12.71}$$

In this model, $\sigma = \sigma_{11,0}$ and thus $d\sigma/dt = 0$ for a creep experiment which implies a constant stress:

$$\frac{d\varepsilon}{dt} = \frac{\sigma_{11,0}}{\eta} - \frac{E\varepsilon}{\eta} \tag{12.72}$$

Integration of this differential equation leads to

$$\varepsilon(t) = \frac{\sigma_{11,0}}{E} \left[1 - \exp\left(\frac{t}{t_{ret}}\right) \right] \tag{12.73}$$

where t_{ret} = retardation time, which is different from the relaxation time. The creep compliance is described by the following function:

$$D(t) = \frac{1}{E} \left[1 - \exp\left(-\frac{t}{t_{ret}}\right) \right] \tag{12.74}$$

According to this model, a sample submitted to a stress $\sigma_{11,0}$ undergoes a deformation (ε) that grows rapidly from 0 (at t_0) at short times (as observed in liquids)

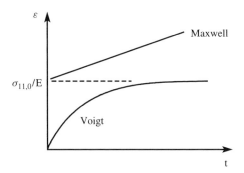

Figure 12.10. Behavior of polymer meeting the requirements of either Maxwell model or Voigt model in a creep experiment.

and then more slowly in a second step. The retardation time (t_{ret}) denotes the time required for the sample to attain $(1 - 1/e)$th ($\equiv 63\%$) of the deformation (ε_f) at the final time t_f if the load is maintained constant. At infinite time, the deformation of the sample is equal to $\varepsilon_0 = \sigma_{11,0}/E$ (see Figure 12.10).

A sample complying with the Voigt model behaves like a viscoelastic solid, and thus like a liquid at short times and like a solid material between $0 < t < t_f$. This behavior makes the Voigt model more relevant than the Maxwell model for describing a creep experiment. If the load applied to the Voigt solid is suppressed at the time t_f, its deformation then reduces and this sample recovers its initial dimension ($\varepsilon = 0$). If the Maxwell model is more appropriate to account for a stress relaxation experiment and the Voigt model to describe a creep experiment, neither of these two analogical models is accurate enough to describe the viscoelastic response of a polymer in its complexity. These models comprise one unique retardation or relaxation time, implying that all chains rearrange themselves in a same lapse of time, which is not realistic.

To overcome these limitations, more elaborate models combining a greater number of springs and dashpots have been proposed with the idea of generalizing the two previously described types of models.

For a relaxation experiment, the generalization of the Maxwell model—that is, a given number of Maxwell elements connected in parallel—implies a relaxation modulus that is the sum of the moduli of all the elements:

$$E(t) = \sum_{i=1}^{n} E_i \exp\left(-\frac{t}{t_{i,relax}}\right) \tag{12.75}$$

Upon associating a Maxwell element and a spring connected in parallel, one obtains the **Zener model**, which describes very satisfactorily the behavior of highly cross-linked polymers. Such behavior is characterized by an instantaneous elasticity followed by a phase of retarded elasticity, with the deformation exhibiting a completely reversible character.

The generalization of the Voigt model corresponds to the connection of a given number of Voigt models. As the regular version, the generalized Voigt model is inappropriate to describe the relaxation of a polymeric material. In the context of a creep experiment, the creep compliance is written as the sum of the compliance of the various Voigt elements:

$$D(t) = \sum_{i=1}^{n} \frac{1}{E_i} \left[1 - \exp\left(\frac{t}{t_{i,\mathrm{ret}}} \right) \right] \tag{12.76}$$

Upon associating a Voigt element and a Maxwell element in series, one obtains the **Burgers model**, which is well-suited to describe the creep behavior of a thermoplastic polymer: the behavior of such a material is characterized by an instantaneous elasticity followed by a phase of retarded elasticity, but the strain retains in this case an irreversible character.

12.3.6. Principle of Time–Temperature Superposition

Leaderman first suggested the existence of time–temperature equivalence in a viscoelastic material. This principle can be illustrated through the case of a sample subjected to a tensile experiment during which the rate of stretching is maintained constant. If this sample is identified with a Maxwell element, the variation of the stress (which is zero at the beginning of the experiment) that it undergoes with time can be written as

$$\frac{d\sigma}{dt} = E \frac{d\varepsilon}{dt} - \left[\sigma \left(\frac{E}{\eta} \right) \right] \tag{12.77}$$

which, after integration, gives

$$\sigma = \eta \frac{d\varepsilon}{dt} \left[1 - \exp\left(\frac{E}{\eta \frac{d\varepsilon}{dt}} \varepsilon \right) \right] \tag{12.78}$$

The profile of the variation of σ versus ε shows a large initial increase which then softens independently of η ($d\varepsilon/dt$), a term that normally remains constant.

In the extreme situation of $\eta(d\varepsilon/dt)$ tending toward infinity, the Hooke law ($\sigma = E\varepsilon$) applies; in contrast, if $\eta(d\varepsilon/dt)$ tends toward 0, σ is equal to zero. At high drawing rates, the sample thus behaves like a rigid material (E is high) and like a rubber (E is low) at low rates. With the help of relation (12.78), the variation of σ versus ε can be calculated for various values of $\eta(d\varepsilon/dt)$, with the two families of curves shown below indicating through their similarity that establishing $E = f(t)$ or $E = f(T)$ are equivalent. The function $E = f(t)$ can thus be transformed into a function $E = f(T)$ and vice versa (see Figure 12.11).

The simplest method to carry out this transformation would be to measure the relaxation modulus [$E(t)$] for constant deformation at various temperatures and to plot its variation as a function of time. For the lowest temperatures, very long times

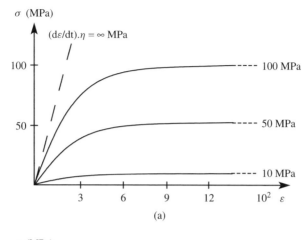

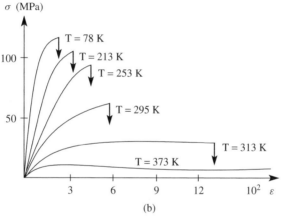

Figure 12.11. Stress–strain diagram: (a) Variation calculated for a Maxwell body for various values of $\eta d\varepsilon/d\sigma$. (b) Experimental curves obtained for a PMMA sample at various temperatures.

are required to establish a complete curve; on the other hand, at higher temperatures measurements at short times are sufficient. This means that it is practically impossible to plot a complete curve at a given temperature due to the limitations related to measurement times.

However, for measurements arbitrarily carried out at $T = 25°C$, one observes that the value of the relaxation modulus at short times coincides with that measured, for example, at $0°C$ at longer times: this equality of the modulus for different pairs of time and temperature can thus be used to build a whole curve by successive shifts and superpositions and then to establish a so-called *master curve* (see Figures 12.12 and 12.13). The latter is drawn starting from a reference temperature that should be chosen in a judicious way (near the polymer service temperature). To each shift of a curve segment along the time axis is associated a shift factor a_t which corresponds to the time gap with respect to the reference temperature (T_r) (see Section 12.2.2).

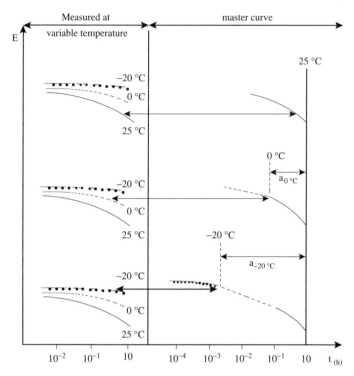

Figure 12.12. Establishing the master curve based on the time–temperature superposition principle.

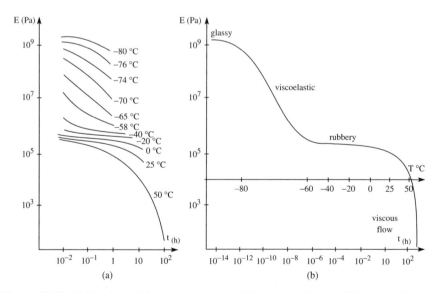

Figure 12.13. (a) Variation of the Young modulus (E) of polyisobutene (PIB) versus time at various temperatures. (b) Master curve for a reference temperature $T = 25°C$.

The time–temperature superposition principle can be written in the following mathematical form:

$$E(T_{r,t}) = E\left(T, \frac{t}{a_T}\right) \qquad (T_r \text{ being the temperature of reference})$$

In other words, a change of temperature is equivalent to multiplying the time scale by an appropriate factor. To be completely rigorous, it is necessary to take into account the variations of volume with temperature and, accordingly, to rewrite the previous relation in the form

$$E(T_{r,T}) = \frac{\rho(T_r)}{\rho(T)} \frac{T_r}{T} E\left(T, \frac{t}{a_T}\right) \qquad (12.79)$$

The same type of expression can be established for the creep compliance

$$D(T_{r,T}) = \frac{\rho(T_r)}{\rho(T)} \frac{T_r}{T} D\left(T, \frac{t}{a_T}\right) \qquad (12.80)$$

and for variables derived from dynamic measurements.

In addition, Williams, Landel, and Ferry proposed a relation describing the variation of a_T as a function of $(T - T_r)$; insofar as T_r is suitably selected the following relation can be relevant:

$$\ln a_T = -\frac{C_1(T - T_r)}{C_2 + (T - T_r)} \qquad (12.81)$$

where C_1 and C_2 are two constants depending on the polymer and on the selected reference temperature. From measurements carried out on 17 polymers at different reference temperatures, they proposed a "universal" curve of the variation of a_T as a function of $(T - T_r)$ (see Figure 12.14). The same authors also showed that if the reference temperature (T_r) selected is the glass transition temperature (T_g), then C_1 and C_2 can be taken equal to 17.4 and 51.6. These constants actually vary little with the nature of the polymer under consideration, and thus the expression (12.81) becomes

$$\ln a_T = -\frac{17.4(T - T_g)}{51.6 + T - T_g} \qquad (12.82)$$

which is called the **WLF relation**.

Another situation to consider is the response of polymers to a dynamic stress applied near thermal transitions.

As previously shown, the storage modulus (E') characterizes the rigidity of a viscoelastic material. It corresponds to the energy retained and restored by the polymer; under the conditions of a recoverable deformation, it is equivalent to Young's modulus (E). In a glassy state—that is, below the glass transition temperature—the modulus takes values in the range of 10^3–10^4 MPa. As previously discussed, this

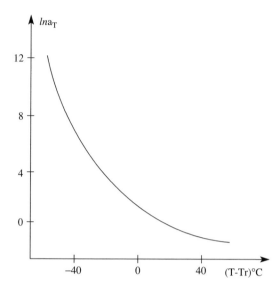

Figure 12.14. "Universal" curve giving the variation of a_T versus $(T - T_r)$ (in °C), according to Williams, Landel and Ferry.

elasticity is of enthalpic origin and thus implies the reversible modification of both interatomic distances and valence angles. The loss modulus (E'') which corresponds to the energy (in the form of heat) dissipated is almost equal to zero as molecular motions are frozen, thus resulting in an absence of internal frictions. Indeed, the chains cannot change their shape in response to the stress imposed.

The expression of the energy dissipated is written as

$$h = \sigma_{21}\dot{\varepsilon} = \eta\dot{\varepsilon}^2 \qquad (12.83)$$

The value of η is very high, but the rate of deformation of the chains ($\dot{\varepsilon}$) is zero. The energy dissipated and hence the loss modulus are then close to 0.

As for the tangent of the loss angle (tan δ), also called the *loss factor*, it corresponds to the ratio of the loss modulus to the storage modulus and is thus a measure of the energy dissipated versus the energy stored. Its value reflects the capacity of a material to absorb a stress in an elastic manner or to dissipate it by internal friction. A zero or low value of tan δ mirrors a purely elastic behavior, whereas a high value corresponds to a pronounced nonelastic response.

When a polymer sample undergoes a transition from the glassy state to the rubbery one upon heating, the polymer segments become more mobile. Their movements, caused by the softening of the material, now follow the stress imposed: the conversion of the internal friction and of the nonelastic deformation into energy is now maximal. Even if the viscosity of the medium tends to decrease, the rate of deformation of the chains becomes significant. Under these conditions, the loss modulus (E'' or G''), which measures the energy dissipated, reaches its highest value whereas the storage modulus decreases considerably by a factor of 1000. At still higher temperature, η decreases to such extent that G'' tends toward 0 even if $\dot{\varepsilon}$ continues to grow.

However, maxima of the loss modulus (E'' or G'') and of the loss factor (tan δ) do not coincide perfectly: as both E' or G' and E'' or G'' moduli change dramatically in this area, the maximum of tan δ is slightly shifted toward higher temperatures compared to that of E'' (or G'') at low frequency of shearing (1 Hz).

The glass transition temperature can then be identified as the temperature at which the loss modulus (E'', G'') or the loss factor reach their maximum value. Dynamic measurements thus provide an efficient means to measure T_g. Other transitions—for instance, second-order transitions—which are related to the movements of the chain substituents or due to certain segments can also be detected by dynamical stimuli.

12.4. MECHANICAL PROPERTIES AT LARGE DEFORMATIONS

12.4.1. Polymers in Practical Situations

Within the limit of small deformations (the case that has been treated until now), a polymer can respond in three different ways to an external mechanical stimulus:

- Instantaneous elastic deformation implying a spontaneous reversibility.
- Time-dependent viscoelastic deformation, implying simultaneously both relaxation and reversibility.
- Viscous and irreversible deformation above T_g, varying with time.

When departing from the limit of small deformations, solid-state polymers ($T < T_g$) lose their capacity to sustain a reversible deformation beyond the proportionality limit. The permanent deformation undergone is said to be *plastic* and the limit of unrecoverable deformations is called the *yield point* (see Figure 12.15d). The plastic deformation of a polymer sample results in a reorientation of its chains on a large scale and can possibly lead to its degradation. It concerns both amorphous and semicrystalline polymers. In the latter, sequences of the amorphous parts and the crystalline lamellae tend to align under the effect of an uniaxial elongation which can reorganize the crystalline zones and even break them upon increasing.

Such a chain orientation in the direction of deformation is accompanied by their plastic deformation; depending upon the proportion of entanglements, it occurs by a disentanglement process. The tensile test is the most convenient way to characterize the mechanical strength of a polymer. Before fracture which can be either fragile or ductile, four different scenarios of the stress–strain behavior can be contemplated, depending on the scale considered.

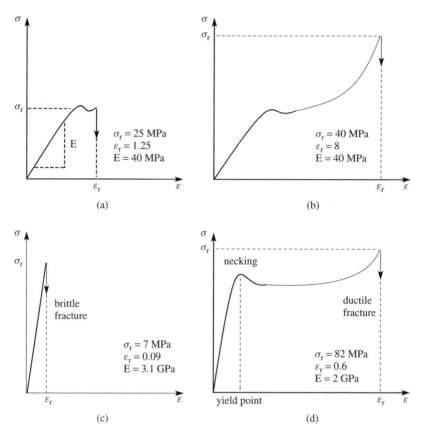

Figure 12.15. Stress–strain curves for four typical situations: (a) soft and strong polymer; (b) soft and ductile polymer; (c) rigid and fragile polymer; (d) rigid and ductile polymer.

A *brittle fracture* occurs in the elastic range and is associated with small fracture elongation and high stresses. The sample—typically fibers or thermoplastics—breaks perpendicularly to the stress direction without flowing.

A *ductile fracture* corresponds to an irreversible plastic deformation; fracture elongations may reach in the latter case several hundred percent and are found in lightly cross-linked elastomers.

Typically, four different situations can be described:

(a) Polymers of low molar mass and thus free of entanglements are characterized by a low elastic modulus (E) and a low tensile strength. Their fracture is generally observed soon after the limit of proportionality of the stress–strain curve (Figure 12.15a);

(b) Elastomers also exhibit a relatively low elastic modulus, but they can undergo high deformations before break. Their deformation is almost elastic until the *break point*; the considerable increase in modulus, which precedes breaking, mirrors an orientation of the sample chains along the deformation axis (Figure 12.15b).

(c) Thermoplastic polymers, such as polystyrene or poly(methyl methacrylate), are prone to a brittle fracture when elongated below their glass transition temperature. Their Young modulus is high and their fracture occurs in the linear part of the stress–strain curve so that their fracture deformation is necessarily small (Figure 12.15c).

(d) Other thermoplastics, such as polyethylene and certain polyesters or poly-amides, are subject to a ductile fracture: beyond the limit of proportionality of the stress–strain curve, chains flow, causing the sample to deform irre-versibly and undergo the so-called *necking* phenomenon. Under the pressure exerted by the clamps of the tensile testing machine, which holds the test specimen, a stress concentration builds up near the clamps, causing a low flow and thus a decrease of the specimen cross section. As chain elements flow, their internal frictions provoke an increase in temperature which, in turn, reduces further the viscosity and favors the flow. Beyond the yield point, the necking phenomenon propagates and the sample deforms under an equal or even lower stress. At the end of the test and before the break, an increase in the stress–strain curve is observed in certain cases, due to the existence of entanglements. The latter behave like physical cross-links and limit the chain motion (Figure 12.15d).

The previously described typical behaviors are also determined by the rate of deformation: a given material can be ductile at low rate of deformation and at high temperature or undergo a brittle fracture at low temperature and at a high rate of deformation.

12.4.2. Physical Damaging

Damages undergone by a polymer during plastic deformation appear macroscop-ically under two forms—that is, through the formation of shear bands or that of crazes (see Figure 12.16).

The existence of shear bands reveals a plastic deformation resulting from the shearing of polymer chains. When the deformation is highly localized, these shear bands can even be seen with a naked eye; in general they form a 45° angle with the direction of stress. Experimentally, compression tests (not elongation ones) favor the formation of shear bands in thermoplastic polymers.

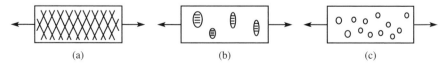

(a)   (b)   (c)

Figure 12.16. Mechanism of damaging during a plastic deformation: (a) formation of shear bands; (b) formation of crazes comprising fibrillae, or (c) constituted by microcavities.

The second damaging mechanism generates crazes inside the material, more particularly in amorphous and even semicrystalline thermoplastic materials during an uniaxial tensile stress. Contrary to cracks (which are, by definition, open fissures without matter between their lips that form and grow from a flow or a defect), the two edges of a craze contain a proportion of matter close to 50%. The interior of a craze indeed comprises either fibrillae (diameter 20 nm) of highly stretched chains and separated by voids or spherical microcavities of 10-nm size, embedded in the polymer matrix. Upon applying an uniaxial stress, crazes will grow perpendicularly to the direction of stretching, with the fibrillae being oriented parallel to this same axis. The formation of crazes comprises two stages, namely, nucleation and growth (terms used by analogy with those of crystallization). The existence of stress concentrations inside the matter—in the form of fissures, surface defects, or inclusions—causes the nucleation of crazes. The latter grow by extension of the voids lying between the polymer fibrillae. The kinetics of deformation affects the role played by the crazes during an uniaxial stress; at a slower rate of deformation, the chains have time to reorient inside the crazes, thus creating microcavities in the whole sample. The propagation of crazes is then slowed and a large plastic deformation is obtained before the final ductile fracture.

12.4.3. Brittle Fracture

In such a case, the sample elongation is small (a few percent) at the break point. The ultimate stress (σ_r) of chains perfectly aligned in the direction of the stress can be easily calculated from the force required to break a chain and its cross section. From data relative to the bond energies (348 kJ/mol and $L = 0.154$ nm for a C–C bond), a theoretical value of σ_r equal to 40.8 GPa is found for polyethylene chains.

In reality, the ultimate strength measured is much lower than the one calculated for an assembly of perfectly oriented chains; this is due to the presence of defects such as cracks in the analyzed sample. The fracture of a material thus originates from the existence, the growth, and the propagation of cracks which are voids; only one of such cracks is enough to cause a macroscopic failure of a sample subjected to a tensile test. One can then speak of a brittle fracture, because of the failure that occurs in the elastic range of the stress–strain curve for samples that were not damaged in a prior plastic macroscopic deformation.

Griffith established the connection between brittle fractures and the presence of cracks. When working on glassy materials, he noticed that small samples show an ultimate stress higher than that of bigger specimens. He concluded that the smaller fracture strength of the latter results from the presence of structural defects in larger number and bigger size. At the origin of the Griffith theory, there is an observation made by Inglis, who found that the presence of a hole inside a material enhances the stresses around its circumference. This observation played a significant role in the understanding of the mechanism of propagation of cracks upon applying a stress σ_a. Inglis treated mathematically the case of a plate-type specimen including an ellipse-shaped hole and subjected to an uniaxial stress: the calculation of the normal stresses (σ_{xx}, σ_{yy}) in the vicinity of the hole shows that the normal stress

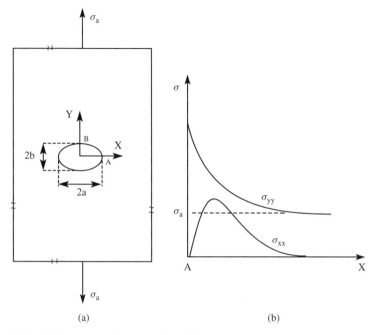

Figure 12.17. (a) Plates containing an elliptic hole whose dimensions are small compared to those of the sample. (b) Variation of normal stresses in the vicinity of the hole, starting from the point A of the sample for an applied stress σ_a.

(σ_{yy}) reaches a maximal value (σ_m) at the focus point (A), and then it decreases to take an average value in the rest of the sample. The σ_{xx} stress in the perpendicular direction increases from zero to take a value higher than the applied stress (σ_a) before decreasing again (Figure 12.17).

The stress is thus maximal at the edge of the sample hole and can be determined using the following ratio:

$$\frac{\sigma_m}{\sigma_a} = \left(1 + \frac{2a}{b}\right) \tag{12.84}$$

A stress concentration factor $K_t = 1 + 2\sqrt{a/\rho}$ can even be defined where ρ, the radius of the ellipse curvature, is equal to b^2/a.

For a circular hole, the stress amplification would be only equal to 3, but, for a highly stretched ellipse ($a/b = 500$), it can reach a factor of 1000. In the presence of a sharp crack, the local stress can reach values capable of breaking carbon–carbon bonds.

12.4.4. Crack Propagation: Griffith's Theory

Cracks can only propagate if the energy of the system decreases and minimizes, and if the local fracture stress is equal to, or even higher than, the theoretical value; this is the basis of Griffith's theory.

The energy balance is expressed as follows: on the one hand, the introduction of cracks and their propagation contribute to lower the elastic energy (U_1) of the system; on the other hand, the formation of new surfaces requires a work (W) that tends to oppose the propagation of cracks. The difference ($U_1 - W$) is thus available for the creation of new crack surfaces; and, on the whole, the energy (U) of the system studied is written as

$$U = U_0 - U_1 + W \tag{12.85}$$

where U_0 is the elastic energy of a plate-shaped specimen without cracks.

As the material considered is elastic, the elastic energy of deformation per volume unit can be identified with $\sigma_a{}^2/2E$ (E being Young's modulus).

Since the plate-shaped specimen has a volume V, U_0 is equal to $U_0 = \sigma_a^2/2EV$.

To calculate U_1, Griffith specified that the introduction of a crack into a two-dimensional plate-type specimen entails in its immediate vicinity i.e., in a volume $\pi a^2 l$ around this crack, a density of elastic energy close to zero; l represents the sample thickness assumed equal to 1. Under these conditions, U_1, which corresponds to the elastic energy lowered by the introduction of a crack, is in first approximation

$$U_1 \approx \frac{\pi a^2 \sigma_a^2}{2E} \tag{12.86}$$

A more rigorous calculation of U_1 based on the precise integration over the crack circumference leads to a slightly different expression:

$$U_1 = \frac{\pi a^2 \sigma_a^2}{E} \tag{12.87}$$

For a thick plate, this same expression becomes

$$U_1 = \frac{\pi(1 - q^2)a^2 \sigma_a^2}{E} \tag{12.88}$$

where q is the Poisson ratio.

Besides, the work (W) required for the formation of the crack surface can be expressed as equal to

$$W = 4a\Gamma \tag{12.89}$$

where Γ corresponds to the surface free energy of the material per surface unit, and $4a$ corresponds to the surface of the two lips of an elliptical crack. On the whole, U, the potential energy of the system, is established as follows:

$$U = \frac{\sigma_a^2}{2E}V - \frac{\pi a^2 \sigma_a^2}{E} + 4a\Gamma \tag{12.90}$$

According to Griffith, the equilibrium is reached when the energy released by the crack propagation (da) is equal to the energy required for the creation of two new surfaces resulting from this growth:

$$\frac{dU}{da} = -\frac{2\pi a\sigma_a^2}{E} + 4\Gamma = 0 \tag{12.91}$$

When the work (W) required to create new surfaces is higher than U—which represents the elastic energy lost in the growth of cracks—dU/da is higher than 0 and there is no possibility of the crack to propagate; on the contrary, when $dU/da < 0$, the crack is unstable and the sample can experience a catastrophic fracture (see Figure 12.18).

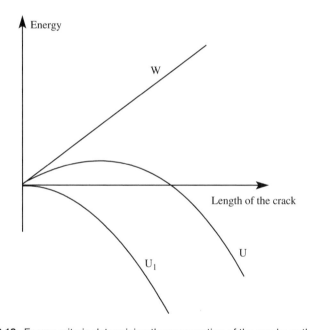

Figure 12.18. Energy criteria determining the propagation of the cracks or their halting.

According to this analysis, the threshold corresponding to equilibrium is associated to a critical stress beyond which the crack is likely to propagate irreversibly:

$$-\frac{2\pi a\sigma_c^2}{E} + 4\Gamma = 0 \tag{12.92}$$

which gives

$$\sigma_c = \sqrt{\frac{2E\Gamma}{\pi a}} \tag{12.93}$$

For a thick plate

$$\sigma_c = \sqrt{\frac{2E\Gamma}{\pi(1 - q^2)a}} \tag{12.94}$$

In the same way and using Griffith's line of reasoning, a crack critical length can be defined for a given stress, beyond which this crack becomes unstable and grows.

Griffith's theory describes satisfactorily the behavior of brittle polymers such as PS or PMMA, and σ_c as well as $a^{-1/2}$ vary in accordance with relation (12.93). However, when these relations are used to calculate Γ, very high values are found (from 200 to 1500 J·m^2), which are well beyond the real surface energies ($\cong 1$ J/m^2). Such a difference is due to the fact that the Griffith theory does not take into account the plastic deformation. In fact, a plastic deformation occurs in these polymers at the bottom of their cracks, which act like energy absorbers and induce a considerable increase of the value of Γ. In this case, H is used instead of Γ, being more universal than Γ; H is defined as the elastic energy released per crack growth unit and is written as

$$H = \frac{1}{2}F^2\frac{dD}{da} \tag{12.95}$$

where D is the compliance of the sample and F is the load applied.

The temperature at which the mechanical stress is applied is the first factor that determines the type of fracture of the material. For $T \ll T_g$ the sample undergoes a brittle fracture, and for $T > T_g$ it undergoes a ductile one. The value of the ultimate stress in a brittle fracture depends not only on the temperature but also on the chain molar mass and orientation; indeed the ultimate stress varies even for T slightly lower than T_g and is independent of the size of the chains only beyond a critical molar mass. The transition between a purely brittle behavior at low temperature and a primarily ductile one at higher temperature occurs in an intermediate zone ($T < T_g$) where the two types of mechanisms of fracture can occur, according to Ludwig's analysis.

The precise temperature (T_d) at which this fragile–ductile transition occurs depends on factors such as

- the rate of deformation (a high rate favors brittle fractures, a slow rate favors ductile fractures),
- the thickness of the sample (greater brittleness in case of thick samples), or
- the presence of a notch in the sample (the latter entailing an increase in the fragile–ductile transition).

All these factors indicate that there is no strict correlation between the fragile–ductile transition temperature and T_g. Beyond T_d and T_g, polymers are prone to a pure ductile fracture, which means that they undergo an irreversible plastic deformation before breaking.

LITERATURE

I. M. Ward, and D. W. Hadley, *An Introduction to the Mechanical Properties of Solid Polymers*, Wiley, New York, 1993.

N G. McCrum, C. P. Buckley, and C. B. Bucknall, *Principle of Polymer Engineering*, Oxford University Press, New York, 1994.

13

RHEOLOGY, FORMULATION, AND POLYMER PROCESSING TECHNIQUES

Thermosetting resins or thermosets apart, polymers are generally processed as melts or as concentrated solutions and occasionally as solids, after the polymerization and the formulation steps.

The melt state usually results from a rise of the temperature of the material above its glass transition temperature or its melting temperature and is reflected in the material fluidity due to the mobility of its constituting entities. **Rheology**, which means the study of flow and deformation, investigates in practice the fundamental and constitutive relations between force and deformation in liquid-like materials, and particularly in those viscoelastic lying between the ideal elastic solid and the ideal viscous Newtonian fluid. Knowledge of the time-dependent dynamics of polymer melts thus constitutes the common base of all the processing techniques.

13.1. DYNAMICS OF POLYMER MELTS

13.1.1. Introduction

The study of the dynamic properties of a system in equilibrium—or near-equilibrium—can be defined as the analysis of the fluctuation of these properties as a function of time. Whether the polymer under consideration is in solution, in the melt or in the solid state, the experimenter can obtain the appropriate information about all the movements undergone by a polymer sample, from totally local ones pertaining to only a few repeating units up to the overall movement of the chain, through the use of a suitable analytical technique.

Organic and Physical Chemistry of Polymers, by Yves Gnanou and Michel Fontanille

Different regimes can be distinguished in the dynamics of an isolated polymer chain that can be monitored using probes sensitive to different time and length scales:

- The long-time regime is generally associated with the Brownian motion of the whole chain.
- The short-time regime corresponds to motions occurring at the level of repeating units.
- The intermediate regimes provide information about the connectivity of the system.

For instance, the solution characterization of polymer conformers (lifetime of about 10^{-10} s) requires Raman spectroscopy whose time scale is about 10^{-13} s, whereas NMR works at a time scale of 10^{-6} to 10^{-7} s.

Another example is that of the self-diffusion and Brownian motion of chains in solution which results in a fluctuation of their concentration on a time scale of about 10^{-3} s. As mentioned in Chapter 6, quasi-elastic light scattering is the ideal technique to determine the diffusion coefficient D from the self-correlation function of the scattered intensity.

13.1.2. Chain Dynamics in the Melt

Above the glass transition temperature (amorphous polymers) or above the melting temperature (semicrystalline polymers), polymers melt and can thus flow.

Chains tend to relax when subjected to a short-time tensile or shear stress:

- Relaxation and recovery of the original state are total for short-time stresses entailing a fully reversible deformation.
- On the other hand, if the stress application time is long enough, the system relaxes and flows simultaneously and the recovery of the original state can only be partial.

For very short-time scales, polymer melts adopt the behavior of a glass, at medium-time scales, it adopts that of a rubber, and at long-time ones, it adopts that of a viscous liquid; in that respect, polymers can be regarded as **viscoelastic liquids**.

As previously mentioned, polymeric materials are processed as melts in most industrial processes, and rheology is the science that studies the deformation and the flow of matter as a function of time.

Two features characterize polymer melts:

- Like their homologs in solution, polymer melts are able to move in their totality by Brownian motion; in the solid state, on the other hand, polymer chains cannot really move due to the very limited motion of subchains. The diffusion of chains in a melt may also entail the cooperative movement of subchains;
- Beyond a critical molar mass (M_c), chains entangle in the melt as they do in the solid state.

13.1.3. Concept of Entanglements

Diluted in a good solvent, polymer chains tend to swell and, according to the Flory theory, their root-mean-square end-to-end distance ($\langle r^2 \rangle_0^{1/2}$) varies with the power 3/5 of their molar mass. In a poor solvent, coils tend to contract and not to entangle; in this case, the end-to-end distance varies as $M^{1/3}$. In θ conditions, as in the melt (i.e., when the chains adopt the so-called "unperturbed" dimensions), $\langle r^2 \rangle_0^{1/2}$ varies as $M^{1/2}$, in accordance with the expression $\langle r^2 \rangle_0 = CXL^2$, where C is a constant that accounts for the correlations of orientation between two successive repeating units, X is the number of these repeating units, and L is their length; in the latter expression, only short-range correlations are taken into account; as revealed from the values of both the viscosity and the elastic modulus (see Figures 13.1 and 13.7), long-range interactions between repeating units carried by two different chains can also occur when chains attain a critical length corresponding to n times ($n > 4$) the persistence length. The interactions that contribute to entangle polymer chains play a role comparable to that of physical cross-links in the sense that the behavior of two entangled chains is interdependent. The comparison between entanglements and cross-links in a network stops here because, unlike cross-linked chains, entangled chains do not form an infinite single macromolecule but can slip one with respect to others and flow. That affects, in turn, the physical behavior of polymer melts.

The existence of entanglements finds its origin in the fact that a sphere of volume $\frac{4}{3}\pi s^3$ containing a polymer chain of radius of gyration (s) can accommodate many others, with the van der Waals volume of the chain considered being equal to XL^3, where L is the length of the repeating units and X is their number. The number (e) of chains that can be accommodated in such a sphere can thus be written as $e = 4\pi s^3 / 3XL^3$. Taking into account the relation between s_0, r_0 (end-to-end distance) and L, one can write $e = AC^{3/2}X^{1/2}$ (where A is a constant equal to $2\pi/9$). Assuming that the effect of the entanglements is felt beyond a critical value e_c, the critical number of repeating units (X_e) or the critical molar mass (M_e) between entanglements are thus proportional to C^{-3} ($M_e \div C^{-3}$) according to this reasoning. Other relations proposing the variations in $M_e \simeq C^{-1}$ have also been considered. This definition is interesting because it can also be applied to concentrated polymer solutions through the introduction of the polymer concentration (C_2) into the expression of e; under θ conditions, one obtains

$$C_2 M_e^{1/2}(C_2) = M_e^{1/2}$$

and in a good solvent

$$C_2 M_e^{4/5}(C_2) = M_e^{4/5}$$

Mechanical as well as rheological properties—that is, those of the material in the solid state and as a melt—actually depend on the presence of entanglements in the sample. Their presence can be detected using one of the three following

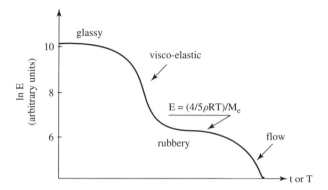

Figure 13.1. Variation of the elastic modulus (E) versus either time or temperature and definition of the molar mass between entanglements (M_e) from the value of E at the rubbery plateau.

methods: through the monitoring of the $E = f(t)$ function giving the variation of the elastic modulus (E) versus time (Figure 13.1) and through the measurement of the compliance or the viscosity.

From the value taken by the elastic modulus at the rubbery plateau, the molar mass between entanglements (M_e) can be indeed deduced using the relation $E = 4\rho RT/5M_e$, where ρ is the density. Plotting the variation of the Newtonian viscosity (η_0) as a function of the molar mass also gives a critical value (M_c) above which entanglements occur (see Figure 13.7), where M_c is roughly equal to 10/4 of M_e (see Table 13.1).

13.1.4. Rheological Behavior of Polymer Melts

To determine the movements of the whole chain and those of subchains, various techniques are accessible and available to the experimenter, including dielectric spectroscopy and mechanical spectrometry. Dielectric techniques are suitable for the study of polymers in a wide range of frequencies (between 10^{-2} Hz and 10^{10} Hz), while the mechanical dynamic characterization of polymers provides access to long relaxation times (>10 s) through creep and stress relaxation tests.

Table 13.1. Critical molar masses between entanglements (M_c) determined from viscosity measurements and molar mass between entanglements (M_e) drawn from the value of the elastic modulus at the rubbery plateau

Polymer	M_e (g·mol^{-1})	M_c (g·mol^{-1})
Polydimethylsiloxane	9,600	24,500
Polyisobutene	5,700	15,200
Polystyrene	13,600	35,000
Poly(α-methylstyrene)	13,000	28,000

The property that permits the characterization of the flow of a liquid is its viscosity. Depending upon the type of stress applied, viscosity may be of either shearing or elongation type. As for the processing techniques, spinning or drawing causes elongation whereas filling induces melt shearing. By definition, the viscosity is the ratio of the stress to the rate of deformation; for an elongation, viscosity is expressed as $\eta_{el} = \sigma/\varepsilon$ and, for a shearing, $\eta = \tau/\dot{\gamma}$.

A liquid whose viscosity is independent of the rate of shearing is said to be Newtonian (η_0). Water, whose viscosity at ambient temperature is equal to 10^{-3} Pa·s, is Newtonian but polymer melts with a viscosity in the range of 10^{10} Pa·s are not Newtonian. Only oligomers and their concentrated solutions undergo a decrease of their viscosity with the rate of shearing, this decrease following a power law $\eta \div (\dot{\gamma})^k$ beyond a critical rate $\dot{\gamma}_c$; k takes negative values and is generally equal to -0.8 (it is worth mentioning that wet sand undergoes an increase of its viscosity with shearing). Below that critical rate $(\dot{\gamma}_c)(0 < \dot{\gamma} < \dot{\gamma}_c)$ the viscosity of polymer melts is Newtonian. In practice, the Newtonian viscosity of polymer melts is obtained by extrapolating the shearing rate to zero ($\dot{\gamma} = 0$) (see Figure 13.2).

Processing techniques such as calendering, extrusion, and injection molding imply very high rates of shearing with a value of about 10^2, 10^5, and 10^7 s^{-1}, respectively. The decrease of η with high values of $\dot{\gamma}$ enables high rates of production. However, the viscosity of polymers can vary with time (Figure 13.3) for a constant rate of shearing, and this has to be taken into account when processing polymers. When shearing induces a decrease of the viscosity, the fluid is said to be *thixotropic*; on the other hand, when viscosity increases under shearing, the fluid is said to be *rheopectic*. For instance, it is essential for paints to be thixotropic, which means highly viscous during their transportation or at rest and reasonably fluid when applied with a brush.

Viscosity measurements can be carried out using three types of viscometers: capillary, rotational, and cone-plate (Figure 13.4). The rates of shearing that can be

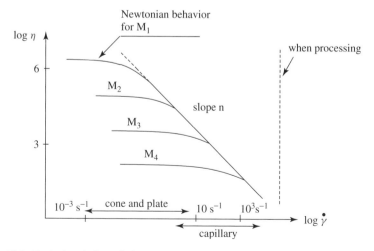

Figure 13.2. Typical variation of viscosity versus the rate of shearing in increasing order of molar masses.

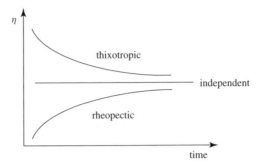

Figure 13.3. Variation of η against time at constant value of $\dot{\gamma}$.

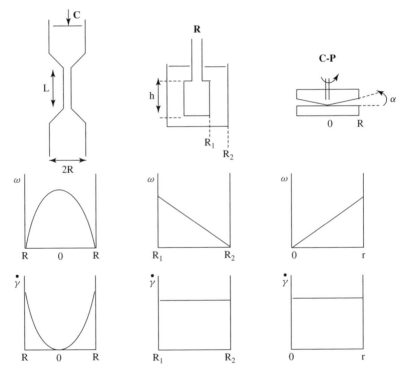

Figure 13.4. Rate profile of the melt (ω) and profile of the rate of shearing $\dot{\gamma}$ in capillary (C), rotational (R) and cone-plate (CP) viscometers.

applied and their profile inside the rheometer vary from one viscometer to another as well as the melt rate profile (ω). Newtonian viscosity is accessible upon using a cone-plate viscometer and other types of viscometers.

13.1.5. Hydrodynamic Theories and Interpretation of the Phenomena

Like the viscosity of all other liquids, that of polymer melts is a function of the nature of intramolecular forces present; their viscosity grows with the polarity of the repeating units and the presence of hydrogen bonds. However, there are two phenomena that occur only in polymer melts:

- During flow, chains slip with respect to each other. When the chains are branched or when they carry bulky substituents, their relative motion is made more difficult, which is reflected in a higher viscosity.
- The molar mass of the chains is a major factor affecting their viscosity.

Chains that exhibit a molar mass smaller than the critical molar mass between entanglements flow as predicted by the Rouse model (see the comprehensive presentation of the Rouse model in Chapter 6 and in the section devoted to the viscosity of dilute solutions). The role given to the solvent in that model is played in the melts by the polymer coils surrounding the chain considered. The scenario described by the Rouse model applies as long as chains do not entangle (Figure 13.5); as the length of chains increases until causing their entanglement, the flow of such melts cannot be described by a sort of Brownian motion that would just consist in chain orientation by the shearing forces applied. Beyond the critical molar mass between entanglements, an individual macromolecule finds trapped itself in a maze of homologous chains that behave as obstacles in its motion. In order to move through these obstacles, that chain has to "reptate" like a reptile from one of its ends to another and avoid any side movement out of the tube formed by segments of other chains (such a chain does not form loops that would hinder its motion). According to this description, which was proposed by de Gennes, Edwards, and Doi, the test chain is thus confined in this tube and is free to move along that tube (Figure 13.5).

The diameter of this tube (d_c), which corresponds to the average distance between entanglements, is much larger than that of the chain: 3.5 nm for polyethylene, 8 nm for polystyrene ($d_c \div X_c$). The tube length and the friction undergone by the chain in its way through the tube are proportional to its molar mass. The polymer motion under such conditions thus occurs through diffusion along the tube center line by anisotropic cooperative movements whose scale remains within d_c (movements perpendicular to the tube center-line and beyond d_c are blocked). The

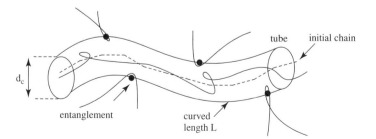

Figure 13.5. Polymer chains confined in an imaginary tube. The obstacles (entanglements) are fixed or labile.

two models, that of Rouse and that of **reptation**, thus differ by their description of both the diffusion and the viscosity of polymer melts. The viscosity of a particle, which reflects its displacement in a liquid in motion, is more difficult to calculate theoretically than its diffusion, which corresponds to the movement of the same particle in a liquid at rest.

The Rouse model predicts for the self-diffusion coefficient (D_2) of a chain, comprising X segments each with molecular friction coefficient ξ_{seg}, a variation inversely proportional to ξ_{seg}:

$$D_2 = k_B T/(X\xi_{seg}) = RT/(\mathcal{N}_a X\xi_{seg}) \tag{13.1}$$

According to this model, D_2 thus varies proportionally to M^{-1} ($D_2 \div M^{-1}$) with $M = M_{seg}X$.

The reptation model leads to a completely different relation. According to the diffusion law, the time (t_r) required for a particle to cover a distance L is written as

$$t_r = L^2/2D_2 \tag{13.2}$$

where D_2 is the self-diffusion coefficient. In the case considered here, the time (t_{rep}) needed for a chain to find its way through a tube of length L_{tub} and come out of it corresponds to

$$D_2 = (L_{tub}^2)/2t_{rep} \tag{13.3}$$

Considering the general expression for the self-diffusion coefficient and the fact that $L_{tub} = XL$, (13.3) becomes

$$t_{rep} = X^3 L^2 \xi_{seg}/2k_B T \tag{13.4}$$

The center-line of the tube is called the "primitive chain," and its radius of gyration is the same as that of the test polymer confined in the tube. Under these conditions, D_2 is expressed as

$$D_2 = \langle s^2 \rangle_0/(2t_{rep}) = k_B T \langle s^2 \rangle_0/X^3 L^2 \xi_{seg} \tag{13.5}$$

Since polymer melts exhibit unperturbed dimensions, XL^2 can thus be assumed equal to $\langle r^2 \rangle_0 = 6\langle s^2 \rangle_0 = 6K^2M$ in virtue of the relation between the radius of gyration and the molar mass (M) or the number of repeating units (X) (see Chapter 5). In other words, the reptation model predicts a dependence of D_2 on the inverse square of the molar mass ($D_2 \div M^{-2}$) above a critical molar mass corresponding to the onset of entanglements:

$$D_2 = \frac{k_B T}{6\xi_{seg}} \frac{1}{X^2} \tag{13.6}$$

Experimentally, an exponent equal to -2 is found for the variation of $\ln D_2$ *versus* $\ln M$ in agreement with the reptation theory (see Figure 13.6).

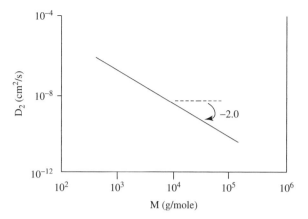

Figure 13.6. Variation of the self-diffusion coefficient as a function of the molar mass for linear polyethylenes at 175° C (according to Pearson and Ver Strate).

For the dependence of the Newtonian viscosity on the molar mass, again the two models (Rouse and reptation) diverge. As established in Chapter 6, the Rouse theory predicts that the viscosity of a fluid is the product of the friction coefficient (ξ_{seg}) times a factor F (see Chapter 6)

$$F = (\rho \mathcal{N}_a/6)(\langle s^2 \rangle_0/M)X \tag{13.7}$$

which takes into account simultaneously the number of segments (or repeating units) and accounts for conformational effects

$$\eta = (\rho \mathcal{N}_a \xi_{\text{seg}}/6)(\langle s^2 \rangle_0/M)X \tag{13.8}$$

According to the Rouse model, η is thus proportional to M, which is indeed verified for all polymers in the domain of low molar masses—that is, lower than the critical molar mass between entanglements. From the observation that the Newtonian viscosity (η_0) of a polymer melt is the product of the shear modulus G times t_{rep}—that is, the time for a chain to emerge from the tube—the reptation model predicts

$$\eta_0 = G t_{\text{rep}} \qquad [\text{with } t_{\text{rep}} = X^3 L^2 \xi_{\text{seg}}/(2k_B T)] \tag{13.9}$$

with the value of the shear modulus being that taken at the rubbery plateau

$$G \div \rho R T / M_e \tag{13.10}$$

where M_e denotes the molar mass between entanglements. After remarking that $k_B \mathcal{N}_a$ is equal to R, the perfect gas constant, that L^2 is equal to $6 \langle s^2 \rangle_0/X$ and X

corresponds to M/M_{seg}, and that M_e is equivalent $4/10$ M_c (see Table 13.1), one obtains for η_0:

$$\eta_0 \div 6\mathcal{N}_a\rho(M_{seg}/M_c)(\langle s^2\rangle_0/M)\xi_{seg}X^3 \qquad (13.11)$$

η_0 varies thus proportionally to the third power of molar masses, namely $\eta_0 \div M^3$. Experimentally, a molar mass dependence of $\eta_0 \div M^{3.4}$ is observed beyond M_c for all polymers; the reason for this small difference between the experimental exponent and that predicted by the reptation model is not understood yet.

The variation of the Newtonian viscosity with the polymer molar mass thus exhibits two regimes (Figure 13.7):

- In the domain of low molar masses, it is the Rouse model which is relevant with a linear dependence of η_0 on M ($\eta_0 \div M$).
- Beyond this domain, the reptation model prevails with a variation of η_0 proportional to $M^{3.4}$.

The transition from one regime to the other occurs precisely at M_c, the critical molar mass above which entanglements occur. Through the image of an imaginary tube and the resistance it exerts on the motion of the test polymer, the reptation theory accounts for the experimental facts more suitably than the Rouse theory; indeed, the viscosity of the polymer increases as a third power of M ($\eta_0 \div M^3$ instead of $\eta_0 \div M$) and its diffusion coefficient decreases as the inverse square of M ($D_2 \div M^{-2}$ instead of $D_2 \div M^{-1}$).

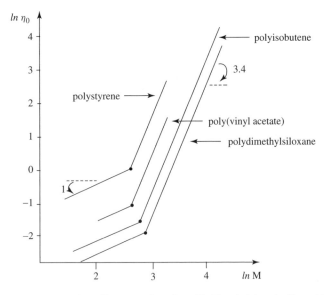

Figure 13.7. Variation of $\ln\eta_0$ as a function of $\ln M$ and determination of M_c.

	Self-Diffusion	Newtonian Viscosity $(\eta_0 = \xi_{\text{seg}} F)$
Rouse	$D_2 = \dfrac{RT}{\mathcal{N}_a \xi_{\text{seg}} X}$	$F = \dfrac{1}{6} \rho \mathcal{N}_a \dfrac{\langle s^2 \rangle_0}{M} X$
Reptation	$D_2 = \dfrac{RT}{6 \mathcal{N}_a \xi_{\text{seg}} X^2}$	$F \div 6 \mathcal{N}_a \rho \dfrac{M_{\text{seg}}}{M_c} \left(\dfrac{\langle s^2 \rangle_0}{M} \right) X^3$

13.2. PROCESSING OF POLYMER MATERIALS

Generally, the raw materials that are delivered by polymerization reactors can-not be directly used or processed without being dried/degassed and eventually compounded. Produced in the form of powders, beads, pellets, granulates, or melts, polymers must be indeed degassed—even washed and dried in the case of beads—before being formulated/compounded; this is a delicate operation that requires the expertise of professionals whose role is to admix additives and modi-fiers so as to adjust the material properties—including mechanical—to the techno-logical requirements of the application. This "formulation" step is generally carried out by the polymer producer, whereas the final processing step of formulated materials requires the expertise and know-how of process engineers. These two operations—formulation and processing—which will be briefly described in the following sections, are essential to produce a material with the desired properties and macroscopic shape.

13.2.1. Polymer Formulation: General Principles

Besides certain basic properties inherent to each polymer and closely related to its molecular structure, there are a number of properties that are "secondary" but can, however, affect the behavior of a material in a given function. These properties can be optimized or modified by means of the formulation—that is, by incorpo-ration of foreign substances that are miscible or immiscible with it into the raw polymer. The technical terminology distinguishes between additives and modifiers according to the proportion of these substances in the final material. **Additives** are used in low proportion (typically less than 5%) and do not affect the mate-rial mechanical properties, whereas **modifiers** such as **plasticizers** or **fillers** are added in higher proportions and even sometimes in majority. These substances are physically dispersed in the polymer matrix in order to modify certain characteris-tics without affecting the basic molecular structure of the polymer as well as its essential properties.

Foreign substances can also be involuntarily present in the polymer (catalytic residues, various impurities, monomers that are not polymerized, etc.) and bring about undesired effects.

13.2.1.1. General Characteristics of the Additives. Intentionally added to the raw polymer, additives can be classified according to their action as processing aids or functional auxiliaries.

Except in special cases, the additives are expected to exhibit their characteristics and their effects irrespective of external constraints. To this end, they must be somehow bound to the matrix in which they are dispersed through sufficiently strong interactions in order not to be eliminated adventitiously. They must be stable over time, and additives that are unstable and could be degraded due to the effect of temperature, light, oxidation, and hydrolysis are to be avoided.

The migration of additives—and hence their possible elimination—depends on the strength of polymer–additive interactions and on their molar mass, especially when they are organic substances (evaporation of volatile molecules); their physical state also matters (immiscible solids do not migrate easily). Additives with low vapor pressure and that afford homogeneous blends because of a negative interaction parameter (χ_{12}) are always preferred.

Another characteristics that is essential when employing an additive is the absence of toxicity:

- At the time of its formulation (which is generally carried out at high temperatures and can thus provoke its possible volatilization or its thermal degradation into toxic side products).
- During usage, in particular in applications such as food packaging.

Toxicity is defined by a toxicity index and by standards that permit its quantification.

Macromolecular compounds exhibit dielectric properties that can be strongly affected by the presence of additives. The electric conductivity can indeed be increased either directly or because of the capacity of certain additives to fix the water that is present in their environment.

In the case of solid additives, microcavities can be formed in the bulk of the material; water and oxygen are trapped, modifying certain properties of the material. By inducing crack formation, they can have considerable effect on the mechanical properties at the macroscopic level.

Incorporation of additives is generally performed by kneading the raw polymer above its softening (glass transition or melting) temperature. In the case of polymers obtained in solution, the mixing step is sometimes carried out in the liquid state. Powders are often mixed as such and then extruded in the form of pellets.

13.2.1.2. Additives Used as Processing Aids. **Heat stabilizers** are required for certain polymers (PVC, PMMA, etc.) that are thermally unstable; their incorporation into these polymers permits to operate and process the latter at higher temperatures, which, in addition to increasing the processing pace, induces a better surface finish. Among stabilizers one distinguishes:

- Primary stabilizing agents—for instance, antioxidants—which stabilize directly the polymer against the thermo-oxidizing degradation.

- Secondary stabilizing agents that transform the peroxides formed into more stable harmless compounds.
- Chelating agents that trap transition metals present in catalytic residues.
- Stabilizing agents for halogenated polymers whose role is to neutralize the acids formed by degradation.
- Synergistic systems (A + B) with better stabilizing effects (often unexplained) than those due to A and B considered independently.

External lubricants are used to lower the forces of friction and thus prevent the premature wear of specimens coming into contact one with another. These additives must not be miscible with the polymer so as to migrate spontaneously to the surface of the objects processed. To this end, either hydrocarbons and silicones forming micron-thick layers or additives developing monomolecular layers are often utilized; lamellar solids (graphite, MoS_2, etc.) are preferred for extreme conditions.

Internal lubricants are processing aids that are added to improve the flow behavior of matter; in their presence, polymers can be processed at moderate temperatures. For example, fatty esters are utilized to improve the processability of PVC.

Thixotropic additives are mineral particles (MgO, SiO_2) that are characterized by a very high specific surface and an ability to develop interactions with the polymer; by forming a sort of physical gel, they prevent flow and are reserved for materials processed by compression molding or calendering (PA, PVC, UP, etc.).

13.2.1.3. Additives for the Modification of Surface Properties.

Additives that increase the **roughness** at the materials' surface are sometimes necessary to reduce the contact surface between specimens and also their gloss. Such an effect is obtained by introducing small rigid or rubbery particles in the polymer, that helps to easily separate the latter from the metallic walls of the processing machine.

Such a surface roughness is also essential for the manufacture of packaging films to prevent film adhesion; "silica flower" particles of $\phi \sim 10$ nm are utilized to this end.

Anti-static agents are required for strongly hydrophobic polymers, that can be subject to electrostatic charging, in particular when rubbing together two such surfaces. It generates static electricity which can attract dust, jam electric devices, and make plastic films sticky. By incorporation of ionizable compounds that migrate to the surface of the material and form a conducting layer through adsorption of atmospheric water, electrostatic charging can be reduced. Sulfonic-, nitrogen-, and phosphorus-based compounds or poly(ethylene oxide) (PEO) are often utilized to this end.

When an inorganic filler is used, it is often necessary to increase the "wettability" of the inorganic surface for better binding of the polymer; **adhesion promoters** are incorporated in the latter case to enhance the interactions between the polymer and the filler.

For example, phosphorus compounds that are known to chelate metal ions are utilized to improve adhesion between metal surfaces and polymers. Acids and amines are also used for the same purpose. For adhesion on glass, titanates are incorporated to react with the substrate as shown below:

$$
\begin{array}{cccc}
\text{OR} & \text{OR} & \text{OR} & \text{OR} \\
| & | & | & | \\
-\text{Ti}-\text{O}-\text{Ti}-\text{O}-\text{Ti}-\text{O}-\text{Ti}-\text{O}- \\
| & | & | & | \\
\text{O} & \text{O} & \text{O} & \text{O} \\
| & | & | & |
\end{array}
$$

///////// glass or metal /////////

13.2.1.4. Additives for the Modification of Optical Properties.
Polymers containing heterogeneous microphases (semicrystalline polymers, blocks, or graft copolymers, etc.) or nonmiscible polymeric blends are opaque or lack transparency. This is due to the scattering of light (or other electromagnetic radiations) by phases of different refractive index; this phenomenon of scattering also depends on the size of the scattering centers, and its intensity becomes maximum when their size is in the range of the incident radiation wavelength.

To improve the **transparency** of crystalline polymers, one can either:

- reduce the degree of crystallinity (case of PET) but that affects the mechanical properties,
- perform heat treatments that favor crystalline structures of low density (decrease of Δn),
- or limit the size of spherulites.

This last possibility is often preferred because it is easy to carry out. A nucleating agent (alkaline benzoates or finely divided inorganic particles, etc.) is added to the polymer that increases its degree of crystallinity and the number of nuclei but decreases the size of the crystalline zones.

To alleviate the effects due to the thermal degradation of polymers such as PVC, that produces conjugated systems and causes their coloring in amber-yellow, fluorescent additives absorbing in the near-UV range and re-emitting light in the visible range can be incorporated.

Organic dyes that are miscible with the polymer matrix or solid pigments are utilized to color polymers. Dyes offer the advantage of being available in a variety of colors, and their cost is extremely low. Pigments are less efficient and are to be added in a larger quantity; they can be organic, mineral, or metallic and bring about various effects. Depending upon their nature, lamellar pigments can provide either a metallic (Cu, Zn, Al) or a pearl finish (lead carbonate, bismuth oxychloride). The most widely used pigment is carbon black, which, in addition to its strong coloring power, exhibits a marked antioxidant effect.

13.2.1.5. Additives Against Chemical Aging. They improve the resistance of polymeric materials to chemical aging and increase their durability. Additives are utilized depending on the targeted application. Against photochemical degradation, either **"total screens"** that absorb all radiations of the solar spectrum (for instance, carbon black) or **"compatible screens"** that transform photons into thermal energy are utilized to protect polymeric materials. Examples are derivatives of 2-hydroxybenzophenone which transform into a quinoidic structure upon photochemical excitation and return to the initial state with emission of heat:

"Quenchers" are additives that interact with excited groups and deactivate them by dissipating energy through harmless infrared radiations. Some organometallic nickel derivatives function in that way.

To protect certain polymeric materials against hydrolytic degradation, it is necessary to neutralize acids and bases initially present in the medium or produced at the onset of degradation so as to prevent its progress. In condensation polymers additives are incorporated that act as **"buffers."** To protect them against microbiological degradation, fungicides and bactericides are added, particularly in wet environment.

13.2.1.6. Self-Extinguishable Polymers. Several polymers are self-extinguishable—for example, halogenated polymers. Others are flammable and thus require addition of flame retardants such as halogen derivatives (alkyl halides, etc.), phosphorus derivatives (aryl-, chloro-alkyl phosphates, etc.), or antimony oxide.

13.2.2. Plasticizers

They are added in variable proportions (from a few percent up to 50%) to basic polymers to lower the glass transition temperature of the resulting material (see Section 11.2.5). In addition to characteristics such as their miscibility with the polymer to be plasticized and the free volume they introduce, plasticizers must not counteract any of the intrinsic features of the polymeric material such as its stability with time, its absence of toxicity, its absence of odors, and so on. For poly(vinyl chloride), whose plasticization is essential, dioctyl phthalate* (DOP) is the plasticizer of choice (see Section 11.2.5.3). For other thermoplastics, tricresyl phosphate

---

*In fact, they are various highly branched alkyl phtalates that give large free volume.

is a good candidate because, in addition to its plasticizing role, it is effective as flame retardant. Diesters (2-ethylhexyl sebacate, adipate and azelate) and many oligomeric aliphatic polyesters are also utilized. To increase their miscibility with PVC, certain unsaturated vegetable oils or alkylated esters of fatty acids are epoxydized, thereby becoming excellent plasticizers.

13.2.3. Fillers and Reinforcing Fillers

Fillers are organic or inorganic compounds that are nonmiscible with the polymer matrix and are added to the latter in large proportions in order to improve the economics of expensive polymers and/or their mechanical properties.

- **"Extenders"** are added to the polymer matrix to lower the cost of the resulting material. They are generally particulate mineral species such as carbonates, sulfates, silicates, hydroxides, and so on, that develop only weak molecular interactions with the polymer;
- **"Reinforcing fillers"** are incorporated to improve the mechanical properties of the original polymer by strongly interacting with its molecular groups. Carbon black is often utilized for this purpose. Added to polydiene elastomers, the latter considerably improve the abrasion resistance in addition to stabilizing the polymer.

Both types of fillers increase the elastic modulus of the material, increase its hardness, and lower its shrinkage.

Reinforcing fillers are often fibers or films that exhibit better mechanical and thermomechanical properties than the polymer matrix and develop strong molecular interactions with it. Upon mixing such fillers with polymers, materials called **"composites"** are obtained. Due the single orientation of fibers, the improvement in the mechanical properties is observed in one direction. To overcome this drawback, nonwoven fibers and fabrics can be utilized as bi-dimensional reinforcing fillers.

13.2.4. Techniques of Formulation (Compounding)

An efficient and homogeneous dispersion of the various components entering in a blend requires powerful and efficient compounding techniques. For pulverulent or low-viscosity liquid additives, regular mixers are well-suited. It is different when the compounding/formulation has to be carried out in the melt, which requires specific techniques because of its high viscosity.

The first technique consists of using a **twin screw extruder** to mix, formulate, and finally extrude polymer melts according to a principle further described in Section 13.3.3. In the latter case, the single screw is replaced by two counter-rotating screws whose opposite rotation ensures the total homogenization of the material. Such a double-screw machine is represented in Figure 13.8. This technique is particularly well-suited to the continuous production of granulates from extruded tubes.

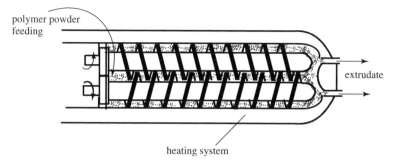

polymer powder
feeding

extrudate

heating system

Figure 13.8. Diagrammatic representation of the principle of an extruder with two counter-rotating screws.

For discontinuous compounding, another technique is preferred; for instance, the plasticization of a polymer such as PVC that is transformed during processing from a pulverulent state to that of melt with a viscoelastic consistency needs techniques such as **calendering** (Section 13.3.4). The latter consists of heated rolls revolving at different speeds against each other and capable of working-in the polymer and its auxiliaries. This compounding technique serves to simultaneously process sheets with a certain thickness.

Another technique consists of kneading the whole components between two counter-rotating roll mills; the operation can be repeated several times but has to be completed by the granulation of the resulting material. The two latter techniques can also be adapted to continuous production.

13.3. MAIN TECHNIQUES OF POLYMER PROCESSING

It is beyond the scope of this textbook to describe comprehensively the complex techniques of polymer processing and process engineering. It is, however, essential to provide polymer chemists with a minimum knowledge about the equipments that are utilized to transform raw materials into "semi-finished" or "finished" products. All processing techniques are based on and take advantage of the monodimensional plastic flow of polymeric materials above their glass transition temperature or their melting point, a feature that was described in the first part of this chapter. The processing (curing) of thermosetting polymers, in particular, requires that the operation be finished before reaching the "gel point," since the shape of the object is not deformable beyond it.

13.3.1. Compression Molding

This is one of the oldest techniques since it was used at the beginning of the twentieth century for the processing of bakelite resins. It is well-suited for the molding of thermosetting prepolymers (phenolic resins, amino resins, unsaturated polyesters, etc.) because it enables the control of the heating time before molding and the completion of the processing before reaching the "gel point."

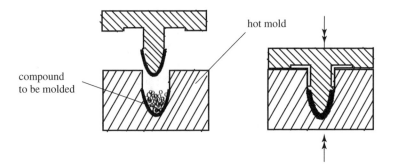

Figure 13.9. Principle of compression molding.

Heating molds are used in which the raw materials to be molded are introduced in slight excess (possibly after preheating); these molds are laid out between the plates of a heating press and compressed. The pressure applied is in the range of 40 MPa and the closing times are in the range of a few tens of seconds, which correspond to the time required to reach the optimal cross-linking. Figure 13.9 illustrates the compression molding technique with a mold that functions like a piston-cylinder system and thus allows the control of the pressure applied; the mold is designed in such a manner that the prepolymer in excess can flow out after filling all the vacant space available.

This technique is well-suited to the manufacture of cylindrical containers of low thickness. A variant of this technique, called "molding by compression transfer," is widely used for the processing of thermosetting resins. It involves the preheating of the material to be molded in a simple mold ("pot of transfer"), permitting its fast thermal transfer; then the matter of appropriate fluidity is introduced under pressure into the compression mold, and the curing of the thermosetting resin occurs. In this case, the amount of matter transferred into the mold should not be in excess.

13.3.2. Injection Molding

At first sight, this technique is close to molding by compression transfer, but it applies primarily to thermoplastic materials, which means that:

1. The matter to be injected can be maintained at the optimal temperature as long as necessary.
2. The molded object has to be cooled in the mold before being ejected.

The two essential parts of the equipment required for injection molding are the mold and the injection molding machine.

Figure 13.10 gives the principle of operation of an injection molding machine whose size closely depends on the mass of matter that the mold can accomodate. After the compounding/formulation step followed by that of granulation, the matter is introduced in the form of pellets in the injection molding machine through its hopper. Processing temperatures must be high enough to permit a good flow and

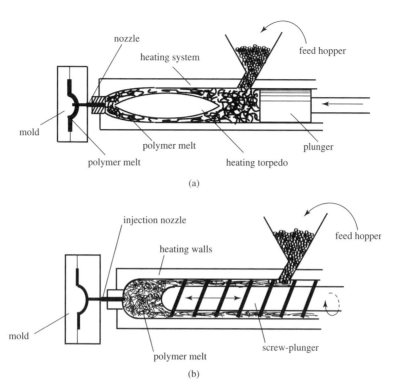

Figure 13.10. Injection molding machine: (a) Plunger–cylinder system; (b) screw–barrel system.

injection into the mold. After-pressure is applied until the specimen injected in the mold is solidified.

Two different processes can be used to make the pellets flow. In the system combining a cylinder and a plunger (Figure 13.10a), the pellets are melted through contact with the heated walls of the injection molding machine; the presence of a *"torpedo"* (plunger) improving the heat transfer through an increased contact surface. When the suitable temperature is reached, matter is injected into the mold through a sprue by a lateral movement of the piston, which is acting as a ram.

A more effective process for melting polymer pellets consists of a screw-barrel system instead of a cylinder–plunger system (Figure 13.10b). In this process, the matter to be molded melts not only from its contact with the heated walls but also from the internal friction imposed by the shape of the screw, with its rotation compressing and transporting the polymer toward the metering zone of the barrel. When enough polymer melt has been conveyed in that zone, the rotation of the screw can be stopped and the latter used to act as a piston (ram) to inject the polymer melt into the mold. The simplest molds consist of two parts; one is fixed to an injection molding machine, with the injection nozzle of the latter allowing the polymer melt to flow to the center of the closed mold. The contact between mold and injection machine through the nozzle helps to maintain the pressure of injection. The mold also comprises a second removable part that forms along with the first part its imprint mold. Upon melt flowing, the air filling the mold is expelled through vents designed to this end. High injection speeds impose a fast cooling of the mold before ejection of the molded article by opening and separation of the two parts of the mold. For this purpose, a cooling fluid is circulated in the two parts of the mold whose form and dimensions can vary considerably; in their largest dimension, molded bodies can exceed 2 meters.

Injection blow-molding is a blow-molding of a "parison" into closed and often complex-shaped molds. This parison is heated and blown by air pressure into this mold to form hollow bodies such as bottles. A derived sophisticated process called "stretch-blow molding," which causes an orientation of polymer chains, leads to an enhancement of the mechanical properties. This process is particularly suitable for the production of bottles (PVC, PET, etc.)

13.3.3. Extrusion of Polymers

Extrusion is one of the most used processing methods. It is well-suited to the production of tubes, pipes, sheets, and filaments. The polymer melt is extruded with extremely high pressure through the nozzle of a die whose cross-sectional shape determines those of the resulting parison. Extrusion applies preferentially to thermoplastic materials, but it can also be adapted to the processing of thermosetting resins.

An extruder with its screw-barrel system has three complementary functions:

- Feeding from a hopper in which the pellets are stored,
- Compacting and melting of the pellets upon heating,
- Extrusion through the die by exerting pressure.

These three functions are fulfilled concomitantly through the use of appropriate screws. The energy required by the process is provided by the rotation of the screw in the barrel, and the pitch and the diameter of the screw are the two parameters that are varied to obtain the expected extrusion.

Many combinations and machine designs have been described as fulfilling these three functions, but the most standard extruder includes a screw of conical form and a constant pitch. Such a screw is schematized in Figure 13.11. It consists of three zones, each one corresponding to the preceding functions of feeding, compacting, and extruding. Because the rotation of the screw and the mechanical energy associated with it do not provide enough thermal energy to melt the polymer pellets, the barrel has to be equipped with heaters to furnish the complementary energy.

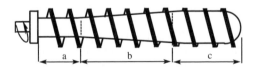

Figure 13.11. Screw of extruder: (a) Feeding section for pellets; (b) compacting section for the matter and melting; (c) extrusion section.

The whole equipment is represented in Figure 13.12.

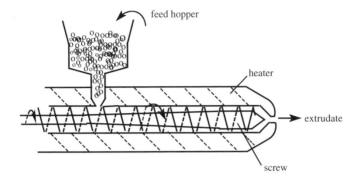

Figure 13.12. Representation of an extruder with a conical screw.

Extrusion-blowing is a process derived from extrusion that is currently used for the production of films and that of hollow bodies; the film blowing process consists of sending compressed air into an extruded tube through a blow head. The air blown expands the tube and produces a tubular film of large diameter. This film is flattened by the upper rollers before being wound by the side winder. The majority of films are manufactured in this manner—in particular, those of polyethylene. Figure 13.13 represents such a film-blowing continuous process.

The extrusion-blowing of hollow bodies also requires pressurized air, but this process is discontinuous. A tube (or parison) is extruded into an open mold. After closing the mold, a blow mandrel (not shown in Figure 13.14) is inserted into the parison head through which pressurized air is blown; this causes the parison to adopt the shape of the mold; the blow mandrel is then lifted and a cutting blade separates the hollow body from its top, ejecting the molded body upon opening

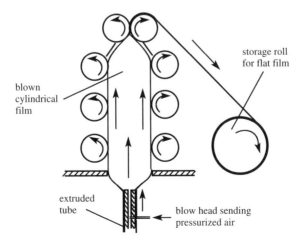

Figure 13.13. Principle of film extrusion-blowing.

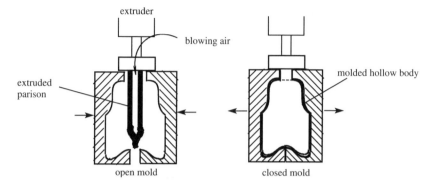

Figure 13.14. Principle of blow molding of hollow objects.

the mold. This process is used for the manufacture of containers of all capacities (from a few cubic centimeters up to few cubic meters) (Figure 13.14).

13.3.4. Calendering

Calendering is used not only to produce semifinished products, but also to mix, formulate, and homogenize polymers with auxilaries when the quantities to be treated are not too large.

The principle of this process is schematized in Figure 13.15. It is suited for the production of sheets, films, and plates of variable thickness. The surface of the calendered sheet can be sculptured through the last cylinder. If calendering is used to mix and homogenize the components of a formulation, the calendered sheet is cut and re-injected in the calenders.

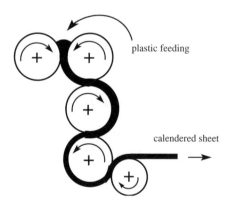

Figure 13.15. Principle of calendering.

Calendering is widely used for the formulation and the production of PVC sheets.

There are several other processing techniques that are specific to the material to be processed and to the geometry of the specimen designed; these techniques are more delicate to reconcile with a continuous production and are thus economically less significant than the preceding ones. For instance, one can mention:

- Molding by casting of a fluid prepolymer,
- Rotational casting (not very different in its principle from molding by casting),
- Thermoforming at temperature lower than T_g (or T_m for semicrystalline polymers),
- Molding by quenching a form in a solution or a fluid medium.

13.3.5. Techniques of Spinning of Artificial and Synthetic Polymers

Natural fibers are still today economically significant; in addition to their relatively low cost, they offer comfort and good aesthetics, which make them irreplaceable.

The production volume ($\sim$3 million tons) of **artificial fibers** (mainly based on cellulose) is stagnant due to their high production cost.

To date, **synthetic fibers** retain a great part of the textile market with an annual production of 30 million tons, which represents 60% of the totality of textile fibers; this dominant position is due not only to their excellent characteristics and performances but also, for the majority of them, to their low production cost. The improvement brought about by the textile industry to the manufacturing processes of these fibers, along with the facile transformation of raw polymeric materials into filaments or into staple fibers of variable length to eventually obtain fabrics or nonwoven materials of high performance, is the main reason for the success of synthetic fibers.

The transformation of raw polymeric material into fibers or continuous filaments requires the melting of the polymer and its spinning under nitrogen through a spinneret. Three different spinning processes can be distinguished, with the choice of one over the others being dictated by the fluidity of the polymeric material and/or its thermal stability at temperatures compatible with a high speed of production.

13.3.5.1. Melt-Spinning. Because it is the cheapest process, it is the most widely used. It implies that the polymer be in the melt state and chemically stable under these conditions. The melt is sent under pressure through a spinneret comprising between several tens and several thousands (for production of tons) of narrow holes of 50 to 100-μm diameter, which affords as many continuous filaments. The latter are pulled off at high speed whereby they cool below their melting point and solidify while forming a yarn. The latter is stretched up to 400–500% whereupon polymer chains and their crystalline domains are oriented along the fiber axis, which further increases their resistance. The principle of this process is represented in Figure 13.16. These monofilaments can be assembled together into multifilament yarns as mentioned above or cut as staples.

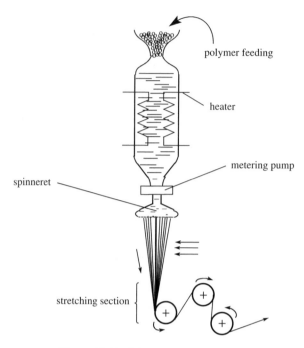

polymer feeding

heater

metering pump

spinneret

stretching section

Figure 13.16. Principle of melt spinning.

Polyolefins (EP and PP) and also polyamides (PA) and poly(ethylene terephtha-late) (PET) are the main polymers processed by melt spinning.

13.3.5.2. Dry Spinning. This process requires that the polymer be soluble in a solvent that can subsequently be evaporated, which limits its possibilities. Its principle is schematized in Figure 13.17. This process is mainly used for the pro-duction of cellulose fibers, for PVC fibers, and for the spinning of certain aramides that are soluble in highly polar solvents and exhibit suitable thermostability. In the production of cellulose acetate-based fibers, the polymer is dissolved in acetone to form a *collodion* that is extruded through the spinneret holes. Upon drawing, the major part of the solvent evaporates and the fibers solidify; due to the very high viscosity of the medium, the chains retain their orientation along the fiber axis. The solvent is recovered by condensation and is reused.

13.3.5.3. Wet Spinning. This wet spinning process is appropriate for thermally labile polymers that cannot be melted without degradation and for which no volatile solvent exists. Its principle is shown schematically in Figure 13.18. The collodion is coagulated in a bath containing, in addition to a nonsolvent, various additives to be incorporated in the fiber.

This process is largely used for acrylic, modacrylic (see section 15.3.5.3), and viscose fibers; the most commonly used solvent in the case of acrylonitrile-based fibers is dimethylformamide (DMF). Fibers of poly(*p*-phenyleneterephthalamide) are also prepared by this process; the corresponding collodion is actually a solution

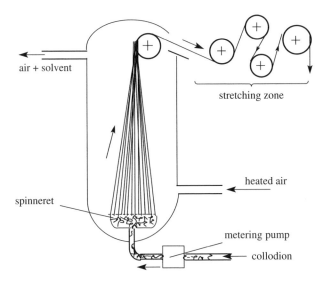

Figure 13.17. Principle of dry spinning.

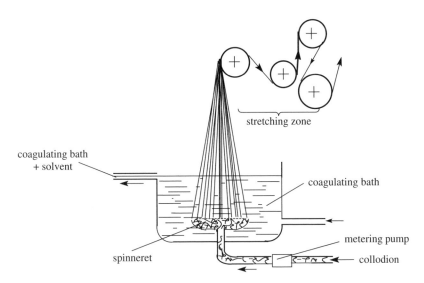

Figure 13.18. Principle of wet spinning.

of polymer in concentrated sulfuric acid, which implies that appropriate solutions to the corrosion of equipments be found.

By simple modification of the shape of the die holes, these three techniques can be adapted to the production of films and membranes. In such cases, the monoaxial stretching of fibers is replaced by a biaxial drawing that improves the two-dimensional mechanical properties.

LITERATURE

C. W., Macosko, *Rheology: Principles, Measurements and Applications*, Wiley, New York, 1994.

J., Dealy and K., Wissbrun, *Melt Rheology and its Role in Plastics Processing*, Van Nostrand, New York, 1990.

N. G., Mc. Crum, C. P., Buckley, and C. B., Bucknall *Principles of Polymer Engineering*, Oxford Science Publication, New York, 1997.

R. Gächter and H. Müller (Eds.), *Plastics Additives Handbook*, Carl Hanser Verlag, Munich, 1993.

A. I. Isayev (Ed.), *Injection and Compression Molding Fundamentals*, Marcel Dekker, New York, 1987.

14

NATURAL AND ARTIFICIAL POLYMERS

Natural macromolecules play a considerable role in the vegetable and animal worlds as well as in the mineral one. Molecular arrangements, which determine the constitution of a large variety of materials and systems of mineral origin, are similar in many respects to those found in organic macromolecules. Nature has exploited the various structural possibilities offered by the mineral macromolecular state, exhibiting, in particular, a close relationship between physical and mechanical characteristics on the one hand and dimensionality on the other hand. Three-dimensional carbon diamond and two-dimensional carbon graphite exhibit completely different properties. Soft sulfur shows a remarkable elasticity related to its linear structure of particularly high molar mass. In silicates and aluminosilicates the tetravalence of the silicon atom is responsible for the formation of single- or double-strand linear chains as well as bi-dimensional layers (micas) or three-dimensional systems.

There is no question of denying the importance of natural polymers of mineral origin. Nevertheless, they will be overlooked in this textbook in favor of natural organic polymers (also called *biopolymers*) that originate from vegetable or animal sources, some of them being even found in both worlds.

14.1. NATURAL RUBBER AND ANALOGS

Rubber was used by American Indians of Pre-Columbus America. They had discovered that the sap of the tree *cahuchu* (which means "wood that cries"), now called

Organic and Physical Chemistry of Polymers, by Yves Gnanou and Michel Fontanille
Copyright © 2008 John Wiley & Sons, Inc.

hevea brasiliensis and growing in the Amazonian forest, is an emulsion (latex) whose coagulation provides an elastic gum. It was used, just as in modern times, for the proofing of fabrics and the production of luxury goods. Its production at the end of nineteenth and the beginning of the twentieth centuries made the fortune of the Manaus City in Brazil. However, production is nowadays shifted to Southeast Asia and Equatorial Africa.

The annual world production is approximately 7 million tons, primarily coming from Malaysia, Indonesia, and Thailand. It represents about 35% of the total production of elastomers (natural and synthetic). The productivity was considerably improved, thanks to the genetic modification of *hevea* resulting in an output of almost 5 tons per hectare per annum.

14.1.1. Molecular Structure

Natural rubber (abbreviation: NR) is a polyisoprene that can be considered as originating from the polymerization of 2-methylbuta-1,3-diene (called isoprene) and contains more than 99% of 1,4-*cis* monomeric units as well as traces of 1,2 and 3,4 units:

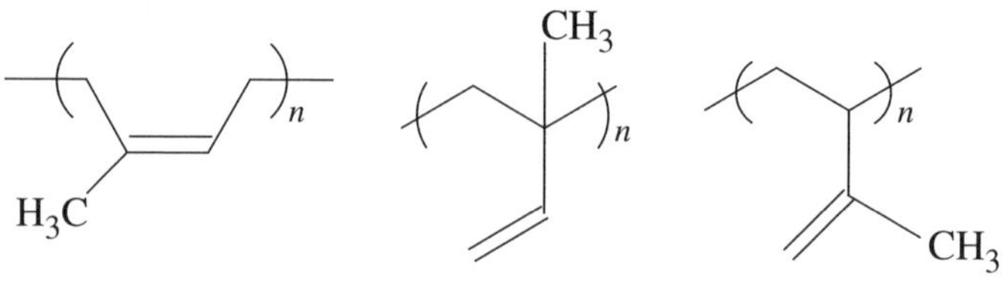

| 1,4-*cis* units | 1,2 units | 3,4 units |
| > 99% | traces | traces |

Molar masses of original rubber are generally very high, from $\overline{M}_n = 2 \times 10^5$ to 3×10^6 g·mol^{-1}, but vary depending on the varieties of trees used. Often molar mass distributions are bimodal.

14.1.2. Formation

The reaction mechanism converting $\{CO_2 + H_2O\}$ into polyisoprene *via* photosynthesis is extremely complex and is only roughly shown in the following scheme:

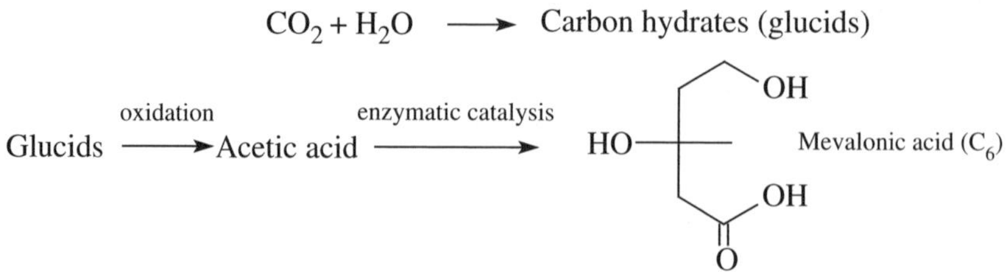

$$CO_2 + H_2O \longrightarrow \text{Carbon hydrates (glucids)}$$

then, under the action of the anenosine triphosphate,

Mevalonic acid ⟶

Mevalonic acid diphosphate

which gives an isopentenyl diphosphate by decarboxylation

$P_2O_6^{3-}$

from which 1,4-polyisoprene is obtained by successive condensations. It is recovered as aqueous emulsion called "latex," and it coagulates spontaneously when exposed to air.

14.1.3. General Characters of Natural Rubber

Like all polydienes, the polyisoprene chain is highly flexible. This property, which relates to the weakness of molecular interactions, maintains this polymer in an amorphous state and is responsible for its marked elastomeric character. However, strain deformations are totally reversible only after vulcanization (cross-linking) of the corresponding material (see Section 9.3). The glass transition temperature of NR is very low ($-72°C$).

When stressed in a monoaxial direction, natural rubber crystallizes, thus causing a heating of the material since the enthalpy of crystallization is equal to $4.4\,kJ \cdot mol^{-1}$. The crystallization generates a helical structure close to a transplanar conformation with two monomeric units per helix turn. The stress relaxation causes the melting of crystalline zones that is an endothermic phenomenon easy to reveal.

Due to its high content in carbon–carbon double bonds, natural rubber is very sensitive to oxidation by molecular oxygen. The resulting degradation occurs by a random breaking of the chains (see Section 9.4). This phenomenon is utilized in combination with a mechanical malaxation to reduce the molar mass of the original latex. This operation is performed to reduce the viscosity and thus to make easier the compounding of the material by allowing an easier incorporation of the various additives and fillers that are essential for its utilization. Thus, before use, the natural rubber should be stabilized against thermo-oxidizing degradation by incorporation of antioxidants.

> **Remark.** Carbon black is a filler that increases the abrasion resistance. It also plays a role of stabilizing agent against oxidation and is of particular use in tire industry.

However, a controlled oxidation is sometimes utilized to reduce the molar mass, thus leading to a viscous liquid that can be used as precursor of networks after vulcanization. If the latter is achieved to the extreme, it becomes a hard and breakable material that has been known and utilized since the middle of the nineteenth century (it was discovered by Charles Goodyear) under the name "ebonite."

14.1.4. Domains of Applicability

Although applications of NR are extremely numerous, the greatest part of the production is utilized by the tire industry. After blending with various synthetic elastomers, this polymer becomes indispensible due to its "sticking" properties, its adherence onto various surfaces (metals or polymers), and its damping properties. It can be easily blended with various fillers (carbon black, silica, etc.), leading to complex systems whose safety, comfort, and abrasion resistance are largely appreciated.

Nevertheless, approximately one-third of the production of NR is utilized in industries other than the tire industry. NR is required due to its mechanical characteristics, more particularly due to its high extensibility.

14.1.5. Gutta-percha and balata

They are also natural polyisoprenes and are isomers of natural rubber. They are extracted from sheets of plants in the family of *sapotaceae* (Southeast Asia and Equatorial America). The polymer chains are constructed from repeating units of 1,4-*trans* isoprene structure. Contrary to natural rubber, they partially crystallize spontaneously according to two conformations: one is 2_1 helical (α-form), the other one is totally transplanar and is equivalent to a 1_1 helix (β-form). The mechanical properties of these materials are determined by the values of the transition temperatures. Amorphous zones undergo a glass transition at $-60°C$ and the two crystalline forms (α and β) melt at $64°C$. The corresponding polymers are thus thermoplastics and acquire (after vulcanization) a highly elastic behavior beyond $70°C$.

Most of the applications of these materials are related to their adhesive properties; the main one is the coating of underwater cables. They also find an application in the manufacture of golf balls.

14.2. POLYSACCHARIDES AND THEIR DERIVATIVES

They are the macromolecular equivalents of simple natural glycosides. They can be chemically modified to give artificial polymers whose economic importance has been well known for more than a century.

14.2.1. Cellulose

It is a particularly important natural polymer since it is the main constituent of the vegetal mass; for this reason, it is produced by Nature in tens of billion of tons every year. It is present in an almost pure state ($>90\%$) in cotton fibers and is also one of the essential components of wood (with lignin). Cellulose is probably the first polymer that found industrial applications since it is present in all vegetable textile fibers.

14.2.1.1. Structure. From a chemical point of view, it corresponds to a regular linking of D-glucopyranose units. From a crystallographic point of view, it contains two monomeric units per fiber period. Thus it can be considered as resulting from the repetition of cellobiose entities,

Nonreducing chain end   Reducing chain end

Glucopyranose units adopt a more stable chair-like conformation.

As for all linear polymers, the monomeric units localized at the chain ends are different from internal units. In the above representation, these terminal units are drawn: one has a reducing character (hemi-acetal), whereas the other one does not.

The molar mass of natural cellulose is about 1.5 to 2.0×10^6 g·mol^{-1}. The treatments performed during its extraction appreciably reduce this value. In pure cellulose, the molecular structure is almost perfectly regular and the percentage of irregular units is about 0.1%. In the original state, cellulose is mixed with hemicelluloses that are branched polysaccharides with saccharide units possessing various molecular structures.

The β-$(1 \rightarrow 4)$ bonds linking the cellulose-constituting units induces a rigidity of the chains; combined with the high regularity of the linking and the development of interchain hydrogen bonds, they favor a particularly high degree of crystallinity ($X = 0.70$). The formation of strongly cohesive crystalline zones is responsible for most of the physical and chemical properties of this polymer.

Depending on its origin and the treatments undergone, cellulose crystallizes in four main structures denoted I, II, III, and IV along with secondary structures. Natural cellulose is generally of type I, but it can be easily transformed into cellulose II. These two structures are monoclinic.

In form I ($a = 0.935$ nm, $b = 0.79$ nm, $c = 1.03$ nm and $\beta = 96°$) the chains that constitute the unit cell are 2_1 helices in parallel position, whereas in form II ($a = 0.80$ nm, $b = 0.90$ nm, $c = 1.03$ nm and $\beta = 117°$) they are in antiparallel position.

14.2.1.2. General Characteristics. All the properties of cellulose are closely correlated with the high proportion of hydrogen bonds that produce an interchain linking.

In spite of the presence of many highly reactive functional groups such as hydroxyls, cellulose is poorly reactive. Interchain molecular interactions (hydrogen bonds) are strong and ensure the main part of the cohesive properties while preventing the penetration of reagents. The breaking of these interactions is the precondition of any reaction. The ways to achieve such a breaking are given in Section 14.2.2, which deals with cellulose derivatives.

Cellulose is not water-soluble but is strongly hydrophilic. This property is responsible for the great comfort exhibited by cellulose-based fibers and by the corresponding fabrics. Under normal conditions of use, cellulose may contain up to 70% of loosely bound water. The partial replacement of polymer–polymer interactions by hydrogen bonds between cellulose and water causes a plasticization of the resulting material and thus a lowering of its mechanical characteristics.

Whereas the tensile strength of highly crystalline and dry cellulose fibers can reach 700 MPa, it can lose up to 30% of its value in wet atmosphere.

Still due to the strong cohesion of this material, cellulose is insoluble in most organic solvents. Only some highly polar mixtures such as N,N-dimethylacetamide/lithium chloride, N-methylmorpholine/water, $Cu(OH)_2$/ammonia, trifluoroacetic acid/alkyl chloride, calcium thiocyanate/water, and ammonium thiocyanate/liquid ammonia are solvents of cellulose. In spite of the potential applications of such solutions, they are exploited relatively little due to their high cost.

The high degree of crystallinity of cellulose makes difficult the measurement of its glass transition temperature. The latter is located beyond 200°C but is impossible to measure accurately since cellulose degrades thermally above 180°C. Obviously, the melting point is not accessible since its value is much higher than the degradation temperature.

As all polymers that contain oxygen atoms in the main chain, cellulose is sensitive to hydrolysis. For example, in acidic medium, a random breaking of the glycosidic oxygen bonds occurs, and species of low degree of polymerization, including glucose, can be obtained from

Natural cellulose ($\overline{X}_n \sim 10^4$):
Technical celluloses ($200 < \overline{X}_n < 1000$)
Hydrocelluloses ($30 < \overline{X}_n < 200$)
Cellodextrins ($10 < \overline{X}_n < 30$)

In addition to acid hydrolysis, cellulose can also undergo both alkaline and enzymatic degradations.

14.2.1.3. Regenerated Cellulose.

A way to solubilize cellulose, other than the direct route presented above, involves the chemical transformation of hydroxyls followed by the solubilization of the corresponding artificial polymer and the regeneration of the primary polymer. The most important method using this principle consists of treating cellulose with soda to transform a high proportion of OHs into ONa groups. Alkali-cellulose thus obtained is soluble in carbon disulfide and reacts with this solvent to give cellulose xanthate:

Na–O–(C$_6$H$_9$O$_4$)-

Monomeric unit of alkali-cellulose $\longrightarrow$

+

S=C=S

$$S=C\begin{cases} O–(C_6H_9O_4)- \\ S\text{-}Na \end{cases}$$

Monomeric unit of cellulose xanthate

The cellulose is then regenerated as either a fiber or a film by neutralization of the medium with sulfuric acid. This regenerated cellulose is known under the name of *viscose rayon*. It is utilized for the production of textile fibers which are in great demand and are utilized for the manufacture of hydrophilic films—in particular, in biomedical engineering (e.g., dialysis membranes). These materials have a degree of crystallinity much lower than that of original cellulose, and thus their mechanical characteristics are lesser than those of the original material. They are interesting because they can be processed in the form of films by conventional spinning and extrusion techniques.

14.2.1.4. Domains of Application of Cellulose.
Original cellulose is mainly utilized as textile fibers (cotton, flax, hemp, etc.). Their annual production reaches 20 million tons.

Extracted from wood (of which it represents ∼50% of the content) by delignification, it becomes the main constituent (∼80%) of paper whatever the method utilized for the treatment of the paper pulp.

Cellulose can also be regenerated from solution, the xanthate method being, by far, the most utilized. This regeneration can be made in the form of wires (rayon) used in textile industry (∼2 million tons) or as films for very diverse applications.

14.2.2. Cellulose Derivatives

They are artificial polymers that retain the skeleton of the primary cellulose and whose hydroxyl functional groups are transformed under the action of various reagents. The general principles of this chemical modification were presented in Chapter 9. From a general point of view, the properties of these cellulose derivatives are highly affected by the nature of the ester introduced, the degree of polymerization, and, especially, by the residual hydroxyl group content; their total transformation considerably lowers the cohesion of the resulting material and drastically modifies the derived properties.

14.2.2.1. Cellulose Nitrates (CN).
They are the source of the oldest thermoplastics, directly obtained from Nature (see Chapter 1), and were used in first instance to manufacture *celluloid* (camphor-plasticized cellulose nitrate) and then "artificial" silk as well as supports for photographic films. These applications were given up due to safety considerations but others appeared which still justify their significant production.

The nitration of cellulose utilizes an attack of hydroxyls by a nitro-sulfuric mixture to give nitric ester (cell-ONO$_2$) with a maximum degree of substitution

(D.S.) equal to 2.8. The properties of these materials are closely related to their D.S., measured and evaluated by the nitrogen content which is equal to 14.14% by weight for D.S. $= 3$. For example, the cellulose nitrate used to prepare celluloid has a D.S. equal to 1.85, which corresponds to $\sim$10.8% of nitrogen. The higher the hydrophilicity of cellulose, the lower the D.S. On the other hand, the higher the solubility in usual organic solvents (acetone, esters, etc.), the higher the D.S. This last property is exploited in the manufacture of varnishes for various uses by dissolution in solvent mixtures.

An essential characteristic of cellulose nitrates is their capability of undergoing thermally breaking to give nitrogen, nitrogen oxides, carbon dioxide, carbon monoxide, and water. This spontaneous reaction requires a high activation energy and is self-catalyzed by the decomposition products. The manufacture of explosives (nitrated cotton) is based on this property.

14.2.2.2. Cellulose Acetates (CA). Acetylation of cellulose is obtained by reaction of the natural polymer with acetic anhydride. The reaction is catalyzed by sulfuric acid. However, to obtain derivatives of high D.S. ($>$92%), it is advisable to operate in the presence of a diluent. When the diluent is a solvent of cellulose acetate—for instance, acetic acid—the cellulose is gradually swollen by the solvent as substitution proceeds, the latter being catalyzed by mineral acids (Lewis or Brønsted acid). This process is called acetylation in the homogeneous phase.

Acetylation in heterogeneous phase (catalyzed by mineral acids) is so called when the diluent is not a solvent of cellulose acetate. Toluene or carbon tetrachloride are such liquids. Under such conditions the original fibrillary structure is reasonably well-preserved because there is less degradation of the constituting chains. Cellulose acetates with $1.6 <$ D.S. < 2.0 are soluble in many solvents (acetone, esters, chlorinated solvents) and can be plasticized by alkyl phosphates or phthalates to give thermoplastic materials exhibiting a good impact resistance.

Cellulose acetate is mainly utilized pure to produce textile fibers exhibiting a medium tensile strength (up to 60 MPa), and providing a great comfort (highly water-absorbent) and aesthetics to the manufactured fabrics. The same principles of manufacture are utilized to prepare cellulose acetate films. In solution, this artificial polymer is utilized in the varnish industry.

The annual world production of cellulose acetates with various degree of substitution is in the range of one million tons.

14.2.2.3. Mixed Cellulose Esters. Only those containing acetate units and corresponding to terpolymers of cellulose, cellulose acetate, and a second ester are worthy of interest. Actually, the situation is even more complex since 3 degrees of substitution may be possible, along with 3 sites of substitution with 3 functional groups for each monomeric unit.

Mixed esters are obtained by reaction between cellulose ester (cellulose nitrate or acetate) and another acid or an anhydride, generally organic. Cellulose acetonitrates, acetopropionates, and acetobutyrates are utilized as films and varnishes.

14.2.2.4. Cellulose Ethers. They are prepared by reaction of alkyl chlorides or their analogues with alkali-celluloses that are much more reactive substrates than native cellulose itself. The reaction pathway to obtain carboxymethyl cellulose (CMC) is given below as an example:

$$\text{Cell–OH} \xrightarrow{\text{NaOH}} \text{Cell–ONa} \xrightarrow{\text{Cl–CH}_2\text{–COO}^-\text{Na}^+} \text{Cell–O–CH}_2\text{–COO}^-\text{Na}^+$$

The resulting product is not only hydrophilic but also totally water-soluble. This property and the capability of this polymer to form aggregates, which increase considerably the viscosity of the corresponding aqueous solutions, result in a wide variety of applications (paper industry, cosmetics, pharmaceutical, food, etc.).

Methyl celluloses (MC) are obtained by reaction of methyl chloride with alkali-cellulose and, depending on their D.S., derivatives having different properties of solubility are obtained. With D.S. $= 1.5$, MC are water-soluble; then for higher D.S. they are soluble in organic solvents.

Ethyl celluloses (EC) are prepared according to the same method as MC. They are not water-soluble, and their applications are mainly in the field of thermoplastic materials.

Hydroxyalkyl ether cellulose [hydroxyethyl- (HEC) and hydroxypropyl cellulose (HPC)] are prepared in a different way. They use the capability of the alkoxide groups of alkali-cellulose to undergo a nucleophilic substitution when reacted with oxiranes:

$$\text{Cell–O}^-,\text{Na}^+ + \underset{O}{\triangle} \longrightarrow \text{Cell–O–CH}_2\text{–CH–O}^-,\text{Na}^+ \quad (\text{CH}_3)$$

Sodium salt of hydroxypropylcellulose

$$\text{Cell–O–CH}_2\text{–CH–OH} \quad (\text{CH}_3)$$

HPC

Depending upon the nature of alkylene group, the derivatives are more (HEC) or less (HPC) water-soluble. Solubility in organic solvents is reversed. However, in both cases, they are strongly hydrophilic polymers whose applications are mainly in relation with this property.

14.2.3. Starch and Its Derivatives

14.2.3.1. Origin. Starch is the main constituent of certain seeds, certain fruits, and tubers. In seeds and tubers, its content varies from 40% to 70%. It is easy to deduce that its main application is food for humans and animals. For its industrial applications, which correspond to approximately 40 million tons annually, it is extracted from cereal seeds (corns, rice) and from potato tubers.

14.2.3.2. Structure of Starch. Although the general formula of this polysaccharide of vegetable origin is identical to that of cellulose $(C_6H_{10}O_5)_n$, their physical and physicochemical properties are completely different. The basic constitutive unit is a D-glucopyranosyl group, but its configuration is different in starch with respect to that in cellulose:

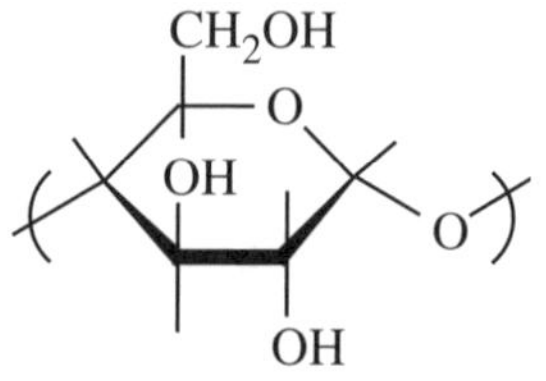

The repeating unit shown corresponds to one of the two units found in cellulose. Its configuration prevents the optimal development of interchain hydrogen bonds and favors the formation of hydrogen bonds with water which is thus included in the crystal lattice.

In addition, it has been shown that starch is, in fact, made up of two families of macromolecular compounds present in variable proportions, depending on the origin:

- **Amylose**, present in minority, which consists of linear chains containing 500–1000 glucopyranosyl groups.
- **Amylopectin**, which is made of branched chains whose monomeric units are of the same type but have irregular units at the branching points. In linear sequences, the monomeric units possess 1,4-links (as in amylose):

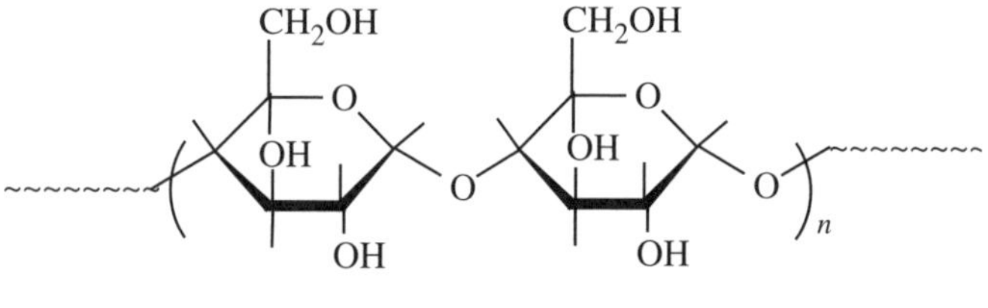

Branching can occur from hydroxyl groups:

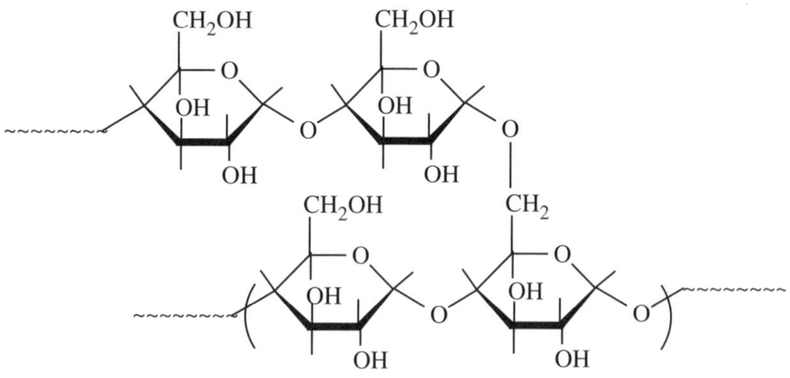

The molar mass of amylopectin chains can reach $50-70 \times 10^6 \ \mathrm{g \cdot mol^{-1}}$.

As concerned the conformational structure, linear sequences are able to crystallize in a helical geometry comprising 6 glucopyranyl residues per helix turn (helix 6_1) corresponding to a fiber period $c = 1.069$ nm.

Helices are assembled by pairs to give double helices (12 monomeric units per turn) that crystallize in the monoclinic system. The degree of crystallinity of starch depends on its origin and varies from 20% to 50%. Amylose is not highly crystalline, and the long linear sequences of amylopectins are responsible for most of the crystallinity.

14.2.3.3. General Characteristics of Starch.

The properties of starch are closely related to the existence of hydrogen bonds between the two strands of the double helix forming the crystalline zones. Interchain interactions ensure most of the cohesive properties of the system. However, due to the molecular structure, certain polymer–polymer interactions cannot be established and the cohesive energy density is definitely lower than that of native cellulose.

Hydroxyls that do not participate in the cohesion of the system strongly bind to molecules of water, and it is impossible to eliminate the latter without complete destruction of the crystalline lattice. Thus, although starch is insoluble in water at ambient temperature, it swells in hot water without total solubilization. Indeed, amylopectin chains are very long and give entanglements that form physical gels which can be irreversibly deformed under mechanical stress. These gels are thus thermoplastic materials whose temperature of creep under stress can be modulated by varying the water content.

A total hydrosolubility is obtained in water containing alkaline metal hydroxide.

Due to its lower cohesive energy density, the reactivity of starch is higher than that of cellulose, and similar methods are used to synthesize esters and ethers.

14.2.3.4. Starch Derivatives.

The most important ester is starch acetate; it is obtained according to the same method as that leading to cellulose acetate. The gradual substitution of acetate groups for hydroxyls decreases the hydrophilicity of modified starch; even at low degrees of modification, the hydrosolubility in hot water disappears and products with D.S. > 1.5 become soluble in organic solvents.

Among ethers, only hydroxyalkylethers and alkylammonium ethers are produced in an industrial scale. They are obtained by reaction of their chlorinated derivatives with starch in alkaline medium (alkali-starch). These derivatives have a variable hydrophilic/hydrophobic balance depending on the nature of the ether and the degree of substitution. It is thus possible to adapt their properties to each application.

14.2.3.5. Domains of Application of Starch and its Derivatives.

Apart from its direct utilization in food applications, extracted starch is an important industrial product, due to its hydrophilicity, a property that can be used in many respects. Thus, it can serve as viscosifying agent in human or animal food and in the pharmaceutical industry. It is also utilized in the manufacturing processes of papers and paperboards as additive in concretes or as finishing material in the textile industry.

Its capability of being degraded under the effect of biological agents in out-door media is used to induce the biofragmentation of polyolefins or in the industry of packaging (expanded biodegradable starch shaped after plasticization by water). Applications in the chemical industry are numerous, and it can be regarded as the natural polymer that is the most widely utilized by industry as an additive.

Concerning starch derivatives, their domains of applications is sensibly the same as those of other natural polymers, but their hydrophilicity could be modulated by the partial chemical modification of hydroxyl groups. Like cellulose derivatives, they can even be used as plastics.

14.3. LIGNIN

After cellulose, the most widely found natural polymer of vegetable origin on earth is lignin. Indeed, present to an extent of approximately 20% in the constitution of lignocellulosic materials, it can be estimated that its annual production by Nature is higher than 1 billion tons. Lignin thus generated many studies, but the various problems induced by its utilization are far from being solved. This is due to two main reasons:

- First, for its extraction and its subsequent valorization, the three-dimensionality of this polymer requires a partial degradation, which is difficult to control.
- Second, the extreme complexity of its molecular structure can be only represented by an average composition that can vary with the vegetable species from which it is extracted.

A possible molecular structure of an element of the network is represented on the next page. When separated from cellulose by partial degradation during the manufacturing process of paper from wood, lignin is mainly used as fuel in paper industry. Before this ultimate stage, it would be interesting to use it as material, and many attempts were performed in this respect.

14.3.1. Structure of Lignin

A three-dimensional polymer consisting mainly of di- and trisubstituted phenyl-propane units can be defined only by its average content of a certain number of molecular functions or groups. Among these various functional groups, hydroxyls and methoxyls are prominent in lignin, and the methoxyl content is generally utilized to identify the origin of lignocellulosic materials and the vegetable species from which it emerges. Lignin also contains carbonyls and unsaturations, phenols and carbohydrates which ensure a good compatibility with cellulose in wood. The scheme represented on the next page gives only a rough idea of the structural complexity of this polymer. It is important to stress that hydrogen bonds developed with cellulose result in the formation of a composite material with semi-interpenetrated network structure whose excellent properties are well-known.

14.3.2. Extraction of Lignin

Due to its cross-linked structure, lignin can be extracted from wood only by breaking up the initial network and a deterioration of its structure. Presently, industrial lignins ($>50 \times 10^6$ tons per annum) are species exclusively obtained from a chemical treatment used in the manufacture of paper pulp or cellulose fibers.

Delignification of wood is carried out in either acidic or basic conditions and in the presence of sulfur in various forms. It results primarily from a rupture of $-C-O-C-$ bonds.

- Kraft process: the wood shavings are treated at $170°C$ under pressure during a few hours in a reactor containing an aqueous solution of soda and sodium sulfide. The resulting hydrolysis allows extraction of a black liquid whose lignin components are recovered by precipitation through modulation of the concentration and the pH.
- Lignosulfonate process: the wood is treated by a sulfite (sodium, calcium, ammonium, or magnesium sulfite) which generates SO_2 *in situ*. The latter reacts with lignin simultaneously and brings about an acid hydrolysis, which induces the degradation of the network and generates highly water-soluble lignosulfonates that can be separated from cellulose.
- Although less important, many other processes can be used—in particular, the flash self-hydrolysis that results from the explosion of shavings of wood impregnated with steam under pressure.

The product resulting from these extractions appeared as a dark-brown solid whose molar mass and properties depend on the conditions of the network fragmentation ($\overline{M} = 10^4$ to 10^6 g·mol^{-1}).

14.3.3. Valorization of Lignin

It is mainly used as combustible in paper industry. However, certain more valorizing applications can be found. Lignin is utilized as a filler in blends with certain thermoplastic polymers (polyolefins, PVC, rubbers, etc.), with the presence of phenol groups ensuring a marked antioxidizing effect. The high percentage of hydroxyls also contributed to use lignins as polymer precursors (macromonomers, functionalized precursors) to give formo-phenolic resins, polyurethanes, or polyesters.

14.4. PROTEIN MATERIALS

Without entering into the chemistry of the processes of life, it is worth stressing the importance of certain proteins resulting from either the vegetable or the animal worlds that are used in industry. Textile fibers, wool, and silk are of great and continuing interest, and scientists have copied them in inventing polyamides.

14.4.1. Structure of Proteins

It can be considered that these compounds are the products of the polymerization of α-amino-carboxylic acids:

$$H_2N\text{-*CH-CO}_2H \longrightarrow \left(\text{NH-*CH-CO}\right)$$
$$\quad\quad\;\; \underset{(A)}{|} \quad\quad\quad\quad\quad\quad\quad\;\; \underset{(A)}{|}$$

The term protein is employed for compounds exhibiting molar mass $> \sim 10^4 g \cdot mol^{-1}$, and the term polypeptide is reserved for the shorter chains. The *C-marked carbon atom is unsymmetrical (except for $A = H$) and always has absolute [S] configuration (indicated as L by biochemists).

In Nature, the variety of side groups **A** (20 different **A** leading to 20 different "residues") imparts a complexity to the molecular structure of natural proteins, whose extent arises from the combination of 20 different "comonomers" in variable proportions (Table 14.1).

The level of structure described as **primary** corresponds to the distribution of the 20 protein residues along the macromolecular chain. To indicate the arrangement of the various comonomeric units in the polypeptide sequences, it is necessary to give an abbreviation to each residue, corresponding to the first letters of the amino acid. For example, the $-Arg-Gly-Asp-$ sequence is known to exhibit antithrombogenic properties that are used in biomedical engineering.

> **Remark.** The increase in both the sensitivity and the precision of the techniques of characterization allows the identification of increasingly long sequences, and it is convenient to indicate each residue by only one letter. Thus, the arginine–glycine–aspartic acid sequence is also indicated by RGD.

The average composition of a given protein and the sequential arrangement of the constituting residues depend on its origin *i.e.* the species from which it results

Table 14.1. Designation and structure of the 20 natural protein "monomers"

Aspartic acid	$HOOC-CH_2-CH(NH_2)-COOH$
Glutamic acid	$HOOC-CH_2-CH_2-CH(NH_2)-COOH$
Alanine	$H_2N-CH(CH_3)-COOH$
Arginine	$H_2N-C(NH)-NH-(CH_2)_3-CH(NH_2)-COOH$
Asparagine	$H_2N-CO-CH_2-CH(NH_2)-COOH$
Cysteine	$HS-CH_2-CH(NH_2)-COOH$
Glutamine	$H_2N-CO-CH_2-CH_2-CH(NH_2)-COOH$
Glycine	H_2N-CH_2-COOH
Histidine	$Imidazolyl-CH(NH_2)-COOH$
Isoleucine	$CH_3-CH_2-CH(CH_3)-CH(NH_2)-COOH$
Leucine	$(CH_3)_2CH-CH_2-CH(NH_2)-COOH$
Lysine	$NH_2-(CH_2)_4-CH(NH_2)-COOH$
Methionine	$CH_3-S-(CH_2)_2-CH(NH_2)-COOH$
Phenylalanine	$C_6H_5-CH_2-CH(NH_2)-COOH$
Proline	$Pyrrolidyl-COOH$
Serine	$HO-CH_2-CH(NH_2)-COOH$
Threonine	$HO-CH(CH_3)-CH(NH_2)-COOH$
Tryptophane	$Indolyl-CH_2-CH(NH_2)-COOH$
Tyrosine	$HO-C_6H_4-CH_2-CH(NH_2)-COOH$
Valine	$(CH_3)_2CH-CH(NH_2)-COOH$

and also, to a lesser extent, on the individual species that has produced it. For example, the composition of keratins, which are the essential components of wool, hair and feathers is different for each of these entities and varies with the species that produce them and also with the individuals that compose these species.

The **secondary** structure is considerably affected by the optimal development of hydrogen bonds that develop between C=O groups and amide functional groups.

The alpha-helix corresponds to the structure shown in Figure 14.1, with two possible directions of notation corresponding to right-handed and left-handed helixes.

In proteins, mostly right-handed helixes are found. It is difficult to utilize the concept of fiber period for such helices since the residues located in identical positions can differ by their side substituent R. However, to find two residues located in the same position along the chain axis of this helix, one has to move across five turns

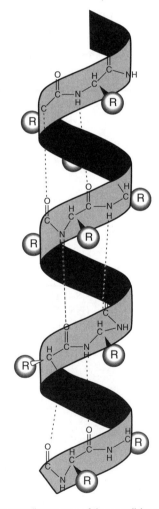

Figure 14.1. Alpha-helix corresponding to one of the possible secondary structures of proteins.

corresponding to 18 residues: there are 18_5 helixes corresponding to a repeat period of 2.7 nm. Figure 14.1 shows (dashed lines) how interchain hydrogen bonds between C=O and N–H groups are formed and induce a high stability of the resulting helical structures.

Many proteins exhibit a secondary β-structure corresponding sensibly to the chain under total extension. Due to an odd number of atoms per residue in the main chain, the crystalline period contains 2 residues (2_1 "flat" helix) with a fiber period (c) equal to 0.70 nm. Hydrogen bonds develop along two dimensions (Figure 14.2), thus inducing the formation of layers whose cohesiveness is highly anisotropic.

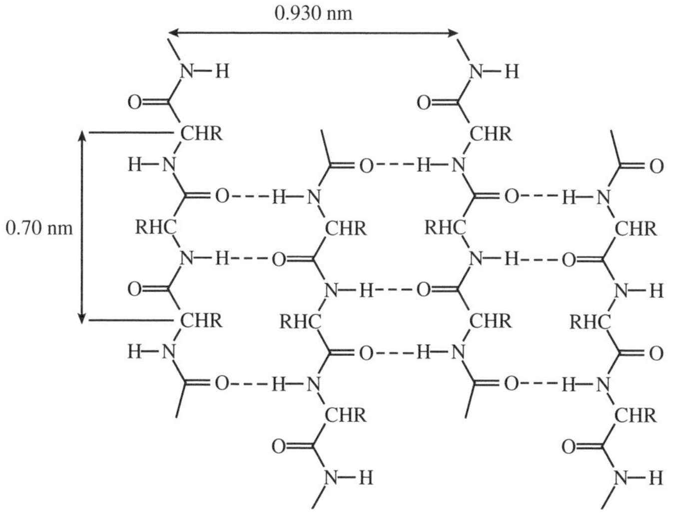

Figure 14.2. Beta secondary structure of proteins.

Under mechanical stress, it is possible to transform an α-helix into a β-helix. The reversibility of the deformation makes this phenomenon close to rubbery elasticity. Under certain conditions, this phenomenon can be made irreversible. For example, under the effect of reducing systems such as thioglycolic acid, the disulfide bridges linking cysteine residues of keratin can be broken and the chains extended by application of an uniaxial stress. Through an oxidizing process the disulfide bridges can be reestablished and a new conformation fixed in a two-dimensional network.

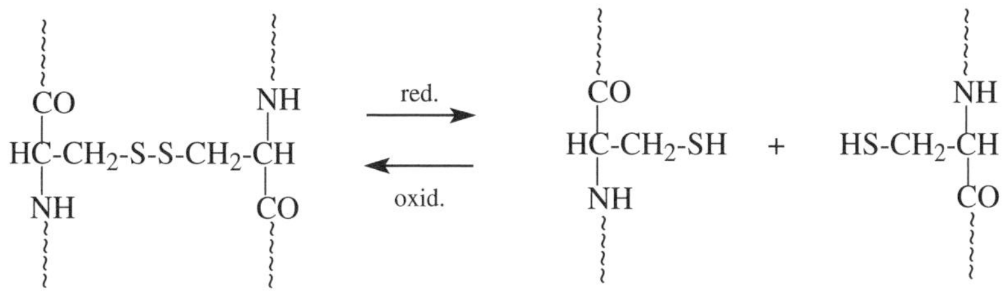

Due to the constitutive dissymmetry of monomeric units, polypeptide assemblies of chains in β-conformation can utilize either parallel or antiparallel chains. Indeed, the optimal development of hydrogen bonds is essential to the formation of these structures.

The **tertiary** structure of proteins reflects a still higher level of organization and corresponds to a preferential arrangement of relatively long ordered sequences. Such structures have been characterized by means of crystallographic methods and, as previously mentioned, mirrors the maximal thermodynamic stability of the protein considered. An example of tertiary structure is presented in Figure 14.3.

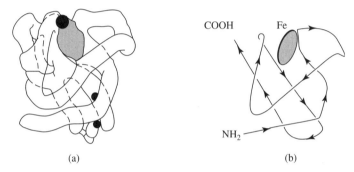

(a)   (b)

Figure 14.3. Tertiary structure of myoglobin identified by X-ray diffraction. (a) Global structure of the macromolecule. (b) Orientation of the various constituting sequences.

The denaturation of proteins is a conformational transformation of the macromolecule that induces a loss of its specific properties. It can result from either a rise in temperature, a change in pH of the medium, a mechanical stress, or a chemical action. The denaturation is primarily the result of a transformation from an α-helix to a β-structure which instantaneously modifies the tertiary structure.

14.4.2. Several Protein Materials

Wool is mainly constituted of keratins—that is, proteins whose main residues are derived from leucine, serine, cysteine, glutamic acid, and arginine. The relatively high proportion of cysteine is responsible for the presence of disulfide links and confers three-dimensionality to these proteic materials that exhibit remarkable reversible deformations. However, when wool is stretched during a short lapse of time in the presence of hot water or hot steam and then relaxed, the macromolecular chains that were initially in β-conformation fold up into partial α-helixes and contract according to a process known as felting. The comfort of wool fibers for textile applications ($\sim$1.5 million tons) is primarily due to their hydrophilicity that results from the presence of polar groups along the chains.

Silk is mainly constituted of a protein excreted by *bombyx mori*. It contains a high proportion of glycine (44%), alanine (26%), and serine (13%) units. With a fiber period $c = 0.695$ nm, it exhibits a characteristic β-conformation including two residues per fiber period. Such a structure is responsible for a high elastic modulus and for a rather low reversible elongation (stretching at break ~15%). Its tensile strength is equal to ~0.5 GPa.

Other proteic materials have also an industrial application, but their importance is considerably lesser than that of wood and silk. However, in the future the situation may change if the use of the biomass and the biodegradation of materials become major objectives.

For example, it is possible to obtain a material from the casein of milk. It is still produced in small quantities under the name of galalith. Casein is extracted from whey by precipitation in acidic media and, after drying, can be molded by hot compression. Treated by formaldehyde, it acquires a hydrophobic surface that prevents it from swelling in aqueous media. From casein, textile fibers can also be manufactured by spinning from an aqueous alkaline solutions and made insoluble in water by formaldehyde treatment.

LITERATURE

K. Kamide, *Cellulose and Cellulose derivatives*, Elsevier, Amsterdam, 2005.

J. Park, *Science and Technology of Rubber*, Elsevier, Amsterdam, 2005.

Kirk-Othmer (Ed.), *Encyclopedia of Chemical Technology*, Wiley-Interscience, New York, 1996.

H. F. Mark, N. M. Bikales, C. G. Overberger, and G. Menges (Eds.), *Encyclopedia of Polymer Science and Technology*, 2nd edition, Wiley, New York, 1989.

15

LINEAR (MONODIMENSIONAL) SYNTHETIC POLYMERS

By definition, these polymers are obtained by polymerization of bivalent monomers and have a finite molar mass. They can be either linear or branched and are soluble in solvents that can break the molecular interactions ensuring their cohesion. Moreover, they are thermoplastics if their softening temperature is lower than their temperature of thermal decomposition.

They correspond to the major part of synthetic polymers, and their annual world production exceeds 130 million tons with about 36% in the United States alone.

Various families of polymers will be presented not only due to their economic significance but also due to their intrinsic characteristics.

15.1. POLYOLEFINS

Olefins (or alkenes) are unsaturated aliphatic hydrocarbons having the general formula $H_2C=CR_1R_2$. The corresponding polymers $-(CH_2-CR_1R_2)_n-$ do not possess polar groups, and their cohesion is thus closely dependent on intermolecular distances and, consequently, on their degree of crystallinity. By modulating the latter, it is then possible to obtain a wide variety of materials from highly cohesive ones (that could be used as textile fibers) to highly deformable ones (that could be used as elastomers). In spite of the extreme variety of the possible molecular structures, only monomers corresponding to R_1 and $R_2 = -H$ and $-CH_3$ (ethylene, propylene, isobutene) are utilized in the production of polymers to a substantial extent; however, poly(but-1-ene) and poly(4-methylpent-1-ene) have attained the industrial level.

Organic and Physical Chemistry of Polymers, by Yves Gnanou and Michel Fontanille
Copyright © 2008 John Wiley & Sons, Inc.

15.1.1. Polyethylene and Its Copolymers

Acronym: PE

Molecular formula: $-(CH_2-CH_2)_n-$

IUPAC nomenclature: poly(methylene)

This is the most important synthetic polymer. Its annual world production is estimated at about 60 million tons in 2006.

15.1.1.1. Monomer. Ethylene (which according to IUPAC rules should be called *ethene* and the corresponding polymer, from source-based rules, *polyethene*) is a gas ($T_b = -104°C$) obtained from the thermal cracking (free radical process) of oil products. The initial reaction is a homolytic rupture of the covalent bonds of hydrocarbons that generate primary free radicals; but the subsequent reactions are extremely varied (H abstractions, additions, decompositions, and isomerizations of radicals, etc.) and lead to a complex mixture that must be fractionated. Ethylene can also be produced either by dehydration of ethanol

$$C_2H_5OH \longrightarrow C_2H_4 + H_2O$$

or from propylene *via* the "triolefinic" process, which uses a metathesis reaction:

$$2\ H_3C-CH{=}CH-CH_3 \underset{}{\overset{cat.}{\rightleftarrows}} H_2C{=}CH_2 + H_3C-CH{=}CH-CH_3$$

For economic reasons, only high-purity ethylene is used for polymerization; this purity is necessary for the processes based on the catalysis by coordination.

15.1.1.2. Methods and Processes of Polymerization. In 1933, scientists of Imperial Chemical Industries (ICI) succeeded in performing the **free radical polymerization** of ethylene while operating at very high pressures (150–300 MPa). The process led to industrial production in 1939. It is still used, with the polymerization being initiated using either an organic peroxide or molecular oxygen at a temperature between 140°C and 180°C. The resulting polymers are specifically named by using the acronym LDPE, which stands for "low-density" polyethylene.

Polymerizations are carried out in continuous flow either in stirred reactors (autoclaves) whose volume is in the range of one cubic meter or in less bulky tubular reactors ($\sim0.4\,m^3$) in which the pressure may be higher than in autoclaves. The monomer conversion is only 15–20% for each passage in the autoclaves, but is a little higher (25%) in tubular reactors. It is important to emphasize that ethylene is in a supercritical state under the temperature and the pressure of such polymerizations; the corresponding fluid has a density close to 0.6 and is used as solvent for the PE formed. It is thus a polymerization in bulk as defined in Section 8.5.12.

The **coordination polymerization** of ethylene is more and more utilized because it allows the production of polymers with a better control of the structure than that of PE obtained by free radical polymerization. More particularly, linear polymers indicated by the acronym HDPE ("high-density"polyethylene) as well as copolymers with other olefins can be obtained. The latter are particularly important because they generate materials of definitely differentiated characteristics by varying only the comonomer content.

"Phillips" catalysts based on supported chromium oxide are still widely used to produce HDPE (see Section 8.8.4). Nevertheless, the discovery of coordination catalysts by Ziegler in 1953 revolutionized the production of polyethylene. Indeed, the catalytic systems based on titanium halides and alkylaluminum offer many advantages relative to the processes (polymerization under moderate pressure) as well as the properties of the resulting polymers. The most widely used catalytic systems consist of $TiCl_4$ and $Al(C_2H_5)_3$, the product of the reaction being supported on $MgCl_2$ (see Section 8.8.2). They give extremely high outputs (up to 500 kg of PE per gram of Ti), which allows the suppression of the polymer "washing," a phase required to eliminate the catalytic residues. A great variety of techniques are used to carry out this coordination polymerization: high pressure ("bulk"), solution in an aliphatic hydrocarbon, "gas-phase" process and suspension in a diluent. Each of these techniques should be adapted to the production of polymer in particularly high quantities. The molar masses are controlled by transfer to molecular hydrogen.

Metallocenes are able to initiate the polymerization of ethylene and also its copolymerization with other α-olefins to produce copolymers (see Section 8.8.3). The efficiency of these catalysts is close to unity. They afford very high outputs which may give these catalytic systems a promising future.

15.1.1.3. General Characteristics of Polyethylenes.

Due to their symmetry, the linear sequences of polyethylene are highly crystallizable. They are arranged in planar zigzag and are assembled according to an orthorhombic symmetry close to a hexagonal system. The fiber period corresponds to only one monomeric unit ($c = 0.254$ nm).

The melting point of the best arranged crystalline zones is $135°C$. The noncrystalline sequences undergo the glass transition phenomenon at $-110°C$. This transition (known as "γ") corresponds to the motion of short sequences (3–4 methylene groups) and is observed in all types of PE. It is admitted that PE presents a second transition phenomenon at $-20°C$ ("β" transition), which is related to the motion of longer sequences and that cannot be observed in highly crystalline polymers. The degree of crystallinity of polyethylenes closely depends on their structure; it can vary from 30% to 70%, depending on whether the proportion of branches (or comonomeric units) is high or low. This degree of crystallinity is generally evaluated by the density which varies between 0.92 and 0.97 for homopolymers and can be reduced up to 0.88 for linear copolymers (LLDPE stands for "linear" low-density polyethylene).

A particular case is that of linear PE with very high molar mass ($\overline{M}_w > 3 \times 10^6$ g·mol^{-1}) whose crystallization can be partially inhibited ($d = 0.94$) not by the proportion of branches (which is low) but due to the very high viscosity of the medium.

The high cohesion energy density of the crystalline zones is responsible for the low solubility of polyethylene: it is insoluble in all solvents at ambient temperature and is soluble only at high temperature ($T > 80°C$) in certain hydrocarbons (decahydronaphtalene, etc.), aryl halides (o-dichlorobenzene, trichlorobenzene, etc.) or ketones, esters, and ethers carrying big alkyl groups (diamyl ether, etc.). The insolubility of PE at room temperature requires the development of techniques operating at high temperature (size exclusion chromatography, etc.) for its structural characterization in solution.

Due to its paraffinic structure, PE exhibits a marked hydrophobic character and a high chemical inertia. Its resistance to thermo-oxidizing degradation is in close relationship to its degree of branching because tertiary hydrogen atoms are more sensitive than secondary ones to the attack of molecular oxygen.

Once processed (molded objects, films, fibers, etc.), polyethylene can be crosslinked *in situ* either by homolytic decomposition of peroxides or by electron or γ beams in order to lower its creep under stress.

15.1.1.4. Various Types of Polyethylene and Copolymers.
There is a wide variety of materials obtained from the (co)polymerisation of ethylene whose physical and mechanical characteristics are quite different.

Conventional **LDPE** ("low density") obtained by radical polymerization is presently still very important since it forms nearly 30% of the total current production of polyethylenes. It is a highly branched homopolymer due to intra- (majority of short branches) or intermolecular (long branches) transfer reactions occurring during polymerization (see Section 8.3.6). Its degree of branching is measured by the number of methyl groups per 1000 carbon atoms. It is about 20–30 with a clear prevalence of short branches (4–6 carbon atoms). Its density varies from 0.915 to 0.925, depending on the polymerization conditions. Mass average molar masses of LDPE are in the range $1-2 \times 10^5$ g·mol^{-1} with a dispersity index (D_M) varying from 4 to 12. Such high values are due to the high proportion of short chains that play an important role as plasticizers for long chains and determine the fluidity of the material in the molten state.

LDPE is translucent and even transparent when processed in thin films.

Contrary to the polymerizations mentioned above, homopolymerization by coordination catalysis ("Ziegler," "Phillips," etc. catalysts) leads to polymers that are almost perfectly linear and thus highly crystallizable. They exhibit a very high density (**HDPE**) since their high degree of crystallinity ($\sim$70%) confers upon them a volumic mass that can reach 0.97 g·cm^{-3}. Moreover, "Phillips" HDPEs carry an unsaturation at the chain end which results from a spontaneous transfer reaction.

The mass average molar masses ($\overline{M}_w$) of commonly produced HDPE are in the range of 10^5 g·mol^{-1}, whereas $\overline{M}_n$ are definitely lower due to a strong heterogeneity of chain lengths. In addition, certain HDPE having very high molar mass (from 1 to 5×10^6 g·mol^{-1}), named UHMWPE;* are obtained by Ziegler–Natta catalysis in absence of transfer agents; they exhibit specific mechanical behavior. In spite of their high stuctural regularity, they crystallize with difficulty ($d = 0.94$) due to

*UHMWPE: ultra high molecular weight polyethylene.

the high viscosity of the medium. Due to same reason, they cannot be processed by the usual techniques but by sintering. They are characterized by an excellent abrasion resistance, a high chemical inertia, and very good frictional properties.

HDPEs are more cohesive than LDPE; they are translucent but not transparent (even at low thickness) as their crystalline zones cause light scattering. Applications of HDPE are not very different from those of LDPE, although the mechanical characteristics of two materials are clearly different (Table 15.1).

The annual production of HDPE reached 28 million tons in 2006. With a consumption of 4.0 kg *per capita*, it is the third-largest plastic commodity material in the world after poly(vinyl chloride) and polypropylene.

Ziegler catalysts are also able to copolymerize ethylene with higher olefins (propylene, butene, etc.), and these copolymers acquire an increasing importance. In particular, conventional LDPE is gradually replaced by **LLDPE**[†] in its numerous applications. In fact, LLDPEs are copolymers generally obtained by using "Ziegler" catalysts. In addition to their improved processability compared to that of the corresponding conventional LDPE, "linear" ones with extremely variable degrees of crystallinity can be obtained in a single reactor and upon using the same catalytic system. Indeed, the only change in the ratio of comonomer allows the production of PE with density varying from 0.89 to 0.95. The most widely used comonomers are propylene and butene; the degrees of "branching" usually lie between 20 and 60 substituents per 1000 carbon atoms. Production of LLDPE represents 14 million tons.

Copolymers with a high proportion of co-α-olefin (generally propylene) are totally amorphous and, due to this reason, exhibit elastomeric properties after cross-linking (vulcanization). There are two types of these copolymers which contain 15–40% of propylene units. The first contain only units of both comonomers (EP copolymers), whereas the second (EPDM) contain in addition a few units resulting from the incorporation of a nonconjugated diene (for example, dicyclopentadiene, 5-ethylidenenorbornene, hexa-1,4-diene, or 7-methylocta-1,6-diene). The cross-linking of EP copolymers is obtained either by treatment with an electron beam or by generation of free radicals *in situ* generated by the thermal decomposition of a peroxide. In the case of EPDM, the incorporation of a diene in the chain consumes only one double bond; the second is a side group that can be used for

Table 15.1. Main mechanical characteristics of polyolefins

Polyolefin	Elastic Modulus (MPa)	Stress at Break (MPa)	Strain at Break (%)
LDPE	150	15	500
LLDPE	250	20	200–900
HDPE	800–1200	35	200–800
UHMWPE	200–600	35	200–500
PP	1300	40	400

[†]LLDPE: linear low-density polyethylene.

a conventional vulcanization (see Section 10.3). As compared to polydienes, these elastomeric copolymers exhibit an excellent resistance to oxidation.

Ethylene can also be copolymerized with polar monomers in order to widely modify the characteristics of the corresponding materials. The comonomers are most often (meth)acrylic monomers or vinyl acetate, with the latter being the most used for the production of EVA copolymers (ethylene/vinyl acetate). EVAs generally contain about 20% mass of comonomer and are very interesting due to their adhesive properties.

After neutralization by a metal cation, copolymers with acrylic acid give materials that behave like thermolabile cross-linked systems.

Copolymers can also be obtained by a chemical modification of homopolymers. Thus, chlorinated or chlorosulfonated polyethylenes are prepared by chemical modification of PE, for the copolymerization of ethylene with the corresponding "comonomers" is impossible to carry out.

15.1.1.5. Fields of Application.
Film packaging is the one of privileged fields of application of polyethylene. Low-density PE is widely used, but HDPE also has some applications in this field. These films are obtained by the extrusion-blowing process (see Section 13.3.3). PE is also utilized for films of agricultural use.

Whatever may be its type, polyethylene is also used to obtain semi-finished products by extrusion process (pipes, sheaths of cables, etc.) as well as various objects by extrusion-blowing of hollow bodies or by injection molding. Depending on the desired mechanical characteristics, PEs having variable density are utilized, with the low-density PE being characterized by a remarkable impact strength. For applications in cable-making, PE is generally cross-linked after extrusion.

EP and EPDM copolymers are used as synthetic elastomers in all sectors of the rubber industry due to their high chemical inertia and low tendency to aging.

HDPE can be stretched to give monofilaments that are utilized in the manufacture of ropes. Its paraffin touch restricts its use in textile industry. The drawing of linear PE with high molar mass can lead to fibers having very high elastic modulus.

15.1.2. Isotactic Polypropylene

Acronym: PP (or iPP to differentiate it from syndiotactic polyproylpene which is appearing on the market and is indicated by sPP)

Molecular formula: $-[CH_2-CH(CH_3)]_n-$

IUPAC nomenclature: poly(1-methylethylene)

Polypropylene (or polypropene) prepared by free radical polymerization is a low molar mass atactic polymer, and it is not much significant product from economical point of view.

Isotactic PP, which was discovered by Natta and was obtained by polymerization of propylene using Ziegler catalysts, is a product economically significant since its annual world production exceeds 30 million tons.

15.1.2.1. Monomer. Like ethylene, propylene (propene) is a gas ($T_b = -48°C$) obtained from the cracking of oil products. The monomer must be free from impurities so as to undergo polymerization by coordination catalysts; indeed, impurities (polar molecules, dienes, etc.) would prevent its coordination on transition metal. Thus propylene is to be very carefully purified before its use.

15.1.2.2. Methods of Polymerization. Polymerizations by radical and electrophilic additions on the monomer double bond are possible, but, due to occurrence of transfer reactions, they only lead to atactic oligomers.

Isotactic polypropylene was obtained for the first time by Natta (see Section 8.8.2). The use of the original Ziegler catalytic systems led to the formation of a significant fraction of atactic polymer which had to be eliminated to obtain a material showing a high degree of crystallinity. Now it is known that this atactic PP was produced by *aspecific* sites of $TiCl_3$ crystals. An important improvement in the stereoregularity of PP was obtained by poisoning the most acidic sites that correspond to *aspecific* ones by addition of Lewis bases (ethers, tertiary amines) to the catalytic system. This allowed attainment of high degrees of isotacticity (*mm* triads content >98%). As improvements in the isospecificity were simultaneously accompanied by a fall in the activity, considerable improvements had to be made on this point. This was achieved thanks to, in particular, the supported catalysis on $MgCl_2$. At present, supported systems contain, in addition to the active catalytic system, two Lewis bases (one internal and one external) to attain rates of *mm* triads close to 99% and activities (see definition in Section 8.8.2) of several hundreds of grams of PP per gram of Ti per h pe MPa.

Although at present most of the production of iPP uses Ziegler–Natta catalysis, one can expect that metallocene-containing systems will be more used. Their high efficiency and the possibilities they offer in the fine control of the tacticity of poly(α-olefins) make them increasingly attractive. In particular, syndiotactic polypropylene (sPP) could be obtained under industrial conditions. This material is different from iPP.

Whatever may be the catalytic systems used nowadays, their productivity and stereospecificity are such that it is useless to proceed for the elimination of the atactic chains and even the catalytic residues.

15.1.2.3. General Characteristics of Isotactic Polypropylene. The high content in stereoregular sequences in iPP makes this polymer highly crystallizable. The resulting regular conformational arrangement is a 3_1-type helix (3 monomeric units per helix turn), which corresponds to a periodicity along the fiber axis $c = 0.650$ nm. In a crystalline lattice, this 3_1 helix can be arranged according to three different positionings indicated by α, β, and γ arrangements whose occurrence depends closely on the heat treatments applied on iPP. The alpha form is the most common structure and corresponds to a monoclinic system.

Crystallization is spontaneous at ambient temperature since $T_m = 170°C$ and $T_g = -8°C$. The maximum rate of crystallization is at about $110°C$. The degree of crystallinity lies between 0.4 and 0.6.

Like PE, iPP is particularly hydrophobic due to its paraffinic nature. Its oxidation resistance is definitely lower than that of the PE because hydrogen atoms carried by tertiary carbons are sensitive to the action of molecular oxygen (see Section 10.4.1).

Original iPP is a highly cohesive polymer in the crystalline state (see Section 15.1), a property resulting from low intermolecular distances in the crystalline phase. It is much more impact-sensitive than PE, particularly at temperatures below ambient. This can be due to a relatively difficult flow of the chains under sudden stress near the glass transition temperature, in relation with their helical structure.

15.1.2.4. Improvement of the Impact Strength of Polypropylene. For
many years, the development of iPP was slowed by its low-impact strength. Thus, the solution of this problem became prioritised, and research in this field was inspired by the methods found for the development of "high-impact polystyrenes" (HIPS).

The most interesting method, which involves the mixing of iPP (before extrusion) with a polyolefinic elastomer (EP or EPDM), causes the mechanical homolytic breaking of the chains and the random recombination leading to the formation of ill-defined block copolymers. Emulsifying properties of the latter are, however, sufficient to finely disperse EP or EPDM elastomers in the iPP matrix and thus to considerably increase the impact strength of the corresponding materials. It was recently established that certain LLDPE with high butylene content are miscible with iPP and can advantageously be used for the same objective.

15.1.2.5. Fields of Application. Since the problems related to its high brittleness were solved, iPP became one of the most significant thermoplastic for the manufacture of molded objects and bi-oriented films (food packaging, castings for automotive engineering, etc.). It has a great capability of adaptation with respect to the totality of the processing techniques. Extruded under the shape of mono-oriented film, it can be cut out in strips that have high elastic modulus and tensile strength. They can be used for the industries of woven bags, strings, and ropes, as well as in the carpet industry (underlayers).

Its mechanical characteristics open it the field of textile industry — in particular, that of the carpets and fitteds carpet for which it gives an excellent quality/cost ratio.

15.1.3. Polyisobutene (Butyl Rubber)

Acronym: PIB

Molecular formula: $-[CH_2-C(CH_3)_2]_n-$

IUPAC nomenclature: poly(1,1-dimethylethylene)

Actually, it is not a homopolymer but a copolymer with a low isoprene comonomer content (about 1%). Unsaturations of the latter are used only for the vulcanization of the material after its processing.

In spite of the high regularity of its molecular structure, this polymer cannot crystallize spontaneously. It can thus only be used as elastomer since its glass transition temperature is lower than the ambient temperature.

Its world production is on the order of one million tons.

15.1.3.1. Isobutene Monomer (or 2-Methylpropene).

It is a gaseous hydrocarbon ($T_b = -7°C$) that results from petrochemistry. The electron donor effect of the two methyl substituents of the double bond increases the electron density of the latter and enhances its sensitivity to electrophilic addition reactions. It is thus a monomer that is particularly suited to be polymerized by cationic means.

15.1.3.2. Methods and Process of Polymerization.

The cationic copolymerization of isobutene with isoprene is carried out in solution through a flow process in a halogenated hydrocarbon ($CHCl_3$, CH_2Cl_2, etc.) at very low temperature in order to restrict the extent of transfer and termination reactions. The initiator utilized is a complex Lewis acid resulting from the addition of very low quantities of water on $AlCl_3$. Incorporation of isoprene units in the chains is 1–4 (60% *trans* and 40% *cis*).

The maintainance of the reactor at low temperature ($-95°C$) in spite of the high enthalpy of polymerization is ensured by a circulation of liquid ethylene or ammonia. The polymerization is total and quasi-instantaneous and the rate of polymerization is thus controlled by the rate of introduction of the monomer into the reactor. PIB is insoluble in the reaction medium and precipitates gradually when appearing. It is recovered by continuous filtration.

15.1.3.3. General Characteristics of PIB.

Due to the low content in isoprene, the chains are constituted by long regular sequences of isobutene monomer units since cationic polymerization of this symmetrical monomer generates only "head-to-tail" placement. Consequently, this polymer is highly crystallizable, but crystallization is prevented by the high mobility of the chains and their interdistance due to the steric effect of methyl groups. Molecular interactions are weak and the resulting specific cohesion cannot counteract the effect of thermal agitation. The polymer is thus totally amorphous and since its $T_g = -73°C$, it exhibits a marked elastomeric character at room temperature after vulcanization through unsaturated isoprene units (approximately one cross-link in the network per 250 carbon atoms).

On the other hand, chain orientation resulting from a unidirectional stretching favors the crystallization in 8_5 helical conformation and orthorhombic assembly. The melting point of these crystalline zones maintained under stretching is 45°C.

In addition to its very high reversible extensibility after vulcanization, the main property of this material is its impermeability to gases, a property that determines its applications.

Resistance to aging is satisfactory since PIB does not have tertiary hydrogen atoms and contains only a small fraction of residual unsaturations after vulcanization.

15.1.3.4. Applications of Butyl Rubber.

They are mainly in connection with the impermeability to gases, with the major application being the manufacture of tire tubes or their substitute—that is, the manufacture of tubeless tires. As secondary applications, one also finds the application in the manufacture of sealing compounds (homopolymers), joints, and various coatings.

15.2. POLY(CONJUGATED DIENES)

Although these polymers are mainly utilized as three-dimensional elastomers after vulcanization, they are produced as linear polymers. For this reason they are classified in the category of monodimensional polymeric materials. Indeed, in spite of the presence of two double bonds in the monomer molecules, their polymerization leads to linear chains with preservation of one unsaturation per monomeric unit.

All polydienes are characterized by a high mobility of their backbone and by weak molecular interactions; that explains their relatively low glass transition temperature and their incapability to crystallize spontaneously, even when their structural regularity is high. Natural rubber, which is 1,4-*cis*-polyisoprene, was discussed in Chapter 14 along with other natural polymers.

15.2.1. Polybutadiene

Acronym: BR (butadiene rubber)

Molecular formula: $-(C_4H_6)_n-$

Depending on the selected method, the polymerization of conjugated dienes can lead to various isomers of the monomeric units shown hereafter; depending on whether the residual double bonds are localized either laterally or in the main chain, one can obtain "polyvinyl" (iso- and syndio-) tacticity or geometrical type isomerisms, respectively, that are presented hereafter. They exhibit different chemical and physicochemical properties. Actually, polybutadienes contain variable proportions of each one of these structures and can thus be assimilated to statistical copolymers whose properties vary continuously with respect to their composition.

Type of Monomeric Unit	IUPAC Nomenclature
1,2-	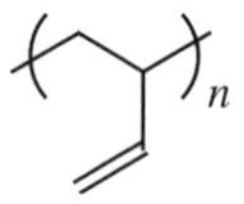 Poly(1-vinylethylene)
1,4-*cis*-	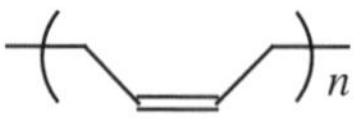 Poly(Z-but-2-enylene)
1,4-*trans*-	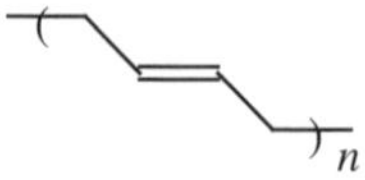 Poly(E-but-2-enylene)

15.2.1.1. Monomer. Buta-1,3-diene is obtained from steam-cracking of oil products, and the C_4 fraction contains approximately 60% of this monomer. Due to its electronic structure, this conjugated diene is sensitive to all types of active centers and can thus be polymerized by all main methods of chain polymerization. It is important to note that this monomer is generally bivalent; but under certain experimental conditions, it becomes tetravalent (vulcanization).

15.2.1.2. Methods of Polymerization. Radical polymerization is the oldest method among those presently used. This polymerization is generally carried out in emulsion and at low temperature ($5°C$) in order to favor 1,4 isomerism. Initiation is thus carried out by means of water-soluble mineral redox systems (ferrous salts/potassium persulfate, etc.). The free radical polymerization initiated by hydrogen peroxide (H_2O_2) produces hydroxytelechelic oligomers ($\overline{M}_n \sim 2500\,g\cdot mol^{-1}$). For such functionalization, one hydroxyl group results from the dissociation of the initiator whereas the second one results from a transfer reaction to the initiator. Actually, the hydroxyl functionality of this telechelic oligomer is slightly higher than 2 as a consequence of side reactions, and its curing with a bivalent coupling reagent leads to a network.

Coordination polymerization in solution by means of Ziegler–Natta systems allows the preparation of elastomers with excellent properties since the corresponding polymer chains can contain more than 95% of 1,4-*cis* units; this isomerism is most often required due to the corresponding mechanical characteristics of the material.

Anionic polymerization initiated by butyllithium in hydrocarbon solution is also used for industrial purpose. It produces BR with preferential 1,4-type units (50% 1,4-*cis*, 40% 1,4-*trans*, and 10% 1,2).

For certain specific applications, it is interesting to prepare polymers of overall 1,2-type isomerism. Anionic polymerization allows this by simple addition of a solvating agent (tetrahydrofuran, tertiary diamine, etc.) to the reaction medium.

Copolymers are prepared either by radical emulsion polymerization (SBR) or by anionic polymerization in cyclohexane solution (SBR statistical and SBS block copolymers).

15.2.1.3. General Characteristics. Even if produced with a high structural regularity, 1,4-*cis*-polybutadiene does not crystallize without the assistance of a mechanical constraint. Its glass transition temperature is closely dependent on the content of various isomers. The situation is similar to that of statistical copolymers, and the value of T_g can be calculated through the relation of Gordon–Taylor (see Section 11.2.5) by using the values of T_g given in Table 15.2. Statistical copolymerization with styrene is equivalent to terpolymerization.

After vulcanization, the sequences of highly stereoregular (*cis* or *trans*) 1,4-homopolybutadiene can crystallize under uniaxial constraint. The corresponding melting points are given in Table 15.2. As with all polyunsaturated hydrocarbons, the mobility of polybutadiene chains is high and confers excellent elastomeric properties to the corresponding materials after vulcanization. This polyunsaturation

Table 15.2. Transitions temperatures of various isomers of polybutadiene

Isomer	$T_g(^\circ C)$	$T_m(^\circ C)$
Isotactic poly(vinylethylene)	—	120
Syndiotactic poly(vinylethylene)	—	154
Atactic poly(vinylethylene)	-10	—
Poly(*cis*-butenylene)	-102	6
Poly(*trans*-butenylene)	-60	96

induces sensitivity to oxygen and the polymer can be used as material only after stabilization by using antioxidizing agents.

The low cohesion energy density of BR is responsible for its solubility in many solvents: aliphatic and aromatic hydrocarbons, tetrahydrofuran, ketones, ethers, and esters carrying heavy hydrocarbon groups. The vulcanization causes its insolubility, with the preceding solvents producing a swelling whose intensity depends on the cross-linking density.

15.2.1.4. Polybutadiene Derivatives.
Copolymers of butadiene are economically more important than the homopolymer. Polybutadiene is used to prepare high-impact polystyrene (HIPS) and ABS. These materials are described in Section 15.4.2, which is devoted to styrene copolymers.

Copolymers with a high content of butadiene are elastomers.

Statistical copolymers (SBR) prepared in emulsion by free radical initiation contain 75% of butadiene monomeric units distributed statistically along the chains ($r_b = 1.4$ and $r_s = 0.6$) with a prevalence of BB dyads.

SBR prepared by anionic polymerization in solution exhibits a quite different distribution of the comonomeric units due to very different reactivity ratios and a higher content of 1,4-*cis* units to the detriment of 1,2 units. When anionic statistical polymerization is carried out in batch and in absence of polar additives, the chains formed are similar to tapered block copolymers with one block mainly constituted of polybutadiene, then a tapered block with an increasing content in styrene and a block containing almost exclusively styrene units.

Butadiene–acrylonitrile copolymers (NBR) are also elastomers. Depending upon the targeted properties, they have a variable content (15–50%) of acrylonitrile (AN). The free radical reactivity ratios of both comonomers are lower than 1, thus favoring an alternation of the constituting units. The strong molecular interactions existing between the acrylonitrile units distributed along the chain considerably modify the mechanical and the physicochemical characteristics of the material. The main characteristics of NBR elastomers is their resistance to swelling in hydrocarbons, which determines their applications. NBR elastomers are prepared by emulsion free radical copolymerization.

Thermoplastic elastomers (SBS) can also be obtained from butadiene and styrene. Their synthesis is described in Section 10.3.1. In these three blocks copolymers, the two extreme PS blocks represent 25% mass of the polymer and are dispersed as hard nodules in a matrix constituted of the polybutadiene block (see Section 5.4.5). In this case, each phase retains most of its proper thermomechanical characteristics, and the entire assembly formed is equivalent to a physical gel.

These copolymers have the drawback of a significant creep for temperatures higher than the ambient one; indeed, a certain miscibility of polybutadiene with the polystyrene phase, although weak, appreciably lowers the glass transition temperature of the latter and thus also its cohesion.

15.2.1.5. Applications of Polybutadienes. The tire industry utilizes large quantity of polybutadiene and derived copolymers. However, in connection with the high level of the volume of production of high-impact polystyrene (HIPS) (see Section 15.2.3), most of the polybutadiene produced is consumed in this application as an intermediate product.

SBS thermoplastic elastomers are used as "pressure-sensitive" adhesives and in all fields of application of elastomers with processing techniques usually intended for thermoplastics.

Hydroxytelechelic oligomers prepared by free radical means (average valence slightly > 2) are used to form polyurethane elastomeric networks by reaction with diisocyanates (which often contain a small proportion of trivalent molecules).

15.2.2. Synthetic Polyisoprene

Acronym: IR (isoprene rubber)

Molecular formula: $-(C_5H_8)_n-$

Whereas natural rubber (see Section 14.1) is almost exclusively constituted of 1,4-*cis* units, the polymerization of isoprene, as in the case of polybutadiene, leads to polyisoprene with various structures resulting from the different possibilities of polymerization:

Type of Monomeric Unit		IUPAC Nomenclature
1,2-		Poly(1-methyl-1-vinylethylene)
3,4-		Poly(1-isopropenylethylene)
1,4-*cis*-		Poly(Z-2-methylbut-2-enylene)
1,4-*trans*-		Poly(E-2-methylbut-2-enylene)

15.2.2.1. Monomer. Contrary to butadiene, which is a gas at ambient temperature, 2-methylbuta-1,3-diene is a liquid ($T_b = 34°C$). It is a by-product of the production of ethylene obtained from steam-cracking of oil products. It can also be prepared either by catalytic dehydrogenation of pentenes or by dimerization of propylene. The world annual production of isoprene monomer is in the range of one million tons.

15.2.2.2. Techniques of Polymerization. As for butadiene, the electronic structure of isoprene allows this monomer to be polymerized by cationic, anionic, or coordinative manner; radical polymerization can also be performed, but it leads to many transfer reactions on methyl group.

Only polymers with a high 1,4-*cis* units content are interesting for their industrial applications. Ziegler–Natta polymerization (by $TiCl_4/i$-Bu_3Al) gives rise to a structure very close to that of natural rubber (NR); however, as with NR, the resulting material contains a small fraction of three-dimensional polymer resulting from the tetravalence of a fraction of the monomer. The catalytic systems containing rare earth metals are more selective and generate gel-free polymers. The 1,4-*trans* isomer can be obtained by using VCl_3 based Ziegler–Natta catalysts.

The polymerization of isoprene can also be initiated by butyllithium in solution in a light hydrocarbon (isopentane, etc.). It leads to an elastomer constituted of perfectly linear chains; although qualified as "low *cis* content" by its users, it contains more than 90% of 1,4-*cis* units. Anionic polymerization is also used to produce styrene–isoprene–styrene (SIS) thermoplastic elastomers (analogs of butadiene-containing SBS) which contain a central elastomeric block of polyisoprene and two extreme hard polystyrene blocks. Like SBS, SIS are prepared by coupling of "living" diblock copolymers by means of $(CH_3)_2SiCl_2$ (see Section 9.3.1). For the same reason as that invoked for SBS thermoplastic elastomers (TPE), they creep slightly under constraint when the temperature is higher than that of the ambient one.

The cationic polymerization of isoprene is used only for the preparation of copolymers with high isobutene content (PIB).

15.2.2.3. General Characteristics of Polyisoprene. As a homopolymer with a high 1,4-*cis* units content, IR is not very different from natural rubber (NR) and it is thus produced only in the countries that are not supplied in NR. Polyisoprenes prepared by anionic polymerization exhibit low dispersities.

1,4-*trans* polyisoprene is crystalline ($T_m = 60°C$ and $T_g = -60°C$); it exhibits the characteristic properties of semicrystalline polymers whose T_g is lower than the ambient temperature.

15.2.2.4. Applications. 1,4-*cis*-polyisoprene is mainly utilized in the tire industry along with natural rubber. As SIS thermoplastic elastomer, it is used -like its equivalent butadiene– in the compounding of pressure sensitive adhesives.

1,4-*trans*-polyisoprene has quite specific applications (golf balls, etc.), but the low level of its production makes it expensive; it is thus little used apart from some specialized applications with high added value.

15.2.3. Polychloroprene [Poly(2-chlorobutadiene)]

Acronym: CR (chloroprene rubber)
Molecular formula: $-(C_4H_5Cl)-$

It is obtained by the radical polymerization of a conjugated diene and thus the corresponding polymer chain contains the various isomers similar to those of polyisoprene:

Type of Monomeric Unit		IUPAC Nomenclature
1,2-		Poly(1-chloro-1-vinylethylene)
3,4-		Poly[1-(1-chlorovinyl) ethylene]
1,4-*cis*-		Poly(Z-2-chlorobut-2-enylene)
1,4-*trans*-		Poly(E-2-chlorobut-2-enylene)

The IUPAC designation of these monomeric units is similar to that given for polyisoprenes. Annual world production of polychloroprene is approximately 500,000 tons.

15.2.3.1. The Monomer

Initially, it was obtained by the catalytic dimerization of acetylene, followed by an addition of HCl:

$$2\ HC{\equiv}CH \xrightarrow{\text{CuCl}} CH_2{=}CH{-}C{\equiv}CH \xrightarrow{\text{HCl}}$$

2-Chlorobuta-1,3-diene
(chloroprene)

At present, it is produced by the chlorination–dehydrochlorination of butadiene according to the following reaction pathway:

It is a liquid with $T_b = 59°C$.

15.2.3.2. Methods and Processes of Polymerization.

For economic reasons, only emulsion radical polymerization at 40°C is utilized for its production; this relatively low temperature leads to both a high content in 1,4-*trans* units and a low content in polymeric gel (resulting from side reactions). The initiators utilized are redox systems.

Certain processes involve copolymerization of chloroprene with small amounts of a given comonomer in order to modify some of the basic characteristics of the material. Thus, copolymerization with 2,3-dimethylbutadiene reduces the tendency to crystallize. Methacrylic acid introduces carboxylic functional groups along the polymer chain and thus improves the adhesive properties onto metals and polar surfaces. Copolymerization with acrylic diesters of diols leads to a prevulcanization of the material, and so on.

15.2.3.3. General Characteristics of Polychloroprene.

Under the conditions of polymerization, the high percentage of 1,4-*trans* units (90%) favors a tendency toward crystallization (orthorhombic system) which is inconvenient for elastomeric properties; the degree of crystallinity exceeds 10% and is still higher when the material is stored at low temperature.

The degree of crystallinity has an inverse relationship with the presence of molecular irregularities (1,2, 3,4, 1,4-*cis* units and irregular dyads), which, however, do not exceed 20% of the monomeric units.

The glass transition temperature is −45°C, whereas the melting zone (T_m) is in the range between 50°C and 60°C. The maximum rate of crystallization is observed at 5°C.

Polychloroprene is compatible with all other elastomers; it is thus often used in blends.

15.2.3.4. Applications.

Polychloroprene finds applications due to its durability, good adhesive properties, and fireproof character. This material is widely used for the molding of flexible objects, the manufacture of flexible tubes, and sealing joints such as those required for the formulation of adhesives.

The annual world production is approximately 500,000 tons.

15.3. VINYL AND RELATED POLYMERS

They correspond to a family of polymers whose main chain consists of sequences of two carbon atoms resulting from the addition on the ethylene double bond:

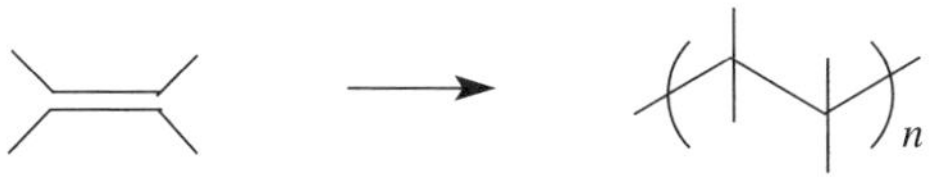

However, the name **vinyl polymer** is generally more restrictive since it is reserved for those resulting from the polymerization of a vinyl group carrying a substituent other than an alkyl group:

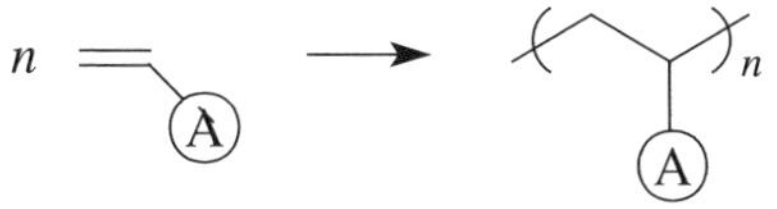

Vinylidene polymers can be included in this important family:

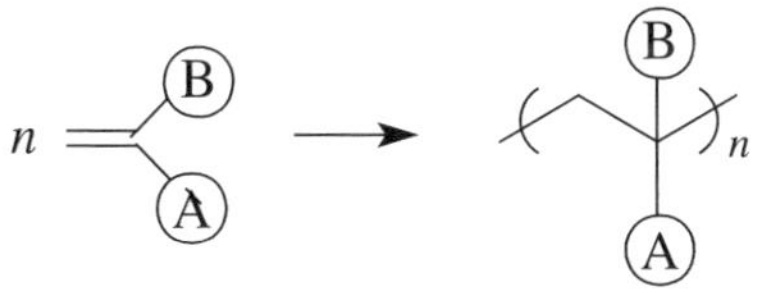

and even **tri- and tetrasubstituted ethylene polymers**:

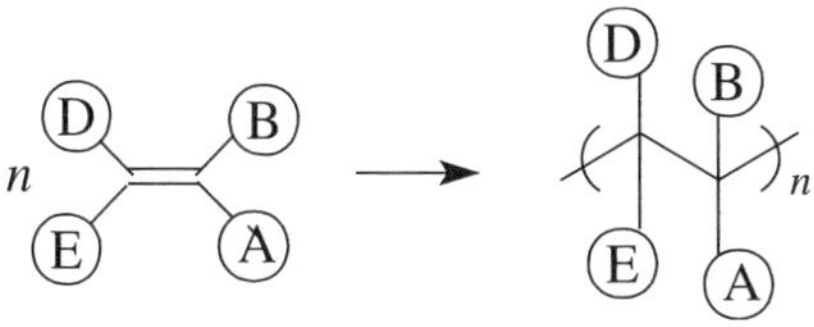

All these polymers exhibit a glass transition temperature higher than room temperature, and most of them are amorphous. Due to either the size of their substituent or their molecular symmetry, some of them are semicrystalline. They have a definite melting point, and according to its value they exhibit a mechanical behavior more complex than that of amorphous polyvinyls. Depending upon the nature of substituents A and B, the physicochemical properties of polymers of this family can be very different, thus enabling them to be used in many fields. Their overall economic importance is considerable but variable depending upon their structure; it is not possible to make an exhaustive presentation of them.

15.3.1. Polystyrene and Its Derivatives

Acronym: PS

Molecular formula:

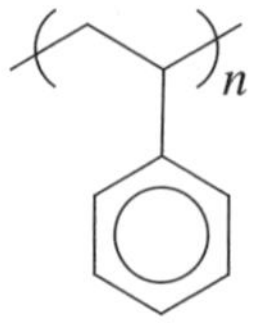

IUPAC nomenclature: poly(1-phenylethylene)

Polystyrene is one of the most common commercial polymers. Its basic properties can be modulated by copolymerization and by the varied presentations of its semi-finished products. The level of its annual world production and that of its derivatives exceeds 18 million tons, and its growth rate remains high. From a scientific point of view, polystyrene homopolymer is regarded as the reference polymer by physicists as well as by chemists and physical chemists. Thus it was comprehensively studied, and its properties are particularly well known.

15.3.1.1. Monomer. It is prepared by dehydrogenation of ethylbenzene, which itself is obtained from an alkylation of benzene by ethylene:

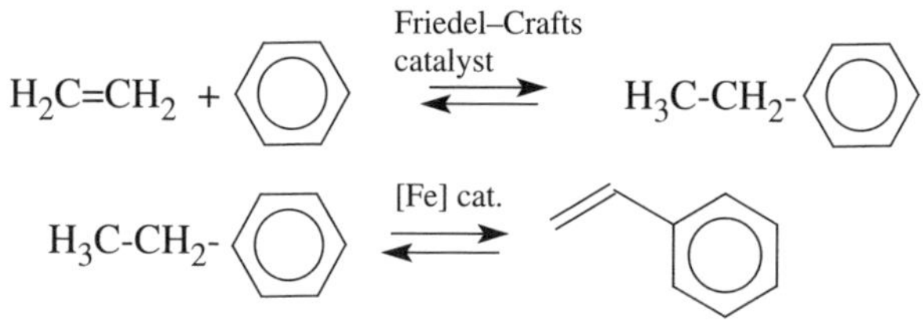

It is a liquid whose boiling point is 145°C. It is slightly water-soluble (0.03% at 25°C), and it should be stored in the presence of a stabilizer (*t*-butylcatechol) to avoid its undesirable radical polymerization.

Its annual world production is about 20 million tons, from which 85% are used for the production of styrene polymers—that is, polymer materials with a majority of styrene units.

15.3.1.2. Techniques and Processes of Polymerization. The conjugation of the vinyl group with the phenyl ring causes the polarization of the ethylene double bond depending upon the nature of the antagonistic active center. Consequently, styrene can be polymerized by all the methods of chain polymerization. However, at the industrial level, most of the production of polystyrene results from a radical process.

The most current processes use the capability of styrene to be polymerized spontaneously in bulk by the sole effect of temperature (see Section 8.5.3). However, more and more organic peroxides are used as initiators, which allows to operate at a lower temperature. In addition, the use of a radical initiator considerably lowers the content of dimer and trimer species that are inconvenient from any point of view.

Polymerization proceeds without transfer to polymer and thus leads to fully linear chains, which appears to be a drawback with respect to the mechanical properties of the resulting material.

The polymerization is carried out according to a flow process with a maximum conversion of 75%. Then, the residual monomer is eliminated by "devolatilization" from the corresponding syrup and the extracted monomer is recycled. The "bulk" processes are often regarded as "solution" processes since the yield does not exceed 75%, with both residual monomer and ethylbenzene that it contains being recycled; they are to be used as co-solvents and monomer in a new polymerization.

Polystyrene can also be prepared by suspension polymerization. Initiators being utilized are also organic peroxides, azo compounds, or sometimes a mixture of both. This technique offers the advantage of affording calibrated PS pearls which are well-fitted for the technique of expansion with pentane (see Section 15.3.1.5).

A combination of the two preceding techniques can also be carried out while starting from a bulk polymerization and then by transferring the resulting syrupy liquid into a reactor intended for suspension polymerization in order to augment the conversion.

Attempts for industrial processes using anionic polymerization in bulk and at high temperature are in progress. The corresponding materials offer many advantages as compared to the radical PS: better thermal stability due to perfectly regular sequences, absence of oligomers, absence of any residual monomer, and better control of molar masses.

As far as the preparation of high-impact polystyrene (HIPS) is concerned, the techniques and processes used are not much different from those allowing the preparation of PS homopolymer (called "crystal" PS); however, instead of starting from pure styrene monomer, the polymerization is carried out in the presence of polybutadiene (BR) solubilized in styrene monomer. The mechanism of the fine dispersion of BR elastomer particles in polystyrene matrix is described in Section 5.4.5.

Industrialization of coordination polymerization of styrene using metallocenes and leading to syndiotactic polystyrene is in progress.

15.3.1.3. Styrene Copolymers.
Only those containing mainly styrene (SBR elastomers are excluded) are included in this group.

The most important copolymer is HIPS. It was mentioned on several occasions (see above and Section 5.4.5); due to its impact strength definitely higher than that of conventional PS "crystal," many applications of styrene polymers are opened. It contains 15–25% of polybutadiene prepared by radical polymerization, finely dispersed in the polystyrene matrix and covalently bound to polystyrene chains to form graft copolymers. The method of preparation implies that the polybutadiene chains on which PS grafts are anchored play the role of an emulgator and are blended with a PS homopolymer matrix. These copolymers form approximately one-third of the total production of styrene polymers (including copolymers). As compared to PS "crystal," they have impact strength 5–10 times higher, whereas their only negative property (for most of them) is the lack of transparency due to their heterogeneous morphology.

Radical statistical copolymerization with acrylonitrile leads to SAN. They contain about 25% of the polar comonomer units homogeneously distributed along the chain because the reactivity ratios of each comonomer are lower than unity. These copolymers show an increase in the density of cohesive energy and thus an improvement in their mechanical properties as compared to those of PS homopolymers (for example, stress at break is 40 MPa for PS and 70 MPa for SAN). As compared to polystyrene, SAN also offers a higher glass transition temperature and a better resistance to solvents—hydrocarbons in particular.

It is possible to combine the strong density of cohesive energy of SAN and the impact strength of HIPS by copolymerizing styrene and acrylonitrile in the presence of polybutadiene. The resulting morphologies are similar to those of HIPS in all aspects, but the dispersing phase, which is constituted of SAN, is much more cohesive than that of PS homopolymer. Acrylonitrile–butadiene–styrene terpolymers (ABS) are thus regarded as technical polymers.

From an economic point of view, the other copolymers of styrene are much less important than those above. Copolymerization with α-methylstyrene leads to an increase in the glass transition temperature of the corresponding polymer. Copolymerization with acrylic esters allows the generation of polar sites along the nonpolar polystyrene chains (ionomers and others). Almost perfectly alternating copolymers can be obtained by copolymerization of styrene and maleic anhydride by a radical process.

Finally, although their production is very low, styrene–divinylbenzene copolymers are three-dimensional systems that afford applications with high added value.

Unsaturated polyesters (UP) primarily consist of styrene, but they will be discussed in Chapter 16 along with other three-dimensional polymers.

15.3.1.4. General Characteristics of Styrene Polymers. Polystyrenes should have molar mass in the range 1×10^5 and 2×10^5 g·mol^{-1} (the molar mass between entanglements is approximately 18,000 g·mol^{-1}) to attain optimal mechanical characteristics. PS materials are always branch free and methods of polymerization which could generate branches, are sought by producers because the mechanical characteristics of the corresponding thermoplastic materials might be somewhat improved. Both homopolymers and statistical copolymers based on styrene are transparent but fragile to impact which limits their applications. In the form of HIPS or ABS, they have definitely improved impact strength, but generally they lose their transparency. Their glass transition temperature (T_g) is slightly lower than 100°C; but for T higher than 70°C, PS can undergo an important creep under constraint. Copolymerization induces an improvement of the thermomechanical properties.

Whereas stress at break of "crystal" PS is about 40 MPa at 25°C, the two-dimensional orientation as sheets causes an increase in the value of this property up to 120 MPa.

15.3.1.5. Applications. As semi-processed products, PS is used in the form of rigid foams usually utilized in the packaging industry and in the insulation buildings. These expanded PS foams (density ~ 0.05) are obtained by swelling PS pearls with pentane followed by a heating that causes the simultaneous evaporation of the inflating agent and the softening of the PS material. For reasons of safety, producers tend to replace pentane by carbon dioxide.

Polystyrene structural foams (density $= 0.5-0.7$) can also be produced. Contrary to expanded PS foams, they keep satisfactory mechanical characteristics that are compatible with the manufacture of various objects.

"Dense" styrene polymers and copolymers are used in all fields of packaging (for food in particular) and mechanical engineering (electric household appliances, automotive engineering, etc.) as well as for the realization of a multitude of low-size objects.

The bad image of "crystal" polystyrene (even up to the recent past) has also affected other thermoplastic materials causing for instance much damage to the image of PMMA. Ps is often chosen due to its low cost and its easy processing but was often used wrongly.

15.3.2. Poly(vinyl chloride)

Acronym: PVC

Molecular formula:

IUPAC nomenclature: poly(1-chloroethylene)

It is a commercial polymer having a major economic significance since its annual world production reaches 33 million tons, and its *per capita* consumption in industrialized countries exceeds 13 kg/year. As thermoplastic polymer, it is utilized in two forms: one, rigid, corresponds to polymer containing only a small proportion of additives, while the other one, flexible, has plasticizer content up to 50%.

15.3.2.1. Monomer (VCM). It is a gas $(T_b = -14°C)$ that can be prepared by addition of HCl to acetylene:

$$HCl + HC\equiv CH \longrightarrow H_2C=CHCl$$

Presently, however, widely used methods involve the direct chlorination of ethylene by electrophilic addition catalyzed by $FeCl_3$

$$H_2C=CH_2 + Cl_2 \xrightarrow{FeCl_3} Cl-CH_2-CH_2-Cl$$

followed by a pyrolysis of 1,2-dichloroethane

$$Cl-CH_2-CH_2-Cl \longrightarrow HCl + CH_2=CHCl$$

or the oxychloration of ethylene, catalyzed by $CuCl_2$:

$$2\,CuCl + 1/2\,O_2 \longrightarrow CuO, CuCl_2$$

$$2\,HCl + CuO, CuCl_2 \longrightarrow 2\,CuCl_2 + H_2O$$

$$H_2C{=}CH_2 + 2\,CuCl_2 \longrightarrow 2\,CuCl + Cl{-}CH_2{-}CH_2{-}Cl$$

The two methods can be combined, with the second one using the HCl produced by the first one. Vinyl chloride monomer is extremely toxic and should be handled only under conditions of high safety.

Its world production is about 30 million tons; it is primarily intended for the preparation of PVC and various copolymers.

15.3.2.2. Preparation of PVC. It is important to note that poly(vinyl chloride) is insoluble in its proper monomer. PVC is prepared only by radical polymerization of VCM through techniques applicable to chain polymerization: bulk, solution, suspension, emulsion.

Bulk polymerization is important at the economic level; this process involves use of two successive reactors under pressure; in the first one the polymerization of VCM proceeds up to 10% yield, and a beginning of precipitation occurs. In the second reactor, the polymerization is continued under different conditions in order to favor the precipitation of high-porosity particles and to reach 25–30% conversion. The polymer is recovered by sifting and the residual monomer is recycled.

> **Remark.** This bulk polymerization is actually a "dispersion" polymerization in the sense which was given to this term in Section 8.5.12c.

Processes using suspension polymerization are successful because the resulting polymers are easier to process than those obtained from bulk polymerization. They produce PVC that is easy to plasticize. Suspension polymerization are carried out under pressure, with liquid VCM being dispersed as 15 to 30 μm droplets in poly(vinyl alcohol)- or methyl cellulose-containing water. After its appearance, the PVC formed gradually precipitates inside these droplets in the form of ~1 μm particles. Polymerization is carried out up to 90% conversion. After recovery of the polymer particles, the residual monomer is extracted in order to reduce its content as low as 10 ppm.

According to the same principle, microsuspensions can be carried out by addition of surfactant in high proportion.

Emulsion polymerization is used to produce either very fine particles or latexes that are directly usable for various applications.

15.3.2.3. General Characteristics of PVC. Radical polymerization of vinyl chloride monomer favors the formation of syndiotactic sequences and, depending on the temperature of polymerization, the content in *rr* triads varies from 60% at 50°C to 20% at 70°C. This tendency to syndiotacticity at lower temperature is responsible for the formation of small-size crystallites ("discs" of $\sim$1 nm thickness and $\sim$4 nm diameter), which are zones of strong cohesion and high thermo-mechanical stability. They consist of chains organized in orthorhombic system and are the "cross-links" of a physical network that resists the flow of the material up to a temperature much higher than the glass transition temperature ($T_g = 80°C$). The proportion of these crystallites can reach 20% at the end of the polymerization but does not exceed 5% on cooling from the fluid state. The presence of these physical cross-links makes delicate the processing of PVC and compels the manufacturer to use temperatures at which partial thermal degradation cannot be avoided. There is generation of polyconjugated sequences that are responsible for an inconvenient coloration of the corresponding materials. The crystallinity of PVC can be reduced by copolymerization of VCM with another vinyl monomer—for example, vinyl acetate (VA).

The relatively low degree of crystallinity is not only due to the low stereoregularity but also due to the presence of many irregular sequences. They are responsible for a low thermostability that considerably limits the applicability of this material (see Chapters 3 and 9). Poly(vinyl chloride) releases HCl on heating at temperatures higher than 50°C. Such behavior makes difficult the valorization of its waste by thermal degradation.

Like all halogenated polymers, PVC exhibits a high density ($d = 1.39$), with most of its mass being due to the chlorine atom.

PVC burns under the effect of a flame but does not propagate combustion: it is flame-resistant.

PVC is sparingly soluble in common solvents. The most common solvent used to give "collodions," which can be used as adhesives to connect objects in PVC, is tetrahydrofuran.

In spite of its low intrinsic solubility, PVC can be plasticized by various compounds; among them, most common are alkyl phosphates and phthalates. These plasticizers are equivalent to heavy solvents that exhibit a negative interaction parameter (χ_{12}) with PVC; they strongly interact with the C–Cl groups of the polymer, through their ester functional groups. Moreover, plasticizers should present a low volatility in order not to be eliminated gradually by evaporation. They are used in proportions up to 50% by weight in order to lower the glass transition temperature (see Chapter 11) by reducing the density of cohesive energy. The corresponding materials are easily deformable and hence they have many applications—in particular, as substitutes of leather.

The morphology of PVC grains prepared in bulk or in suspension shows a strong porosity that favors the plasticization.

To allow an easy processing of PVC and the improvement of its resistance to aging, poly(vinyl chloride) has to be blended with many additives. In particular, organic bases are incorporated to trap HCl released during processing, the thermal degradation being self-catalyzed by acids.

To improve the resistance to strain deformation at temperatures higher than ambient, PVC can be overchlorinated. This operation is carried out by reaction of chlorine gas on PVC powder in a fluidized bed; it involves partial substitution of hydrogen atoms by chlorine atoms. The glass transition temperature of the resulting polymer can then be increased above $100°C$.

Incorporation of mineral fillers such as finely powdered calcium carbonate can be used to reduce the cost of PVC and to increase its temperature of strain deformation.

15.3.2.4. Fields of Application.
Although use of all halogen-containing substances is disputed, poly(vinyl chloride) still has a high rate of growth.

In its rigid form, it is widely used for the manufacture of pipes that are easily assembled by means of a PVC-based collodion in tetrahydrofuran. This application is related to its low cost, its chemical inertia, and its very good resistance to aging at ambient temperature. It is also widely used for food packaging (water, etc.).

In the last few decades, PVC has found many applications in buildings (door and window frames), where it is appreciated for its fireproof character.

PVC fibers are obtained by dry spinning from a collodion in tetrahydrofuran; they are used to manufacture imputrescible and car-fireproof furnishing fabrics. Finally, the electrical insulating character of PVC opens the market of small electrical equipments to it.

Plasticized PVC has a level of production comparable to that of rigid PVC. It is a material completely different from the precedent; it is widely used for the manufacture of molded objects which should be deformable or have resistance to impact. Films produced by calendering have the appearance of leather and compete with it for certain inexpensive applications (furniture, clothing, etc.). It also is widely used in construction—in particular, for floor coverings.

15.3.3. Other Halogen-Containing Vinyl Polymers

All polymers presented in this section have a common molecular structure, which induces a semicrystalline morphology. They are different from other vinyl polymers whose mechanical properties are similar to those of their amorphous state. The partial crystallinity of these halogenated polymers has to be considered with their glass transition temperature, which is lower than the ambient temperature for most of them. Moreover, they have at least one halogen atom in each monomer unit and hence they are flame-resistant.

15.3.3.1. Poly(vinyl fluoride)

Acronym: PVF

Molecular formula:

IUPAC nomenclature: poly(1-fluoroethylene)

This polymer is obtained by radical suspension or emulsion polymerization under a slight overpressure because the monomer is a gas ($T_b = -72°C$). Polymerization occurs with transfer to polymer, which generates many branches.

In spite of its atacticity and other structural defects at the molecular level (many irregular sequences), poly(vinyl fluoride) is semicrystalline. Indeed, the small size of the fluorine atom enables it to crystallize in orthorhombic system according to a planar zigzag, whatever the configuration of tertiary carbon atoms. Defects in the molecular structure, however, affect the value of the degree of crystallinity, which can vary from 25% to 55%. The melting zone of PVF is dependent on the degree of crystallinity and is in the range of 190–200°C.

The amorphous fraction of PVF undergoes the glass transition phenomenon at various temperatures, depending on the localization of the chain segment under consideration. The most important transition occurs at about −18°C.

As observed in all fluorinated polymers, PVF exhibits solution properties rather different from those of other vinyl polymers. Its solubility parameter ($\delta = 24.5\,\mathrm{MPa}^{1/2}$) corresponds to weak interactions with most nonpolar or slightly polar organic solvents and also corresponds to a marked hydrophobia as observed in all fluoro polymers. PVF is insoluble in all solvents at room temperature. PVF is a very cohesive polymer as reflected by its mechanical properties. Because the carbon–fluorine bond is particularly strong, PVF is much more thermally stable than PVC.

In the absence of molecular oxygen, thermal degradation occurs only at 450°C. Poly(vinyl fluoride) is used to manufacture biaxially stretched films and coatings, which exhibit a strong chemical inertia and an excellent resistance to aging. Films are prepared by extrusion of a polymer swollen by polar solvents followed by their evaporation.

15.3.3.2. Poly(vinylidene chloride)

Acronym: PVDC

Molecular formula:

$$\left(\!\!\begin{array}{c}\text{Cl}\\[-2pt]\diagdown\diagup\!\!\diagdown\!\!\underset{\text{Cl}}{\overset{|}{\text{C}}}\end{array}\!\!\right)_{n}$$

IUPAC nomenclature: poly(1,1-dichloroethylene)

Monomer, $H_2C{=}CCl_2$ (VDC), is a liquid ($T_b = 32°C$) that is obtained by dehydrochlorination of 1,1,2-trichloroethane in the presence of soda. In spite of the steric hindrance brought about by two geminal chlorine atoms, this monomer polymerizes very easily by free radical process. In the absence of inhibitor, it polymerizes spontaneously during storage under the effect of atmospheric oxygen.

For industrial production, the polymerization is carried out in bulk, in emulsion or in suspension. Bulk polymerization, like that of vinyl chloride, gradually leads to precipitation of polymer after its appearance (dispersion polymerization). Emulsion polymerization is initiated by redox systems at relatively low temperature ($\sim50°C$). Suspension polymerization (like that in bulk) is initiated by means of organic peroxides (lauryl peroxide, etc.). The vinylidene chloride is also copolymerized with vinyl chloride to give a material whose T_g is higher than that of PVC.

Due to steric hindrance of the two chlorine atoms, poly(vinylidene chloride), which does not possess an asymmetrical carbon atom, exhibits a regular placement of monomeric units. Because its glass transition temperature is $17°C$, it crystallizes spontaneously at ambient temperature. The atomic radius of the chlorine atoms prevents crystallization in planar zigzag, and the chains thus adopt a helical conformation with two monomeric units per helix turn (helix 2_1) which are organized in a monoclinical system. The degree of crystallinity is high, and it can reach 80%. The melting point of the crystalline zones is in the range of $200°C$. The density of this polymer, like that of all other chlorinated polymers, is high; it depends on the degree of crystallinity, but it is about 1.90. PVDC is insoluble in all currently available solvents at ambient temperature, thus reflecting the strong density of cohesive energy of the crystalline zones.

PVDC has two physicochemical characteristics that determine its applications.

Its first service property determines its main application: poly(vinylidene chloride) is more impermeable to gases and liquids than most of other polymers due to its high density and degree of crystallinity. Indeed, the fraction of free volume is low and does not allow the passage of small molecules. This barrier property is used in food packaging as films, molded containers, and coatings.

As any halogen-containing polymer, PVDC is flame-resistant. This second property is used in the production of textile fibers for furnishing fabrics.

However, one should not underestimate the thermal degradation that proceeds by HCl elimination (see Section 3.2.3).

Chain orientation during spinning causes a considerable increase in tenacity which passes from 50 MPa for an unstretched PVDC to more than 300 MPa for an oriented fiber.

15.3.3.3. Poly(vinylidene fluoride)

Acronym: PVDF

Molecular formula:

IUPAC nomenclature: poly(1,1-difluoroethylene)

Many analogies exist between PVDF and PVDC at the structural level. However, the strong energy of C–F bonds reduces the possibility to form branches, and thus this polymer is mainly linear. It is obtained by radical polymerization of vinylidene fluoride monomer (VDF) in emulsion or in suspension. Because the monomer is a gas ($T_b = -84°C$), the polymerization is carried out under a pressure of about 10 MPa and at $\sim 80°C$. However, temperatures and pressures higher or lower than this one can also be used, with the choice being determined by the use of a given initiator—that is, by a redox system or a peroxide.

PVDF is a highly crystalline polymer, and its degree of crystallinity ranges from 35% to 70%. It exhibits different regular conformations, with the most frequent being a helical 2_1 conformation of tg^+tg^- type. The melting zone is in the range $160–180°C$, and the glass transition temperature is around $40°C$.

PVDF exhibits an excellent thermal stability and a good chemical inertia.

Due to its high degree of crystallinity, PVDF exhibits excellent mechanical characteristics. The value of its tensile strength at break (σ_r) can reach 100 MPa. However, its most interesting physical property is closely related to its crystalline structure. The high electric moment of C–F groups and their preferential orientation in the crystallline zones induce particular electrical properties: di-, piezo-, pyro-, and ferroelectric properties, depending on the crystalline structure of the unit cell.

These properties are widely used in many applications; due to its high permittivity and its low dielectric losses, PVDF is widely used for the insulation of electric and electronic devices. For example, it can be used as insulating layer in condensers. It is also useful as thin film electret in transducer technology. Its ferroelectric properties are used in microphones, loudspeakers, and so on.

Poly(vinylidene fluoride) is thus classified in the category of specialty polymers, although its mechanical properties resemble those of technical polymers. Because of its high cost it is used in high-value added applications.

15.3.3.4. Polytetrafluoroethylene

Acronym: PTFE

Molecular formula:

IUPAC nomenclature: poly(difluoromethylene)

PTFE is a well-known polymer due to the large number of its applications even though the volume of its production is relatively low, about 20,000 tons *per annum*. It is a high cost material whose considerable economic interest is only due to some of its properties that are unique in the field of polymers.

The monomer ($CF_2=CF_2$) is a gas ($T_b = -76°C$) that is obtained by pyrolysis of difluorochloromethane. This reaction generates many by-products, and the process requires a careful purification of the raw monomer before use. Polymerization is carried out by free radical means according to two techniques that are difficult to carry out. Each one of them allows the preparation of PTFE under a particular morphology and takes into account the total insolubility of the polymer in the reaction medium. These techniques operate in dispersed media:

- In absence of surfactants and by simple stirring in water under pressure, the polymerization initiated by peroxides leads to the dispersion precipitation of the resulting polymer followed by a coagulation in the form of irregular grains.
- In the presence of surfactants, the resulting particles are finer and remain in suspension as long as a process of coagulation does not occur; the product resemble latex.

In polytetrafluoroethylene, the energy of C–F bond is very high (485 kJ/mol) and is responsible for perfectly linear chains. This structural property is due to the absence of transfer reactions to polymer which would require the breaking of this bond. This structural regularity is responsible for a strong tendency to crystallize (see Section 5.2.1) in a triclinic system below $19°C$ and hexagonal above it. The degree of crystallinity of PTFE powders and grains obtained from the polymerization is always higher than 90%: it is the most crystalline among synthetic polymers. However, after melting, PTFE chains find it difficult to organize in the crystalline state, and the degree of crystallinity falls to 50–75% according to the thermal cycle. In spite of the weakness of molecular interactions, the melting point of this polymer is high ($T_m = 327°C$) because the entropy variation during melting is particularly low (see Chapter 11). The glass transition temperature is about $30°C$.

PTFE is insoluble in all currently available solvents even at high temperatures. It is nonflammable and exhibits a high chemical inertia: it reacts only with alkali metals and soda at high temperature, as well as with fluorine and certain fluorinating agents.

Its density is equal to 2.15 for a degree of crystallinity $\sim$50%. PTFE has a particularly low friction coefficient that induces highly requested self-lubricating properties. It is a very effective electrical insulator, a property related to its marked hydrophobic character.

The rigidity of PTFE chains avoids the use of the usual techniques of processing because the viscosity of the molten bulk is very high at $380°C$, and it undergoes thermal degradation at higher temperatures. Thus it is impossible to inject or extrude and specific techniques of processing had to be found.

Starting from PTFE grains, objects of variable size can be manufactured by compression molding. This operation is carried out in two stages:

- The first one involves packing PTFE grains under a pressure varying from 20 to 50 MPa according to the size of the object to be realized at a temperature in the range 50–100°C, in order to approach the true density and to obtain an object of final form with a beginning of cohesion.

- The second stage is that of sintering; it is performed according to a complex thermal cycle that brings the object up to $380°C$. This stage is accompanied by an important dilatation of the object corresponding to a decrease in the degree of crystallinity.

PTFE grains can also be processed by "extrusion" of the powders followed by compression. This technique corresponds to a continuous compression molding.

As far as processing of fine powders resulting from coagulation of the suspensions is concerned, addition of a lubricant and various additives improves the flow of the mixture through a die and leads to profiles of small section. Then the lubricant is vaporized and the extrudate is sintered as previously described for granular PTFE.

The products obtained from suspension polymerization can be directly utilized for the coating of various supports; they are also used for the production of fibers by spinning from a mixture with cellulose xanthate which is then carbonized before the sintering of PTFE.

There are many applications of PTFE that use one or more of its specific properties: self-lubrication, chemical inertia, thermostability, and anti-adhesive capacity. It also finds applications in mechanical engineering, chemical and textile industries as well as in electrical engineering.

15.3.3.5. Poly(chlorotrifluoroethylene)

Acronym: PCTFE

Molecular formula:

IUPAC nomenclature: poly(chlorotrifluoroethylene)

The molecular structure of this polymer corresponds to the maximal steric hindrance for a polymer. The rigidity of the chains is increased compared to that of PTFE; and, in spite of the atacticity of the polymer obtained by radical polymerization in emulsion or suspension, PCTFE is highly crystalline. Its melting zone is about $211-216°C$. Its degree of crystallinity is approximately 50%. Its glass transition temperature is $85°C$ due to the high rigidity of its chains.

It shares numerous properties with PTFE: high density, chemical inertia, thermostability, flame-resistance, and anti-adhesive capacity, but it also exhibits good mechanical characteristics at very low temperatures. It also exhibits a basic property that differentiates it from PTFE: processability by the usual techniques of extrusion, injection, and compression molding. Due to its high cost, it is a technical polymer that is used only for high-value added applications. Its annual world production does not exceed 500 tons.

15.3.4. Poly(vinyl acetate) and Its Derivatives

Acronym: PVA or PVAc

Molecular formula:

IUPAC nomenclature: poly(1-acetoxyethylene)

15.3.4.1. Monomer (VA). It is an enol ester that is obtained either by addition of acetic acid onto acetylene in the presence of a mercury, zinc, or cadmium salt (catalyst),

$$HC{\equiv}CH + H_3C{-}COOH \xrightarrow{\text{cat.}} H_2C{=}CH{-}O{-}CO{-}CH_3$$

or through oxidizing addition of acetic acid onto ethylene catalyzed by palladium chloride:

$$H_2C{=}CH_2 + H_3C{-}COOH \xrightarrow[\text{PdCl}_2]{O_2} H_2C{=}CH{-}O{-}CO{-}CH_3 + H_2O$$

It is a liquid ($T_b = 73°C$) that is slightly water-soluble ($\sim$2% at 50°C), a specific property among vinyl monomers.

15.3.4.2. Polymerization of VA. This can be carried out by radical means most often in emulsion but also possibly in bulk, in solution, and in suspension. The widely used initiators are mineral or organic peroxides as well as redox systems. Due to the partial hydrosolubility of this monomer, the model suggested by Smith–Ewart for emulsion polymerization is not applicable to this monomer.

15.3.4.3. General Characteristics. PVAc is an amorphous atactic polymer for which the glass transition temperature is around 30°C, which corresponds to a significant change of its mechanical properties at a temperature slightly higher than ambient. Thus, only polymers having a very high molar mass can be used as structural materials. In addition to a certain hydrophilicity, poly(vinyl acetate) is soluble in many polar organic solvents (esters, ketones, higher alcohols, halogenated hydrocarbons, carboxylic acids). As a counterpart, it is insoluble in aliphatic and aromatic hydrocarbons. Its capacity to absorb small amounts of water determines its electrical and dielectric properties. For example, its permittivity, which is equal to 3.3 in a dry state, can reach 10 in an atmosphere at 100% relative humidity. In addition, this absorbed humidity plays the role of plasticizer lowering the glass transition temperature up to the ambient one.

The thermal stability of PVAc is relatively low; at temperatures higher than 150°C, it releases acetic acid with a process similar to that of thermal degradation of PVC and thus leads to colored polyconjugated chains. The chemical reactivity of PVAc is that of aliphatic esters. In particular, it is sensitive to hydrolysis in either acidic or basic medium to give poly(vinyl alcohol). One of its essential physicochemical characteristics is its capacity to adhere on surfaces that exhibit high surface energy. Its adherence to cellulose, wood, glass, metals, mineral fillers, and polar-group-containing polymers is very good; this property determines the number of its applications.

15.3.4.4. Vinyl Acetate-Based Copolymers and Derived Polymers. Vinyl acetate units transmit their intrinsic adhesivity to all VA-containing polymers. Thus, vinyl acetate is copolymerized with ethylene, acrylic monomers, and so on. Its reaction behavior can be predicted through parameters $Q = 0.026$ and $e = 0.22$ (see Section 8.5.11).

Through partial acidic or basic hydrolysis, poly(vinyl acetate) leads to copolymer-like derivatives whose hydrophilicity is directly related to the proportion of hydroxyl groups. Hydrolysis of copolymers gives terpolymers whose adjustment of the composition allows a simultaneous optimization of several properties. The "total" hydrolysis of PVAc gives poly(vinyl alcohol) (PVAL), which is obtained from latexes of PVAc, primarily by transesterification in acidic or basic medium,

as well as by direct hydrolysis. Gradually with the generation of hydroxyls along the chains, the viscosity of the reaction medium greatly increases due to interchain H bonding. The characteristics of the resulting polymer are closely dependent on the type of catalysis.

Main properties of PVAL are determined by the high density of hydroxyls that are very reactive groups and are able to give hydrogen bonds. Thus, the solubility parameter of PVAL, which reflects its density of cohesive energy, is equal to 26 MPa, a value much higher than that of other vinyl polymers.

PVAL is water-soluble, and it interacts strongly with all the groups able to accept H bonds. Aqueous solutions are very viscous because interchain H bonds are highly favored.

Due to the small size of OH groups, the crystallization of PVAL can occur whatever its tacticity. It occurs according to a transplanar conformation and in a monoclinical system with a fiber period equal to 0.55 nm. Its degree of crystallinity is about 60%, and its glass transition temperature is 85°C. Its melting point can only be estimated at 260°C as PVAL undergoes thermal degradation at temperatures higher than 150°C. In addition to its water-solubility, which is, however, reduced by the tendency to crystallize and which is sensitive to the presence of salts, PVAL is soluble in highly polar solvents (DMSO, DMF, glycols, etc.) at high temperature. It is totally insoluble in hydrocarbons as in most of nonpolar solvents. In the absence of moisture, it exhibits barrier properties with respect to molecular oxygen, which are much better than those of all other polymers.

The degradability of poly(vinyl alcohol) can be considered under two aspects. One is negative and is related to thermal degradation according to a reaction pathway common to the whole of vinyl polymers: first of all, there is a dehydration that generates a polyconjugated polymer

The second aspect, which is more beneficial, is related to its biodegradability. Indeed, PVAL is the only really biodegradable vinyl polymer because it is completely transformed into CO_2 and H_2O under the effect of multiple bacteria and according to a complex mechanism.

As far as its chemical reactivity is concerned, poly(vinyl alcohol) exhibits the properties of secondary alcohols; in particular, it generates intra- or intermolecular acetals, thus allowing a decrease in its surface hydrophilicity by reaction with aldehydes:

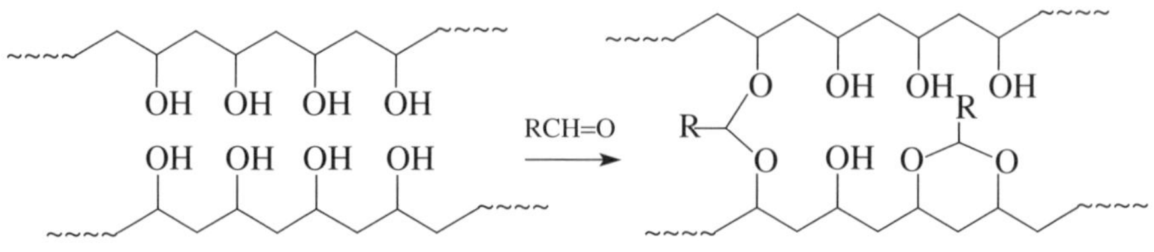

The total chemical modification of PVAL leads to a poly(vinyl acetal) in which the majority of the acetal functional groups are of cyclic type:

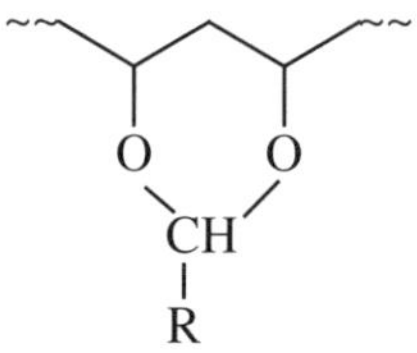

However, the molecular structure of these poly(vinyl acetal)s is ill defined because they contain hydroxyls and residual vinyl acetate monomeric units. The most important are poly(vinyl formal) and poly(vinyl butyral). Due to their chain stiffness, their thermal stability is definitely higher than that of the corresponding vinyl polymers.

15.3.4.5. Applications of Poly(vinyl acetate) and Its Derivatives.
Poly-(vinyl acetate) homopolymers and high-VA-content copolymers are mainly used as emulsions. They are part of many compounds due to their adhesive properties. They are also the main component of vinyl paints; indeed, their glass transition temperature is close to ambient, and their adhesive properties are particularly well-fitted to this application.

PVAc is widely used in paper industry for the coating of papers and paperboards. Textile industry also uses PVAc as a binder in nonwoven materials and for the coating of natural and synthetic fibers. It is also used as additive in bitumens and mortars. High-VA-content copolymers are used for the same application.

Annual world production of poly(vinyl acetate) and VA-containing copolymers is 2.5 million tons.

Due to its hydrophilicity and its other characteristics, poly(vinyl alcohol) (PVAL) has applications different from those of PVAc. It is the basis for the preparation of poly(vinyl acetal)s and, in addition to its use in adhesive formulations and for the final coating of textile fibers or papers, it is introduced as additive in aqueous solutions as viscosifying agent, in particular for suspension polymerizations.

An unexpected application, taking into account its hydrosolubility at high temperature, is the manufacture of textile fibers and papers: the high hydroxyl density brings tenacity ($\sigma_r = 125\,\mathrm{MPa}$) and comfort while the surface hydrofugation is obtained by its formalization. This is obtained by coagulation of the fibers in an aqueous solution of formaldehyde (see Section 13.3.5.3: Wet Spinning). The total annual production of poly(vinyl alcohol) is about 1.5 million tons.

15.3.5. Acrylic and Methacrylic Polymers

It is a large family whose members exhibit a wide variety of properties and whose economic significance are increasing considerably.

They are derived from

Acrylic acid 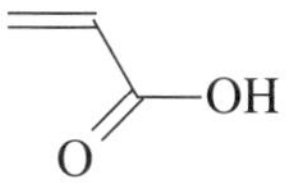   and

Methacrylic acid

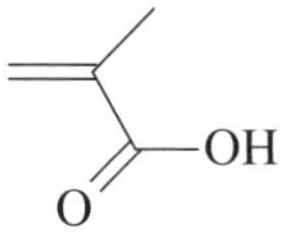

and can be found as polymers from acids, esters, amides, and nitriles. Although some of these derivatives can be polymerized by anionic means, (meth)acrylic polymers are mainly obtained by free radical polymerization. Except for polyacrylonitrile, they are amorphous and their mechanical characteristics are thus highly affected by the value of their glass transition temperature.

From Table 15.3, it can be pointed out that T_g of poly[(meth)acrylic ester]s are highly affected by the size of the side chain, in spite of the fraction of free volume that it brings to the system.

15.3.5.1. Poly(acrylic acid) and Poly(methacrylic acid)

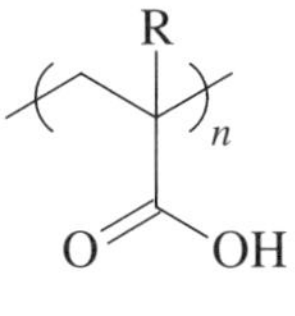

with $R = H$ or CH_3

Table 15.3. Glass transition temperature (in °C) of atactic poly[(meth)acrylic ester]s

$-R$	$CH_2=CH-COOR$	$CH_2=CH(CH_3)-COOR$
CH_3	6	105
C_2H_5	-24	65
$n\text{-}C_3H_8$	-45	35
$n\text{-}C_4H_9$	-50	20
$iso\text{-}C_4H_9$	-43	53
$tert\text{-}C_4H_9$	43	107
cyclohexyl	16	104
benzyl	—	54

Acrylic acid monomer is obtained by catalytic oxidation of propylene in two steps:

$$CH_2=CH\text{-}CH_3 \;+\; O_2 \longrightarrow CH_2=CH\text{--}CHO \;+\; H_2O$$
Acrolein

then

$$CH_2=CH\text{-}CHO \;+\; 1/2\,O_2 \longrightarrow CH_2=CH\text{--}COOH$$
Acrylic acid

Methacrylic acid can be prepared from isobutylene either by a similar process (see hereafter the preparation of methyl methacrylate) or by hydrolysis of methyl methacrylate by steam water.

Due to the hydrosolubility of the corresponding monomers and polymers, these two acids can be polymerized either in bulk or in aqueous solution or even in reverse-dispersed medium (suspension or emulsion). For the processes using aqueous solutions, the concentrations in monomer varies from 20% to 30%, the polymerization being initiated by water-soluble free radical generators (ammonium persulfate, alkali persulfate/bisulfite, persulfate/ferrous salt, etc.). The resulting polymers are atactic with a tendency toward syndiotacticity.

The properties of these polyacids are closely related to the presence of carboxylic acid functional group: sensitivity to pH, hydrosolubility, cross-linking by dehydratation, strong interactions with polar surfaces, and so on. However, due to its molecular structure, poly(acrylic acid) (PAA) exhibits more water absorption capability than its methacrylic counterpart (PMAA), and it is difficult to obtain it in a completely anhydrous form.

When heated to 200°C, a dehydratation of PAA *m* dyads occurs to form cyclic anhydrides:

This dehydratation does not occur between *r* dyads.

The glass transition temperature of PAA is 110°C. This value can be considerably increased by the presence of the above anhydride functional groups along the chains, which increases the rigidity of the structure.

When free from anhydride interchain links, PAA is completely water-soluble. It is always soluble in slightly basic water, and most of its applications are related to this property: thickening agent of aqueous solutions, agent to improve suspensions and dispersions, flocculating agent, and ion exchange resins. Due to its capability to interact with metal surfaces through its carboxylic groups, PAA is also used in certain adhesive formulations.

The world production of these two polyacids is about 300,000 tons *per annum*, with acrylic resin being the most important.

15.3.5.2. Poly[(meth)acrylic ester]s. Only poly(methyl methacrylate) is of major economic significance.

Acronym: PMMA

Molecular formula:

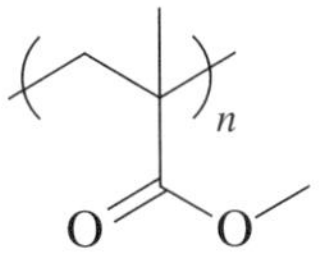

IUPAC nomenclature: poly[1-(methoxycarbonyl)-1-methylethylene]

The most economic and widely used process for the production of MMA monomer involves the reaction of acetone with cyanhydric acid according to the following reaction pathway:

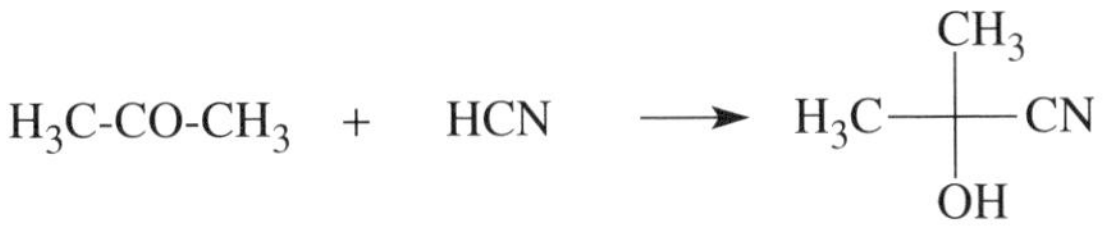

$$H_3C\text{-}CO\text{-}CH_3 \quad + \quad HCN \quad \longrightarrow \quad H_3C\underset{OH}{\overset{CH_3}{\big|}}CN$$

Then

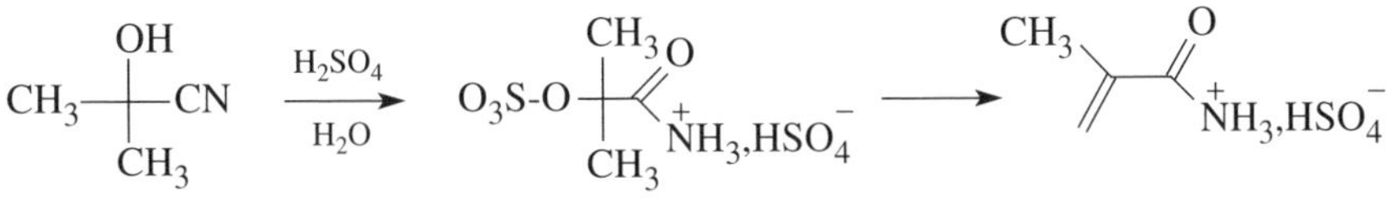

This salt is finally esterified by methanol:

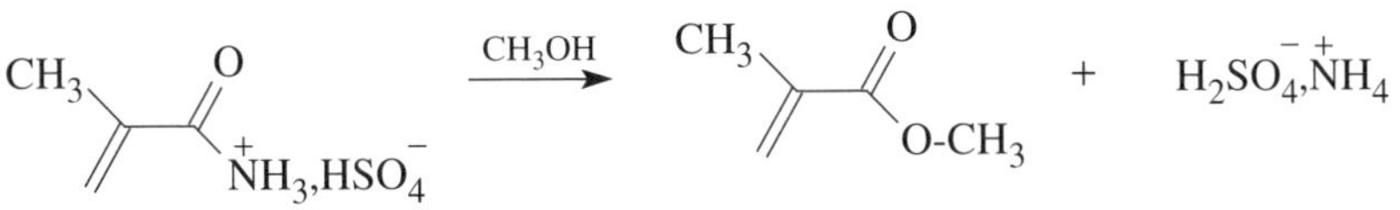

MMA can be also obtained by oxidation of isobutylene

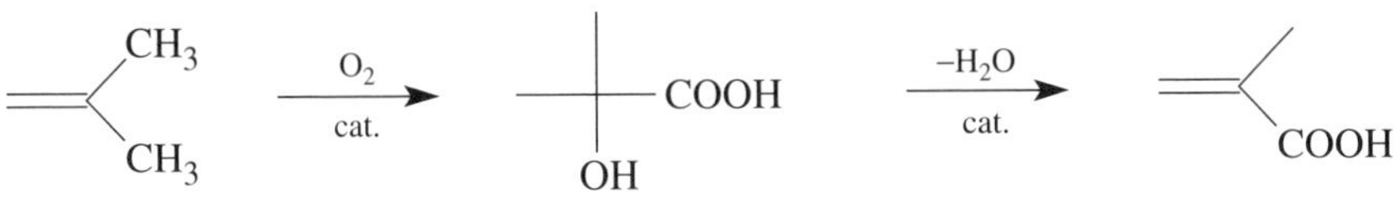

Methacrolein

followed by an esterification of methacrylic acid by methanol:

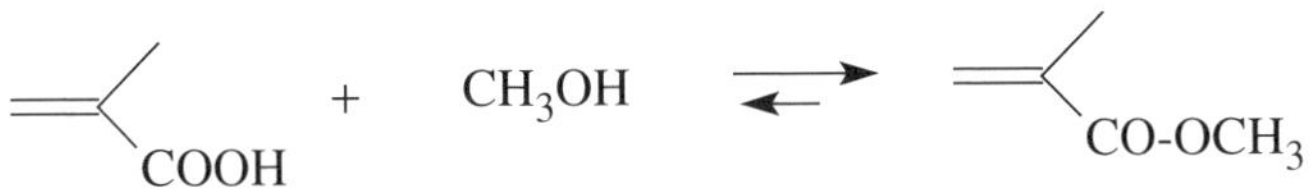

MMA

Although it can be polymerized by anionic process (see Chapter 8), PMMA is almost exclusively produced by radical polymerization in suspension or in bulk. It should be noted that this polymerization is retarded [as is the case for all (meth)acrylic esters] or inhibited by the presence of molecular oxygen (O_2).

PMMA is atactic with a tendency toward syndiotacticity. It is highly regular with an almost perfect head-to-tail placement, but the termination occurs by combination thus generating fragile irregular head to head dyads; the breaking of the C–C bond by a rise in temperature gives free radicals that are able to cause the total chain depolymerization as the ceiling temperature is relatively low (220°C).

The glass transition temperature of this amorphous polymer is very dependent on tacticity. For isotactic polymers its value is 65°C, whereas for syndiotactic ones it is located at about 140°C. For commercial PMMA it is about 105°C and many attempts directed toward an increase in its syndiotacticity were made in order to improve its behavior at high temperatures. PMMA has a better resistance to hydrolysis than that of its acrylate equivalent. It is a fairly cohesive material ($\delta = 18.8\,MPa$) whose stress at break can reach up to 75 MPa.

PMMA is the most transparent among the common polymers; in the range 360 to 100 nm it absorbs only slightly electromagnetic radiations and its transmission coefficient is higher than that of mineral glass. Unfortunately, its low scratch resistance limits its applications in the field of optical intrumentation.

Applications of PMMA are related to its high transparency, its satisfactory mechanical characteristics, and also its easy processing. It is widely used in glazing and in optics. It is also a highly appreciated material in biomedical engineering. In addition to motor oils, it generates visco-statics properties which result from the drastic modification of its solubility with temperature.

The annual world production of PMMA is higher than two million tons.

Acrylic ester homopolymers or statistical copolymers, whose annual world production is higher than 1 million tons, are used in the fields of paints in emulsion, adhesives, textile fiber processing, and paper industry. These applications are in close relationship with their glass transition temperature, which is relatively low. Poly(methyl acrylate) is much more stable against hydrolytic attack by acids and bases than its isomer poly(vinyl acetate). The poly(butyl acrylate) exhibits an elastomeric character which induces multiple applications.

15.3.5.3. Polyacrylonitrile

Acronym: PAN

Molecular structure:

OR

IUPAC nomenclature: poly(1-cyanoethylene)

Acrylonitrile monomer is obtained by catalytic oxidation of propylene in presence of ammonia:

$$\text{CH}_2=\text{CH}-\text{CH}_3 \ + \ \text{NH}_2 \ + \ 5/4\,\text{O}_2 \ \xrightarrow{\text{cat.}} \ \text{CH}_2=\text{CH}-\text{CN} \ + \ 5/2\,\text{H}_2\text{O}$$

Acrylonitrile (AN) is a liquid boiling at $77°C$. It is only used as precursor of (co)polymers. Its world production is about 3.5 million tons. AN is polymerized only by radical means, although its anionic polymerization can be contemplated. All techniques can be used, but it is important to emphasize that the monomer is water-soluble at a concentration of about 10% at $70°C$ and that its polymer (PAN) is insoluble in the monomer.

The most widely used technique of polymerization is dispersion polymerization from an aqueous solution of the monomer; this means that the initial system is homogeneous and the polymer precipitates as soon as it is formed. The process is initiated by water-soluble generators of free radicals (persulfates, mineral redox systems, etc.).

Polymerization in emulsion is mainly used to prepare copolymers.

Acrylonitrile is one of the rare monomers that are polymerized in solution through a radical process. The solvent used is dimethylformamide (DMF) or dimethylsulfoxide (DMSO), and initiators are then azo compounds or organic peroxides. This technique is used only when the resulting concentrated polymer solution (collodion) serves for the production of textile fibers.

Due to the marked polarity of cyano groups ($\mu = 3.9$ D), PAN is an extremely cohesive polymer, thus explaining its particular structural and physicochemical properties.

In spite of its atacticity, PAN is a semicrystalline polymer with a degree of crystallinity close to 50%. The chain conformation in the crystalline zones is not completely regular but gives cylinder-like helixes of about 6 nm diameter. This value is obtained by differential enthalpy analysis with an ultra-fast heating rate because PAN is thermally degradated at $200°C$. As far as the glass transition temperature is concerned, its value is controversial, but most specialists have detected the corresponding signal between $80°C$ and $100°C$.

The melting of PAN is impossible to reach since the corresponding temperature is much higher than the temperature of decomposition; this considerably limits the possibilities of its processing. Homopolymers are utilized exclusively for the production of textile fibers (acrylic fibers), which are obtained by the spinning from collodions that are concentrated solutions in dimethylformamide, dimethylsulfoxide, or even dimethylacetamide because PAN is insoluble in currently available organic solvents. Dry spinning can be used; but due to the high temperatures necessary to evaporate these solvents, this process leaves a considerable proportion of residual solvent in the fiber. Wet spinning process with coagulation of the extruded filament in an aqueous solution of mineral salts is more commonly utilized.

The tensile strength of acrylic fibers ($\sigma_r = {\sim}0.25$ N/tex or 0.29 GPa) is lower than that of PET or polyamides 6 or 6,6 (${\sim}0.8$ GPa), but equivalent to that of cotton and higher than that of wool.

> **Remark.** The textile industry uses the *tex* as unit to measure the fineness of a fiber. It is the mass in grams per 1000 meters of wire. This unit replaced the denier, which was the mass in grams per 9000 meters of wire. The justification of these units is due to the fact that the section of synthetic fibers is seldom circular, with the shape of the die allowing the adaptation of the geometry of the fibers to the optimal required properties; it is thus difficult to characterize their fineness by their diameter.

PAN fibers are moderately comfortable because their capability to absorb water is not very high in spite of the presence of strongly polar groups. The latter preferentially develop polymer–polymer interactions even in the presence of water.

This drawback is partially compensated by the possibility of obtaining extremely fine fibers (up to 0.13 tex corresponding to a diameter equal to ${\sim}12\,\mu$m) that induce a great smoothness of the corresponding fabrics.

The improvement in the intrinsic comfort of these fibers is obtained by copolymerization with comonomers in order to improve certain properties. Thus, the increase in the capability of water absorption can be obtained by copolymerization of acrylonitrile with a hydrophilic monomer (acrylic acid, hydroxyethyl acrylate, etc.). Tinctorial properties are improved by incorporation of monomers that decrease the crystallinity and bring chemical functional groups interacting with the dyes (ester functions, etc.) and so on.

Actually, PAN homopolymer fibers are seldom used as textile fibers. Common fibers known as "acrylics" contain small proportions of another comonomer.

When the comonomer content exceeds 5%, certain properties of the fibers are largely modified and they are then called "modacrylic" fibers.

These modifications are primarily related to the solubility, the capability of water absorption, and the thermal stability. The comonomers being used are numerous: vinyl pyridine, vinyl acetate, methyl acrylate, vinyl chloride, vinylidene chloride, and so on.

PAN homopolymer fibers are also used to produce carbon fibers utilized in composite materials. The process involves a controlled thermal degradation that begins with an intramolecular cyclization which leads to an intensely colored polyconjugated polymer which stabilizes the fiber. Then, depending on the processes, the fiber is heated at temperatures varying from $1200°C$ to $3000°C$ in order to completely eliminate the hydrogen and nitrogen atoms as well as H_2, HCN, N_2, NH_3, and other, more complex molecules and to lead to a carbon fiber of graphite type.

The production of polymers mainly constituted of acrylonitrile is about 3 million tons. Acrylonitrile is also used as minor comonomer with styrene (SAN and ABS copolymers) and butadiene (NBR) (see Section 15.1.3).

15.3.5.4. Polymers of Acrylamide and Its Derivatives

Acronym: PAAM

Molecular structure:

IUPAC nomenclature: poly[1-(aminocarboxy)ethylene]

This monomer ($CH_2=CH-CO-NH_2$) can be obtained by biocatalyzed hydrolysis of acrylonitrile. It is water-soluble in all proportions, and it is polymerized only by radical process so as to obtain the aforementioned polymer. Consequently, it is polymerized either in aqueous solution (20% solution) or in reverse suspension or in water-in-oil emulsion.

Remark. However, the anionic polymerization of this same monomer is possible, but it leads to a completely different polymer as a consequence of a rearrangement of the position of the nucleophilic species:

Poly(β-alanin)
corresponding to polyamide-3 (PA-3)

The polymer obtained from radical polymerization is an amorphous solid with $T_g = 188°C$, this high value being due to the presence of hydrogen bonds. It is completely water-soluble and also in strongly polar solvents (DMF, glycol, etc.). On the other hand, it is insoluble in most of other organic solvents.

Aqueous solutions can be hydrolyzed in acidic medium to give an equivalent of {acrylic acid/acrylamide} copolymers whose composition depends on the experimental conditions.

All applications of PAAM are related to its water-solubility. Its principal field of application uses its capacity to aggregate impurities contained in aqueous suspensions. It is thus primarily used as flocculating agent for treatment of waste water and ores. Another application of polyacrylamide is related to the coating of paper to which it simultaneously confers a barrier effect and a hydrophilic character that are required for printing. PAAM with high molar mass are also used as viscosifying agents to facilitate oil recovery or as industrial thickeners.

By copolymerization with N,N'-methylene-bis-acrylamide, which is a tetravalent monomer, acrylamide gives a cross-linked polymer that is used as superabsorbent polymer and also for the water removal of underground constructions.

Annual world production of polyacrylamide is about 250,000 tons.

Polymers derived from N,N'-alkylated acrylamide monomers exhibit properties completely different from those of polacrylamide. The absence of hydrogen bonds easily explains these differences.

15.3.6. Various Vinyl Polymers

15.3.6.1. Poly(vinyl ether)s

Molecular formula:

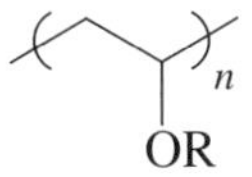

OR

Depending upon the nature of R, they exhibit a wide variety of properties. They are obtained either by radical or cationic polymerization of the corresponding monomer. Under certain conditions, the process can be a "controlled" and/or stereospecific polymerization (see Sections 8.4 and 8.7.5).

The main application of these polymers is related to their capability of undergoing polymerization by photochemical irradiation. They are used as coatings, the polymerization of di- and tetravalent mixtures of comonomers being initiated by irradiation. Alkylvinyl ether homopolymers have a T_g lower than ambient and, generally, they are not used as structural materials. They are used in mixtures, in the field of adhesives, as additives for thermoplastics, as lubricants and as elastomers. The most important is poly(methyl vinyl ether).

There are many copolymers produced at the industrial level. The most used comonomers are maleic anhydride, N-vinylcarbazole, and tetrafluoroethylene.

15.3.6.2. Polyvinylpyrrolidone

Acronym: PVP

Molecular formula:

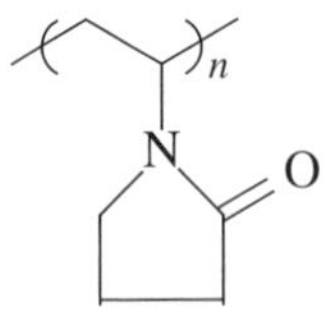

The monomer is a water-soluble liquid that can be polymerized by a free radical means either in bulk or in aqueous solution. The widely used initiators are hydrogen peroxide (H_2O_2) possibly with photochemical activation as well as organic peroxides and azo compounds.

This polymer is a highly hygroscopic solid whose glass transition temperature is 175°C. Due to this high value, the processing of this polymer by conventional techniques normally used for thermoplastics is difficult because it undergoes thermal degradation before reaching a sufficient fluidity. The solution properties of PVP are very peculiar; indeed, this polymer is soluble in many fairly polar organic solvents (alcohols, halogenated hydrocarbons, tertiary amines, hydrocarbons/alcohols mixtures, carboxylic acids, etc.) and also in water. This ambivalence is related to the simultaneous presence of imide and methylene groups in PVP.

Polyvinylpyrrolidone is thus utilized as either aqueous or organic solutions in a multitude of applications. Many of these applications are also related to the chelating properties of this polymer. Thus, it forms complexes with molecular iodine (I_2) and can thus be used as a reservoir of this molecule whose disinfecting properties are well known. It also gives strong interactions with natural and synthetic dyes and thus facilitates their anchoring on textile fibers by complexation with the corresponding polymers.

The interactions of polyvinylpyrrolidone with anionic surfactants are used to lower of the critical molecular concentrations; it is thus often used as additive with such compounds.

The copolymerization of polyvinylpyrrolidone with hydrophobic monomers allows the modulation of its hydrosolubility and its adaptation to the application contemplated: cosmetics, adhesives, binders, encapsulation of galenic compositions, and so on. Finally, by copolymerization with tetravalent monomers (diallylic....), polyvinylpyrrolidone affords aqueous gels that are widely used for biological cultures.

15.3.6.3. Other Vinyl Polymers.

There are many **other vinyl polymers** in addition to those described previously. For most of them, their economic significance does not justify the attribution of a heading: polyvinylcarbazole, polyvinylpyridines, polyvinylfuran can be mentioned. It happens that the economic significance of a monomer is higher than that of the corresponding polymer because the former is more often used as minor comonomer in an important copolymer and not as a homopolymer.

15.4. ALIPHATIC POLYETHERS AND RELATED POLYMERS

These polymers are characterized by a high mobility of the chains in the melt state and also by the possibility of developing hydrogen bonds with water; the latter is responsible for a marked hydrophilicity of these polymers, which largely determines their use in applications.

15.4.1. Poly(ethylene oxide)

Acronym: PEO

This polymer is obtained by ring-opening polymerization of ethylene oxide

by anionic, anionic-coordinated, and cationic processes (see Chapter 8). In general, these methods of polymerization lead to dihydroxy-ended perfectly linear polymers. Rare earth metal-containing initiators are the only ones that lead to PEO with very high molar masses (several millions).

PEO is a highly crystalline polymer whose helical conformation, due to intramolecular dipolar interactions, is not very stable because its melting point (variable with the molar mass) reaches an asymptotic value at 65°C. Amorphous zones undergo the glass transition phenomenon, but the corresponding temperature is also very much dependent on both the molar mass and the degree of crystallinity; its maximum value is $T_g = 17°C$ for $\overline{M}_n = 6000\,\mathrm{g\cdot mol^{-1}}$ and a maximal degree of crystallinity.

PEO is hygroscopic and water-soluble at temperatures higher than ambient, giving solutions whose viscosity is dependent on the nature of the end-groups; thus, α,ω-dihydroxylic PEO gives more viscous solutions than those of their monohydroxylic equivalents due to the development of H bonds between chain ends. Only oligomers are completely water-soluble at ambient temperature. The derivatives with high molar mass are only soluble at low concentration (a few %), with the higher concentrations leading to elastic gels. In general, PEO has a marked preference for polar solvents.

Applications of PEO are in close relationship with their hydrosolubility, their very low toxicity, their low melting point, and their capacity to give interactions with polar surfaces. They are widely used in the biomedical field and in industry as viscosifying agents of aqueous media. They are also used as flocculating agents and in number of adhesive formulations. On cross-linking by irradiation, they are also used in agriculture as superabsorbents. In the years to come, one can predict an important utilization as dissociating medium of lithium salts in batteries with ionic conducting polymers. One of the main applications of PEO is as intermediates in the manufacture of polyurethanes, as diol precursors.

15.4.2. Poly(propylene oxide)

Acronym: PPrO*

As compared to PEO, the interest of this polymer primarily lies in its lesser hydrophilicity and its amorphous character. Indeed, resulting from the polymerization of propylene oxide, it carries methyl groups on each monomeric unit. This

---

*PPO would be the logical acronym, but the latter is also used for the designation of poly(phenylene oxide). That is why it is preferable to use PPrO in order to avoid ambiguity.

substituent increases the hydrophobic character of the polymer and hinders its crystallization due to its irregular atactic placement.

Remark. Tertiary carbon atoms that are responsible for the atacticity initially exist in the monomer molecules. However, certain stereoselective polymerization by giving polyR and polyS chains from [R] and [S] monomer molecules can be isospecific.

The polymerization of propylene oxide, like that of its ethylene oxide lower homolog, can be performed by anionic (NaOH or KOH), cationic (BF_3,Et_2O), or coordinative (R_2Zn/ROH) polymerization.

It is important to preserve the formation of terminal hydroxyl functional groups, and certain processes use additives (pluriols) to increase the number of OH groups per chain.

PPrOs are mainly used as precursors for polyurethanes to which they bring their low glass transition temperature ($T_g = -58°C$) and the high mobility of their chains.

Their annual production exceeds the million tons.

15.4.3. Polytetrahydrofuran. [or poly(oxytetramethylene)]

Acronym: PTHF or PTMEO

In addition to its use as solvent, THF can be cationically polymerized to give a dihydroxyl polyether. For example, when polymerization is initiated by a sulfonic acid, one obtains

Hydroxytelechelic PTHF with a high structural regularity can thus be obtained, which gives semicrystalline polymers ($T_m = 60°C$ and $T_g = -84°C$). These polymers are only intermediates since their almost exclusive application is as polyurethane precursors in which they bring, compared to PEO and PPrO, their lesser sensitivity to the hygroscopy of the surrounding medium. Most of elastomeric polyurethanes are based on this precursor, and a great part of their production is used to produce spandex fibers. Recently, they have been used as diols for the production of polyester thermoplastic elastomers. Their annual production approaches 20,000 tons.

15.4.4. Polyacetals

Due to the difficult classification of these polymers in another family, polyacetals are assimilated to aliphatic polyethers, although their physical and chemical characteristics are completely different from those of the preceding polymers.

The most important polyacetal is obtained from the polymerization of either formaldehyde or trioxane, its cyclic trimer. This polyacetal is called poly(oxymethylene), and its acronym is POM. Trioxane can possibly be copolymerized by cationic process with other heterocycles, particularly ethylene oxide, leading to polymer chains having structural regularity less than that of the conventional POM and yielding less crystalline and less cohesive materials.

The polymerization of formaldehyde is carried out by nucleophilic addition onto carbonyl groups, with the initiator used being a weak base:

$$B^-,Met^+ \; + \; H_2C{=}O \quad \longrightarrow \quad B\text{-}CH_2\text{-}O^-,Met^+$$

$$\Big\downarrow H_2C{=}O$$

$$B\text{www}OH \; \xleftarrow{\; H^+ \;} \; B\text{-}CH_2\text{-}O\text{-}(CH_2\text{-}O)_n\text{-}CH_2\text{-}O^-,Met^+$$

When pyridine is utilized as base, active zwitterionic centers are formed—that is, cyclic ion-pairs. POM gradually precipitates in the reaction medium after its formation.

Cationic polymerization of formaldehyde is also possible:

$$A^-,H^+ \; + \; O{=}CH_2 \quad \longrightarrow \quad A^-,\overset{-}{H}\text{-}\overset{+}{O}{=}CH_2 \quad \longleftrightarrow \quad H\text{-}\overset{+}{O}\text{-}CH_2,A^-$$

$$\Big\downarrow n \; CH_2{=}O$$

$$H\text{-}(O\text{-}CH_2)\text{-}O\text{-}CH_2\text{-}OH \quad \xleftarrow{\qquad} \quad H\text{-}(O\text{-}CH_2)_n\text{-}O\text{-}\overset{+}{C}H_2,A^-$$

As with other cyclic ethers, cationic polymerization of trioxane proceeds through an oxonium cation as active center:

$$\text{wwww-}\overset{+}{O},\, A^- \Big\langle \begin{array}{c} {-}O \\ {-}O \end{array}$$

Commercial polymers have a molar mass in the range $50{,}000{-}100{,}000 \, g{\cdot}mol^{-1}$. Poly(oxymethylene) are crystallizable polymers whose regular chain conformation is helical with a 9_5 helix inducing a marked rigidity which is responsible for most of its physical characteristics. The degree of crystallinity is high and can exceed 80% after annealing at 150°C. The melting point is 175°C, but the value of the glass transition temperature is disputed because two signals are detected: that located at 13°C would correspond to the true T_g, whereas that at 73°C would correspond to the motion of shorter sequences. Due to its strong cohesive energy ($\delta = 22.4 \, MPa^{1/2}$), POM is a technical polymer exhibiting high mechanical characteristics with a tenacity of 70 MPa and an elastic modulus of 3.2 GPa. It is a polymer exhibiting an excellent durability. Polyoxymethylenes (and their copolymers) are used along with polyamides and polyesters. Their principal field of application is the manufacture of components molded by injection and used in mechanical engineering.

The annual production of POM and copolymers reaches 500,000 tons.

15.5. LINEAR CONDENSATION POLYMERS

Several families of polymers constitute this category of thermoplastics, but two are particularly important: polyesters and polyamides. These polymers are generally obtained by step growth polymerization using reactions whose mechanism was described in Chapter 7. Polyesters and polyamides obtained from chain polymerization by ring opening of heterocycles will also be presented in this chapter, with the two methods used to obtain the same material.

15.5.1. Aliphatic Polyesters

For his first experiments in polycondensation, Carothers utilized aliphatic diacids and diols. The poor mechanical properties of the corresponding condensation polymers prevented further investigations on aliphatic polyesters. However, polyesters acquired a renewed interest when relationships between molecular structure and physical—particularly mechanical—properties were precisely established.

Aliphatic linear polyesters correspond to two possible general structures

$$-\!-[CO\text{-}(CH_2)_x\text{-}O]_n\!-\!- \qquad \text{and} \qquad -\!-[CO\text{-}(CH_2)_x\text{-}CO\text{-}O\text{-}(CH2)_y\text{-}O]_n\!-\!-$$

"Unsymmetrical"   "Symmetrical"

whose properties are closely related to the values of x and y—that is, to the density of the ester functional groups, the molecular groups ensuring the cohesion of the system. Due to the preferential intramolecular cyclization during the polycondensation reaction, the first terms of the series cannot be obtained by polycondensation; they are prepared by chain polymerization of the corresponding lactones or lactides as follows:

Table 15.4. Designation and melting temperature of some aliphatic and aromatic polyesters

Formula of the Monomeric Unit	Designation of the Polyester	T_m (°C)
$-(CO-CH_2-O)-$	Polyglycolate or polyglycolide	225
$-[CO-(CH_2)_2-O]-$	Poly(3-hydroxy-propanoate) or polypropiolactone	122
$-[CO-CH(CH_3)-O]-$	Poly(2-hydroxypropanoate) or polylactate or polylactide	215
$-[CO-CH_2-CH(CH_3)-O]-$	Poly(β-hydroxybutyrate)	180
$-[CO-(CH_2)_5-O]-$	Polycaprolactone	55
$-[CO-C_6H_5-O]-$	Poly(p-hydroxybenzoate)	>350
$-[CO-(CH_2)_7-CO-O-(CH_2)_6-O]-$	Poly(hexamethylene sebacate)	38
$-[CO-C_6H_4-CO-O-(CH_2)_2-O]-$	Poly(ethylene terephthalate)	267
$-[CO-CO-O-(CH_2)_2-O]-$	Poly(ethylene oxalate)	172
$-[CO-C_6H_4-CO-O-(CH_2)_4-O]-$	Poly(butylene terephthalate)	234

Homopolyesters are semicrystalline materials with a melting point that increases with the decreasing molar mass of monomeric units (see Table 15.4). This occurs in relation with a concomitant increase in the density of ester functional groups and an increase in cohesive energy. "Symmetrical" polyesters are generally obtained by polycondensation between a diacid (adipic, azelic, etc.) and a diol (ethylene glycol, propylene glycol, etc.), whereas "unsymmetrical" ones result generally from the polymerization of the corresponding lactones or lactides.

For high values of x (or x and y), the "density" of cohesive ester groups is low and the corresponding materials exhibit poor mechanical characteristics; they are thus not used as structural materials in a pure state. On the other hand, dihydroxytelechelic oligomers are very important diols and essential precursors for the production of polyurethanes.

Another application of these polyesters is related to their miscibility with halogenated polymers: they are used as polymeric internal plasticizers when used along with PVC.

The situation is different for polyesters with a short alkylenic chain for which the density of cohesive energy is high. Polyglycolide (PG), $-(CO-CH_2-O)_n-$, is obtained by polymerization of glycolide

at high temperature, using either tin salts (tin octoate, tin chloride dihydrate, etc.), Lewis acids, or various organometallic compounds as initiators. Its mechanical characteristics are as good as those of polylactide (PL), $-[CO-CH(CH_3)-O]_n-$, obtained from lactide:

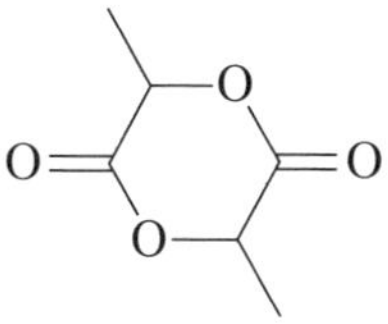

Their cohesive energy is very high, and they are soluble only in highly polar solvents such as hexafluoroisopropanol or hexafluoroacetone sesquihydrate. These two homopolymers are crystalline (if lactide is optically pure): for PG, $T_m = 225°C$ and $T_g \sim 35°C$; their copolymerization in variable proportions allows the control of their degree of crystallinity as well as their hydrophilicity. The principal application of these copolymers (PGL) is in the biomedical field; they are used as biodegradable materials, with their biodegradability being mainly controlled by their capability to absorb water.

Poly(β-hydroxybutyrate), $-[CO-CH_2-CH(CH_3)-O]_n-$, is a polyester obtained through biochemical techniques using bacteria. It is a perfectly regular isotactic polymer and is thus highly crystalline ($T_m = 180°C$ and $T_g = 5°C$). It is copolymerized with its ethyl analog in order to reduce its degree of crystallinity and thus to increase its impact strength. The major property of this polymer is its capability of being biologically degraded.

Other polyesters of this type also exhibit this property but have not yet reached a significant industrial development:

Polydioxanone	$-(CO-O-CH_2-CH_2-O)_n-$
Poly(trimethylene carbonate)	$-(O-CO-O-CH_2-CH_2-CH_2)_n-$
Polycaprolactone	$-[CO-(CH_2)_5-O]_n-$

They are obtained by polymerization of the corresponding heterocycle.

Poly(propylene fumarate), $-[CO-CH=CH-CO-O-CH_2-CH(CH_3)-O]_n-$, is also biodegradable. It is obtained by polycondensation (diacid onto diol or cross-esterification).

15.5.2. Aromatic Polyesters

Examination of Table 15.4 shows that rigid groups present in the backbone of the chains are required for the cohesion of the corresponding polymers. The investigation of this structure–property relationship led to the discovery of an important family of polymeric materials by Dickinson and Whinfield (ICI) in 1942.

15.5.2.1. Poly(ethylene terephthalate) (PET)

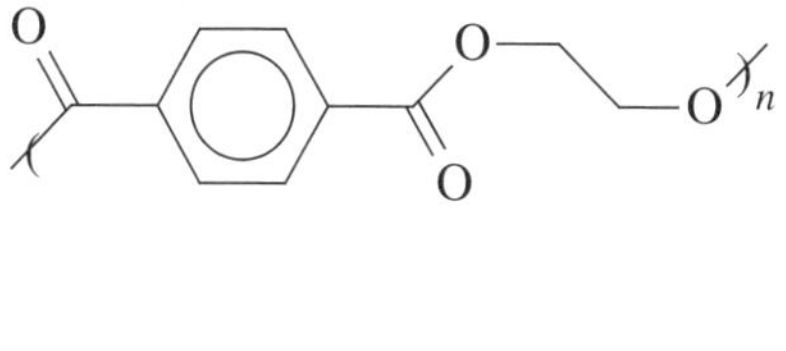

It is obtained either by slow direct esterification or, mainly, by catalyzed cross-esterification between dimethyl terephthalate and glycol (see Section 15.3.6).

Dimethyl terephthalate (DMT)

is thus an important intermediate obtained by esterification of terephtalic acid, which itself is obtained by the controlled oxydation of p-xylene. Optimal mechanical properties of the polymeric material require the preparation of high molar mass PET. For this purpose, DMT is purified by recrystallization before cross-esterification.

Glycol is prepared by oxidation of ethylene; the resulting ethylene oxide is then hydrolyzed to obtain ethylene glycol (or ethane-1,2-diol).

Polymerization is carried out in bulk in two distinct steps. First of all, an oligomerization is carried out (by esterification or cross-esterification) in order to eliminate most of the by-products of condensation (water or methanol); this operation is catalyzed by metal derivatives (Ti, Mn, Ca, Zn, Sn, etc.). In the second step, an elimination of either residual water or methanol is accomplished by intense stirring under vacuum at high temperature. It is important to point out that all the processes being used generate diethylene glycol ($HO-CH_2-CH_2-O-CH_2-CH_2-OH$) during polyesterification. This molecule can, in turn, be incorporated in the chains by copolymerization. Its content lies between 2% and 15% in the final polymer.

To attain optimal properties, the molar masses of PET should be in the range $25{,}000 < \overline{M}_n < 50{,}000 \, g \cdot mol^{-1}$. Due to the regularity of its molecular structure, poly(ethylene terephthalate) is a highly crystallizable polymer. Under suitable conditions of temperature, it crystallizes in a triclinic system ($a = 0.46\,nm$, $b = 0.59\,nm$, and $c = 1.08\,nm$) with only one monomeric unit per fiber period.

In spite of the high volume of the monomeric units, its cohesive energy is high ($\delta = 20.5\,J^{1/2}cm^{-3/2}$), thus explaining the high value of the melting point ($T_m = 264°C$). The glass transition occurs between $65°C$ and $80°C$, with such a wide range being explained by the effect of the molar mass, the water content, and the degree of crystallinity on this property. For dry and amorphous PET with high mass molar, $T_g = 70°C$.

The rate of crystallization of PET from the molten state is relatively slow (see Chapter 11). Since its glass transition temperature is higher than the ambient one, it can be obtained and preserved in a totally amorphous state. Its degree of crystallinity can thus be modulated according to the desired characteristics required for its application. As thermoplastic structural material, PET is often used in a "quenched" amorphous state.

In order to obtain PET in a semicrystalline state, it should be kept in the hot mold for the time required for its crystallization; its physical properties are thus highly dependent on its thermal history. As textile fiber, PET is highly oriented by drawing and it is maintained at its temperature of maximal crystallization before being cooled. Its degree of crystallinity depends on the content of diethyleneglycol units being incorporated as well as the degree of drawing, whereas the chain orientation with respect to the fiber axis depends only on the degree of drawing which is about 600%.

The major part of the production of PET is utilized for the manufacture of textile fibers. Their annual world production is about 18 million tons, which corresponds to two-thirds of the production of synthetic textile fibers. These fibers are used either alone or mixed with cotton or wool. However, the "tinctoriability" of PET fibers is low and, in a pure state, dyeing is obtained by dispersion of the dye in amorphous zones of the semicrystalline material. Another solution to solve this problem involves replacement of pure PET by various copolymers that introduce reactive functional groups along the chains simultaneously with the reduction of the degree of crystallinity; the latter point is detrimental to the mechanical properties.

Copolymerization is also used to obtain fibers whose melting point is considerably lowered, which is useful for manufacture of nonwoven fabrics.

Poly(ethylene terephtalate) is used more and more as technical polymer. However, apart from its use in the textile industry, its principal application lies in the manufacture of bottles. For this application, the mechanical characteristics are improved by a two-dimensional drawing during processing (extrusion-blowing).

The annual total production is about 18 million tons. PET is thus part of the family of important industrial polymers.

15.5.2.2. Poly(butylene terephthalate) (PBT)

With respect to its molecular structure, this polyester differs from the precedent only by the presence of two additional methylene groups in the monomeric unit. The method of preparation exhibits considerable analogies with the preceding one. Side reactions are rarer, which allows the attainment of higher molar masses more easily. However, this small difference in the molecular structure induces definitely different physical properties. Indeed, the incorporation of two methylene groups substantially increases the mobility of the chains and their crystallizability. Thus, contrary to PET, PBT is spontaneously semicrystalline and is used in this state as thermoplastic material.

Its crystalline state leads to two structural forms (α and β) in the triclinic system. The fiber period corresponds to one repeating unit in total extension for the structure observed after drawing. In the crystalline form corresponding to a disoriented polymer, the methylenic chain acquires a left-hand conformation that corresponds to a shorter fiber period.

Thermomechanical properties reflect the lower density in cohesive groups as compared to PET. Thus, the melting point is lower ($T_m = 224°C$) and, due to same reason, the effect of atmospheric moisture on the mechanical characteristics is limited. The high mobility of PBT chains and the melting point lower than that of PET make this polymer easier to process than PET; it is mainly processed by injection at a temperature $\sim 250°$C.

PBT is often mixed with short glass fibers, which simultaneously allows an increase of its stress at break (tenacity) and its elastic modulus.

Its annual world production is approximately 200,000 tons and its growth rate is high.

15.5.2.3. Poly(bisphenol A carbonate) (PC)

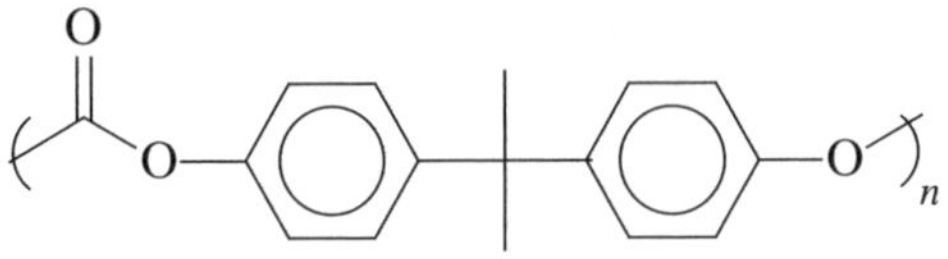

Among the various diols that were tested for polycarbonate synthesis, the one resulting from bisphenol A (BPA) was found to be a material exhibiting interesting mechanical characteristics. Thus, it corroborates the general idea that the introduction of bisphenol A groups into a macromolecular chain largely improves the mechanical properties of the resulting material.

Among the various methods that can be utilized to produce PC, processing through interfacial polycondensation is widely used. The hydrosoluble precursor is the disodic salt of bisphenol A. It reacts with phosgene (Cl_2CO) solubilized in a hydrophobic solvent that is generally a chlorinated solvent (CH_2Cl_2, $CHCl_3$, C_6H_5Cl, etc.):

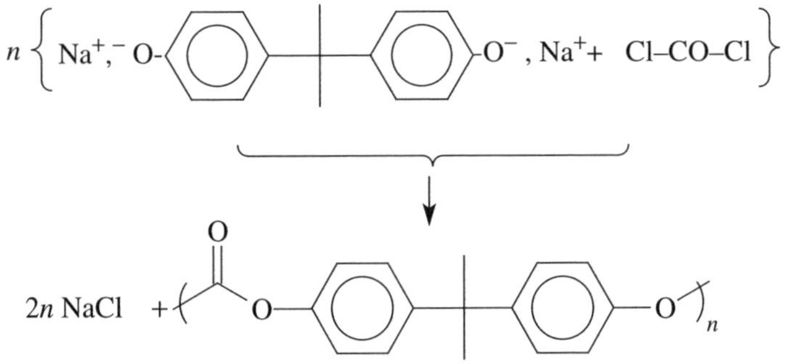

Another method for the polycondensation involves the reaction between $COCl_2$ and bisphenol A in methylene chloride solution in the presence of pyridine to trap the HCl produced.

The transesterification with phenyl carbonate is also possible but is more difficult to transpose to an industrial level than the preceding methods.

Depending upon the method being used, it is possible to obtain PC whose molar masses vary from 2×10^4 to 2×10^5 g·mol^{-1}. However, for injection molding, the best adapted molar masses are in the range $2-3 \times 10^4$ g·mol^{-1}.

Poly(bisphenol A carbonate) shows a perfectly regular structure. However, it is unable to crystallize spontaneously. Indeed, the rigidity of the chains due to the presence of phenylene rings restricts their possibility of folding up in the molten state. By prolonged annealing, it is, however, possible to crystallize it and then its melting temperature is 260°C.

PC is only used in an amorphous state, thus it gives a completely transparent material whose $T_g = 150$°C. This high temperature causes the great rigidity of the chains. Poly(bisphenol A carbonate) resists the attack of most of the chemicals; it has also a high thermal stability.

The mechanical characteristics of PC are good, but its most remarkable specific property is resilience. Its impact strength measured under equivalent conditions is approximately 10 times higher than that of PET, 30 times higher than that of PMMA, and 300 times higher than that of mineral glass. Its applications are thus mainly based on this property and on its transparency: car industry, electrical engineering, packaging, compact disks, and so on. PC is also frequently used in blends with other thermoplastics.

Its world production is 2 million tons, and its annual growth rate approaches 10%.

15.5.2.4. Other Aromatic Polyesters. Poly(cyclohexyldimethylene terephthalate) (PCT)

As in case of PET, the rate of crystallization of PCT is very slow. Only a long annealing allows a partial crystallization ($T_m = 290$°C and $T_g = 80$°C). Its molecular structure can be considered as a copolymer between *cis* and *trans* isomers with respect to cyclohexylene, and the transition temperatures of completely either *cis* or *trans* polymers are slightly higher than those given above. PCT is interesting due to its thermal stability, its thermomechanical properties, and its low water absorption. It is mainly used in electronics industry, but it can also be used as structural material in mechanical engineering.

Poly(ethylene naphthalene dicarboxylate) (PEN)

It is obtained by polycondensation between ethylene glycol and either naphthalene-2,6-dicarboxylic acid or its dimethyl ester. Its high thermomechanical characteristics and its excellent durability opened markets for it in the fields of electronics, industrial textiles, and food packaging (hot drinks and sparkling beverages).

Liquid crystal polyesters (LCP) are generally derived from *p*-hydroxybenzoic acid:

They are thermotropic liquid crystals obtained from aromatic homo- or copolyesters, which exhibit particularly high thermomechanical characteristics while preserving excellent impact strength up to very low temperatures. Taking into account their high cost, they are utilized only for high-valued applications, particularly in electronics industry.

15.5.3. Aliphatic Polyamides

They were the first linear condensation polymers produced on an industrial scale by the end of the 1930s. Contrary to their polyester analogs, they exhibit excellent thermomechanical characteristics. This difference is mainly due to their tendency to form hydrogen bonds that increase their density of cohesive energy. They are indicated by initials PA followed by one (or 2) number(s) referring to the number of carbon atoms in the main chain constituting the monomeric unit.

PA followed by one single number (PA-*x*) results theoretically from the polycondensation of an α-amino,ω-carboxylic acid:

$$n \; H_2N-(CH_2)_x-COOH \longrightarrow [-HN-(CH_2)_x-CO-]_n$$

Actually, these PA are generally prepared by chain polymerization of the corresponding lactam. Nevertheless, they are presented here with condensation polymers.

Aliphatic PA indicated by two numbers [PA-x, y + 2] are generally prepared by polycondensation between a diamine whose number of carbon atoms is given by the first number (*x*) and a dicarboxylic acid whose structure determines the second number (*y* + 2):

$$n \left\{ H_2N-(CH_2)_x-NH_2 \; + \; HOOC-(CH_2)_y-COOH \right\}$$

$$\downarrow$$

$$-[NH-(CH_2)_x-NH-CO-(CH_2)_y-CO]-_n$$

Remark. In the United States, it is common practice to indicate polyamides by the term "nylon," which is, in fact, the commercial name of PA produced by DuPont de Nemours Company.

Polyamides have certain common chemical, physical, or physicochemical characteristics, primarily resulting from their capability of developing hydrogen bonds. So, the zigzag planar structure allows the maximal stabilization of the crystalline state, and the fiber period depends on the value of $(x + y)$. The rate of crystallization of aliphatic polyamides is high and, although their glass transition temperature is higher than ambient, it is very difficult to quench them in the amorphous state from the molten state by fast cooling. In addition, after quenching, they spontaneously tend to increase their degree of crystallinity when heated above the ambient temperature. Their density of cohesive energy is high ($\delta_{PA-6,6} = 28 \, J^{1/2} cm^{3/2}$). It is closely dependent on the even or odd number of carbon atoms of the repeating unit. It is easy to explain this phenomenon using the following schemes, which compare the the extent of hydrogen bonds in PA-6 and PA-7. The melting point of PA-7 (223°C) is logically higher than that of PA-6 (215°C) in spite of a lower density of cohesive amide functional groups.

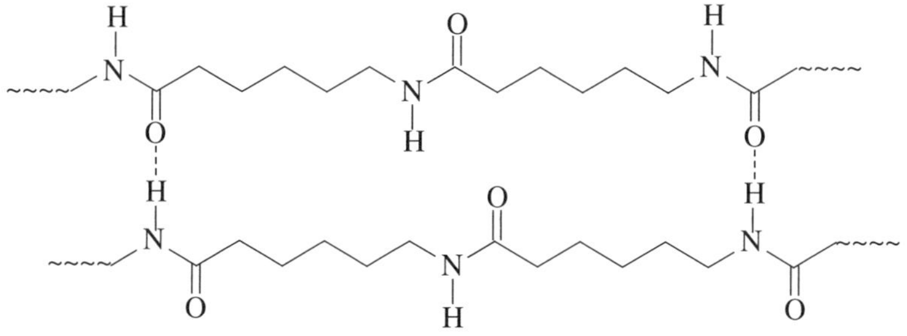

PA-6

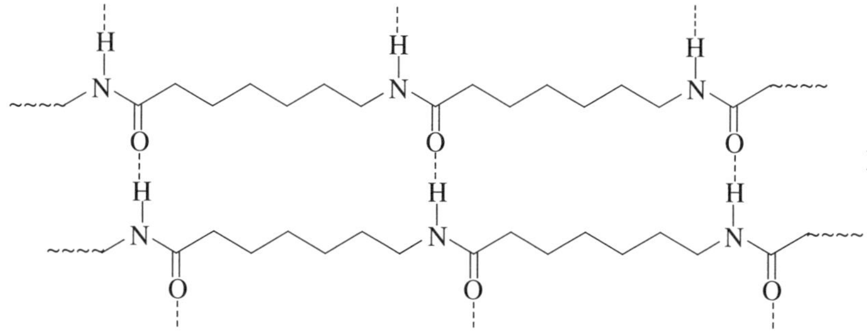

PA-7

By replacing interchain H bonds, the interactions established with water molecules deeply affect number of properties. In fact, water plays the role of a plasticizer that swells the material, diminishes its cohesion, and lowers its T_g. This effect is considerable because the density of amide functional groups is high. Thus, from the examination of Table 15.5 related to PA-x, one can see that the capacity of water absorption decreases gradually with the number of methylene groups $(x - 1)$ separating two amide functional groups. If melting points are also considered, they decrease with the same structural parameter but inside two series corresponding to an even and an odd number of carbon atoms, respectively. In the same way,

Table 15.5. Melting temperature and capability of water absorption of polyamides-x, at 100% relative humidity (RH)

$$-[CO-(CH_2)_{x-1}-NH]_n-$$

x	$T_m(°C)$	% H_2O
3	340	25
4	260	—
5	258	13.5
6	215	9.5
7	230	5.0
8	200	3.8
9	209	3.4
10	188	2.0
11	190	2.8
12	180	2.7

mechanical characteristics reflect the effect of the density of H bonds inside the material. A similar evolution of the properties can be easily interpreted in the series of PA-x,y.

At ambient temperature polyamides are soluble only in highly polar solvents: phenols, formic acid, fluorinated alcohols, and mineral acids. Thus, their structural analysis in solution is difficult, particularly the determination of their molar mass by end group titration. Their solubilization in commonly available solvents (tetrahydrofuran in particular) can be obtained by alkylation of NH groups, which suppresses H bonds and thus lowers the strong cohesion of these materials.

All polyamides are suitable for being hydrolyzed in proportion of their water content. Depending on the relative orientation of the dipoles formed by amide functional groups, PA may or may not exhibit piezoelectric properties, *i.e.*, generation of an electric signal under mechanical constraint. PAs that have all their dipoles directed in same direction (PA-7, PA-9, etc.) leads to a marked piezoelectric effect whereas the alternation of orientations (in PA-6 for instance) cancels this effect.

Many polyamides or copolyamides can be synthesized. Actually, only a small number of them reached a significant industrial development. Polyamides 6 and 6,6 share ~85% of total annual market of 7.5 million tons (4 million tons for PA-6 and 2.5 million tons for PA-6,6). The three quarters of the production (4 million tons) are used for the manufacture of textile fibers, the rest is utilized as thermoplastic technical polymer.

15.5.3.1. Polycaprolactam (PA-6)

Molecular structure:

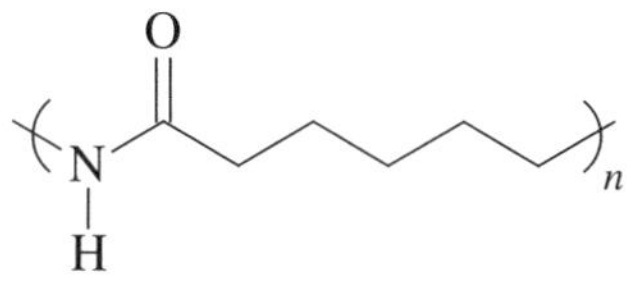

IUPAC designation: poly[imino(1-oxohexamethylene)]

Caprolactam monomer,

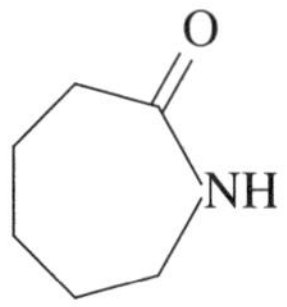

can be prepared either from toluene or from benzene. For example, in the latter case we have

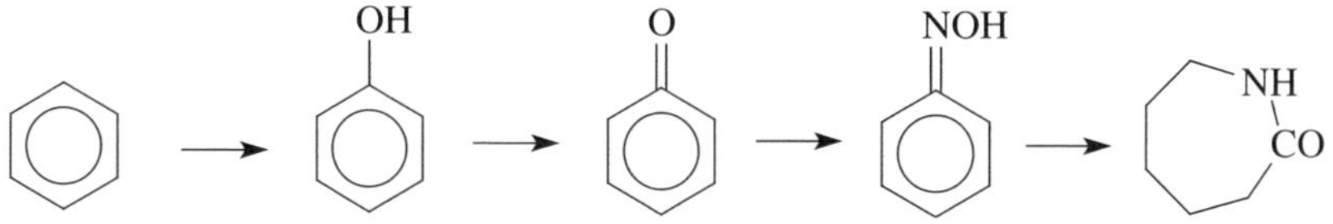

After purification, caprolactam is polymerized by two different methods.

The first, which is well known, involves the hydrolysis of caprolactam in order to generate ε-aminocaproic acid whose polycondensation leads to PA-6 having very high molar mass if the free water is eliminated at high temperature by vacuum pumping. This polymerization "catalyzed" by water is generally carried out continuously.

The second method uses the anionic chain polymerization of the heterocycle, whose mechanism (complex) was presented in Section 8.6.4. The activated monomer sodium lactamide is used as initiator. This method allows the preparation of statistical copolymers—for example, with lauryllactam, which is the monomer molecule of PA-12.

PA-6 crystallizes in the monoclinical system with a fiber period of 1.72 nm corresponding to the chain in total extension. The degree of crystallinity is about 50%; the melting point of PA-6 is 215°C and its glass transition occurs at 52°C.

Its mechanical characteristics are excellent due to its strong cohesive energy ($\delta = 28\,\mathrm{J}^{1/2}/\mathrm{cm}^{3/2}$). Its stress at break ($\sigma_r$) reaches 80 MPa after chain orientation by drawing, and its initial elastic modulus is equal to 2.8 GPa. Like all other polyamides, PA-6 has a remarkable reversible elongation, a property that is used for its applications in textile industry. Strain at the yield point is about 10–15%.

In addition to its main application, which is the production of textile fibers, PA-6 is used as structural and technical thermoplastic material in many sectors of the mechanical engineering. Spinning from the melt is used to process PA-6 fibers. The manufacture of various objects can use all common processing techniques of thermoplastics.

15.5.3.2. Polyhexamethyleneadipamide (PA-6,6).

Molecular structure:

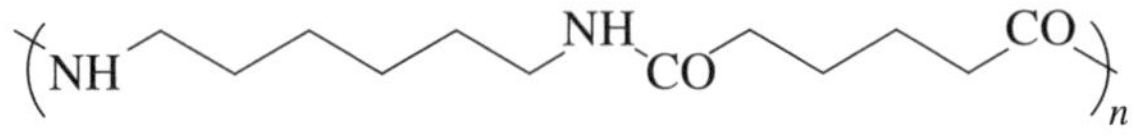

It is obtained by direct polycondensation between adipic acid and hexamethylene-diamine. The following scheme gives one of the main methods of preparation of these two comonomers, which are, moreover, important reaction intermediates:

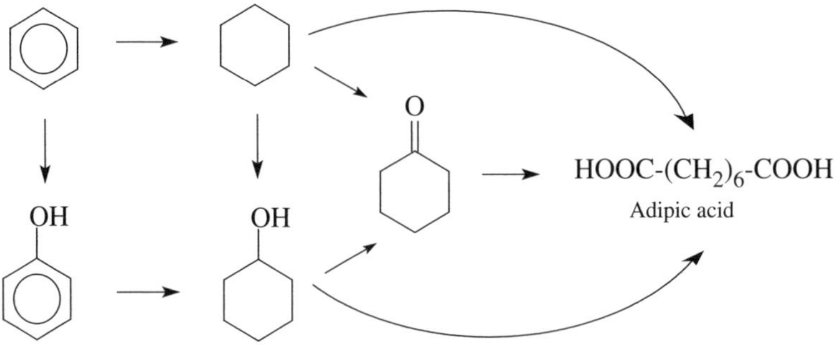

$$HOOC\text{-}(CH_2)_6\text{-}COOH \longrightarrow NC\text{-}(CH_2)_6\text{-}CN \longrightarrow H_2N\text{-}(CH_2)_6\text{-}NH_2$$

Adiponitrile   Hexamethylene diamine

Initially, both comonomers react in aqueous solution at neutral pH and lead to a salt called "nylon 6,6 salt,"

$$\left\{ \begin{array}{c} H_3N^+\text{-}(CH_2)_6\text{-}NH_3^+ \\ {}^-OOC\text{-}(CH_2)_4\text{-}COO^- \end{array} \right\}$$

which can be obtained in the pure state with a perfect stoichiometry by recrystallization in water. In addition to the hydrosolubility of this salt, it is much less oxidizable than hexamethylenediamine; thus it does not generate colored by-products that would be inconvenient for most of the applications of the derived materials. For the polymerization, the solution of the salt is concentrated to 80% at 160°C; then the steam is pumped in order to allow the equilibrium to shift toward amide formation.

The desired molar masses $(\overline{M}_n)$ are about 20,000 g·mol^{-1}.

PA-6,6 crystallizes in the triclinic system with a fiber period of 1.72 nm. Its degree of crystallinity is about 50%. Its melting point T_m is 260°C and its glass transition temperature is 57°C, which is slightly higher than that of PA-6. Due to its molecular structure, the development of interchain H bonds is maximal, thus inducing mechanical characteristics slightly higher than those of PA-6.

Other physical and physicochemical characteristics are not very different from those of PA-6, and consequently the fields of application are also the same.

15.5.3.3. Other Aliphatic Polyamides. They are mainly polyamides having longer polymethylene sequences. Their marketing was dictated by the need of technical polymers whose mechanical characteristics are less sensitive to the hygrometry of the ambient conditions than those of PA-6 and PA-6,6. It concerns PA-6,10, which is obtained by the polycondensation between hexamethylenediamine and sebacic acid [HOOC–$(CH_2)_8$–COOH], PA-11 and PA-12.

PA-11 results from the polycondensation of 1-aminoundecanoic acid (molecule whose raw material is castor oil).

PA-12 is obtained by polymerization of the corresponding lactam, which itself is obtained from the trimerization of butadiene.

These polymers are less cohesive and thus have a glass transition temperature and a melting temperature lower than those of PA-6 and PA-6,6. Their mechanical characteristics are also slightly weaker but depend much less on the relative humidity (see Table 15.5). These polymers are not utilized for textile applications. They are used for manufacture of monofilaments (fishing nets, cords for musical instruments, ropes) and also as technical polymers for the surface coating and the molding of various objects that are required to resist moisture.

15.5.4. Aromatic Polyamides (Aramides)

As for polyesters, the introduction of an aromatic moiety into polyamide chains considerably changes their physical and physicochemical characteristics. Due to their molecular structure, aromatic polyamides combine together structural regularity, stiffness of the chains, and compacity of phenylene rings. This results in a very strong cohesion of the corresponding materials that exhibit exceptionally tough mechanical characteristics.

Aramides are prepared by polycondensation using the Schotten–Baumann reaction. This reaction, which uses the selected isomers of phthaloyl chloride and phenylenediamine (or its chlorohydrate), is carried out at low temperature (from 0°C to −40°C) in order to avoid side reactions; it is carried out in an amide solution (dimethylacetamide, N-methylpyrrolidone, tetramethylurea, etc.) to which mineral salts are added.

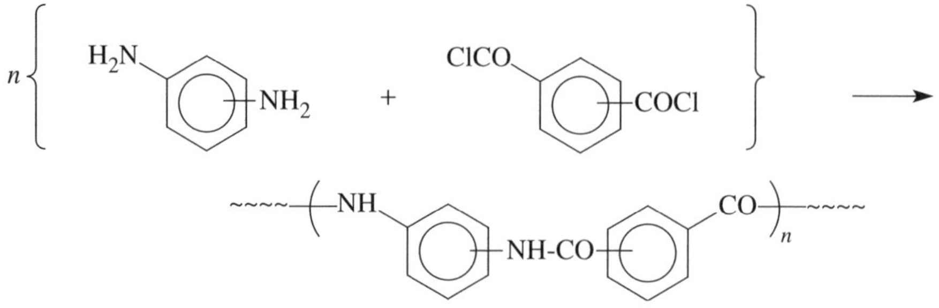

Aramides can also be prepared by interfacial polycondensation but in this case the molar masses obtained are lower than those attained in the solution process.

Due to their chain stiffness and their particularly high melting points, aramides cannot be processed by usual techniques applicable to thermoplastics. As in the case of polyacrylonitrile (PAN), they can only be utilized as fibers, which are obtained from the corresponding collodions either by dry spinning or wet spinning processes.

In addition to the high level of their mechanical properties (in particular their elastic modulus and stress at break), which are preserved up to temperatures higher than 200°C, aramide fibers exhibit an excellent durability of their properties even under extreme conditions. They resist very well to most of chemicals except strong acids. Their low combustibility (LOI $= 28-30$) makes them irreplaceable as safety materials.

> **Remark.** The combustibility of a polymer is measured by the limit proportion of molecular oxygen in a gas mixture O_2/N_2 for which its combustion is not propagated (LOI, limit oxygen index).

Many aramide structures were synthesized and studied. Several of them were produced industrially but at present only two are significantly developed. They are poly(p-phenyleneterephthalamide) and poly(m-phenyleneisophthalamide). Due to different molecular symmetry, these two aramide fibers have different specific physicochemical and application properties.

15.5.4.1. Poly(p-phenyleneterephthalamide) (PPD-T)

Molecular structure:

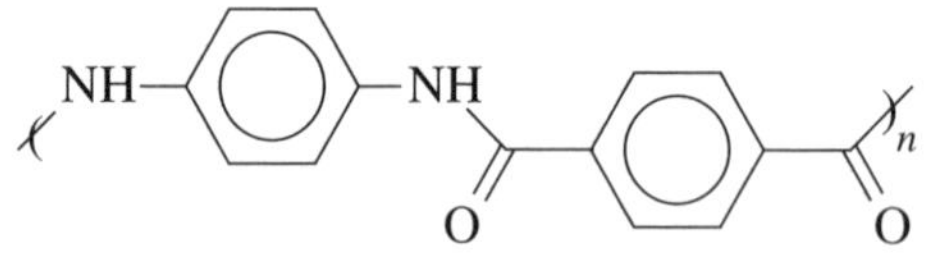

The molecular symmetry and the rigidity of the chains of this aramide are responsible for the mesomorphic structure of this polymer which exhibits a lyotropic character. The trademark of this well-known commercial fiber is Kevlar®. It is obtained by wet spinning process from a collodion in concentrated sulfuric acid, which induces many problems related to the corrosion of the equipment. The coagulation of lyotropic solutions followed by drawing leads to a highly crystalline polymer. The chains crystallize in a monoclinical system with two repeating units per fiber period which corresponds to $c = 1.28$ nm. The estimated melting point of PPD-T is $T_m = 550°$C, but at this temperature the polymer quickly undergoes degradation. The glass transition occurs at about $360°$C. The density of this material is equal to 1.45 and it depends on the degree of drawing.

The tenacity (stress at break) of PPD-T can reach 2.8 GPa with an initial elastic modulus equal to 120 GPa.

The applications of PPD-T fibers are in the fields where very high mechanical characteristics able to be preserved in a wide range of temperature are required. The fabrics containing this aramide are used in the fields of clothing and of safety cables and industrial fabrics. However, the main application is in the field of composite materials as reinforcement of strongly cohesive matrices (polyepoxy or others).

15.5.4.2. Poly(m-phenyleneisophthalamide) (MPD-I)

Molecular structure:

This material is mainly known under the trademark Nomex®. It was the first aramide produced at the industrial level. As compared to PPD-T, its molecular symmetry is lesser and does not induce the mesomorphic character. Consequently, its solubility is slightly higher, thus allowing the processing of the corresponding fibers by dry spinning from collodions in polar organic solvents. For economic reasons the spinning can be carried out from the solution in which it was prepared.

The mechanical characteristics of MPD-I are definitely weaker than those of PPD-T. It can, however, reach a tenacity of 0.6 GPa with an initial elastic modulus of 20 GPa. For this reason, applications are found in the fields where its chemical inertia, self-extinguishability, and excellent behavior at high temperatures are required.

In spite of the cost of their development, aramides reached a relatively high level of industrial production, and these materials became essential in many high-value-added applications.

15.5.5. Linear Polyurethanes

General molecular formula: $-(O-R-O-CO-NH-R'-NH-CO)_n-$

The generic acronym of polyurethanes is PUR, but those that are made of linear chains are often indicated by TPU (thermoplastic polyurethanes).

The urethane function, $-O-CO-NH-$, also called carbamate, is generated by reaction of a hydroxyl group with an isocyanate function (see Section 8.4.3).

As reflected by its molecular structure, this functional group combines the inherent properties of amide and ether functional groups. The former induce a high density of cohesive energy through the development of H bonds, whereas the latter favor the free rotation around the bonds, thus inducing a certain mobility of the polymer chains and the deformability of the corresponding materials.

15.5.5.1. Structural Polyurethanes.
Generally, thermoplastic polyurethanes consist of segmented copolymers (copolymers with multiple and short blocks) with alternation of rigid sequences which bring cohesion to the material and flexible sequences which confer its deformability. Generally, the rigid sequences result from the step co-oligomerization of a short diol (ethylene glycol, propylene glycol, etc.) with a rigid aromatic or cycloaliphatic diisocyanate, for example:

Diphenylmethanediisocyanate (MDI)   Dicyclohexylmethanediisocyanate (DCI)

These oligomers are prepared by reaction of an excess of diisocyanate in order to be functionalized by NCO end-groups. The molar mass of these sequences is only a few hundreds of $g \cdot mol^{-1}$. These rigid sequences are step copolymerized with α,ω-dihydroxy oligomers of either a polyether [poly(ethylene oxide), poly(oxytetramethylene), etc.] or a polyester [poly(butylene adipate), etc.] type whose density of cohesive energy and glass transition temperature are low. These oligomers, whose molar mass is a few thousand $g \cdot mol^{-1}$, play the role of flexible segments and afford the deformability to PUR material. The structure can thus be schematically presented as follows:

PUR

The heterogeneous morphology of these systems, where each phase preserves its specific properties, is an essential condition for the emergence of their predicted properties. Such PUR materials exhibit mechanical properties approaching those of thermoplastic elastomers with an elastic modulus of only 5–10 MPa at 100% strain (the corresponding creep is about 3%) and an elongation at break reaching 400%. On the other hand, if "long" diols are replaced by "short" diols, rigid phases prevail and give a highly cohesive material possessing a high elastic modulus.

Thus, the mechanical characteristics of thermoplastic polyurethanes are versatile, depending on the nature of the diol used to build them. When heated, isocyanate-ended polyurethanes can be cross-linked, with the corresponding reaction being either a trimerization

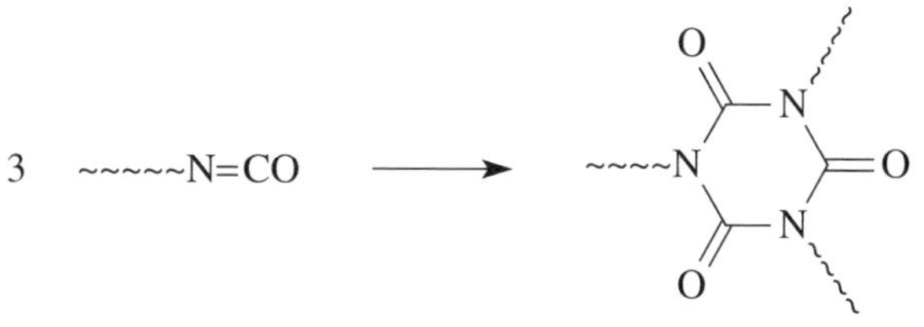

$$3 \quad \sim\sim\sim\sim\sim N{=}CO \quad \longrightarrow$$

or the formation of an allophanate by reaction of a terminal isocyanate with the acidic hydrogen of a urethane functional group:

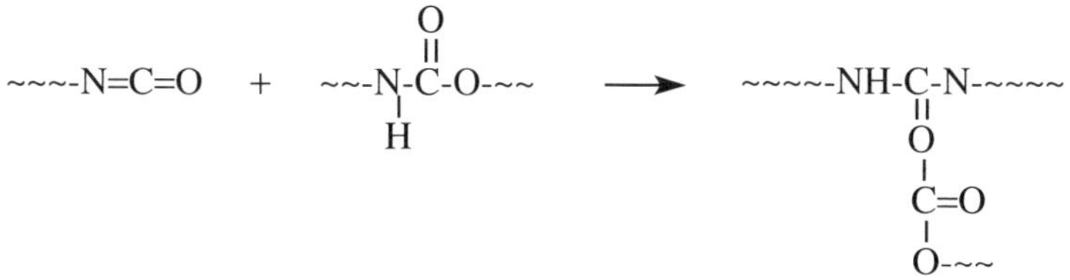

TPU thus exhibit adjustable properties, which explains the variety of their applications: they are produced as profiles, sheets, films, pipes, snap rings, and so on, and are usable for all applications that simultaneously require deformability and high stress at break. Among their specific fields of application, one finds the glass industry (as adhesives) and biomedical engineering.

Their annual volume of production is rather low (approximately 20,000 tons) due to their high cost.

15.5.5.2. "Spandex" Fibers.

It is the name of polyurethane-based elastic fibers; these are materials experiencing a very significant development in the textile industry. These "spandex" fibers have an initial molecular structure close to that of previously described TPU, but their preparation is performed out of stoichiometry in order to limit their molar mass and to produce two-ended chains carrying isocyanate functional groups.

The processing of these fibers is generally carried out by dry spinning from a collodion in either dimethylformamide or dimethylacetamide. Simultaneously, a chain extension proceeds by reaction with a bivalent molecule. The latter can be either a diol or a diamine; in this latter case, a biuret group is formed:

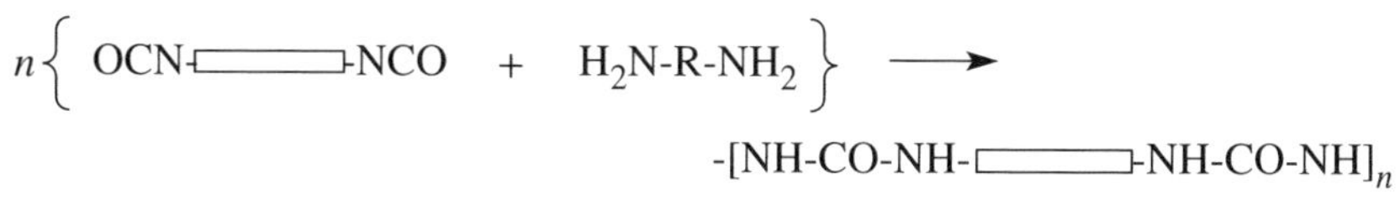

$$n\left\{ OCN{-}NCO \quad + \quad H_2N{-}R{-}NH_2 \right\} \longrightarrow$$

$$-[NH{-}CO{-}NH{-}{-}NH{-}CO{-}NH]_n$$

Spinning can also be performed by either extrusion from the molten state or through a wet process. Sometimes, the chain extension is carried out through a reaction with trivalent reagents in order to form a three-dimensional network inducing a total reversibility of the strain deformations. Along with rubber fibers, "spandex" fibers are the only elastomeric fibers produced at an industrial level in large volume. Like polydienic fibers, they can undergo up to 600% elongation but PUR fibers have a better tenacity, abrasion resistance, tinctorial properties, and, especially, resistance to chemical ageing (resistance to oxidation). "Spandex" fibers are never used alone but always blended with some other fibers: cotton, wool, PET, and so on.

15.5.6. Other Technical and Specialty Linear Condensation Polymers

These polymers correspond to materials whose performances justify their high cost. There is a great number of such polymers exhibiting varied structures and widely used due to their specific properties. Some are technical polymers while others ones are specialty polymers; only the most important ones will be presented here.

15.5.6.1. Polyimides (PI). Only aromatic polyimides are experiencing a significant development. They combine chain stiffness and a high density of cohesive energy. They are obtained by reaction of a dianhydride with a diamine, with the polymerization being performed in two steps according to the following scheme:

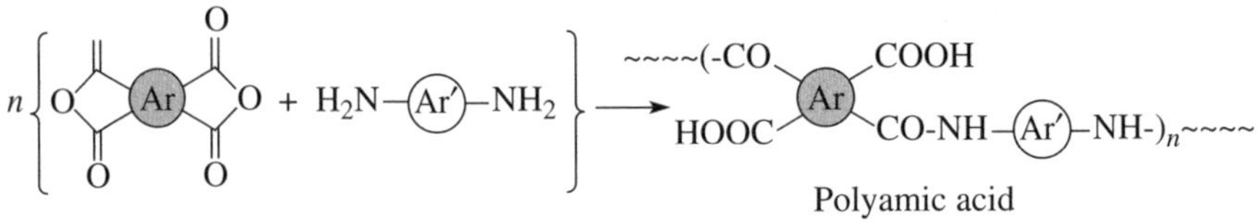

Polyamic acid

then, by heating

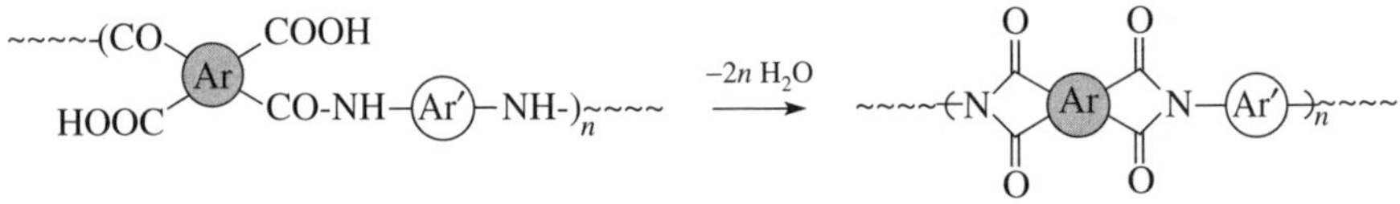

The first step is carried out in solution in a polar solvent (N-methylpyrrolidone, N,N'-dimethylacetamide, dimethylformamide). Then the solvent is removed in order to allow dehydration by heating, initially at the boiling point of the solvent, then at 280–300°C. The reaction is carried out under stoichiometric conditions and the control of the molar masses is accomplished by adding the required quantity of a monovalent reagent (degree of polymerization limiter) such as phthalic anhydride, maleic anhydride, or nadic anhydride. With the last two degree of polymerization limiters, the condensation generates an unsaturation at the chain end which can be possibly used later on to create cross-linking.

Depending on the nature of the dianhydride and the diamine, the resulting polyimide exhibits variable characteristics in addition to common basic properties: thermomechanical and chemical thermostabilities, along with chain stiffness that is responsible for a high viscosity. However, due to the diversity of the dianhydrides and the diamines available, the potential choice is broad, but it is much less at the economic level. Indeed, the processability, the solubility of the resulting polymer, and the cost of the precursors result in seeking compromises at the expense of the thermostability. The main dianhydrides and diamines used are shown in Table 15.6. In stepwise polymerization, their combination leads to polyimides whose glass transition temperature is very high and melting point is beyond the temperature of decomposition. Consequently, their processing is generally carried out from poly(amic acid), and a subsequent heating leads to the formation of polyimide.

Another method is sometimes utilized for the preparation of poly(aromatic etherimide)s. It involves reaction between dinitrobisimide precursor and alkali bisphenolate by aromatic nucleophilic substitution:

$$n\left\{Met^+, {}^-O{-}\boxed{Ar}{-}O^-, Met^+ \ + \ O_2N{-}\boxed{Ar'}\ \overset{O}{\underset{O}{\diagup\!\!\!\diagdown}}N{-}R{-}N\overset{O}{\underset{O}{\diagup\!\!\!\diagdown}}{-}NO_2\right\}$$

$$\longrightarrow \quad 2n\ NO_2{}^-, Met^+ \quad + \quad \left(\!\!{-}O{-}\boxed{Ar}{-}O{-}\boxed{Ar'}\ \overset{O}{\underset{O}{\diagup\!\!\!\diagdown}}N{-}R{-}N\overset{O}{\underset{O}{\diagup\!\!\!\diagdown}}\boxed{Ar'}\!\!{-}\right)_n$$

To clarify the views on the subject, some data related to several essential characteristics of polyimides resulting from monomers of Table 15.6 are given in Table 15.7. These values account for the high-level performances of these materials. Their excellent mechanical characteristics are preserved over a very wide range of temperatures due to the strong interactions developed by the carboxy groups of a chain with the nitrogen atoms of another chain. Moreover, polyimides are characterized by a strong resistance to degradation by thermo-oxidation; their melting temperature is such that they can be used up to 300°C or more during hundreds of hours. They are also insensitive to hydrolysis and ionizing radiations. Finally, they are insoluble in most solvents unless their structure is selected to favor their solubility in highly polar solvents (p-chlorophenol, dimethyformamide, etc.) to the detriment of their thermomechanical characteristics.

Several polyimides contain other chemical groups (fluoride, ether, sulfone, etc.) that correspond to the combination of the inherent properties of each of these groups to generate a particular material. Polyimides are widely used as films obtained from poly(amic acid) solutions by evaporation of the solvent and dehydration by heating. In particular, these films are utilized for the realization of membranes that are supposed to work at high temperature (reversal osmosis and fuel cells) as well as for the coating of electronic systems. PI are also components of composite materials and adhesives. The manufacture of objects by sintering, possibly followed by a machining, is another field of application of these high

Table 15.6. Dianhydrides, aromatic diamines, and dinitrobisamide used for the preparation of polyimides

Ref.	Comonomer	Molecular Formula
	Dianhydrides	
1	Pyromellitic anhydride	
2	Dianhydride of 2,3,6,7-naphthalene tetracarboxylic acid	
3	Phthalic dianhydride of bisphenol A	
4	Benzophenone dianhydride	
	Diamines	
5	*m*-Phenylene diamine	
6	4,4′-Diaminobiphenyl	

Table 15.6. (continued)

Ref.	Comonomer	Molecular Formula
8	*p,p'*-bis(*m*-Amino-phenoxy)biphenyl	
9	3,3'-Diamino-benzophenone	
	Dinitrobisimide	
10	1,3-bis(4-Nitro-phthalimido)-benzene	
	Monoanhydrides (for degree of polymenzation Control)	
11	Phthalic anhydride	
12	Nadic anhydride	

performance materials. Finally, certain molecular structures of PI correspond to a softening temperature compatible with processing by conventional techniques of molding or extrusion.

The annual world production of PI is a few thousands of tons; it is a relatively low volume compared to some other technical polymers, but, due to their high cost, PI are intended for very high value added applications (depending on their structure,

Table 15.7. Thermal and mechanical characteristics of some polyimides (formulas of the polymers concerned can be found in Table 15.6)

Structure	$T_g(°C)$	$T_m(°C)$	$\sigma_{r(MPa)}$
1 + 7	250	388	93
4 + 8	246	350	136
3 + 5	215	(amorphous)	105

their cost can vary hundredfold) and are thus important from the economic point of view. Their growth rate is high and reaches up to 20% some years.

15.5.6.2. Aromatic Polyethers [or poly(oxyphenylene)s] and Their Analogs.
It is a family corresponding to various molecular structures. All polymers of this family have a high thermal and chemical stability as well as excellent mechanical characteristics. Besides that, they have a much better processability than that of pure polyimides due to the deformability of the chemical bonds which lowers their viscosity in the molten state. Certain polyimides take advantage of this property, and the structure,

which corresponds to an alternation of imide and ether groups, permits a satisfactory compromise between thermomechanical characteristics and processability.

Only those polymers containing ether functional group are considered as **aromatic polyethers**. The most important is the polymer obtained by the oxidizing polycondensation of 2,6-dimethylphenol

with the reaction being catalyzed by copper salts or amines. The presence of two methyl groups on the phenylene moiety considerably increases the rigidity of the chain, since for 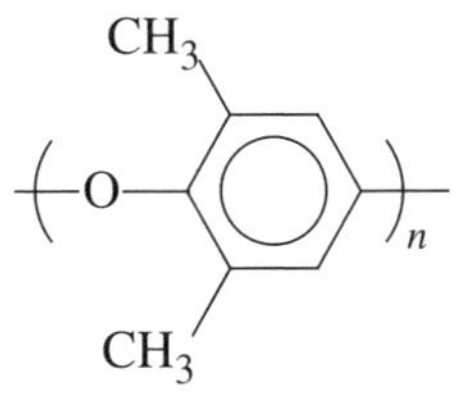 $T_g = 211°C$

whereas for 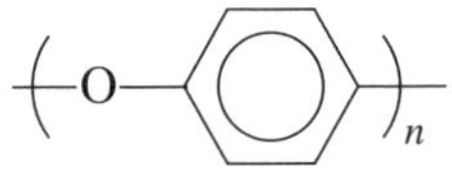 $T_g = 82°C$ only

This high rigidity of the chains makes the processing of this material very difficult in a neat state. However, its total miscibility with polystyrene confers it an unexpected property, which allows the preparation of blends with better processability. They are indicated by the term "modified PPO."

Aromatic polyether can also consist of a copolymer obtained from the simultaneous polymerization of di- and trimethylphenol:

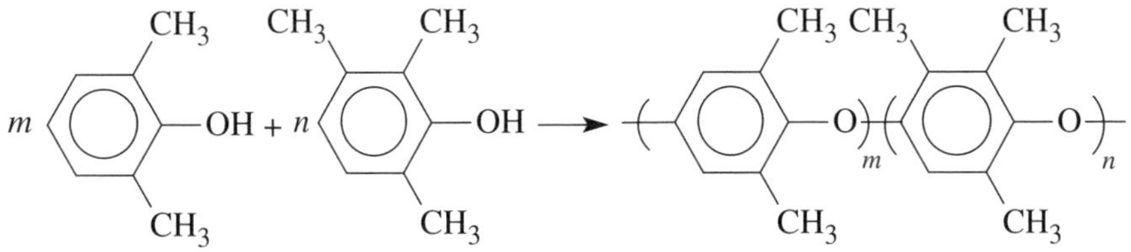

The most important commercial polymer is a blend of poly[oxy-(2,6-dimethylphenylene)] with high-impact polystyrene (HIPS). The corresponding material exhibits variable characteristics, depending on the PS content. It is a widely used technical polymer in mechanical engineering. Indeed, it exhibits a good impact resistance at very low temperatures in addition to its good thermal and mechanical properties. Its marked electrical insulating character even in wet atmospheres finds applications in electric and electronic industries.

Polyarylethersulfones combine the flexibility of ethers and the strong cohesion of sulfone groups. They can be prepared using two methods. The first one utilizes polycondensation in solution between a sodium diphenoxide and a dichlorinated derivative, for example:

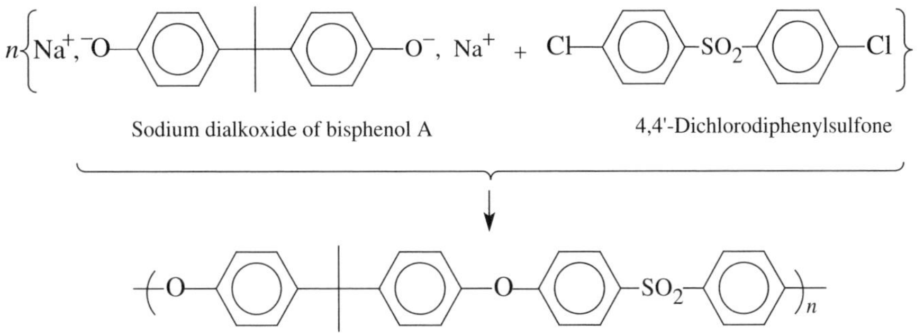

Sodium dialkoxide of bisphenol A   4,4'-Dichlorodiphenylsulfone

The second one is a Friedel–Crafts reaction between a sulfonyl dichloride and an aromatic ether catalyzed by small amounts of a Lewis acid ($FeCl_3$, $SbCl_5$, etc.):

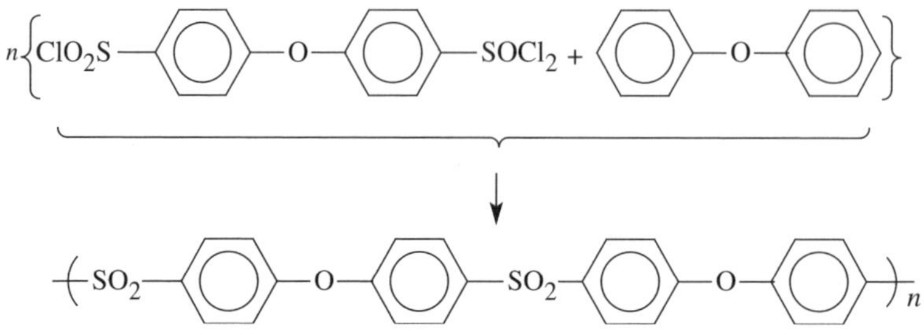

Depending on the nature of initial diphenols, polymers are obtained whose properties are variable but nevertheless exhibit common characteristics:

T_g: 185–220°C;

σ_r: 70–85 MPa;

Young's modulus: ~2.5 GPa;

Elongation at break: 40–75%

Other properties that have found practical applications are: impact strength resistance, fatigue resistance, high resistivity, and low permittivity as well as good resistance to hydrolysis. Due to the said properties, these materials are utilized for the manufacture of electric and electronic components. Finally, their combustibility is low, which provides a particularly high level of safety in use.

Polyaryletherketones are also part of the family of aromatic polyethers. They can be obtained either by nucleophilic substitution of monomers that carry the two antagonistic functional groups (the process is schematically represented as follows)

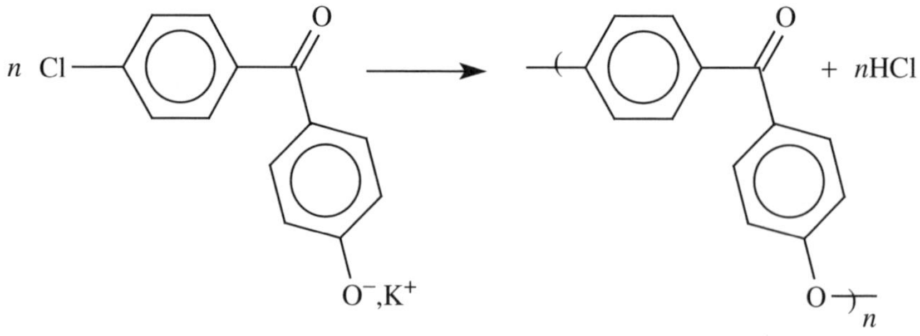

Poly(1,4-oxyphenylenecarbonyl-1,4-phenylene)

or by dehydrofluorination:

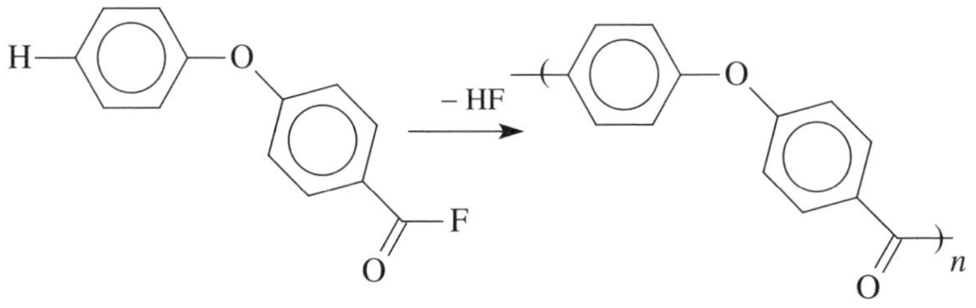

The most important polyetherketone is shown below; it contains two ether functional groups per monomeric unit, thus explaining its name: polyetheretherketone (PEEK):

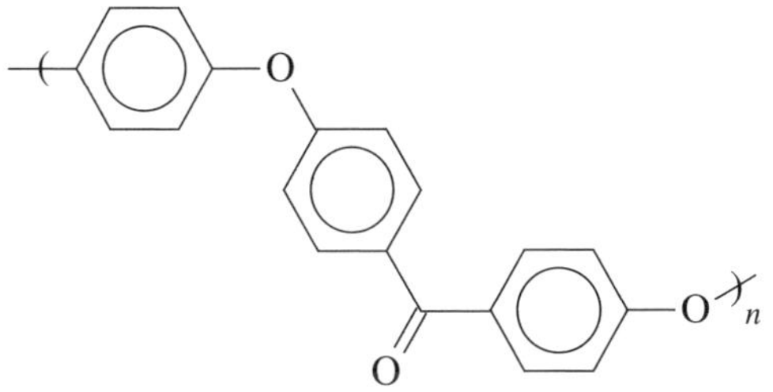

This polymer has a $T_g = 135°C$ and a melting temperature $T_m = 335°C$. Its stress at break reaches up to 90 MPa, and its Young's modulus is equal to 4 GPa. This material exhibits an excellent thermostability and is mainly used after filling with either glass or carbon fibers.

Aromatic polysulfides are analogs close to polyethers, and poly(phenylene sulfide) (PPS) is an important technical polymer. It is generally prepared by nucleophilic coupling by means of sodium sulfide

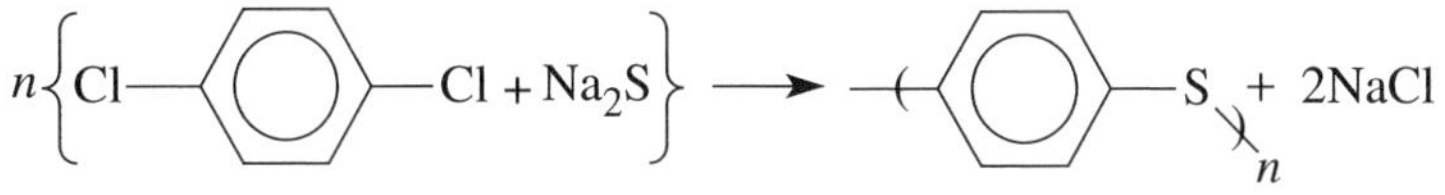

but many other methods can be used to obtain it, for example:

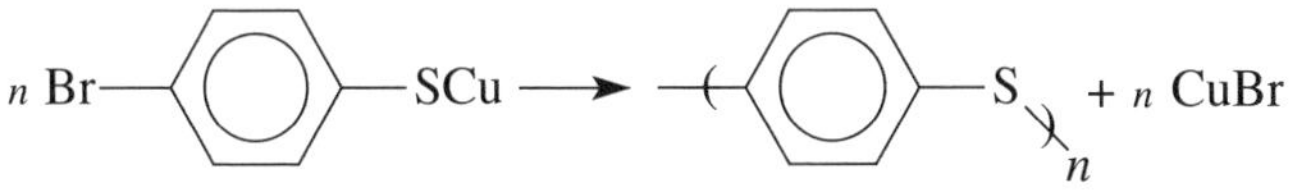

The very high thermal stability of this material is due to the high energy of the constituting bonds as well as to the low mobility of its chains. It is a material that can be used uninterrupted at $150°C$ when it is reinforced by carbon fibers or glass fibers. Its glass transition temperature is $85°C$, and it crystallizes slowly beyond this temperature ($T_m = 285°C$).

Its physical and mechanical characteristics justify its high cost. Its annual world production is about 30,000 tons.

LITERATURE

J. Brandrup, E. H., Immergut, and E. A., Grulke, *Polymer Handbook*, 4th edition. Wiley, New York, 1999.

J. A., Brydson, *Plastic Materials*, Butterworths-Heinemann, Oxford, UK, 1999.

O. Olabisi (Ed.), *Handbook of Thermoplastics*, Marcel Dekker, New York, 1997.

A. K., Bhowick and H. L., Stephens (Ed.), *Handbook of Elastomers*, Marcel Dekker, New York, 1988.

M. Lewin and E. Pearce (Ed.), *Handbook of Fiber Chemistry*, Marcel Dekker, New York, 1998.

S. Fakirov (Ed.), *Handbook of Thermoplastic Polyesters*, Wiley VCH, Weinheim, 2002.

Kirk-Othmer (Ed.), *Encyclopedia of Chemical Technology*, 4th edition, Wiley, New York, 1996.

H. F., Mark and N. M., Bikales, C. G. Overberger, and G. Menges. (Eds.), *Encyclopedia of Polymer Science and Technology*, 2nd edition, Wiley, New York, 1989.

16

THREE-DIMENSIONAL SYNTHETIC POLYMERS

Whatever the type of polymerization—chain or stepwise process—only the polymerization of monomers or prepolymers exhibiting an average valence ($\overline{v} = \sum_i N_i v_i$) higher than 2 (see Section 7.3) are considered in this chapter.

"Vulcanized" elastomers, in which the three-dimensional structure is obtained by reaction of functional groups carried by the monomeric units of linear chains, are also excluded from this category of polymers.

16.1. SATURATED POLYESTERS (ALKYD RESINS)

The name "alkyd resins" clearly distinguishes these polymers from thermoplastic polyesters (PET, PBT, PC, etc.) as well as from unsaturated polyesters (UP), which are also part of the family of three-dimensional polymers (see Section 16.2).

Three-dimensional saturated polyesters are condensation polymers obtained from the reaction of polyols with carboxylic polyacids, their anhydride, or their esters.

There is a wide variety of possible structures based on the use of glycerol ($v = 3$), trimethylolpropane ($v = 3$), pentaerythritol ($v = 4$), and sorbitol ($v = 6$) or their mixture as polyol, reacting not only with phthalic anhydride ($v = 2$) but also with pyromellitic dianhydride ($v = 4$), citric acid ($v = 4$), trimellitic acid ($v = 3$), or their mixture as polyacid. The most important resins are those obtained from the polycondensation of o-phthalic anhydride with glycerol, and the resulting network is represented hereafter.

Organic and Physical Chemistry of Polymers, by Yves Gnanou and Michel Fontanille
Copyright © 2008 John Wiley & Sons, Inc.

This polymer may potentially have a high density of cross-links since the average valence of the prepolymeric stoichiometric mixture glycerol/phthalic anhydride is equal to 2.4. However, due to the inaccessibility of a fraction of reactive sites, it is extremely difficult to reach such an ideal structure, and the true cross-link density of the network thus depends on the stoichiometric balance and on the reaction conditions.

The preparation of prepolymeric mixtures using mono- or bivalent monomers as well as dihydroxylic oligomers permits to adapt of the structure of the network to the required properties.

In addition, intramolecular reactions can reduce the average valence of the pre-network and afford monovalent species; for example, glycerol can react with phthalic anhydride to give

which then reacts as a mono-alcohol.

In a first step, these alkyd resins are prepared to a limited extent of conversion in order to afford a viscous liquid which can be molded by casting in a selected mold. Under these conditions, primary hydroxyl groups of glycerol react preferentially, which allows to consume the comonomers without crossing the gel point; in other words, glycerol proceeds initially as a divalent monomer. In a second step, the alkyd resin is hardened by heating in the mold at about 250°C and in the presence of esterification catalysts, taking advantage of the trivalence of glycerol.

The resulting materials exhibit an excellent resistance to solvents. Their mechanical properties, in particular, their impact strength can be improved by incorporation of various plasticizers.

However, the most important application of alkyd resins is not in the field of structural materials but in the field of paints and varnishes. For this purpose, it is advisable to formulate the initial network precursor in such a manner that it is completely soluble in organic solvents and natural or synthetic oils. This is obtained by replacing a fraction of phthalic anhydride with long chain fatty acids through triglycerides. When the oils used as solvent are not siccative (castor oil, for example, is not siccative), the coating should be hardened in a furnace at 250°C as in the case of conventional alkyd resin. However, quite often, these oils are siccative (linseed oil, soy oil, etc.) which means that their unsaturations can react by air oxidation to form a cross-linked network. The simultaneous formation of the two networks affords an interpenetrated system whose properties are well-suited to applications such as paints and varnishes (glycero-phthalic paints).

16.2. UNSATURATED POLYESTERS (UP)

Strictly speaking, at the end of the process, the so-called unsaturated polyesters are not polyesters since they contain a high proportion of polystyrene. They are obtained by radical polymerization of styrene in the presence of low molar mass unsaturated polyester capable of undergoing radical reactions through their C=C double bonds. These unsaturated polyesters, which are precursors of the final polyester network, are obtained by copolycondensation of diols with various anhydrides such as maleic anhydride and ortho- and/or isophthalic anhydrides. The polymerization is performed by simple heating in the presence of usual esterification catalysts. A typical molecular structure is represented below, which corresponds to a stepwise "quaterpolymerization" of maleic anhydride, o-phthalic anhydride, ethylene glycol, and diethylene glycol:

The use of phthalic anhydride as comonomer for the preparation of polyester prepolymer lowers its unsaturation content and reduces the cross-link density of the resulting network. In addition, incorporation of phthalic moieties in the network improves its behaviour at high temperatures.

UP prepolymers possess a molar mass $(\overline{M}_n)$ ranging between 1800 and 2500 g·mol^{-1} and about 5 to 8 double bonds per chain. In order to form a network, they are dissolved in neat styrene whose radical polymerization is most often initiated by a peroxide. Then there is a "copolymerization" between the bivalent styrene ($v = 2$) and the plurivalent unsaturated polyester acting as a comonomer. The average valence of the latter is twice the average number of unsaturations per chain. The reaction mechanism of the process is described below:

The structure and thus the mechanical characteristics of these materials can be fine-tuned through two ways:

- by variation of the nature and/or the proportion of the various comonomers entering in the composition of the polyester precursor and/or
- by a possible partial or total substitution of another vinyl monomer (MMA, etc.) for styrene.

Due to the brittleness of these materials, which is inherent to the high polystyrene content, UP are almost exclusively used as composite matrices; they are generally compounded with glass fibers as fillers. Their economic significance is indisputable; this is mainly due to their easy processing, in particular from pre-impregnated fabrics, which can be rigidified by mixing with mineral powders having high specific area. These pre-impregnated fabrics are stored at low temperature in the presence of polymerization inhibitors because they contain a free radical generator dissolved in the prepolymer. They are utilized in all fields of mechanical engineering. Their annual world production crosses to 2 million tons.

"Vinyl esters" are commonly classified among unsaturated polyester precursors even if their structure is different from that of UP. This classification is due to the similarities of the processing techniques as well as their field of application.

These "vinyl esters" are di(meth)acrylic prepolymers obtained by the reaction of (meth)acrylic acid with diglycidic ether of bisphenol A (DGEBA) (precursor of epoxy resins) (see Section 16.6). The reaction can be catalyzed by amino or phosphonated bases

and the resulting structure of the prepolymer is schematized below:

with $0 < n > 2$–3

These dimethacrylate epoxy resins are tetravalent. They can be copolymerized with styrene or MMA to generate networks whose mesh size is more homogeneous than that of networks obtained from common UPs. Moreover, their excellent mechanical properties, that are inherent to the incorporation of bisphenol A moieties in the chains, give use to materials sought for high-added-value applications. However, their cost restricts their application.

16.3. PHENOPLASTS (PHENOL-FORMALDEHYDE POLYMERS, PF RESINS)

They correspond to the first industrial polymer synthesized since the process was first developed by Baekeland in 1907, and their production started in 1910. These materials are also called "formo-phenolic resins" but it is preferable to retain the term "resin" for the precursor of the network. These prepolymeric resins are obtained by polycondensation of formol ($v = 2$) with phenol ($v = 3$) or some of its analogs.

The use of substituted phenols allows to reduce the average valence of the precursor and thus the cross-link density of the final material. These structural modifications are mirrored in varied mechanical characteristics. Among substituted phenols used, one finds

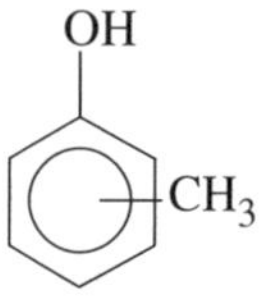

Cresols (o–, m–, p–)   R p-Alkylphenols

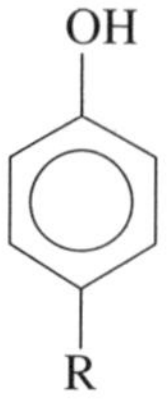

p-Phenylphenol

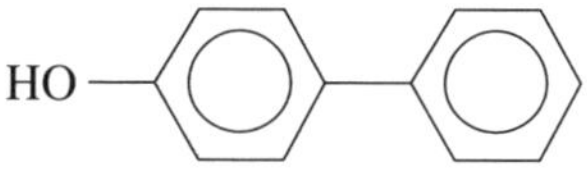

Bisphenol A

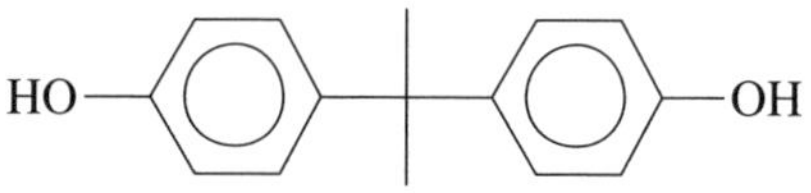

Resorcinol

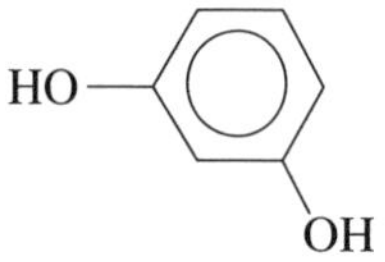

The production of formo-phenolic materials is always carried out in two steps. The first one corresponds to the formation of an oligomer (resin) that is used as prepolymer, and its molecular structure depends on the pH of the reaction medium. The mechanism of these reactions is described in Section 7.4.2.

Novolacs are resins obtained in acidic medium with a [phenol]/[formaldehyde] ratio > 1 (generally ranging between 1.20 and 1.30). They consist of linear or moderately branched condensation products, with the phenolic moieties being connected through methylene bridges. Due to a low formol content, their molar mass is low (about 2000 g·mol^{-1}). They are soluble in polar solvents and can be fluidified upon heating. The following scheme represents a possible structure of an oligomer present in a novolac resin:

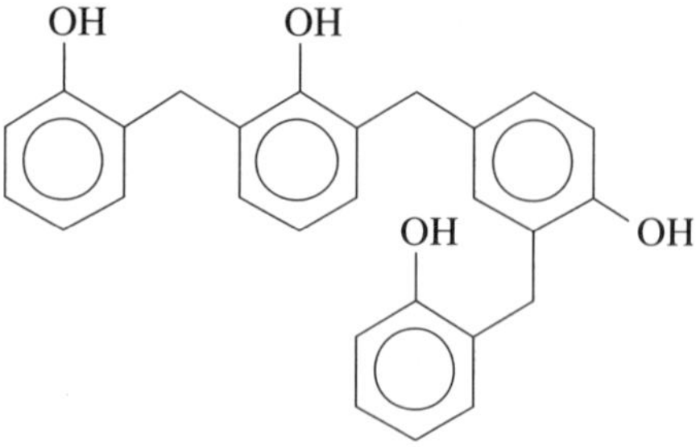

To cause the cross-linking of such a prepolymer, it is necessary to attain the average valence $\bar{v} = 2.40$ corresponding to stoichiometry. To this end, a compound is added to the reaction medium in order to generate formaldehyde by hydrolysis. Generally, hexamethylenetetramine (HMTA) is used for this purpose in the proportion ranging from 5% to 15%, depending on the desired properties (see Section 7.4.2). HMTA is a solid which can be easily dispersed in the reaction medium.

The reaction leading to cross-linking is carried out at about $150-170°C$. It affords materials whose cross-link density is high and which, moreover, exhibit a high density of cohesive energy. The latter is due to the presence of hydrogen bonds in addition to covalent bonding. The theoretical structure of a phenoplast cross-linked up to the maximum of its capability is schematized below:

In addition to their cohesive energy, these networks also exhibit excellent properties such as durability and resistance to burning, good chemical inertia, excellent behavior against moisture, and a marked electric insulating character. Their high cross-link density generates a certain brittleness against impact; consequently, novolac resins are seldom used alone as structural material. Hydroxyls of phenol groups develop strong interactions with a wide variety of fillers (filaments, powders, granules, etc.). They are widely used for the manufacture of electric hardware, in mechanical engineering (transport, household electric appliances, etc.) and in packaging. They are processed by compression and injection molding. Novolacs are widely used as rigid adhesives, for their ability to develop strong interactions with many substrates.

Resols are prepolymers obtained in basic medium in the presence of an excess of formaldehyde. They consist primarily of methylolphenols as well as di-, tri-, and tetraphenolic compounds resulting from the polycondensation of phenol with formaldehyde at its early stage:

etc.

These resol compounds can exhibit different physicochemical characteristics, depending on their structure: they can be solid or liquid, water-soluble or not. The structure of the final material or the kinetics of its formation can be finely tuned. The cross-linking of resols occurs upon heating (without additive) at ~150°C, more often after their neutralization. During molding, the heated matter is first elastomeric and is then called "resitol." Then it hardens quickly to afford "resite." The reactions occurring during this cross-linking are particularly complex, and the precise structure of the resulting materials is ill-defined.

Resols can also be cross-linked by means of either hexamethylenetetramine (as for novolacs) or dimethylamine. The catalytic process of their polycondensation is generally carried out using ammonia. In the processing of resols advantage is generally taken of their solubility in light solvents (water, methanol, etc.). Solutions are used for the coating of surfaces, fabrics (composite materials reinforced by carbon or glass fibers), or wood derivatives (lignocellulosic flours, sawdust, and other particles) to give reconstituted wood. The latter application requires large quantities of phenolic resins. The resols can also be processed in the molten state. Phenoplasts find a wide range of applications: building, transport, electric household appliances, adhesives, paper industry, and so on.

Their world production is about four million tons, which equates to a higher quantity of the materials derived.

16.4. AMINOPLASTS (AMINO RESINS)

There are many analogies between phenoplasts and aminoplasts: these two families use formaldehyde as comonomer, they are prepared from water-soluble precursors, they use the same techniques of processing, and they lead to highly cross-linked polymeric materials that find applications in similar fields. The main difference between phenoplasts and aminoplasts concerns the color of the corresponding materials: whereas phenoplasts are rather intensely coloured to dark brown, aminoplasts can be obtained in colorless form, which permits their use as coating materials, an application seldom accessible to phenoplasts.

The essential quality of aminoplasts is their hardness, which is exceptionally high for a polymer but is detrimental to their impact resistance. They are not very sensitive to hydrolysis and to most of chemical reagents. The differences between the materials of this family are seen to secondary properties. In addition to formaldehyde, aminoplasts are obtained from a second comonomer carrying amino groups. Two molecular compounds are used for this purpose, which correspond to two types of aminoplastic precursors existing on the market:

- Urea-formaldehyde resins make use of urea [$(NH_2)_2C{=}O$], a difunctional tetravalent reagent antagonist to formaldehyde;
- Melamine-formaldehyde resins make use of melamine, a trifunctional hexavalent molecule, to afford highly cross-linked networks:

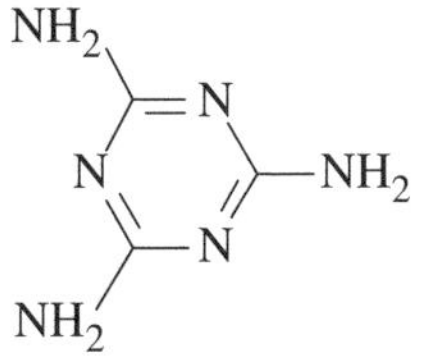

Other amino comonomers can be used, but their proportions in the materials are low compared to those of the two preceding molecules. The reaction mechanisms corresponding to the formation of the precursors and their polymerization were presented in Chapter 7. The precursors are generally prepared in batches from the comonomers in aqueous solution. The structure and the properties of the final material are determined by the composition of the reaction medium, its pH, and the time and temperature of reaction. The first step of the process is the formation of methylol derivatives, and the second step involves their condensation with elimination of formaldehyde or water. An acidic medium does not permit an easy control of the formation of methylol compounds, and polycondensations are generally carried out in slightly basic medium (pH 7.5 to 9.0).

16.4.1. Urea–Formaldehyde Resins (UF)

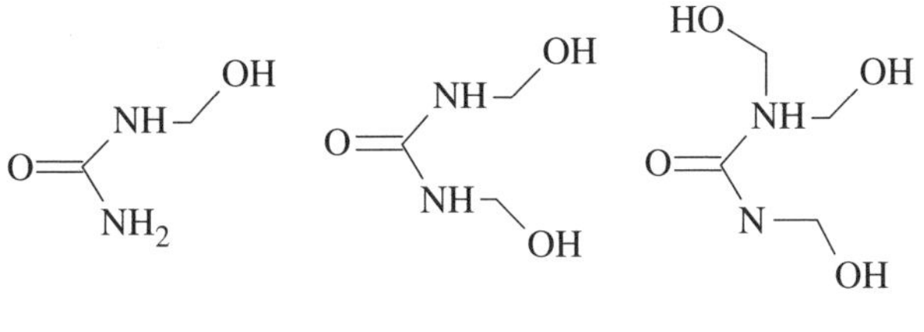

They account for approximately 80% of the production of aminoplastic polymers, their annual production being estimated to 1.2 million tons. The precursor consists of mono-, di-, and trimethylolurea, that are water-soluble and are obtained at pH $= 8-9$ with a molar ratio of formaldehyde/urea equal to ~ 6 and at a temperature of about $50°C$. The purpose of a basic medium is to catalyze the reaction of formation of methylolureas without allowing their condensation. For the polymerization to occur, it is advisable to acidify the reaction medium (pH ~ 5), and, even at ambient temperature, some condensation is inevitable. In addition to the formation of methylolureas, their partial condensation leads to compounds of higher molar mass that are water-insoluble, even those with a degree of polymerization of a few units. Those which remain water-soluble (or methanol soluble since this solvent is often used as reaction medium) increase the viscosity of the solution and give syrup usable for the second stage. This (aqueous or methanolic) syrup is used as coating for fabrics, papers, and various fillers. After partial evaporation of the solvent, the filled prepolymer is heated to give further condensations under pressure and at high temperature.

The following scheme gives an idea of the possible molecular structure of an UF resin containing methylol groups that are unable to move but can strongly interact with the fillers. It is important to note the presence of ether-oxide functional groups obtained from simple condensation:

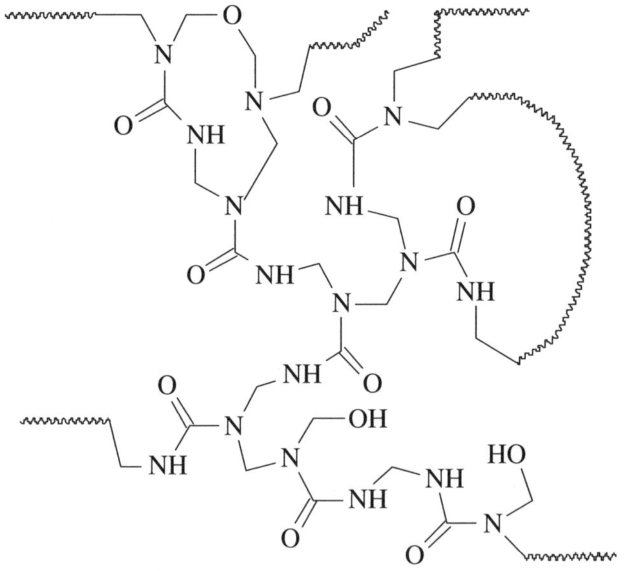

The major application of UF resins, which nearly correspond to 70% of the utilization of these materials, is in the field of adhesives and, more particularly, in the industry of wood and its derivatives (particle boards, plywood...). Compared to the other adhesives used in the same field, UF resins exhibit a moisture sensitivity that restricts their use as materials for interior applications owing to a tendency to slowly release formaldehyde.

16.4.2. Melamine–Formaldehyde Resins (MF)

The reaction mechanisms occurring with systems based on melamine (2,4,6-triamino-1,3,3,5-triazine), a monomer that is synthesized from urea, are not different from those affording UF resins. It is important to mention a higher reactivity of MF resins compared to that of UF resins and the real hexavalence of melamine, which gives rise to the hexamethylol derivative shown below and other methylols by reaction with formaldehyde.

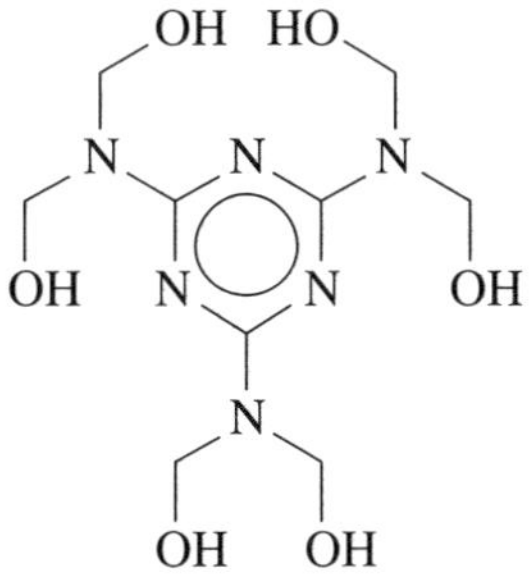

This hexavalence of the prepolymer leads to a network whose cross-link density can be higher than that of UF resins. This network is schematized below. It is obtained by heating the prepolymer syrup under neutral or slightly acidic conditions:

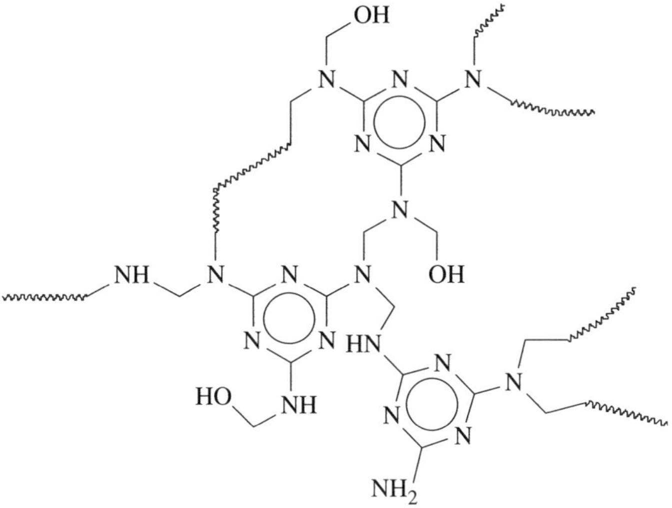

UF precursors can be mixed with MF precursors to give copolymer networks. Due to the difference in reactivity, melamine methylols are consumed first, thus inducing a certain heterogeneity in composition of the resulting network. It is even possible to mix MF resins with phenoplast resins; however, there is no evidence for the existence of covalent bonding between the two networks. The fields of application of MF resins only slightly differ from those of UF ones: adhesives for lignocelluloses derivatives, laminates, varnishes, coating of textile fibers and papers, and ground reinforcement. The annual world production of melamine formol resins is about 250,000 tons.

16.5. POLYURETHANES (PUR)

The name polyurethane is often abusively used because urethane functional groups generally represent only a tiny fraction of the said polymers. Consequently, the properties of the final material depend more on the nature of the prepolymer than on the urethane groups ensuring the links between the chain segments. Indeed, PUR resins are obtained from reactive oligomers, with the generation of the urethane (also called carbamate) functional group serving only for chain extension of diol or diisocyanate prepolymers; the presence of prepolymers of higher valence causes cross-linking.

The reaction mechanism occurring in the formation of polyurethanes was presented in Chapter 7, and the preparation of prepolymers was presented for linear PUR (Chapter 15). The structure of the prepolymers used to prepare three-dimensional PUR is simpler than that of linear ones; indeed, with the former the cohesion of the material is mainly due to the presence of cross-links.

Most three-dimensional polyurethanes are used in the foam industry. Depending on the valence of the precursors, they are flexible or rigid. In general, these foams are obtained not only from polyethers [poly(propylene oxide), poly(ethylene oxide), polytetrahydrofuran] but also from dihydroxylic aliphatic polyesters [poly(ethylene adipate- or succinate), etc.]. Rigid foams are obtained from polyols of higher valence (up to 8); the latter result from the copolymerization of diisocyanates with either macrodiols or simple polyols (glycerol, pentaerythritol, etc.). The rigidity of the resulting foam is thus related to its high cross-link density.

Current diisocyanates used are either aromatic or aliphatic. Toluenediisocyanate (TDI)

is not exactly divalent because it contains a small fraction of triisocyanate formed by substitution of the second ortho position; it is the most used isocyanate.

Highly flexible foams generally use aliphatic diisocyanates for chain extension, for instance, hexamethylene diisocyanate (HMDI),

$$O=C=N-(CH_2)_6-N=C=O$$

whereas, on the contrary, an additional rigidity is conferred to rigid foams by incorporation of aromatic diisocyanate such as diphenylmethane-4,4-diisocyanate (MDI).

$$OCN - \langle\bigcirc\rangle - CH_2 - \langle\bigcirc\rangle - NCO$$

The preparation of foams implies the presence of porogens. Two methods are used for this purpose. The first one involves the *in situ* generation of carbon dioxide by a controlled reaction of water with an excess of diisocyanate in an initial phase. It is a useful solution but expensive due to the cost of isocyanates; it is, however, the method chosen for the production of the flexible foams. It is sometimes also used for rigid foams

$$R-N=C=O \; + \; H_2O \; \longrightarrow \; R\text{-}NH_2 \; + \; CO_2 \nearrow$$

The second method uses a light solvent (trichlorofluoromethane, CH_2Cl_2, etc.), which is vaporized by heating, simultaneously with the exothermic cross-linking (analogy with the formation of expanded PS). PUR foams are processed from parallelepipeds of variable dimensions. These parallelepipeds can either be molded between two plates controlling the volume expansion, or sliced from an expanded volume. Apparent densities of these foams are low:

from 0.016 to 0.030 for free expansion,
from 0.025 to 0.040 for sliced plates.

Applications of PUR foams vary from sound and thermal insulation, to mattress, furnishing, packaging, and so on.

The world production of cross-linked polyurethanes is approximately 9 million tons, taking into account the low density of foams that represents an important volume of material!

16.6. POLYEPOXIDES OR EPOXY RESINS (EP)

The term "polyepoxide" should be preferred to that of "epoxy resin" which is generally used, because there is an ambiguity in the use of the latter term. Indeed, it refers at the same time to the epoxy precursor and the final material itself obtained from the reaction of the former with a hardener. In this textbook, the term "epoxy resin" will be reserved for the precursor whose reaction with a hardener gives rise to a three-dimensional polymeric material:

$$\text{Epoxy resin} + \text{Hardener} \longrightarrow \text{Polyepoxide}$$

Remark. The term "polyepoxide" should preferentially be reserved for polymers obtained from the polymerization of oxiranes [poly(ethylene oxide), poly(propylene oxide), etc.]. These polymers are viewed as aliphatic polyethers, thus limiting the possibility of confusion.

As in the case of the polymers prepared from reactive oligomers (precursors), the properties of polyepoxides are closely related to the nature of the monomer used for the preparation of this precursor. In the case of epoxy resins, the potential monomers are numerous but, in reality, one of them corresponds to more than 80% of the market; it is bisphenol A [or (4,4'-diphenylol)-2-propane], whose reaction with epichlorhydrin was described in Section 7.4.4. This reaction leads to a precursor that is bivalent under the usual conditions of use. The cross-linking of the system imposes the use of a hardener of valence $v \geq 3$, and the network formation can then be schematized as follows (here a tetravalent hardener reacts through its reactive hydrogen atoms):

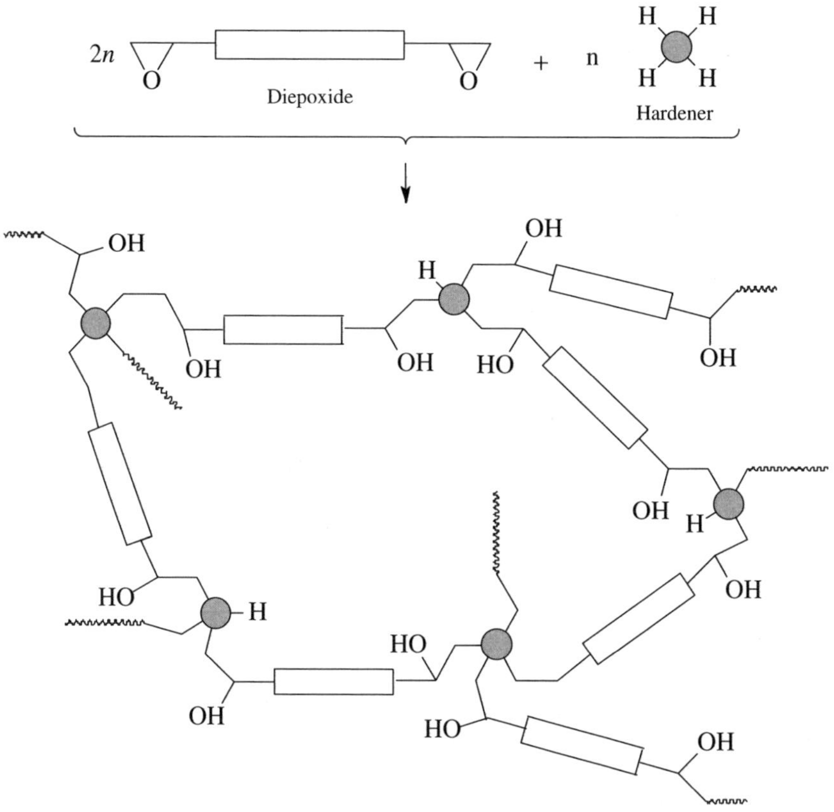

The above scheme shows the large number of hydroxyl groups generated by the reaction. They play an important role with respect to adhesive properties; it also shows that, even under stoichiometric conditions, residual reactive hydrogen atoms may remain unreacted. The variety of the hardeners is much larger than that of the epoxy resins. Moreover, even if they affect only secondarily the properties of the final material, they are not less important because they determine not only certain characteristics but also the conditions of cross-linking.

16.6.1. Epoxy Resins (Precursors)

By far the most important diepoxide precursor is that obtained from the reaction of epichlorhydrin with bisphenol A, called "diglycidyl ether of bisphenol A" (DGEBA); the stoichiometric ratio of the reactants determines the value of x:

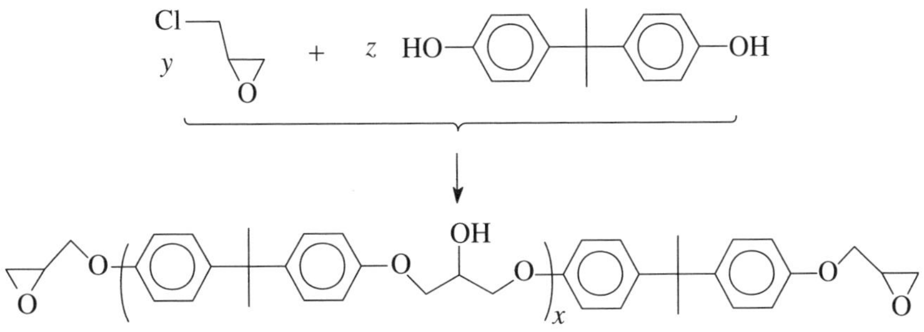

The first term of the series, which corresponds to $x = 0$, is widely used in all formulations. It can crystallize, and its melting point is $41°C$. However, because it is seldom pure since mixed with a small proportion of the derivatives corresponding to $x = 1, 2 \ldots$, it is in supercooled state at room temperature; its crystallization occurs only at temperatures lower than $0°C$. In addition, this precursor brings excellent mechanical characteristics, adhesiveness, thermostability, and good resistance to chemicals and hydrolysis.

DGEBF (obtained from bisphenol F) has a viscosity definitely lower than that of DGEBA but exhibits general characteristics that are close to those of DGEBA.

Remark. The names "bisphenol A" and "bisphenol F" are given, depending on whether these molecules are synthesized according to the following scheme—that is, from acetone

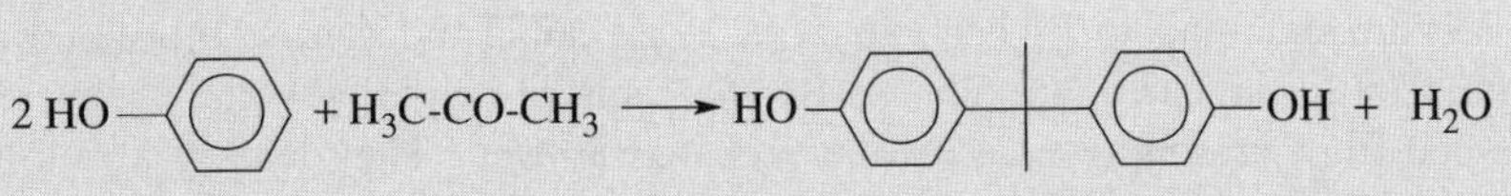

Bisphenol A

or from formaldehyde, respectively:

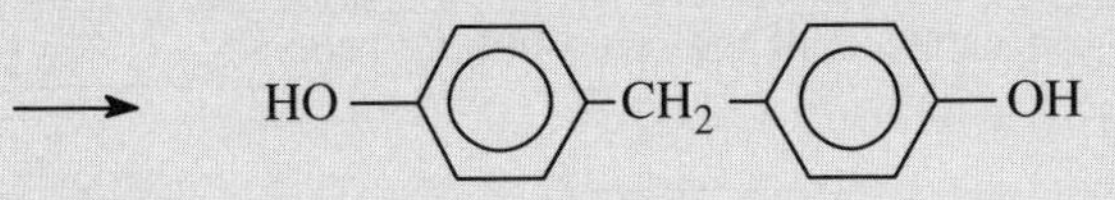

Bisphenol F

To obtain precursor systems with low viscosity, structural elements inducing chain stiffness are to be suppressed; for example, use can be made of diglycidyl ether or other epoxides, such as phenylglycidyl ether (PGE), which is a monovalent model of bisphenol A

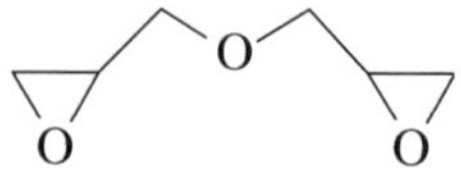

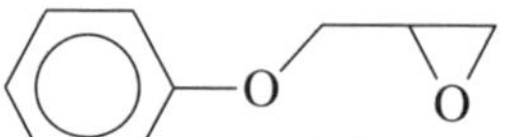

PGE

or butyl glycidyl ether

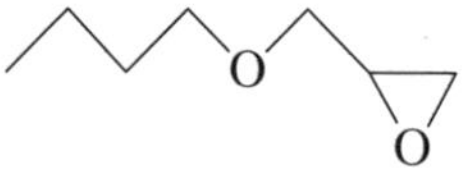

In addition, the two latter epoxides lower the cross-link density of cured polyepoxydes.

On the contrary, certain formulations including polyfunctional resins tend to increase the viscosity of the prepolymeric mixture. This is true for epoxy resins prepared from prepolymeric novolac formo-phenolic resins (oligomerized in acidic medium) (see Section 16.3). These novolac resins are polyphenols that can react with epichlorhydrin to give polyepoxides. Their structure can be schematized by

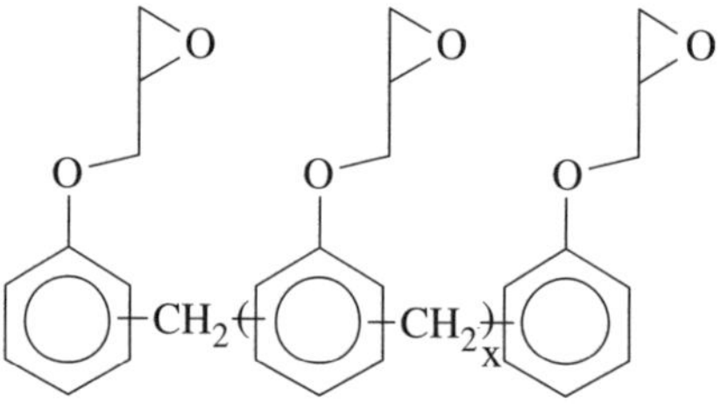

with $2 < x < 5$

Those corresponding to higher values of x are solid and can be processed only at temperature higher than ambient. By the same principle, novolac epoxy resins obtained from cresols ($v = 2$) can be prepared as well. They lead to a family of polyepoxides that exhibit an excellent thermal stability.

An extremely large variety of resins exist which allows the fine-tuning of the processing and/or the characteristics of the final material. For instance, N,N'-tetraglycidyl-4,4'diamino-diphenylmethane that carries tertiary amino groups ensures the self-catalysis of the cross-linking process.

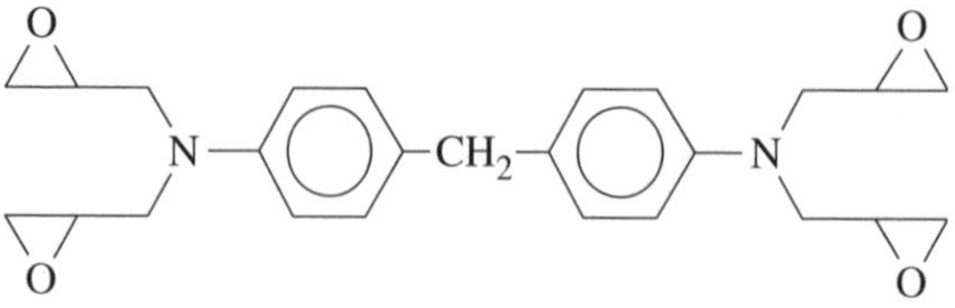

16.6.2. Hardeners

Hardeners are the second main component of the precursory system, essential to the formation of polyepoxide. They can be a mere cationic (Lewis acid) or anionic (tertiary amine) initiator of chain polymerization. Because oxirane reacts as a bivalent functional group, in this case, the diepoxide precursor is tetravalent and gives spontaneously a highly dense cross-linked network:

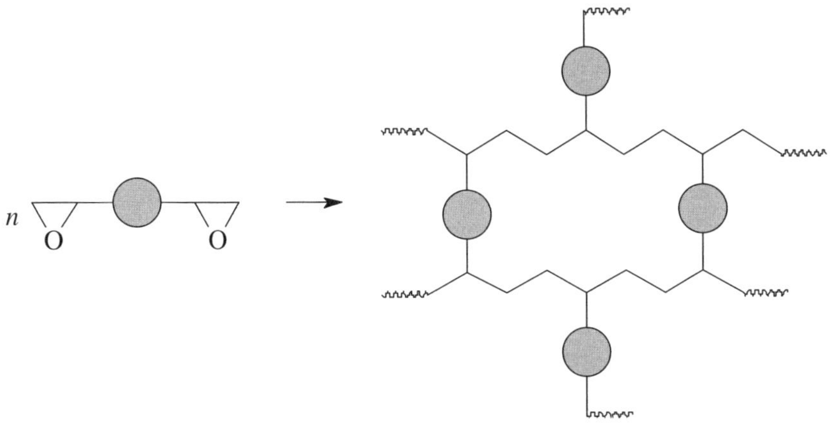

However, epoxy resins are generally cured through step polymerization. In the latter case, oxirane reacting as a monovalent functional group at ambient temperature the hardener should exhibit a valence higher than 2 to cross-link the diepoxide. Apart from epoxide precursors, a wide variety of hardeners exists. This variety arises from the nature of the reacting functional groups and on the architecture of the molecule. Each one brings its specificity and affects the properties of the final material (mechanical characteristics, thermostability, hydrophilicity, etc.). All systems that contain a hydrogen atom able to react with the oxirane group can be used: primary and secondary amines, amides, phenols, thiols, and so on.

The most widely used hardeners are primary amines whose reaction mechanism was presented in Section 7.4.4. This reaction occurs even at ambient temperature with aliphatic amines; it requires higher temperatures when aromatic amines are used. Some of these molecules that are frequently mixed to combine their specific properties are presented below:

$H_2N-(CH_2)_6-NH_2$ Hexamethylenediamine ($v = 4$)

$H_2N-CH_2-CH_2-NH-CH_2-CH_2-NH_2$ Diethylenetriamine ($v = 5$)

$H_2N-CH_2-CH_2-NH-CH_2-CH_2-NH-CH_2-CH_2-NH_2$

Triethylenetetramine ($v = 6$)

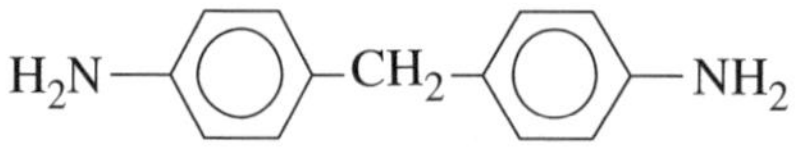

H_2N—⟨O⟩—CH_2—⟨O⟩—NH_2 4,4'-Diaminodiphenylmethane $(v = 4)$

H_2N—⟨O⟩—SO_2—⟨O⟩—NH_2 4,4'-Diaminodiphenylsulfone $(v = 4)$

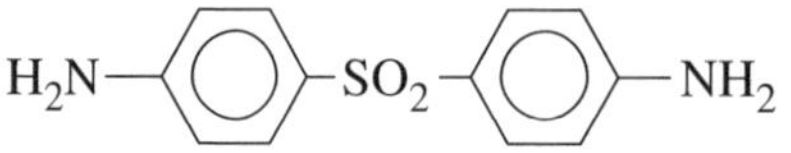

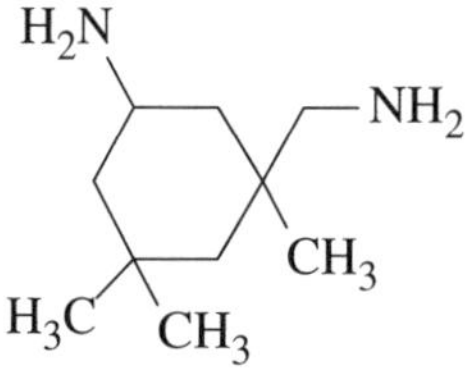

Isophorone diamine $(v = 4)$

$$CH_3 \qquad CH_3$$
$$H_2N\text{-}CH\text{-}CH_2\text{-}O\text{-}(CH\text{-}CH_2\text{-}O\text{-})_nNH_2$$ Poly(oxypropylene)diamine $(v = 4)$

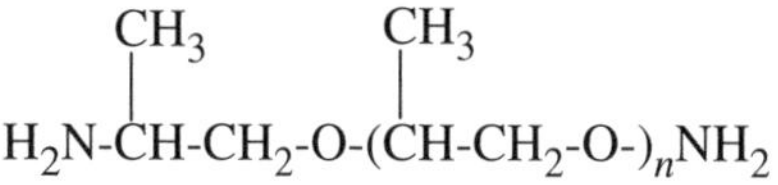

$(CH_2)_7CO\text{-}NH\text{-}CH_2\text{-}CH_2\text{-}NH\text{-}CH_2\text{-}CH_2\text{-}NH_2$
$(CH_2)_7CO\text{-}NH\text{-}CH_2\text{-}CH_2\text{-}NH\text{-}CH_2\text{-}CH_2\text{-}NH_2$ $v = 8$

$CH_2\text{-}CH{=}CH\text{-}(CH_2)_4CH_3$

Mixture in variable proportions

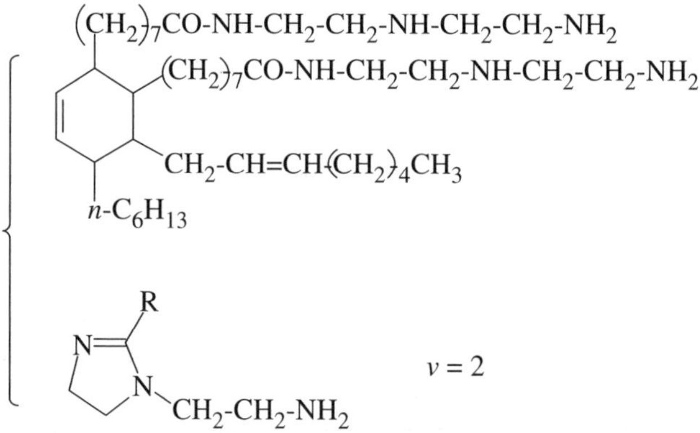

$n\text{-}C_6H_{13}$

$v = 2$

$CH_2\text{-}CH_2\text{-}NH_2$

... named "amino-polyamide"

The curing of epoxy resins can also be obtained (or completed) by using of carboxylic diacid anhydrides. This reaction is a simple esterification of hydroxyls initially present in the resin or formed by reaction with primary amines. Among the widely used anhydrides, one finds

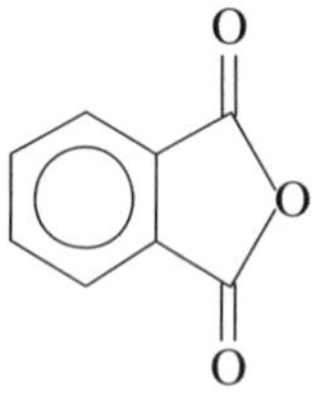

Phthalic anhydride

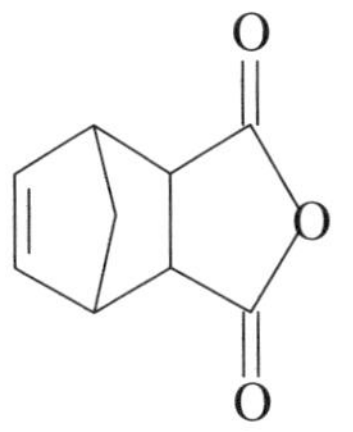

Nadic anhydride

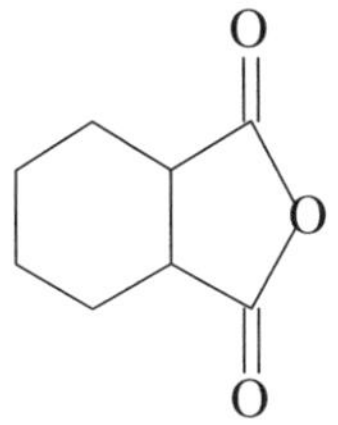

Hexahydrophthalic anhydride

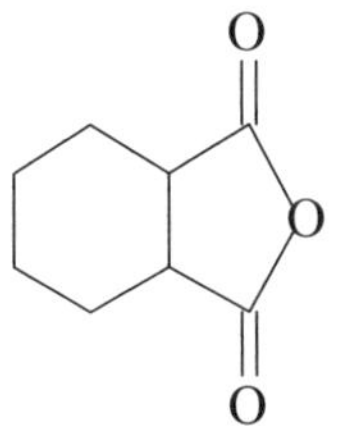

Hexahydrophthalic anhydride

The curing by anhydrides (or the corresponding carboxylic acids) requires high reaction temperatures $(T > 200°C)$ and relatively long reaction times. This draw-back is largely compensated by the improvement of the characteristics of the materials formed. The cross-linking can also be obtained by means of polythiols, with each $-SH$ group being monovalent and reacting according to a mechanism similar to that explained for primary amines.

16.6.3. Formulation of Precursor Mixtures

The simple cross-linking leads to dense and rigid networks that have low-impact strength. To optimize the characteristics of the materials formed, it is advisable to add either softening or hardening agents which can modify the cross-link density of the network, and thus the basic properties. For instance, the cross-linking in the presence of ω-thiolic polysulfides $(v = 1)$ considerably improves the impact strength of polyepoxides. The reduction in the cross-link density of the network can also be caused by the diepoxide itself. For example, oligomers of ethylene–butylene copolymers or of propylene oxide that are end-functionalized with oxirane groups are usually used for this purpose.

16.6.4. Applications

There are three main domains of applications of polyepoxides.

- The first one is that of surface coating, a domain in which EP are used due to their adherence to surfaces of variable nature and also to their hardness, their resistance to chemicals, and their easy processing even if high temperatures are often required for this application. The encapsulation of electronic devices can be classified in this type of application.
- EP resins are also structural materials, but the thickness of the objects to be manufactured is limited due to the difficult elimination of the heat of reaction. This difficulty can be alleviated by incorporation of fillers that are used as thermal balance while reducing the costs. The use of EP as matrices of composite materials with aramide or carbon fiber reinforcers leads to particularly cohesive materials.
- Adherence and strong cohesion are the properties that are needed for applications in the domain of the rigid adhesives. The possibility of processing them at ambient temperature is also a particularly interesting factor for their utilization.

Annual world production of EP resins is close to 800,000 tons.

16.7. POLYSILOXANES (THREE-DIMENSIONAL SILICONES) (SI)

Silicones contain mainly dimethylsiloxane units and are thus generally called poly-(dimethylsiloxane) (PDMS) $-[Si(CH_3)_2-O]-$, whatever the nature of other monomeric units and the dimensionality of the system.

16.7.1. General Characteristics

PDMS containing polymers exhibit certain characteristics that clearly differentiate these inorganic polymers from their organic analogs. First, PDMS have an excellent thermal stability resulting from the strong energy of $-Si-O-$bonds in the main chain and of $-Si-C-$bonds in the side chains:

$-Si-O-$: $535 \, kJ \cdot mol^{-1}$
$-Si-C-$: $370 \, kJ \cdot mol^{-1}$

For comparison the bond energy of C–C bonds is: $\sim 305 \, kJ \cdot mol^{-1}$.

PDMS consist of $-Si-O-$ sequences that exhibit a very low variation of the potential energy of the system with respect to internal rotation angles. This means that the chains exhibit high mobility even at low temperature ($T_g = -120°C$). This property combined with the low density of their cohesive energy ($\delta = 14.9 \, MPa^{1/2}$) favors their use as elastomeric materials. The regularity of the molecular structure induces a certain crystallinity which appears by stretching and corresponds to a melting zone observed at about $-38°C$. For low temperature applications, the crystallinity can be suppressed by introduction of side branches by copolymerization.

Finally, due to the nature of $-Si-O-$ sequences, silicones have an extremely low surface energy that induces anti-adhesive characteristics and a damp-proof character which are used in multiple applications.

One of the main application of silicones is as low-viscosity oils; these oligomers are not considered within the scope of this textbook in spite of their economic significance.

16.7.2. Preparation of Polysiloxanes

Two principal methods are utilized for the production of poly(dimethylsiloxane) homo- and copolymers. The first one concerns the chain polymerization of octamethylcyclotetrasiloxane (indicated by D_4),

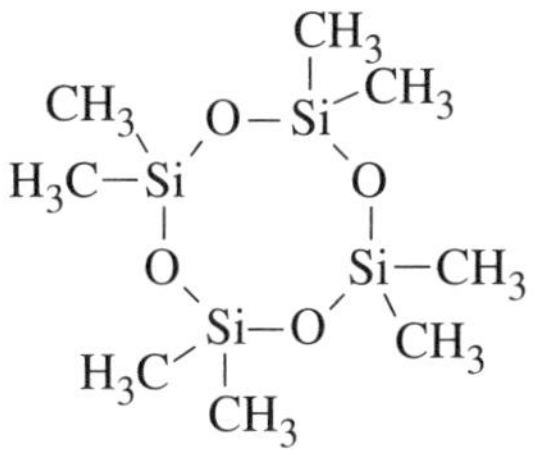

initiated by potassium alkoxide.

Cationic polymerization is required for the polymerization of tetramethylcyclotetrasiloxane and cyclotetrasiloxane bearing acidic H.

The formation of networks from linear siloxane chains requires the presence of reactive groups distributed along the chain by copolymerization. Thus, D4 can be copolymerized with a comonomer carrying vinyl groups to give linear chains with reactive side groups:

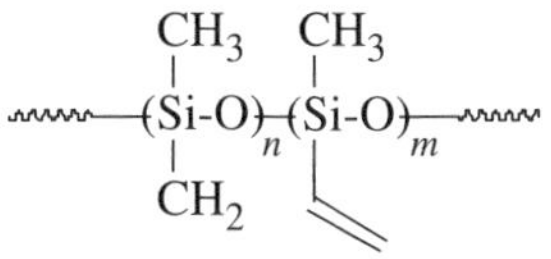

Three-dimensional SI are often prepared by step polymerization using dimethyldichlorosilane $(CH_3)_2-SiCl_2$ as main monomer; similar silane functional groups can be incorporated along the chain or at chain ends. The corresponding reaction mechanism was presented in Chapter 7; the reaction proceeds as follows

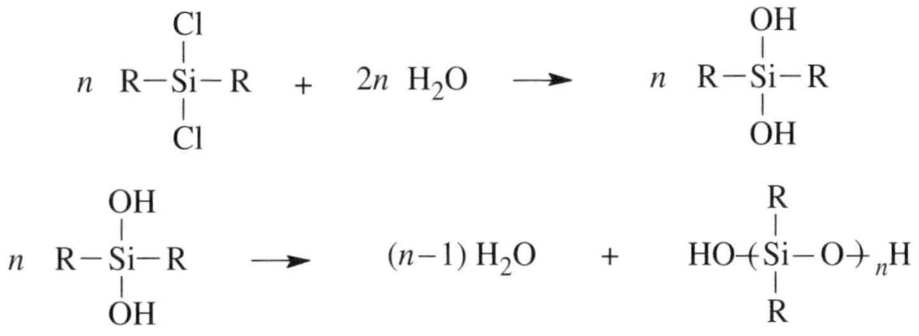

The number of hydrolyzable $-Si-Cl$ groups determines the valence of these monomers. Dichloro monomer provides most of the monomeric units, monochloro generates units located at the chain ends, and trichloro corresponds to the cross-links of the network. Vinyl and phenyl groups are incorporated in the chains from the following trichlorosilanes, respectively:

$$
\underset{\underset{Cl}{|}}{\overset{\overset{Cl}{|}}{Cl-Si}}-CH=CH_2 \qquad \underset{\underset{Cl}{|}}{\overset{\overset{Cl}{|}}{Cl-Si}}-\bigcirc
$$

The cross-linking of silicones is obtained through two methods, depending on whether the process is carried out by either step or chain polymerization.

The first one consists of the copolycondensation of divalent monomers (or possibly its divalent oligomers) with a trivalent monomer. For example, a dichlorosilylated molecule is spontaneously hydrolyzed under the effect of atmospheric water and copolymerizes with a trichlorinated molecule to directly give a network:

The cross-linking of linear chains carrying vinyl groups can be carried out either by radical means using peroxides

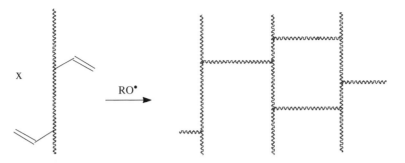

or by hydrosilylation while reacting −Si−H groups with vinyl side groups:

$$\underset{|}{\overset{|}{-}}\text{Si}-\text{CH}=\text{CH}_2 \ + \ \text{H}-\underset{|}{\overset{|}{\text{Si}}}- \ \xrightarrow{\text{Pt}} \ -\underset{|}{\overset{|}{\text{Si}}}-\text{CH}_2\text{-CH}_2-\underset{|}{\overset{|}{\text{Si}}}-$$

The vulcanization of silicone synthetic rubbers at ambient temperature generally utilizes chains of disilanol prepolymers ($v = 2$) reacted with an orthosilicate ($v = 4$):

$$\underset{\underset{\text{CH}_3}{|}}{\overset{\overset{\text{CH}_3}{|}}{\text{HO}-\text{Si}}}\text{-PDMS-}\underset{\underset{\text{CH}_3}{|}}{\overset{\overset{\text{CH}_3}{|}}{\text{Si}-\text{OH}}} \ + \ \underset{\underset{\text{OR}'}{|}}{\overset{\overset{\text{OR}'}{|}}{\text{R}'\text{O}-\text{Si}-\text{OR}'}} \ \underset{\longleftarrow}{\overset{\text{cat.}}{\longrightarrow}}$$

$$\underset{\underset{\text{CH}_3}{|}}{\overset{\overset{\text{CH}_3}{|}}{\text{HO}-\text{Si}}}\text{-PDMS-}\underset{\underset{\text{CH}_3}{|}}{\overset{\overset{\text{CH}_3 \quad \text{O}}{|}}{\text{Si}-\text{O-Si-O}}}\text{-}\underset{\underset{\text{CH}_3}{|}}{\overset{\overset{\text{CH}_3}{|}}{\text{Si}}}\text{-PDMS-}\underset{\underset{\text{CH}_3}{|}}{\overset{\overset{\text{CH}_3}{|}}{\text{Si}-\text{OH}}}$$

with vertical: PDMS-O and O-PDMS

Depending on the cross-link density, either elastomers of variable elastic moduli or rigid materials are obtained.

16.7.3. Applications

In addition to the use of the linear PDMS in the fields of oils and greases, three-dimensional silicones find various applications as either elastomeric materials or rigid resins. Depending on their formulation and their applications, these elastomers are hot- or cold-vulcanized. They are characterized by their high thermal stability, their high resistance, and a relatively good chemical inertia. They are used for the coating of electric systems, the manufacture of flexible molds, and the making of seals and coatings. They are usable at temperatures ranging between -60°C and $+250^{\circ}$C.

Due to the bio-tolerance of these materials, they find wide applications in the biomedical field.

Rigid SI resins correspond generally to structures more complex than those of elastomers. Indeed, a fraction of PDMS methyl groups are replaced by phenyl, vinyl, alkoxy, and so on, groups so as to adapt the properties of the material to a given application. These resins are employed in the field of paintings that should work uninterrupted at high temperatures ($>200^{\circ}$C). They have a particularly high durability at ambient temperature. These resins are also used for the insulation of electric components functioning under extreme conditions and retain their high insulating capacity. They are processed by molding, coating, or impregnation.

Aircraft and electric industries, and biomedical fields are the main utilizers of silicones produced in the world.

LITERATURE

J. P. Pascault, H. Sautereau, J. Verdu, and R. J. J. Williams, *Thermo-setting Polymers*, Marcel Dekker, New York, 2002.

J. Scheirs, T. E. Long (Eds.), *Modern Polyesters*, Wiley, New York, 2003.

E. M. Petrie, *Epoxy Adhesives Formulations*, McGraw-Hill, New York, 2005.

S. Lee (Ed.), *The Polyurethane Book*, Wiley, New York, 2003.

A. Gardziella, L. A. Pilato, and A. Knop, *Phenolic Resins: Chemistry, Applications, Standardization, Safety and Ecology*, Springer, New York, Berlin, 1999.

E. Occhiello, *Polyurethanes: Chemistry and Technology*, Wiley, New York, 2003.

R. G. Jones, W. Ando, and J. Chojnowski (Eds.), *Silicon-Containing Polymers*, Kluwer Academic Publishers, Norwell, Dordrecht, 2001.

INDEX